MASS LOSS AND EVOLUTION OF O-TYPE STARS

INTERNATIONAL ASTRONOMICAL UNION
UNION ASTRONOMIQUE INTERNATIONALE

SYMPOSIUM No. 83
HELD AT VANCOUVER ISLAND, CANADA
JUNE 5–9, 1978

MASS LOSS AND EVOLUTION OF O-TYPE STARS

EDITED BY

P. S. CONTI

Joint Institute for Laboratory Astrophysics,
University of Colorado, Boulder, U.S.A.

and

C. W. H. DE LOORE

Astrophysical Institute, Vrije Universiteit, Brussels, Belgium

D. REIDEL PUBLISHING COMPANY

DORDRECHT : HOLLAND / BOSTON : U.S.A. / LONDON : ENGLAND

Library of Congress Cataloging in Publication Data

Main entry under title:

Mass loss and evolution of O-type stars.

  (Symposium – International Astronomical Union ; no. 83)
  Includes index.
   1. O stars–Congresses. I. Conti, Peter S. II. Loore, Camiel W. H. de
III. Series: International Astronomical Union. Symposium; no. 83.
QB843.012M37   523.8   79–11933
ISBN-13: 978-90-277-0989-9 e-ISBN-13: 978-94-009-9452-2
DOI: 10.1007/978-94-009-9452-2

Published on behalf of
the International Astronomical Union
by
D. Reidel Publishing Company, P.O. Box 17, Dordrecht, Holland

Sold and distributed in the U.S.A., Canada, and Mexico
by D. Reidel Publishing Company, Inc.
Lincoln Building, 160 Old Derby Street, Hingham,
Mass. 02043, U.S.A.

TABLE OF CONTENTS

PREFACE

The organization of this Symposium had its beginnings
at the International Astronomical Union General Assembly in
Grenoble in 1976. The initial "rounding up" of the Scientific Organizing Committee was begun by Drs. Snow and Swings;
most of us who became the eventual organizing committee met
a few times during the Assembly and formulated the essential
outlines of the meeting. Extensive correspondence with all
the committee subsequently established the program. The
idea was to bring together both observers and theoreticians
to discuss the stellar winds and mass loss rates and their
effects on evolutions of O-type stars.

On the observational side, there are now spectroscopic
data from the far UV to the near IR regions concerning the
stellar winds. There is also information about the free-free
emission in the wind from the IR and radio portions of the
spectrum. Fortunately, these different detection methods
give more or less the same mass loss rate for the one star,
ζ Pup,, which has been observed at all wavelengths. One of
the intents of the first three sessions of this Symposium is
to outline the existing data on mass loss rates as it per-
tains to the O-type stars.

While there is some reasonable agreement about the mass
loss rates observed in a few stars, there is not yet a consensus about the physical cause for very extensive stellar
winds that are found. A panel discussion was suggested for
Session 4 as a means by which the conflicting points of view
on this problem could be addressed. The Organizing Committee
made up a series of questions in advance, listed in Session 4,
which it wished the protagonists of the various stellar wind
theories to address. This theme of confronting the existing
data was a reasonable success as may be judged by the contributed papers. There was agreement that the high terminal
velocities observed are caused by radiation pressure, but
apparently heating is present and is not understood. Observations from the forthcoming HEAO-B X-ray satellite will
nicely confirm or eliminate one of the current theories for
explaining the high ionization state.

A few detailed theoretical contributed papers are contained in Session 5, along with some results on binary stars,
our only source of masses. The data are still rather skimpy,
especially for Of stars which appear to be losing mass at the
highest rates.

Sessions 6 and 7 are concerned with the theoretical considerations of evolution of O-type stars when mass loss is
taken into account. Several indirect arguments about mass

loss are also presented there. Perhaps the most interesting
general result is that the subsequent stellar evolution is
very sensitive to the mass loss rate; factor of two differences either have little or no effect, or peel a star down to
a helium core. Unfortunately, the actual mass loss rates
for most stars are not known that accurately.

The final session is concerned with the descendants of
the O-type stars, the WR objects, both from an observational
aspect and from a theoretical one. A major problem, never
really addressed during the Symposium, is the question of
why these stars have mass loss rates appreciably larger than
the most luminous O types; it does not appear possible that
this is only due to radiation pressure.

In the very beginning of organizing the Symposium, it
was decided to limit the subject to the O stars themselves.
There was not enough time to address several related topics
concerned with the mass loss and evolution. The stellar wind
certainly has an effect on the surrounding interstellar medium
due to its pressure effects, leading to stellar "bubbles",
and to the possible initiation of subsequent star formation
in adjacent molecular clouds. It also appears that if O-type
stars lose an appreciable fraction of their mass while core
hydrogen burning, their subsequent behavior as a supernova
and a supplier of heavier elements to the galactic interstellar medium is greatly modified. The role of stellar winds
in the production of X-rays in those close binary systems
with collapsed companions also could not be addressed with
the time available. Any or all of these broad themes resulting from the ubiquitous presence of stellar winds in O-type
stars could well be a topic for future Colloquia or Symposia.

Practically all the invited contributers have presented
camera-ready manuscripts which are contained in these Symposium Proceedings. Slides shown during the meeting were nicely handled by Mr. Ray Carlberg. The discussions were recorded on tape with the able assistance of Mr. R. Scholes; the
participants also usually wrote down what they thought they
said immediately after the session. Although these written
comments form much of the discussions, a substantial fraction
was taken verbatim from the tapes. In all cases, the participants had a final look at what they really had said and
were able to make final corrections. In a few cases the Editors have eliminated extraneous questions or comments, but
the transcipt is otherwise reasonably verbatim. With some
effort the Editors were able to resist listing "laughter" or
"uproarious laughter" at a number of places in the discussions: These are best left to the reader's imagination.

Generally speaking, P.S. Conti was responsible for the preparation of the first half of the Proceedings and C. de Loore for the second part. In JILA, we were ably assisted in the preparation of the discussion sessions from the tapes by Ms. Krog and Ms. Briggs; the camera ready copy of the discussion, and some editorial work on several contributions was typed by Ms. Romey and ably organized by Ms. Volsky. At the Astrophysical Institute of the Vrije Universiteit at Brussels the discussion sessions from the sheets and the tapes was prepared and typed by Ms. Michiels. We would like to thank Dr. Werner Zits for assistance in interpreting parts of the General Discussion following Session 4 and David Van Blerkom for contributing the limericks.

The presence of participants from a number of institutions and countries was greatly aided by financial assistance from the International Astronomical Union and from the National Research Council of Canada. The Scientific Organizing Committee is greatly indebted to Drs. Cowley and Hutchings for the local arrangements of the meeting which was in most pleasant surroundings. The Symposium took place at the Qualicum College Inn, a rustic resort overlooking the beach and the Georgia Straits from Vancouver Island. The weather was invariably pleasant, greatly facilitating the outdoor interactions possible in a clement location. The night before the Symposium began it was noticed by all astronomers that a "close encounter" between Mars and Saturn was occurring in the sky. The Symposium Astrologer, Dr. Tony Hearn, was able to interpret this event as fortelling a few days of "war", followed by a peaceful ending and reconciliation. From the discussion sessions it may be observed how well this prediction was borne out.

The Editors

SCIENTIFIC ORGANIZING COMMITTEE

P.S. Conti (Chairman), A.P. Cowley, J.B. Hutchings, J.M. Marlborough, D.C. Morton, B. Paczynski, T. Snow, A.V. Tutukov, E.P.J. van den Heuvel, J.M. Vreux

LOCAL ORGANIZING COMMITTEE

A.P. Cowley, J.B. Hutchings

LIST OF PARTICIPANTS

D.C. ABBOTT, University of Wisconsin, Madison, USA
Y. ANDRILLAT, Observatoire de Haute Provence, St.Michel,
    France
M.J. BARLOW, Anglo-Australian Observatory, Epping, New South
    Wales, Australia
W.P. BIDELMAN, Warner and Swasey Observatory, East Cleveland,
    USA
G.F. BISIACCHI, Instituto di Astronomia, Universidad de
    Mexico, Mexico
B. BOHANNAN, University of Colorado, Boulder, USA
C.T. BOLTON, David Dunlap Observatory, Richmond Hill, Canada
J. BREYSACHER, European Southern Observatory, Santiago, Chile
R. CARLBERG, University of British Columbia, Vancouver,Canada
L. CARRASCO, Instituto di Astronomia, Universidad de Mexico,
    Mexico
J.P. CASSINELLI, University of Wisconsin, Madison, USA
J.I. CASTOR, Joint Institute for Laboratory Astrophysics,
    Boulder, USA
C. CHIOSI, Instituto di Astronomia, Padova, Italy
P.S. CONTI, Joint Institute for Laboratory Astrophysics,
    Boulder, USA
R. COSTERO, Instituto di Astronomia, Universidad de Mexico,
    Mexico
A.P. COWLEY, University of Michigan, Ann Arbor, USA
D. DEARBORN, Steward Observatory/University of Arizona,
    Tucson, USA
E.L. VAN DESSEL, Royal Belgian Observatory, Brussels, Belgium
C.G. DAVIS, Las Alamos Scientific Laboratory, USA
A. DELGADO, Max Planck Institut für Physik und Astrophysik,
    München, FRG
D. EBBETS, University of Colorado, Boulder, USA
H.J. FALK, University of Western Ontario, London, Canada
C. FIRMANI, Instituto di Astronomia, Universidad de Mexico,
    Mexico
C.D. GARMANY, University of Colorado, Boulder, USA
G. HAMMERSCHLAG-HENSBERGE, Astronomical Institute, Amsterdam,
    Netherlands
S. HEAP, Goddard Space Flight Center, Greenbelt, USA
A.G. HEARN, Sterrewacht Sonnenborgh, Utrecht, Netherlands
H.F. HEINRICH, Astronomical Institute, Amsterdam, Netherlands
T.J. HERCZEG, University of Oklahoma, Norman, USA
E.P.J. VAN DEN HEUVEL, Astronomical Institute, Amsterdam,
    Netherlands
V.A. HUGHES, Queen's University, Kingston, Canada
D.G. HUMMER, Joint Institute for Laboratory Astrophysics,
    Boulder, USA
J.B. HUTCHINGS, Dominion Astrophysical Observatory, Victoria,
    Canada
A.R. HYLAND, Mt. Stromlo and Siding Spring Observatory,
    Australia

S. JEFFERS, York University, Downsview, Canada
M. KLUTZ, Institut d'Astrophysique, Liège, Belgium
P.B. KUNASZ, University of Colorado, Boulder, USA
S. KWOK, York University, Downsview, Canada
S.A. LAMB, University of California, Los Angeles, USA
H.J.G.L.M. LAMERS, Space Research Laboratory, Utrecht,
    Netherlands
P. LEDOUX, Institut d'Astrophysique, Liège, Belgium
E.M. LEEP, University of Washington, Seattle, USA
J.B. LESTER, University of Toronto, Mississauga, Canada
K.C. LEUNG, U.S. Department of Energy, Washington D.C., USA
C. DE LOORE, Astrofysisch Instituut, Brussels, Belgium
L.S. LUUD, Tartv Observatory, Estonia, USSR
B.T. LYNDS, Kitt Peak National Observatory, Tucson, USA
J.M. MARLBOROUGH, University of Western Ontario, London,
    Canada
P. MASSEY, University of Colorado, Boulder, USA
T.J. MAZUREK, University of Texas, Austin, USA
A.F.J. MOFFAT, Université de Montreal, Canada
N. MORRISON, Joint Institute for Laboratory Astrophysics,
    Boulder, USA
D.C. MORTON, Anglo-Australian Observatory, Epping, Australia
E. NASI, Instituto di Astronomia, Padova, Italy
V.S. NIEMELA, Instituto de Astronomia y Fisica del Espacio,
    Buenos Aires, Argentina
P. NOERDLINGER, University of Colorado, Boulder, USA
G. OLSON, Astrofysisch Instituut, Brussels, Belgium
M. OVENDEN, University of British Columbia, Vancouver,Canada
W. PACKET, Astrofysisch Instituut, Brussels, Belgium
S. PARSON, University of Texas, Austin, USA
P. PISMIS, Instituto di Astronomia, Universidad de Mexico,
    Mexico
M. PLAVEC, University of California, Los Angeles, USA
R. POECKERT, Dominion Astrophysical Observatory, Victoria,
    Canada
J. RAHE, Remeis Observatory, Bamberg, FRG
B. ROCCA-VOLMERANGE, Institut d'Astrophysique, Paris, France
D. SCHNEIDER, California Institute of Technology, Pasadena,
    USA
R. SCHOLES, Dominion Astrophysical Observatory, Victoria,
    Canada
W. SEGGEWISS, Observatorium Hoher List, Daun, FRG
P.R. SCHWARTZ, Naval Research Laboratory, Washington D.C.,
    USA
T.P. SNOW, Jr., University of Colorado, Boulder, USA
R. SREENIVASAN, University of Calgary, Canada
R. STALIO, Osservatorio Astronomico di Trieste, Italy
J.P. SWINGS, Institut d'Astrophysique, Liège, Belgium
R.N. THOMAS, Institut d'Astrophysique, Paris, France
A.V. TUTUKOV, Astronomical Council, Academy of Sciences,
    Moscow, USSR

A.B. UNDERHILL, Institut d'Astrophysique, Paris, France
D. VAN BLERKOM, Joint Institute for Laboratory Astrophysics,
    Boulder, USA
D. VANBEVEREN, Astrofysisch Instituut, Brussels, Belgium
J.M. VREUX, Institut d'Astrophysique, Liège, Belgium
A.J. WILLIS, University College London, England
R.E. WILSON, University of South Florida, Tampa, USA
W.J.F. WILSON, University of Calgary, Calgary, Canada
B. WOLF, Landessternwarte Königstuhl, Heidelberg, FRG
S. WOLFF, Institute for Astronomy, University of Hawaii, USA
M. ZEILIK, University of New Mexico, Albuquerque, USA
J. ZIOLKOWSKI, Nicholas Copernicus Astronomical Center,
    Warsaw, Poland

             CAMEO APPEARANCES BY:

C. AIKMAN, Dominion Astrophysical Observatory, Victoria,
    Canada
J. AUMAN, University of British Columbia, Vancouver,Canada
A. BATTEN, Dominion Astrophysical Observatory, Victoria,
    Canada
D. CRAMPTON, Dominion Astrophysical Observatory, Victoria,
    Canada
A. GOWER, University of Victoria, Canada
F.D.A. HARTWICK, University of Victoria, Canada
J. WOODROW, University of British Columbia, Vancouver,Canada

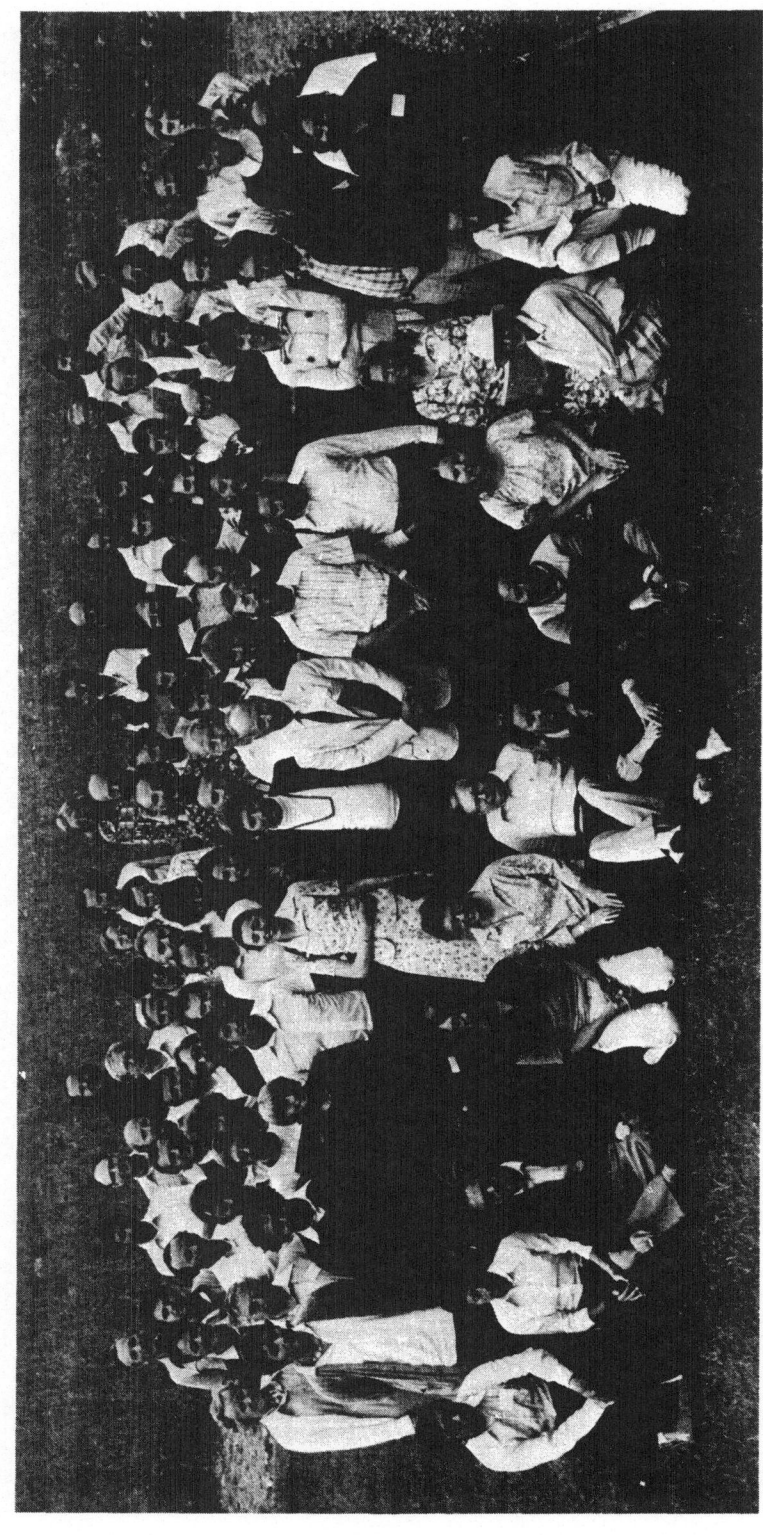

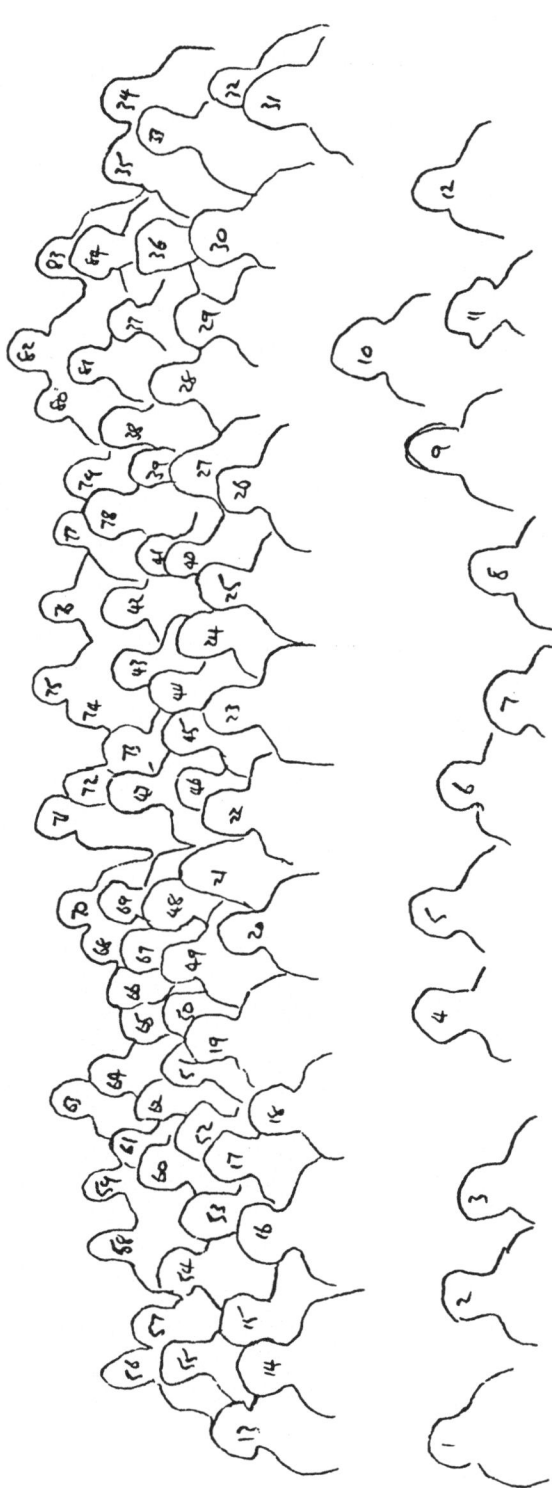

1. Kwok 2. Klutz 3. Bisiacchi 4. Ebbets 5. Wolff 6. Bidelman 7. Niemela 8. Costero 9. Leep 10. Heap 11. Underhill
12. Leung 13. Cowley 14. Rocca-Volmerange 15. Herczeg 16. Carrasco 17. Falk 18. Pismis 19. Snow 20. Hensberge
21. Lamb 22. Hyland 23. Ledoux 24. Schneider 25. Castor 26. Vreux 27. Morrison 28. De Loore 29. Nasi 30. Chiosi
31. Zeilik 32. Sreenivasan 33. Seggewiss 34. Scholes 35. Lynds 36. Conti 37. Willis 38. Vanbeveren 39. Bolton
40. Lester 41. Breysacher 42. Mazurek 43. Delgado 44. Wilson 45. Wilson 46. Andrillat 47. Noerdlinger 48. Andrillat
49. Van Blerkom 50. Massey 51. Poeckert 52. Abbott 53. Schwartz 54. Marlborough 55. Moffat 56. Hughes 57. Carlberg
58. Swings 59. Morton 60. Hearn 61. Parson 62. Luud 63. Jeffers 64. Cassinelli 65. Rahe 66. Garmany 67. Ovenden
68. Olson 69. Barlow 70. Tutukov 71. Van den Heuvel 72. Hummer 73. Packet 74. Ziolkowski 75. Stalio 76. Firmani
77. Wolf 78. Henrichs 79. Van Dessel 80. Kunasz 81. Plavec 82. Hutchings 83. Davis 84. Bohannan

NP Lamers; Dearborn; Thomas

A Belgian astronomer I kneux
Pointed out what some O stars deux
Their spectral lines flicker
He said with a snicker
Which serves all wind theories to screux.

SESSION 1

OPTICAL SPECTROSCOPY

Chairman: E. VAN DEN HEUVEL
Introductory Speaker: J.B. HUTCHINGS

1. J.M. VREUX and Y. ANDRILLAT: H alpha variations in two mass-losing stars.

2. W.G. WELLER and S. JEFFERS: Short term variability of line strengths in some Of and Wolf-Rayet stars.

3. G. HAMMERSCHLAG-HENSBERGE: Mass loss in the spectrum of the O6.5f binary HD 153919.

4. B. WOLF and C. STERKEN: Mass loss of B1 Ia-O supergiants and evolutionary consequences

5. T. NUGIS, I. KOLKA and L. LUUD: On the formation of continuous spectrum and emission line profiles of P Cygni.

6. P. PISMIS: Evidence for non-isotropic mass loss from central stars of some emission nebulae.

7. C.D. ANDRIESSE and R. VIOTTI : Mass loss from Eta Carinae.

8. J. BREYSACHER and M. AZZOPARDI: Wolf-Rayet stars in the Magellanic Clouds.

9. B.T. LYNDS: Stellar outflow: Relative motions of nebulae and Of stars.

THE O STARS: OPTICAL REVIEW*

J.B. Hutchings,
Dominion Astrophysical Observatory
Herzberg Institute of Astrophysics
National Research Council Canada
Victoria, B.C.

1.   INTRODUCTION

I would like to start with a quick overview of the O stars - their significance and role in the galaxy and in astrophysics - just to remind ourselves of why we are here and what we hope to talk about. In Table 1 I show a rough outline of the contribution of O stars to what happens in the galaxy as a whole. Because of their extreme luminosity, they contribute a large fraction of the radiation of the galaxy, while forming a very tiny group of objects and mass. Because of their short lifetime they are a population that has gone through 10^4 generations in the life of the galaxy. Their high mass loss rates may account for a large fraction of the new matter injected into the interstellar medium, and they probably power some significant fraction of the hard X-ray sources in the galaxy, by virtue of the fact that a companion can become a neutron star a) without disrupting the binary and b) while the companion is still a mass losing O star.

TABLE 1. O Stars in the Galaxy

	Each	Total	Fraction of galaxy
stellar mass	30 $M_\odot$	10^7 $M_\odot$	10^{-5}
number	-	3×10^5	10^{-6}
luminosity	$\sim 10^{39}$ erg s	10^{44} erg s	10^{-1}
mass loss	3×10 $M_\odot/y$	1 $M_\odot/y$	0.3?
lifetime	$10^6 y$	-	10^{-4}
hard X-rays	10^{37} erg s	60×10^{37} erg s	> 0.1

*Dominion Astrophysical Observatory Contribution No. 387 =NRC No. 16869

P. S. Conti and C. W. H. de Loore (eds.), Mass Loss and Evolution of O-Type Stars, 3–22.
Copyright © 1979 by the IAU.

Clearly, O stars are important in the galaxy (or any galaxy) and can be seen a long way off. We know roughly what they are doing, and we need to discuss how they got there, how they are doing it, and where they go to.

II. MASS LOSS SIGNATURES

It is generally accepted that one looks at the UV resonance lines to see if a star is losing mass. There, one looks for recessional velocities in excess of escape, and in most cases there they are. However, it has been known for just as long that there are mass-loss indicators in the visual and it is clearly relevant that we consider them here.

First, it is not necessary for matter to be moving in excess of escape velocity to escape, provided it is being pushed. Further, if we see matter moving away from a star and none returning, we may conclude that it escapes. Finally, the escape velocity falls with distance from the photosphere (as R^{-2}) so that at $1R_*$ it is only of the order of 125 km s^{-1}. These points are elementary but often overlooked.

There are several mass-loss indicators in the visual and blue spectrum which are easily seen, and others which require good high dispersion spectra. They are listed below and are mostly self-evident. In general, the more extreme the phenomena the greater the mass-loss, and by calibration with detailed models, it seems that we can detect stellar winds down to $\sim 10^{-7}$ M$_\odot$/year by a careful study of the ground based spectrum.

1. Hα emission present.
2. Hβ emission present, He I λ 5875 emission present.
3. Balmer velocity progression.
4. Velocity excitation-potential relation present.
5. He I λ 4471, Mg II λ 4481, He I λ 4026, C II λ 4267 velocities separate out.
6. He I λ 3888 separates from H8.
7. Hγ emission, other He I emission, further emission lines.

Some words of caution. A Balmer progression may be caused by P Cygni emission which displaces the absorption minimum. A little care and sense will suffice to check this, but in general I would be suspicious of a Balmer progression which shows only in Hα and Hβ, when there is any sign of emission at Hα. The velocity-excitation relation can be confused by 1) not using the Balmer asymptotic value, but some sort of average which includes the highly shifted Hα, β etc. lines and 2) clear deviations from LTE populations, such as occur in N III, Si IV in extreme mass-loss O stars. In these cases there is obviously a contribution to the visible line spectrum from the regions of high velocity where the UV lines are formed, and we are not measuring the acceleration of the inner parts of the atmosphere.

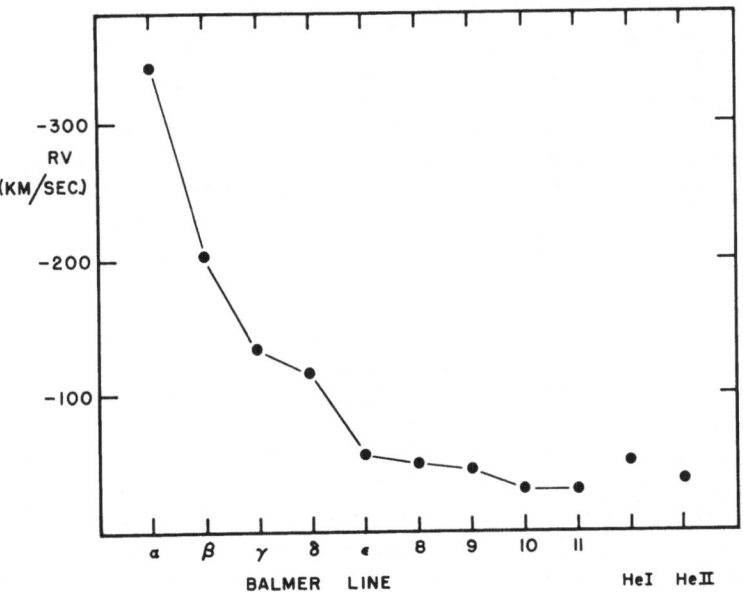

Fig. la). Balmer velocity progression in Of star HD 148937, and mean He I, He II velocities.

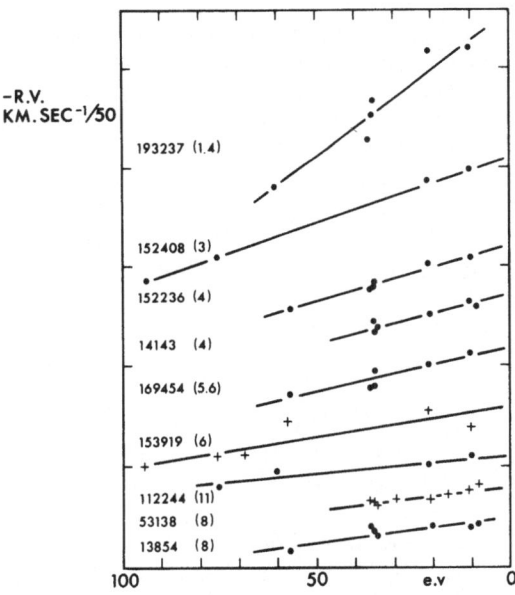

Fig. 1b). Velocity excitation linear slopes for absorption lines in mass-loss stars. HD numbers given and increase slopes of lines in parentheses. Steeper slopes indicate greater mass loss.

Figure 1 shows examples of a Balmer progression and velocity-excitation slopes in mass-loss stars. In the discussions that follow we shall use and refer to these mass-loss criteria as empirical indicators of stellar winds. They should also be of use in a detailed theory of stellar mass-loss, and naturally must complement the UV spectral information. One of the curiosities we must explain is that in the stars with the most obvious visual mass-loss characteristics (e.g. P. Cygni), the UV lines show no P Cyg profiles, and very low velocity shifts.

MASS LOSS RATES

There are two large scale surveys at present - that of Barlow and Cohen (1977) based on I-R data and my own (Hutchings 1976), based on optical spectroscopic data (fig. 2). These deal with 40 and 65 stars respectively of which 23 and 59 respectively are OB stars (i.e. earlier than B5). There are probably systematic errors in each: in the B + C sample, a velocity law derived for P Cyg is applied to all. In view of the unique nature of the P Cyg wind, and new data concerning its distance and mass loss (T.P. Snow; preprint) this is not a good idea. Also, B + C use estimated (i.e. unobserved) values of V for a number of stars, to derive mass loss rates. My own work is a quantitative compilation of mass-loss indications in the spectra, but the mass loss rates are based on a poorly defined grid of a <u>few</u> detailed models. It is clear that this grid needs revision. The result is that I think my rates tend to be high and B + C's tend to be low. The diagram 3 shows the comparison of 18 common stars. The general agreement is encouraging and it is clear at least that we are dealing with rates in the 10^{-5} to 10^{-6} M_Θ/year for the luminous OB stars, with the higher rates occurring in the O stars. In the case of ζ Pup there are now several independent estimates of the mass loss rate which agree to within a factor two, at a mean of 5×10^{-6} M_Θ/year.

IV. EXTREME OF STARS

There are a few Of stars with very pronounced Of characteristics, P Cygni Balmer profiles and emission in most of the N III lines. They also have broad emission bands underlying the Hδ and λ 4630-4690 region, reminiscent of weak W-R bands. There are five such stars in the galaxy which are well studied (see table 2; Hutchings 1976 and references therein) and close investigation shows them all to have very strong stellar winds. They also appear to have very high luminosity and lie in a very small region of the H-R diagram.

It seems reasonable to suppose that they are stars of initially very high mass (>60 M_Θ) which have lost much of their initial mass. The strong N III spectrum could be processed material which has reached the surface (so they are He burning objects?), or a result of peculiar level populations and ionisation in the extreme stellar wind conditions.

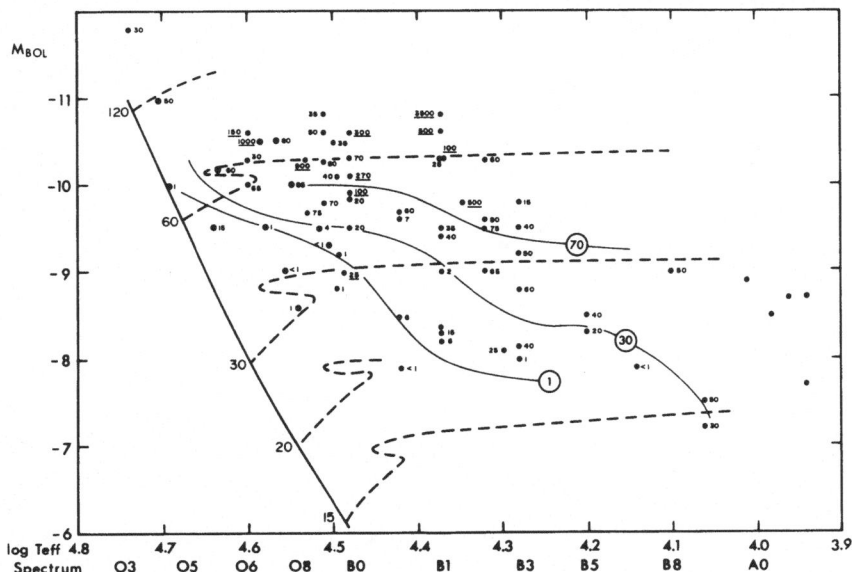

Fig. 2.   Theoretical HR diagrams with conservative mass loss evolution tracks.   Mass loss rates of Hutchings in 10^{-7} $M_\odot/yr$ units and suggested contours in same units.

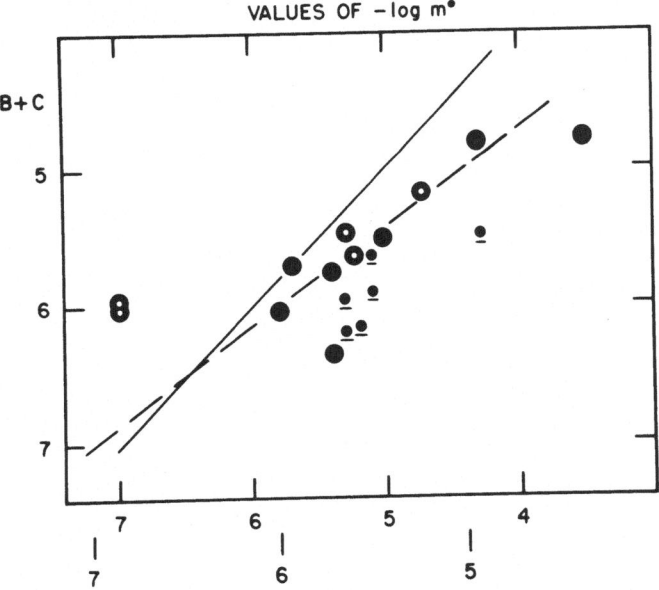

Fig. 3.   Comparison of mass loss rates of Hutchings (JBH) and Barlow and Cohen (B+C).   Open symbols O stars; closed symbols B stars.   Small symbols are stars for which B+C guessed $V_{terminal}$.   Lower JBH scale is revised after recalibration of detailed models and rejection of P Cyg.

Table 2.   5 Extreme Of Stars and 2 which are not

HD	Sp	M_V	$-\dot{m}$ $M_\odot/yr$	RV_{phot} km s^{-1}	$H\beta$ km s^{-1}	M_{BOL}	λ 4686 emis. peak
108	07f	-7.0	2×10^{-5}	-70	-300	-10.5	30%
148937	07f	-7.2	7×10^{-6}	-45	-200	-10.5	15%
151804	09f	-7.2	6×10^{-6}	-40	-145	-10.3	10%
152408	08f	-7.1	4×10^{-5}	-50	-275	-10.3	80%
153919	06f	-6.9	7×10^{-6}	-65	-150	-10.6	40%
66811	04f	-6.4	5×10^{-6}	-10	- 40	-10.3	30%
210839	06f	-6.5	3×10^{-6}	-70	- 90	-10.2	15%

These stars are of special interest as they show a wealth of detail which we can hope to interpret in the theory of stellar mass loss.   The Balmer progressions are very clear (see Fig. 1) and the excitation-velocity relation has a large slope for these stars. Almost every line in the spectrum has a characteristic profile; asymmetry velocity, width and possibly P Cygni emission. I have derived detailed ad-hoc models for the winds of two stars, which, fortuitously or not, indicated high mass loss rates and slow acceleration envelopes, a number of years before they were found reasonable by the real pundits in the field.   If for no other purpose than to initiate discussions, I show the main features of these models in Figure 4.   They are based

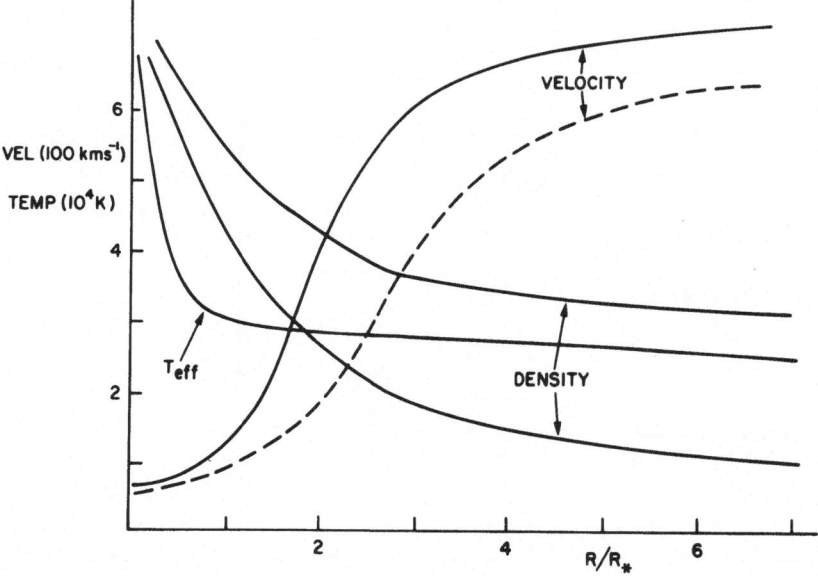

Fig. 4.   Velocity temperature and density fields of HD 152408, from Hutchings (1968).

mainly on fitting observed profiles by calculating ones in an accelerating wind using a simple numerical solution of the transfer equation. These models show too that ionisation (and hence temperature?) increase in the envelope, after an initial drop, as also deduced much more recently by Lamers and Snow (1978).

Two of the stars are binaries, with low mass (and hence evolved) companions, so that we can obtain further information on the fundamental parameters of the O stars. The X-ray star system 4U1700-37/HD 153919 is particularly valuable in this regard, since it eclipses, and the optical to X-ray luminosity is high enough ($\sim$5000:1) that there is little alteration of the O star behaviour by the X-radiation. In this system we find the wind stratification indicated by a relation between K and V_o for different ions (see Table 3). We also see a phase dependent variation in the wind, as aspect and varying tidal distortion in a slightly non-circular orbit vary the surface gravity over the observed disk of the star. We find evidence that the wind is quite sensitive to surface gravity. Similar evidence is found in HD 108, and some less extreme Of stars (e.g. 29 CMa).

TABLE 3.   153919 Line Velocities in km s^{-1}

Absorption	V_o	K	exc*	Emission	V_o	K	I.P.
N IV	− 60	25:	94	He II 4686	18	13	54
He II	− 64	19	76	Si IV 4116	-40	18:	45
O III	− 65	22:	68	C III 5696	-74	16	48
Si IV	− 83	--	57	N III 4640	-77:	15	47
He I	− 87	12	21				
Balmer	− 79	16	10				
H	−110	12:	10				
C IV	−117	22	85				
Mg II	−136	--	16				
Hβ	−150	<10	10				

*I.P. of lower ions + e.p. of line lower level.

V.   PHOTOMETRY AND POLARIMETRY

Since this aspect has not been specifically included in the program I will spend a few moments on it here. Firstly, we know little about the stability of single O stars, beyond a few scattered observations indicating variability at the 0^{m}1 level over periods from

weeks to years. We also know little about the polarisation produced by
extended moving envelopes. Considering what has been learned about Be
star envelopes in this way, I think this would be a valuable observational program.

Turning to binaries, we find that there are 6 eclipsing systems in
the graded catalogue of Koch et al (1973), of 200 entries. In addition
there are several X-ray binaries with good photometry, and a couple of
ellipsoidal systems. Now that good light curve synthesis programs are
available, photometry of interacting binaries can yield important
fundamental data on the stars: temperatures radii, mass-ratios, limb
and gravity darkening, and inclinations and masses for the systems.
These analyses have been particularly fruitful in the X-ray systems
(see Figure 5). This type of analysis has also been pursued by Leumg
and Wilson whose models indicate several contact O star systems. I
find this somewhat worrying, as there are discrepancies with
spectroscopic and other photometric analyses, and this is a point we
may do well to discuss here. If they did exist, what would contact O
stars become?

On the polarisation, it is encouraging that two groups (Koch et
al. and Kemp et al.) are doing observations which, combined with a
simple model, yield information on circumstellar envelopes in O star

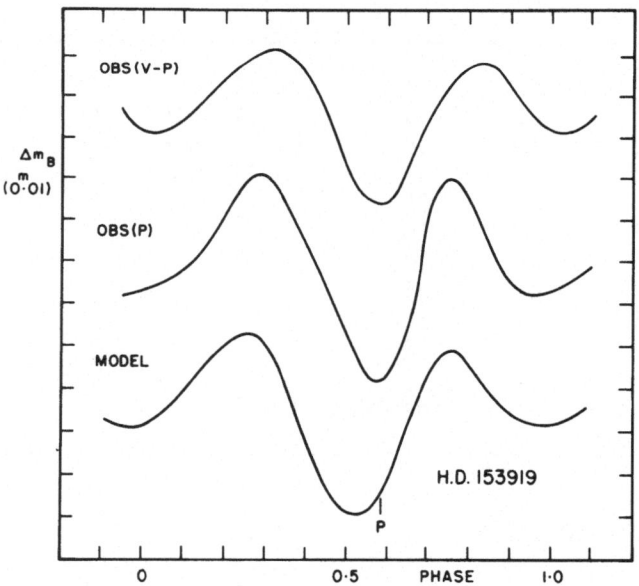

Fig. 5. Light curves of X-ray binary HD 153919. Upper: Van Paradijs
et al. observations. Centre: Petro observations. Lower: model with
e⌁0.05. P marks periastron passage.

binaries, and inclinations for the systems. Clearly, such work is a valuable complement to spectroscopy and should be pursued on as many binaries as possible. Again, a word of caution: the analysis applied to Cyg X-1 yields a value of i (76°) inconsistent with the non-existence of eclipses. A closer look at the situation shows that a more complex model is called for which allows lower values of i. It is clear that we must all beware of the crank-the-handle approach even to very standard looking problems.

A final word about photometry is to emphasise the value of a wide wavelength base. We have the capability of observing from 1000A to several microns, and in the O stars, UV data are particularly sensitive to temperature, reddening, limb darkening and line blanketing. They are well worth getting for the extra information they may yield on these points, and because light curve analysis is much more powerful over a long λ base.

VI. BINARIES AND MASS-LOSS

An important question is that of the effect of a companion on the mass-loss and the disentangling of the Roche lobe overflow versus stellar wind dichotomy, especially in the X-ray systems, where the accretion rate of a neutron star depends on the flow velocity.

The observational picture is as follows. A binary companion does not have a detectable effect on mass loss unless the tidal interaction brings the star close to its critical surface (say within $\sim$10%). I say this on the basis of ground-based data. It may be that UV data from IUE will show effects not detectable in the visible. If we look at the X-ray binaries which have circular orbits (SMC X-1, 1538-52, Cen X-3 - the latter not set studied in sufficient detail) we see no evidence for increased flow towards the companion (see e.g. Hutchings et al. 1977). In fact, we find no clear evidence for a wind $> 10^{-7}$ $M_\odot$/year at any phase. Yet the X-ray luminosity is evidence that mass flow exists. However, it is significant that all OB star X-ray binary orbits show e=0 and $\omega \simeq 0$ (Table 4), and this is qualitatively the effect of increased outflow (by only 3-5 km s^{-1}) along the line of centres (i.e. where g_{eff} is reduced). If this is the explanation we note also that this effect apparently dominates over gravity darkening and T_{eff} variation effects, which should yield spurious ω values of 90° or 270°.

If we look at systems with non-circular orbits, which cause tidal interactions at periastron, we see very clear effects. In HD 187399 and AZ Cas (both B stars), e $\sim$ 0.4, and mass-loss is spectacular at periastron. In the low e>0 stars (Hutchings 1978b) shown in Table 5 we see a continuous variation in the wind, with maximum at periastron. HD 163181 (BO Ia + O) is another example, and we also note that a periodic wind modulation is seen in HD 153919 by Hammerschlag-Hensberge (1978). In this connection it is significant that my light curve analysis shows that e>0 and that this effect once again is maximum at

TABLE 4. X-ray System Parameters
Summary of approximate values

System	X-ray $\underline{e}$	X-ray eclipse	Spectroscopy $\underline{e}$	Spectroscopy ω	$\underline{g}$	light curve $\underline{e}$	light curve	$\underline{i}$	$\dfrac{L_{opt}}{L_x}$
4U1700-37	-	±46°	0.2	330°	~20	0.05	300°	87°	5000
4U0900-40	0.12	±35°	0.2	10°	13	0	0	73°	1300
Cyg X-1	-	-	0.06	330°	~2	0.04	330°	30°-60°	100
Cen X-3	0.0008	±40°	-	-	~15	(0	0)	~90°	100
4U1223-62	-	-	-	-	-	<.05	(90)	<60°	100
LMC X-4*	-	±32°	(0.2	50)	9	-	-	72°	3
SMC X-1*	<0.0007	±26°	(0.3	0)	16	(0	0)	64°	4
4U1538-52	-	±28°	0.2	350°	10	-	-	70°	500

* e = 0 orbits adopted

TABLE 5.   Stellar Winds in Noncircular Binary Orbits

System	$-\dot{m}$ ($M_\odot$ year^{-1})			e
	max	min	max/min	
29 CMa	1.5×10^{-5}	10^{-7}	150	0.09
163181	1.5×10^{-5}	10^{-6}	15	0.08
47129	1.5×10^{-5}	2×10^{-6}	7	0.04
108	4×10^{-5}	10^{-5}	4	?

the phase of periastron. (Note that the spectroscopic e, ω values are once again spurious, and attributable to the permanent tidal deformation of the primary, as mentioned above.)

There is a further indication. In my mass-loss survey, the known binary stars have twice the mean mass loss rate of the average for all objects. This is not highly significant in a small sample, but supports the general picture above. We need more detailed studies of close binaries to clarify the whole position, and quite possibly the UV will provide more clear cut answers.

There is every indication that many O stars are in close binary systems. Thus, in determining their fundamental parameters we must beware of distortions to the velocity curves. Alternatively, we may regard high quality spectra of close binaries as containing information on the structure of the stellar winds.

It is of interest to look at the statistics of O star binaries. Since new binaries are being found and studied all the time I may have missed a few and my numbers will certainly be out of date soon. At present the rough picture is as follows. There are 11 eclipsing systems known, of which two have O star companions, 5 B stars, 1 W-R star, and 3 neutron stars. There are 45 spectroscopic binaries, of which 16 have O star companions, 17 are single-lined, 8 have W-R companions and 4 have neutron stars (or Black holes). This bears out the ideas that massive stars spend very little time between being OB stars and neutron stars, and that the W-R stage is a shortlived but significant part of the evolutionary scheme. Note that the VV Cep systems include stars very close to being O stars and should be studied in the evolutionary scenario. They are probably over-represented because their spectroscopic peculiarities makes them easy to discover.

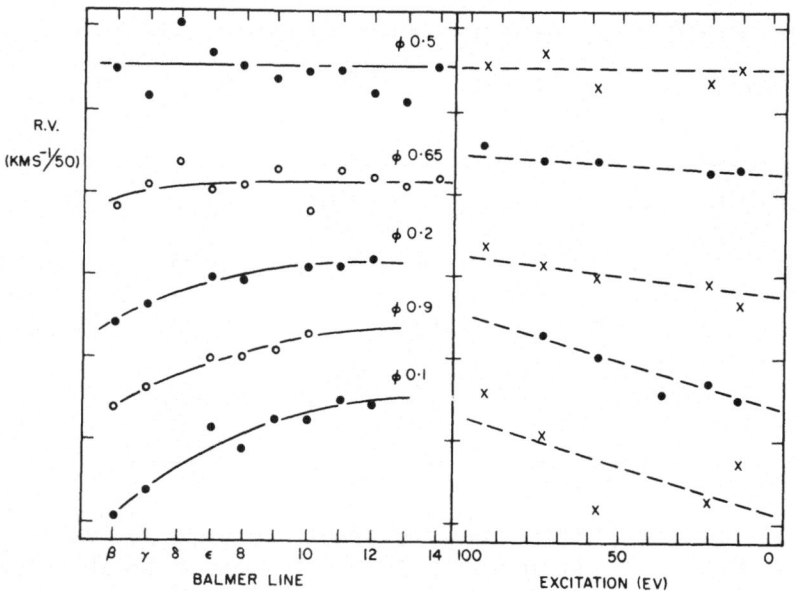

Fig. 6.   Systematic variation of Balmer progression and velocity-excitation slope with phase, in e>0 binary 29 CMa.

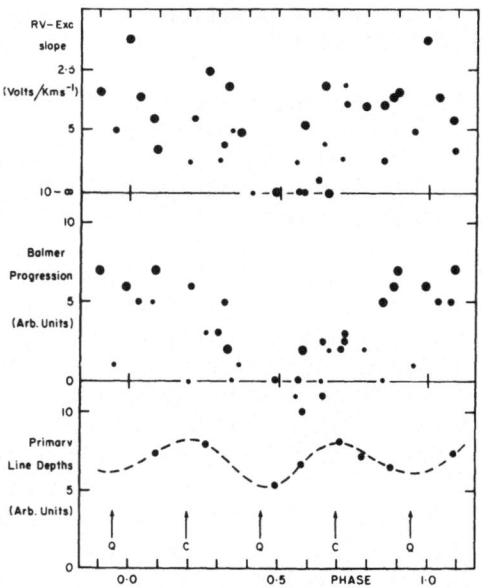

Fig. 7.   Variation of mass-loss rate of primary in e>0 orbit of HD 47129.

VII. MASSES AND DIMENSIONS

The masses of the 0 stars are not well known. Conti and Burnichon (1975) derived M_v and temperatures for a sample of stars of uniform classification and well established distances. Applying bolometric corrections, they showed that they lie on the H-R diagram occupied by evolutionary tracks for stars of mass 20 to 120 $M_\odot$. Masses from spectroscopic binaries are few in number and are generally at the low end of this range. However, they are generally consistent with the picture. Nevertheless, these cases are rare and one can point to uncertainties in every one. Probably the best determined masses are those of the X-ray binaries, but, as pointed out earlier, they are all low for their luminosity, if we believe the above picture. We can brush this off as being the result of extensive mass-loss or exposure to a supernova explosion but we should perhaps still worry a little about this anomaly and the comparison between observational and theoretical H-R diagrams (i.e. do we really know T_{eff}'s and the B.C.'s?).

The X-ray systems provide us with radii and in general the luminosity derived from these, the accepted T_{eff} values and B.C.'s agree well with those derived from distance and reddening determinations (i.e. to within 0^m1 in most cases). The few eclipsing normal systems yield numbers consistent with these too. Thus, the main sequence radii run from $\sim 8R_\odot$ at B0 to $25R_\odot$ at 05, and supergiants in the 15-30 $R_\odot$ range. Figure 8 shows Conti and Burnichon's stars on the H-R diagram, and binary stars whose masses are known.

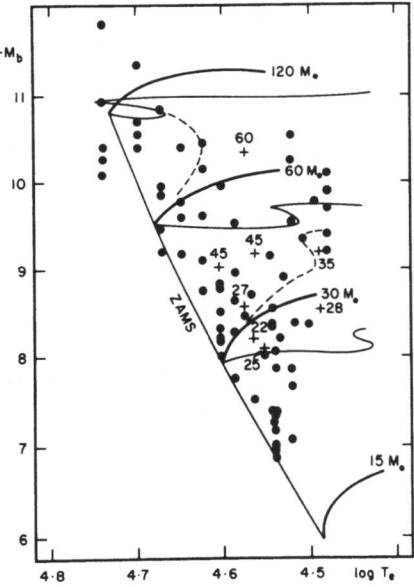

Fig. 8. HR diagram for 0 stars showing apparent agreement of observed and calculated positions. Crosses are binaries for which masses (as marked, in $M_\odot$) are known.

VII. SOME SPECTROSCOPIC MYSTERIES

The optical spectra of O stars contain a few features which are still unidentified - a rare phenomenon in modern astrophysics. The most famous are the emission features at λλ 4485, 4503. These lines are almost always 3-4 A wide and have intensities from 3-10% above the continuum. They have been associated with C III λ 5696 emission but the unidentified lines seem to occur more frequently. It is of interest that these lines have been seen in LMC X-4, where abundances may be different from the galaxy.

A less well known line is an absorption at λ 4726, which has been seen in X Per, HD 47129, HDE 245770, Cen X-3, and the optical primary of 4U1538-52. Three of these are X-ray source companions and all are O8-09.5 stars. The line is similar in profile and strength to He I λ 4713. It is listed by Herbig as an interstellar feature, and its strength correlates with reddening in the above stars.

Next I would point out the broad emission bands seen around Hδ and the λ 4650 region in extreme Of stars like HD 152408, 153919. They are up to 5% in intensity and ∿50 A wide. I guess they originate in the outer parts of the wind, are connected with W-R bands, and arise in the strong Si IV, N III and He II transitions in these regions of the spectrum. Figure 9 shows the bands in the extreme Of stars.

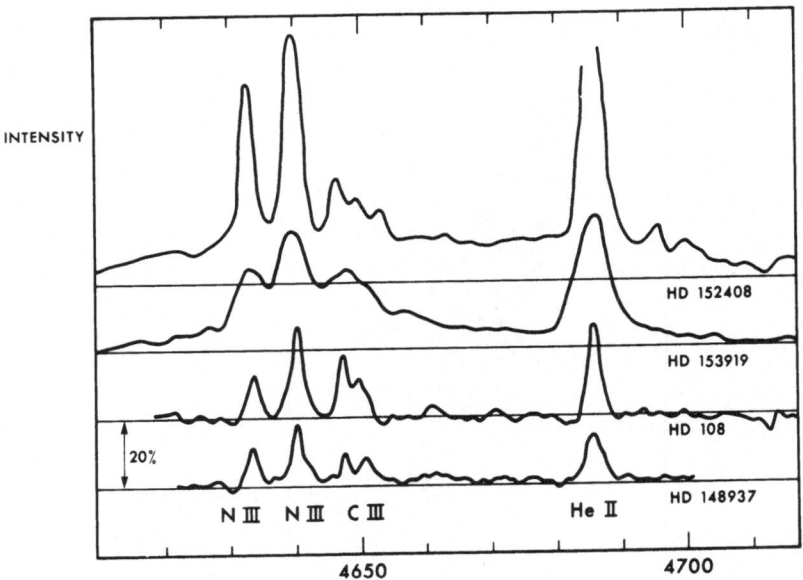

Fig. 9. Tracings of spectra of extreme Of stars, showing broad emission features in HD 152408, 153919.

I want next to mention He I λ 5875.   This line shows a high velocity extra component, at times.   It is seen e.g. in HD 152408 at ℩ -800 km s⁻¹, and in HD 153919 at about this velocity at φ =0.7 (X-ray).   The latter case is often associated vaguely with a "wake" trailing the X-ray emitter, which is immersed in the stellar wind (< 1 R∗ away).   The only similar spectral feature is He I λ 3888 in HD 152408, in which the line is formed through a large depth, accumulating enough absorption near the terminal velocity, to give a second dip.   It is not known why this process should occur in λ 5875.

Finally, there is the mysterious emission line at λ 5300 seen in HD 153919 by Dupree and tentatively identified as Fe XIV.   She claims to have confirmed the observation in this and other O stars.   If it is real (and perhaps it needs further demonstration) its identification may indicate the existence of a hot corona, and/or special line passes.

IX.   ROTATION

The study of rotation in O stars by Conti and Ebbets (1977) showed that the situation is much as might be expected, and similar to the B stars.   There is probably some macroturbulence which adds 30-90 km s⁻¹ to V sin i going from O9 to O3.   The main sequence stars show higher velocities, corresponding to the rapid rotating Be class.   It is interesting to note that the rapid rotators also have high ṁ (10⁻⁵ M⊙/year).   This would imply that rotation enhances a stellar wind by lowering surface gravity.   More work along these lines needs to be done on the Oe stars.   Unfortunately, the high rotation leads to weak lines and difficulties in measuring velocities, so that high quality data would be required.

I would like to emphasise a pet point here.   In the extreme cases a stratified expanding envelope will underestimate V sin i measured from absorption lines.   This leads directly to a controversy over HD 153919 whose absorption line V sin i is much slower than synchronous.   My recent light curve analysis (Hutchings 1978) suggests that rotation must be synchronous (as do circularisation arguments, and V sin i estimates from emission lines).

X.   RAPID VARIATIONS

There has been a great deal written about rapid spectrum variations in OB stars (see Lacy 1977 for critical comment and references).   Claims have been made for changes within a star or from night to night, based on narrow band photometry, photographic spectroscopy, line scanners, and electronic detectors of various types. In nearly every case there is a possibility that the effect is instrumental, or measuring errors, and the experimenters have not made sufficient control observations.   On the other hand, some good experiments have been done and in some cases, very similar results have been obtained by very different techniques.   Later in this volume some very spectacular changes in the B supergiant HD 190603 are reported.

Figure 10 shows changes in the Of emission lines in λ Cep (Hutchings and Walker 1977), on a moderate time scale.

It is known that over periods of years or months spectral changes do occur in O and particularly Of stars, and so do changes of order 0^{m}1 in brightness. A star with a typical stellar wind will replace its entire line forming envelope in about a day, a time shorter than the average rotation period. Thus, inhomogeneities in the flow or modulations in the wind could show up as rapid variations, and indeed seem very probable to me. It is more questionable what we will learn from them, especially as a proper observational job calls for large expenditure of telescope time. Irregular looks at stars and occasional cries of "oh look it's changed" are not getting us very far. We should look at these changes photometrically and spectrophotometrically over a wide range of λ's and spectral lines. This may show us whether brightness or temperature changes occur, and show whether we can follow disturbances propagating outwards in the wind, or stellar rotation, etc.

XI. RUNAWAY STARS

Blaauw (1960) estimated that some 20% of O stars are runaways, based on radial velocity data. Stone (1978) comes to a similar conclusion, estimating their space velocities. Conti <u>et al.</u> (1977a) give 8% on a larger sample and paying more attention to the confusing effects of stellar winds and binary membership. It is clear (Conti <u>et al.</u> 1977b) that in extreme cases, photospheric velocities up to $\overline{65}$

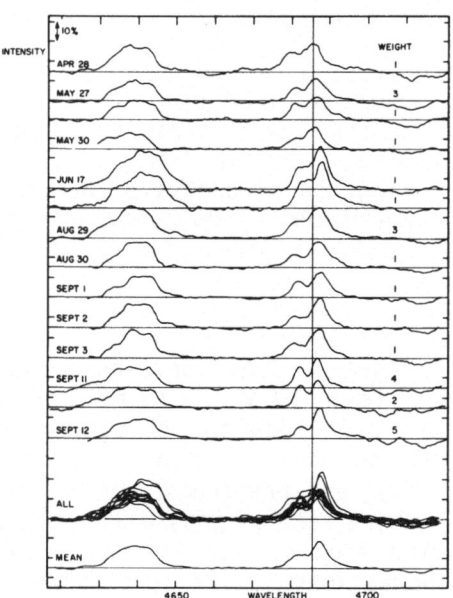

Fig. 10.   Image Isocon scans of Of emission lines in λ Cep, showing short term profile changes in 1976.

km s^{-1} can arise from mass flow. We should also expect in general that stars from a finite area of the galactic plane that are runaway stars with high velocity will show more with positive than negative velocities. It seems to me that there are probably 10% of O stars that are runaways (i.e. receding by > 40 km s^{-1}), and this is probably a direct result of supernova explosions. In this connection it is worth noting the high (-110 km s^{-1} wrt local standard of rest) radial velocity of the O8 primary of the X-ray pulsing source 4U1538-52. Evidently the system can achieve high velocity <u>and</u> remain bound. Note the discussion in this volume on runaway stars.

Perhaps it is worth searching for supernova remnants near to O stars which may have (had) an evolved companion. O stars generally outlive SNR's by 1-2 orders of magnitude, but the chance seems to me not negligible. There is intriguing nebulosity associated with HD 148937, γ Cas, and R148, a candidate star for LMC X-1.

XII. ABUNDANCES

A question which aries in connection with extensive mass-loss is whether the surface abundances are altered by exposing evolved material. This may well occur if 30-50% of the outer layers are removed. Abundance determination however, is complicated by the extreme non-LTE conditons in the envelopes of the stars: particularly the apparent existence of a hot corona-like outer layer in which the high ionised UV resonance lines are generated.

The observational evidence is not compelling either way. However, in some extreme cases, like HD 152408, the N III spectrum is complete, and in emission. This phenomenon is also seen (less markedly) in the X-ray primary of 4U1538-52, which does not show evidence for a strong wind, but which is undermassive by ∿40%. On the other hand, we don't see it in LMC X-4, SMC X-1 or Cyg X-1, which are apparently similar objects. The mass exchange binary HD 163181 has a BO primary which is undermassive by ∿70% and shows strong CNO lines. Its (O type) companion is hard to see but has normal mass. This is the strongest correlation I know of mass-loss and abundance. I don't know of any cases of strong He, except for β Lyr, which is not an O star. It would be worthwhile trying to derive abundances for O and Of stars.

If we consider WN and WC stars to be showing strong abundance anomalies by mass loss, we may expect effects of that magnitude to appear when some 80% or more of the original mass is lost.

XII. THE MAGELLANIC CLOUDS

I have spectra of a number of OB stars in the clouds of which two (Of) are in the SMC and another 4 in the LMC. The brightest of these are M_V -7 and thus compare with high mass loss objects in the galaxy. There are not extreme stellar wind indications in these, though the most likely object has Hβ P Cyg profiles. One star is an Oe star.

The most obvious stellar winds in the clouds are seen in some B supergiants.   This is very preliminary and more careful work needs to be
done before concluding anything about the cloud stars.   It appears that
the distribution on the HR diagram of MC and galactic supergiants is
similar.

CONCLUSION

        This is not the place for summarising.   I hope my remarks have
opened topics for discussion and brought relevant data forward to be
included in the many issues we have to look at in the next few days.

REFERENCES

Blaauw, A. 1960, B.A.N. 15, 265.
Barlow, M.J., and Cohen, M. 1977, Ap. J. 213, 737.
Conti, P.S., and Burnichon, M.L. 1975, A & Ap 38, 467.
Conti, P.S., Garmany, C.G., and Hutchings, J.B. 1977b, Ap. J. 215, 561.
Conti, P.S., and Ebbets 1977, Ap. J. 213, 438.
Conti, P.S., Leep, E.M., Lorre, J.J. 1977a, Ap. J. 214, 759.
Hammerschlag-Hensberge, G. 1978, A. & Ap. 64, 399.
Hutchings, J.B. 1968, M.N.R.A.S., 141, 329.
Hutchings, J.B. 1976, Ap. J. 203, 438.
Hutchings, J.B. 1978, Ap. J. Nov. 15.
Hutchings, J.B., Crampton, D., Cowley, A.P., Osmer, P. 1977,
        Ap. J. 217, 186.
Koch, R.H., Plavec, M. and Wood, F.B. 1970, Univ. of Philadelphia
        Pub. Vol. X.
Lacy, C. 1977, Ap. J. 212, 132.
Lamers, H.J.G.L.M., and Snow, T.P. 1978, Ap. J. 219, 504.
Stone, R.G. 1978, preprint.

DISCUSSION FOLLOWING HUTCHINGS

        Snow: For OB supergiants which show photometric variations which
are non-periodic, what are the typical timescales at these variations?

        Hutchings: These timescales are not well known; to date little
systematic observational work has been done.

        Underhill: The small extra reddening found by Isserstad for the
most luminous O stars in some clusters relative to O stars of the same
subtype but low luminosity and interpreted as circumstellar reddening
may be a result of the fact, see paper by me, Divan, Doazan and Prevot-
Burnichon, that the effective temperatures of luminous OB stars are
significantly lower than those of main-sequence OB stars of the same
subtype.   I have considerable doubt that true circumstellar reddening

exists at a measurable level and can be determined by the method used by Isserstad.

Hutchings: I too suspect some intrinsic reason of this nature. Still the effect needs to be explained one way or the other.

Lamers: You find your lines of equal mass loss rates in the HR-diagram to decrease towards the A-supergiants; whereas Barlow and Cohen's data suggest that the rates are dependent on luminosity only, so these lines should be horizontal. A possible explanation for this discrepancy might be found in the fact that the stars which you use for calibrating your mass loss indicators are all of spectral type around O9 to B1. If you consider the fact that in late-B or A stars the hydrogen lines are formed at lower density than in the B0 stars, you can get the same Balmer-progression in a late-B supergiant with a smaller mass loss rate than in an early-B supergiant. Consequently, your mass loss rates for late-B and A-type supergiants may be overestimated.

R. E. Wilson: What is it about the spectroscopy of the contact or near contact systems which indicates that they may not be in contact?

Hutchings: The sort of information is indication of eccentric orbits with consistent indication of enhanced mass flow at periastron for quite a random distribution of ω values.

R. E. Wilson: Normally spectroscopy doesn't give this kind of information accurately enough.

Hutchings: I worry about photometric solutions for systems where Roche geometry is imposed, e = 0 orbits only are considered, and where we may have photospheric mass flow at tens of km/sec and luminosities near to the Eddington limit (e.g., AO Cas, 29 CMa).

Leung: The spectroscopic eccentricity is not real. The asymmetry in the radial velocity curve is not due to an eccentricity in the orbit but is caused mainly by the effect arising from the tidal distortions of the components [e.g., UW CMa, Leung and Schneider, Ap. J. 222, 924 (1978)].

The luminosity ratio derived from strengths of the spectral line ratio is not dependable, since the mass accreting component most likely has a peculiar atmosphere -- very low density rarified atmosphere.

Hutchings: The effects you mention are calculable and imply ω ~ 270° or 90°. In practice ω is well distributed and derived from the primary spectrum. Luminosity ratios may be suspect. I note that your photometric solutions do not consider e = 0 so I don't see how you know e is not real.

Seggewiss: You showed several diagrams in which the velocity of
the expanding envelope is plotted against "excitation."   If one looks
through the literature we can find different values used for "excitation."   L. F. Smith & L. H. Aller [Ap. J. 164, 275 (1971)] and A. F. J.
Moffat & W. Seggewiss (this Symposium) use the maximum of the lower excitation potential (EP) of the line and the ionization potential (IP)
of the preceding ion.   You used in 1976 (Ap. J. 203, 438) the sum of
the line EP and the IP of the preceding ion, and in 1978 (P.A.S.P. 90,
179; also M. de Groot, B.A.N. 20, 225) the sum of the EP of the line
plus the sum of the IPs of all preceding ions.   I feel that we should
come to an agreement about what should be plotted as "excitation" in
the future.

Hutchings: I agree with you, although the qualitative sense of
what we are showing is not affected.   You do not quote me correctly on
the 1976 reference: In that and all subsequent references I have summed
the IP of all preceding ions.   This is supposed to measure the temperature (monotonically falling) to the work done in arriving at the lower
level of a given absorption line.

Morton: What are your present calibration stars for mass loss
rate?

Hutchings: The stars are the same as I used before, but I think
P Cyg is not typical or trustworthy any more (Snow et al. 1978).   Also
I use ζ Pup and revised numbers for the other stars, taking into ac-
count the work of Barlow & Cohen, Conti, etc.

Morton: Are the broad emission lines similar to the features Bob
Wilson reported 20 years ago?

Hutchings: Probably.

Morton: Do they have the same velocity width as would be expected
by the terminal velocity in the wind.

Hutchings: No.   They are larger by 2 to 3 times the UV line values.

Underhill: The lines in the visible and UV are probably coming
from different optical depths and so may well indicate different outflow velocities.

H ALPHA VARIATIONS IN TWO MASS LOSING STARS

J.M.Vreux
Institut d'Astrophysique, Liège, Belgique
Y.Andrillat
Observatoire de Haute Provence, France

The aim of this paper is to report quite preliminary results obtained with a new multichannel detector put into operation at the Haute Provence Observatory. The observations have been made at the Cassegrain focus of the 1.93 m. The dispersion of the spectrograph was 17 Å.mm^{-1} at H_α and the slit was 300μ wide. The detector was a cooled SIT TV camera tube commercially available under the name of "Nocticon".Details on the observing technique have been given elsewhere (Adrianzyk and al.,1978). We will only point out that all the wavelengths are recorded simultaneously : this recording technique is different from the scanning technique the result of which have been recently reviewed by Lacy (1977).

The first result we will report concerns HD 60848 : (figure 1).

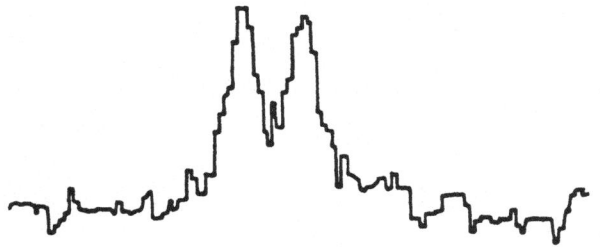

Figure 1. H_α profile in HD 60848 (V = 7.7)
Integration time : 180 sec.

The profile of H_α is double peaked as expected in this Oe star (Conti and Leep,1974), the separation between the peaks is of the order of

P. S. Conti and C. W. H. de Loore (eds.), Mass Loss and Evolution of O-Type Stars, 23-26.

4.3 Å. This profile has been observed during half an hour (interval between two successive integrations : 1 min) and no variation has been detected.

The next result concerns BD +40° 4220. This is a binary (O7+O6) on its way to become a Wolf Rayet (Bohannan and Conti,1976). This 9.1 magnitude star is strongly reddened, and the spectra given in figure 2 have been obtained with an integration time of 200 sec.

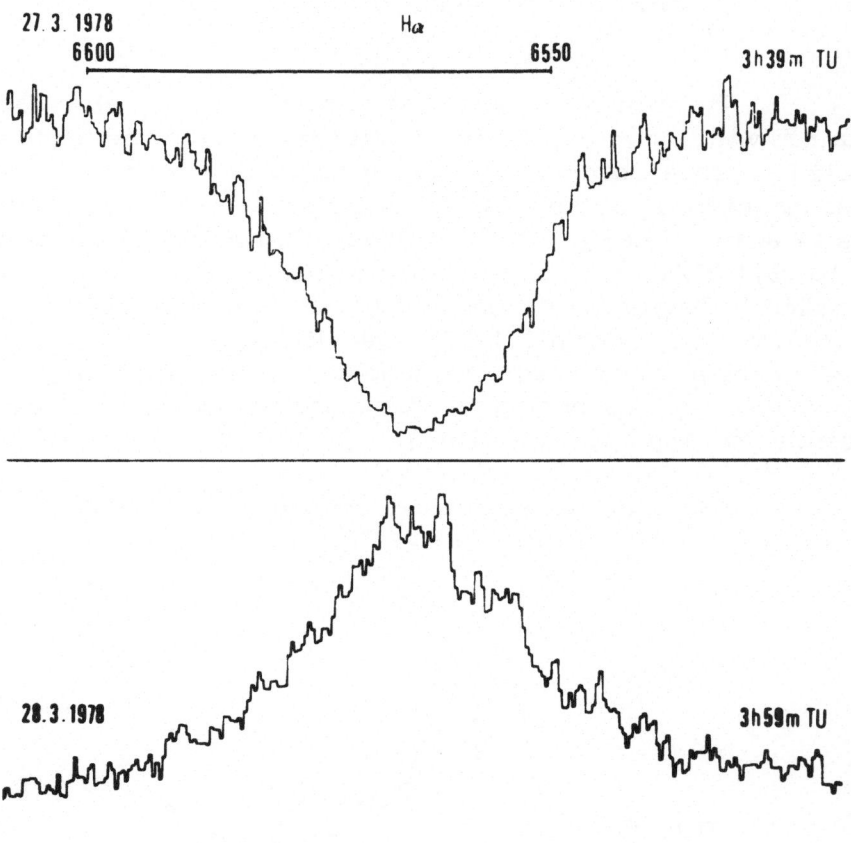

Figure 2. Variations of H$_\alpha$ profile in BD +40° 4220.

Each of them is a sample of the night : during each night the profile was stable during the observations (about half an hour). The emission observed during the second night (28/03/78) is compatible with the few photographic observations of Bohannan and Conti (1976). The result of the first night (27/03/78) is new : the frequency of appearance of these pure absorption profiles will be investigated in the near future due to its importance on the mass loss rate (10^{-5} $M_\odot$ yr^{-1}) derived by Bohannan and Conti (1976).

The last result concerns HD 190603. This is a B 1.5 supergiant with a mass loss rate of 5.10^{-5} $M_\odot$ yr^{-1} (Hutchings, 1976). According to Hutchings (1971) the spectrum of this star is variable "P Cygni type emission is seen at H_β and occasionally at H_γ . There is a definitive activity in both these lines". We have observed this star during about one hour, the integration time of each spectra being of the order of 120 sec. Large variations of the H_α profile on a time scale of a few minutes have been recorded. However these results have been obtained near the saturation level of the detection system and they could be purely instrumental. This point will be checked as soon as possible.

REFERENCES.

Adrianzyk, G., Baietto, J.C., Berger, J.P., Fehrenbach, Ch.,
      Prévot, L., Vin, A., 1978, Astron.Astrophys., 63, pp.279-283.
Bohannan, B., Conti, P., 1976, Astrophys.J., 204, pp.797-803.
Conti, P., Leep, E.M., 1974, Astrophys.J., 193, pp.113-124.
Hutchings, J.B., 1976, Astrophys.J., 203, pp.438-447.
Hutchings, J.B., 1971, in "Colloquium on Supergiant Stars",
      Proceedings of the third Colloquium on Astrophysics held at
      Trieste, ed.: M.Hack, pp.38-47.
Lacy, C.H., 1977, Astrophys.J., 212, pp.132-140.

Editor's Note: Considerable discussion ensued at the meeting concerning
                 the consequences of a "rapid" change in the H_α profile
                 of HD 190603. The participants were not fully aware of
                 the very tentative nature of the result, as presented
                 in this written contribution.

     Note added in proof : variations of H_α in the spectra
of HD 190603 were no longer observed on November 20 and 22
(different instrument, integration time : 600 sec.)

DISCUSSION FOLLOWING VREUX AND ANDRILLAT

Thomas: I agree with Hutchings' caution in his previous paper that such new short term variability observations should be greatly extended. This is being done. But please note that the behavior of the Hα line exhibited in this paper has very strong significance on the question of "deep chromosphere" vs. "extended atmosphere -- cold or hot" models. I assume we will discuss all this in Session 4 tomorrow; but I am glad this paper, these data, by Andrillat and Vreux came so early in this Symposium. As Hutchings knows well, from his own work, the evidence for such variability in Hα -- and other -- emission line profiles is widespread: the whole question is the frequency of occurrence and the amplitude. I assert that current theoretical models tell us nothing -- indeed there is really no theory, at most certain broad thermodynamic characteristics. So -- these kinds of observations are really what is guiding us in modeling.

Conti: How common are the short time scale variations demonstrated in this B type supergiant? Are they typical or atypical for such stars?

Vreux: So far we only have observations for five objects. You have seen the results for three of them. The two others have a noisy spectra: We can only say that there are no dramatic variations. For HD 190603 we have many spectra and the variations you have seen are typical. But for this object we are near the saturation of the system and we have no idea about the amplitude of the variations you have seen; the ordinate of the graph (intensity scale) is not linear and it could be that the amplitude of the variations is quite small. This point will be investigated as soon as possible. As I said these results are quite preliminary and more than anybody here we regret we have not had time enough to perform all the tests and calibrations before this meeting.

SHORT TERM VARIABILITY OF LINE STRENGTHS
IN SOME OF AND WR STARS

W.G. Weller and S. Jeffers
Physics Department and CRESS
York University, Downsview, Ontario

1. INTRODUCTION

In recent years many observations of short time scale variability
in a variety of spectral classes have been published. Some of these
observations relate to phenomena which are reasonably well understood,
as for example the Beta Canis Majoris stars, which are short period
pulsators. In other cases, such as the Of and WR stars, the mechanism
responsible for the variation is not fully understood. Models
proposed to explain this variability include:- the binary hypothesis
which ascribes the variations to fluctuations in gas streams in and
around the members of a close binary system, and:- the pulsation
hypothesis, which ascribes the variation in line strength to the
presence of pulsational instability of massive carbon burning cores of
evolved objects. The choice of the correct model is made difficult by
the lack of an extended set of homogeneous observations. We must ful-
fil two conditions before making this choice. It is necessary to observe
individual objects over extended periods of time at high dispersion to
establish whether or not all WR's are binary systems. It is also
required firmly to establish the temporal nature of the variations,
since short term periodicities would tend to favor the core pulsation
model.
Brucato, (1971), published a photographic survey of line
strengths in nine Of stars, and concluded that there was evidence for
variability in several of his samples. Bhang, (1975), has published
photoelectric line profiles for five WR stars. He similarly reported
variations in the profiles of emission lines, but on longer time
scales. Most of the remaining observations of rapid variability have
been of isolated stars. We report here on an analysis of some data
obtained for a sample of 10 Of and 13 WR stars observed in May, June,
and July 1974. A more complete analysis will be published elsewhere.

2. OBSERVATIONS

The spectra from which we obtain our results were taken with a
SIT Vidicon described by Jeffers and Weller (1976). These spectra

P. S. Conti and C. W. H. de Loore (eds.), Mass Loss and Evolution of O-Type Stars, 27–30.
Copyright © 1979 by the IAU.

cover 150.nm at a dispersion of ·3 nm per channel and instrumental
FWHM of 3 channels. The stellar lines were resolved, even for the
coolest WR subclasses. Each data set comprised from 9 to 20 spectra,
exposed for 30 seconds, with 2 minutes between exposures.

A list of stars observed is as follows:-Of stars HD57060(=29CMa)
HD124314, HD148937, HD150938, HD151804, HD153919(=3U1700-37),
HD164794(=9Sgr), HD175754, HD188001(=9Sge); WN stars HD50896(=EZ CMa),
HD86161, HD92740, HD151932; WC stars HD68273(=Gamma Vel), HD92809,
HD93131, HD136488, HD156385, HD164270, HD165273, HD168206(=CV Ser);
and the WP starHD90657.

Spectral features were reduced in the following way. Two
rectangular "filters" were defined, each centered on the feature to be
measured. The widths of the filters were chosen to be 1.5 and 6.0 nm.
The ratio of the fluxes in these pass bands defines a line strength
parameter which is independent of instrumental and atmospheric effects.
Data were standardised with observations of stars of spectral types
B5 and A0. These stars have absorption lines which are comparable in
strength to the emission lines in the program stars, and can be
assumed to have constant line profiles. Program stars were assumed not
to vary if the ratio of fluxes did not vary any more than that for the
standard stars (.014 mag. peak-to-peam). Figure 1 is a sample plot for
two Of stars indicated by this study to be constant in line strength.
The upper line in each plot is a measure of the flux in a portion of
the continuum. It varies strongly due to the effects of scintillation
and guiding errors. This fluctuation is not seen in the ratio plot.

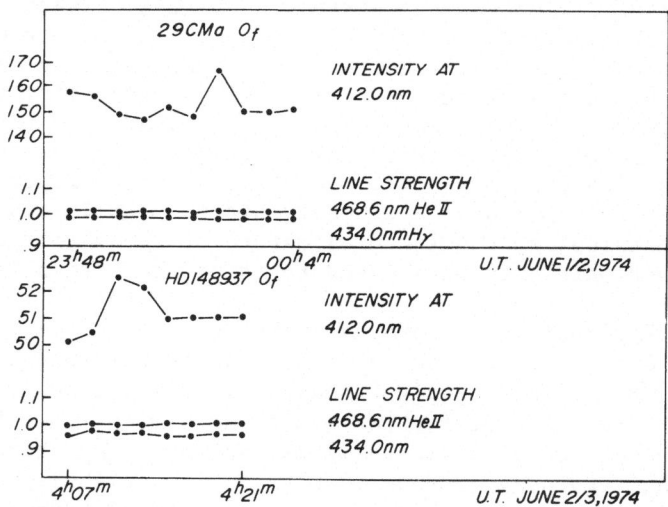

Fig.1 Total intensity and flux ratios for HD148937 and 29 CMa

The nature of the plots is indistinguishable from that of the
standard stars observed. Figure 2 presents data for HD50896 on 2
nights. On both nights both plotted lines show a gradual increase
in strength over the 18 minutes of the observations. On the night of
June 3/4 there is evidence for irregular rapid fluctuations at 486.1 nm
(HI, HeII).

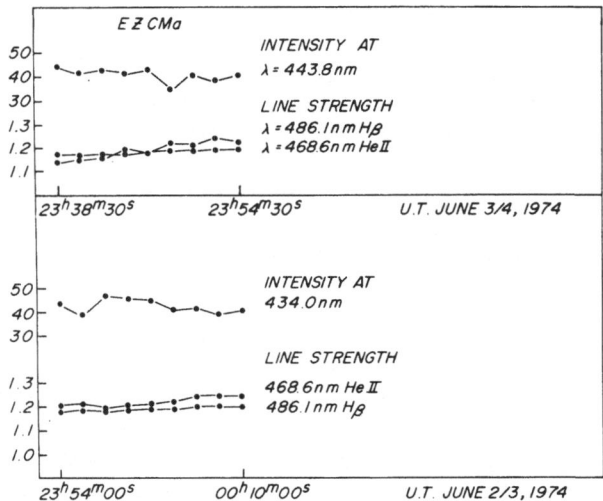

Fig.2   Total intensity and flux ratios for HD50896 on 2 nights

3.   RESULTS AND DISCUSSION

The results of the study are uniformly negative for all observed
Of stars, to the limit of our sensitivity. The WN stars in general
also yield negative results, except for HD50896 a 1.01day binary
(Kuhi, 1967). For the stars of type WC we find activity in several
spectral features. HD68273 (Gamma Vel), and HD92809 both vary at 468.6
nm (HeII), but apparently not at 434.0 nm (HI+HeII) HD156385 yields
marginally positive results at 443.8 nm (CIV), but not at 465.0 nm
(CIII). HD164270 appears to vary at 434. nm, 486.1 nm (HI+HeII) and
465.0 (CIII). No inference of periodic activity can be drawn on the
basis of our data. There appears to be a trend in the following sense:
The later the subclass, the more likely one is to find variability.
The general lack of variability on time scales less than one hour in
both Of and WN stars and the presence of such in the WC stars
provides another observational distinction between the types.

ACKNOWLEDGEMENTS

The work described here was supported by the National Research
Council of Canada. S.J. gratefully acknowledges travel support from
the I.A.U. and the Physics Department of York University.

REFERENCES

Bhang, J.D.R., 1975, M.N.R.A.S., *170*, 611.
Brucato, R., 1971, M.N.R.A.S., *153*, 435.
Jeffers, S., and Weller, W., 1976, AEEP *40*, 887.
Kuhi, L.V., 1967, PASP, *79*, 57.

DISCUSSION FOLLOWING WELLER AND JEFFERS

Thomas: I would agree with your second conclusion: Short-term variability of emission line strengths implies variability in energy input, which implies mechanical energy. But I disagree on the first, that no short-term variability implies radiative dominance. Note that all chromospheric theories based on a time-independent mechanical flux (i.e., generated by statistically-steady mechanical energy input) predict no variability, even though chromosphere-corona are dominated by mechanical energy input.

Underhill: Lack of observation of short-period variations in the intensity of a strong emission line is not evidence that short-period variations in the physical state of the atmosphere do not exist. For Of, WN stars Hα, Hβ and He II 4686 may come chiefly from a large outer "nebula" which may not show irregular variations. However, an irregular input of energy and excitation into a low-lying, relatively dense chromosphere/corona may occur. To see this you must observe a line originating chiefly in that region. The visible spectral range does not offer suitable lines for the expected Te, Ne ranges to be found just outside Of and WN photospheres. The choice of spectroscopic feature which is observed is very critical.

Jeffers: Agreed.

Noerdlinger: You cannot conclude from the presence of variations that there is mechanical energy input. Radiation pressure in lines can cause instabilities leading to mechanical waves. The instability was first discussed by E. A. Milne.

Jeffers: Agreed. Our statements regarding the interpretation of the data should be regarded as speculative.

MASS LOSS EFFECTS IN THE SPECTRUM OF THE O 6.5f BINARY HD 153919

G. Hammerschlag-Hensberge
Astronomical Institute, University of Amsterdam

Abstract. Apart from the known P-Cygni profiles of Hβ, HeI 5876, 4471 in the spectrum of HD 153919, also HeII 3923 and a line at 3759 A which is tentatively identified as OIII show an emission component. These lines indicate that the Of star is losing mass at a high rate. A detailed analysis of the radial velocities obtained from 76 spectrograms indicates that the outflow of the wind is not spherically symmetric: at the side where the Of star faces its companion the velocity of the wind has increased.

The extreme Of star HD 153919 has been identified as the optical counterpart of the X-ray binary 4U1700-37 (Jones et al., 1973; Jones and Liller, 1973; van den Heuvel, 1973). In this paper we will discuss the spectrum of the Of star and its mass loss. The visible spectrum of this star shows many characteristics of a high mass loss rate: several P-Cygni profiles, Hα, HeII 4686 and NIII, CIII 4650 in emission (cf. Hensberge et al., 1973; Conti and Cowley, 1975).

During the last few years we collected 76 blue plates of this star, all obtained with the 1.52 m telescope of the ESO in Chile. An inspection of the intensity tracings of these plates reveals that also some other lines show emission, which was previously undetected. The HeII line at 3923 A shows a broad -though weak- emission component on top of the CaII interstellar line. Fig. 1 shows also another P-Cygni profile which was detected at 3759 A. It is present on all plates. The only possible identification seems to be OIII, although this line appears in absorption in ζPup, an O5f star.

A detailed radial velocity study of the spectral lines on the 76 spectrograms shows that the stellar wind is not spherically symmetric. The assymmetry seems to be due to the deformation of the shape of the Of star by the presence of the compact companion. The mean velocity curve derived for all lines is shown in fig. 2. Each point represents the average radial velocity for one plate. The curve drawn through the points represents the best-fit solution to the data (cf. also

P. S. Conti and C. W. H. de Loore (eds.), Mass Loss and Evolution of O-Type Stars, 31-34.
Copyright © 1979 by the IAU.

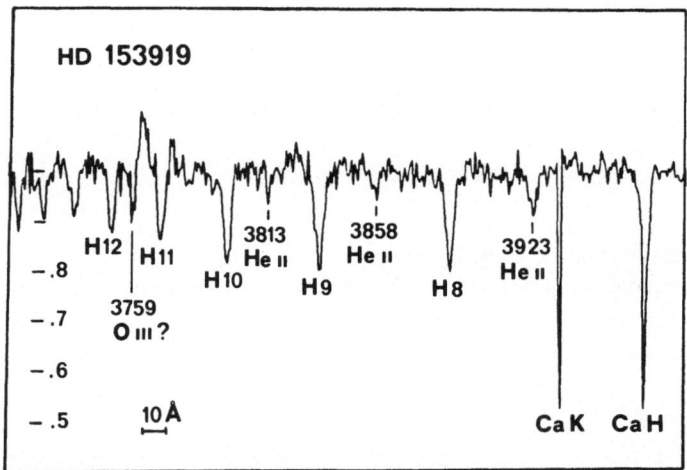

Figure 1. Part of the spectrum of HD 153919.

Hammerschlag-Hensberge, 1978 for more details). For a star with an out-streaming atmosphere as is the case here, one knows that the lines which are formed more outwards in the atmosphere have higher negative velocities than those formed more inwards. There is good evidence that even lines of HeII are formed in a region which is moving outwards. Before calculating the velocity curve, I corrected the velocities of

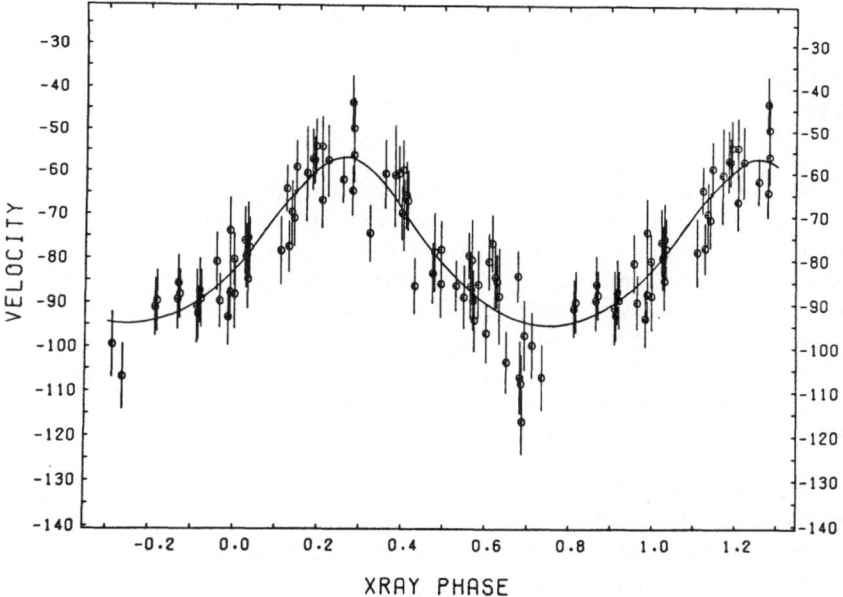

Figure 2. The average radial velocity for all lines plotted against X-ray phase. Phase zero corresponds to mid X-ray eclipse time.

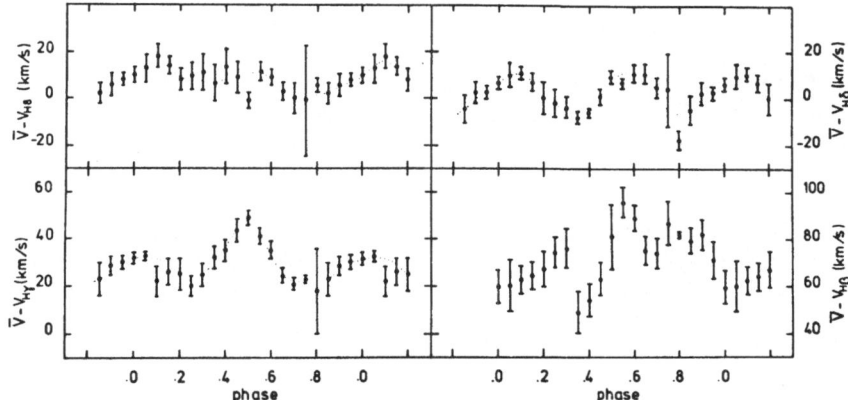

Figure 3. Radial velocity deviations from the mean for some
Balmer lines, plotted against X-ray phase.

all lines for this systematic deviation from the mean velocity. For a
spherically symmetric wind one would expect that this deviation would
be independent of binary phase. But a plot of the deviations gives for
some lines clear evidence for a phase dependence, as fig. 3 shows. For
instance, Hγ shows a clear deviation at phase 0.5, and to a lesser extent also at phase zero, where phase 0.5 is the phase at which the X-
ray source is in front of the Of star. Near phase 0.5 the velocity of
Hγ is systematically more negative. Comparison of the Hγ profile near
phase 0.5 with the profile at other phases shows no difference in intensity or shape. As a consequence the deviation must be due to a
velocity difference in the wind.

The effect can also have implications for the determination of the
binary parameters, e.g. the mass of the binary system. It clearly influences the eccentricity: we have corrected the Hγ-radial velocities
for the just mentioned deviation and after the correction the eccentricity obtained in the orbital solution was reduced from 0.36 to 0.06
(Hammerschlag-Hensberge, 1978). The deviation also influences the velocity amplitude and as a consequence the mass.

REFERENCES

Conti, P.S., and Cowley, A.P.: 1975, Astrophys.J. 200, pp. 133-144
Hammerschlag-Hensberge, G.: 1978, Astron.Astrophys. 64, pp. 399-405
Hensberge, G., van den Heuvel, E.P.J., and Paes de Barros, M.H.: 1973,
    Astron.Astrophys. 29, pp. 69-75
van den Heuvel, E.P.J.: 1973, Int.Astron.Union Circ. Nr. 2526
Jones, C., and Liller, W.: 1973, Int.Astron.Union Circ. Nr. 2503
Jones, C., Forman, W., Tananbaum, H., Schreier, E., Gursky, H., Kellogg,
    E., and Giacconi, R.: 1973, Astrophys.J. 181, pp. L43-48

DISCUSSION FOLLOWING G. HAMMERSCHLAG-HENSBERGE

Seggewiss: Could you please quote the errors of the eccentricities
in your last table? It might be possible that the error of the eccentricity for the corrected RV curve is of the order of the value of 0.06
itself.

Hammerschlag-Hensberge: The error on the eccentricity of the
velocity curve for Hγ is rather large: e ≃ 0.2, because the scatter
in the velocity curve is also large for this system. The intrinsic
scatter in the velocity values is as large as the velocity amplitude
of the orbit.

MASS LOSS OF B1Ia-O SUPERGIANTS AND EVOLUTIONARY CONSEQUENCES

B. Wolf, Landessternwarte Heidelberg-Königstuhl
C. Sterken, Astrophysical Institute, Vrije Universiteit
Brussel

I. Observations

The superluminous B supergiants are loosing mass and their progenitors are supposed to be O stars between O3 and O6 and hence it is justified to talk about B1Ia-O supergiants in this Symposium. Extensive high dispersion spectroscopic observations of four luminous B1 supergiants of the southern hemisphere have been carried out during 1972 and 1975 at ESO, La Silla. Some characteristic data of the program stars are summarized in Table 1.

Table 1: Some characteristic data of the B1 supergiants

Object	Spectral type	M_V	M_{bol}	T_{eff}
HD169454	B1Ia-O	-8.5	-11.0	24200
HD152236	B1Ia-O	-8.7	-11.3	24200
R 116	B1Ia-O	-8.2	-10.7	24200
BD-14°5037	B1.5Ia-O	-7.0:	- 9.5	23150

R116 is the brightest B supergiant of the LMC and hence its absolute visual magnitude is well known which is of special importance for the determination of the mass loss rates.

The most conspicuous spectroscopic features of our programme stars are P Cygni type profiles of several lines, especially of H_α (Sterken and Wolf, 1978a). These P Cygni profiles are variable with time.

Fig. 1 shows the line profile variations of the H_α and HeI $\lambda5876$ lines of HD152236 on two spectrograms taken at a time difference of 30 days in 1973 (note that the interstellar sodium lines are almost identical on both spectrograms which shows that variations due to instrumental effects are negligible).

The program stars are also radial velocity variables (Sterken and Wolf 1978b). The range of the variations which may occur on time scales of days is up to 30 km/sec and is higher than the velocity of sound which is about 20 km/sec in the atmospheres of these stars.

P. S. Conti and C. W. H. de Loore (eds.), Mass Loss and Evolution of O-Type Stars, 35–38.

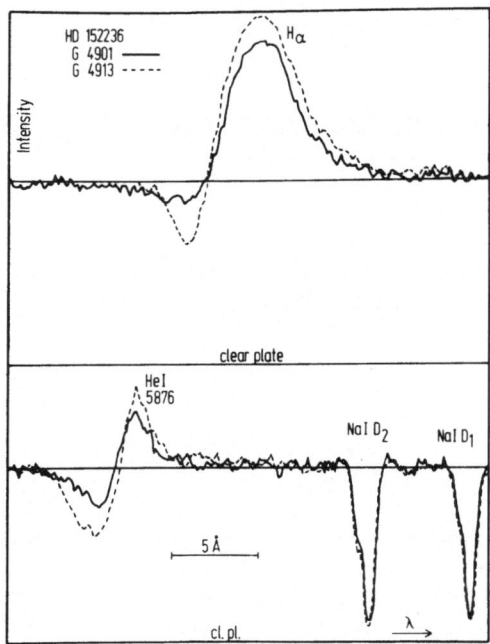

Fig. 1: Line profile variations of the H_α and HeI $\lambda 5876$ lines of HD152236 on two spectrograms taken at a time difference of 30 days in 1973.

II. Mass Loss and Mass Loss Variations

From the H_α P Cygni profiles mass loss rates were determined according to the method suggested by Hearn (1975). In this method a uniform spherical symmetric expanding shell is assumed. The excitation and ionization balance in the shell is determined by radiation from the star. The calculated shell radius is about five times the stellar radius, the electron density in the shell is $n_e \approx 5 \times 10^9 \, cm^{-3}$ and the mean mass loss rates are of the order of $1.5 \times 10^{-6} \, M_\odot y^{-1}$ (see Table 2).

Table 2: Shell radius R_{sh}, mean electron densities n_e, and mean mass loss rates $\dot{M}$ of the program stars

	HD152236	HD169454	BD-14°5037	R116
R_{sh} / R_*	4.9	5.7	5.0	10.4:
$n_e \, [cm^{-3}]$	5.5×10^9	5.0×10^9	6.2×10^9	1.4×10^9:
$\dot{M} \, [M_\odot y^{-1}]$	1.5×10^{-6}	1.7×10^{-6}	3.4×10^{-7}:	1.8×10^{-6}

The observed H_α intensity and profile variations are interpreted in terms of time dependent density variations in the shell, implying variable mass loss rates. The mass loss variations are as high as a factor of two and may occur on a time scale of days. Therefore one has to conclude that a steady state driving mechanism can not provide a complete description of the stellar wind mechanism in extreme B supergiants. Allowance has to be made for a time dependent process.

III. Evolutionary Consequences

The mass loss rate determinations are obviously of interest in connection with the current evolution theories of very massiv stars. In Fig. 2
the evolutionary tracks (cf. de Loore et al. 1978) of mass loosing stars
with initial masses of 60 and 100 $M_\odot$ and with mass loss rates of about
3×10^{-6} and $8 \times 10^{-6} M_\odot y^{-1}$ are shown together with the most luminous galactic stars (taken from Hutchings, 1976). The asterisks denote the po-
sition of the B1 supergiants studied here. Time marks (t_1, t_2, t_3) cha-
racterizing the evolutionary speed during the shell burning phase are
quoted. The assumed mass loss rates for which the evolutionary tracks
were calculated are comparable to the rates found by us (and are comparable to the rates found in stars of previous evolutionary phases by
several authors (cf.e.g. the review article by Lamers and Morton,
1976). The obvious lack of extreme supergiants later than B2 and brighter than $M_{bol} \approx -9$ has been discussed recently by de Loore et al.(1977).
According to the theoretical evolution calculations including mass losses of the above mentioned quantities, this zone of the HRD between t_2
and t_3 should be populated with comparable density as the area to the
left between t_1 and t_2.

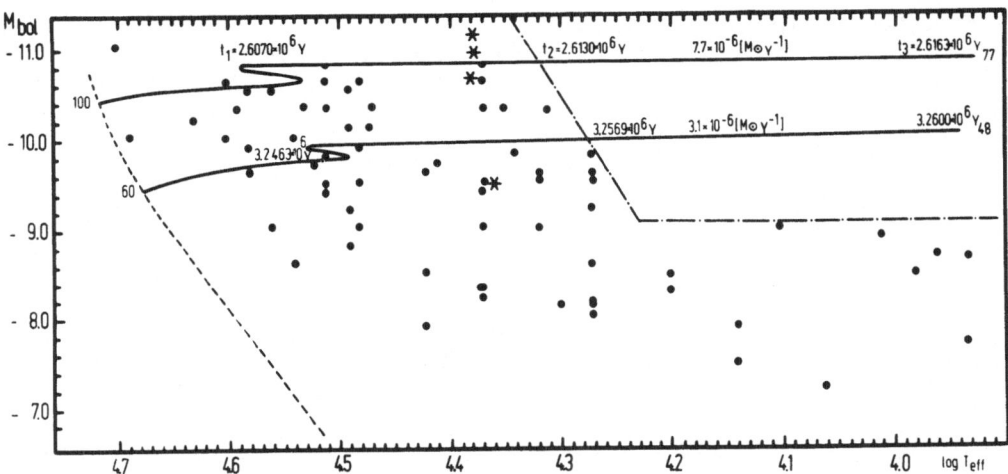

Figure 2: Theoretical HR diagram with the most luminous galactic stars.
          The asterisks denote the position of the B1 supergiants stu-
          died here. The evolutionary tracks (de Loore et al. 1978) of
          mass loosing stars with initial masses of 60 and 100 $M_\odot$ are in-
          cluded. For more details see in the text.

A possible explanation may be that all stars on the parts of the evolutionary tracks beyond $t_2 \approx 3 \; 10^6$ are variables. Objects occasionally
found in just this zone of the HRD are the Hubble-Sandage-variables.
These very luminous variables of spectral type A to F (Hubble and Sandage, 1953) with absolute photographic magnitudes around -10 (Rosino

and Bianchini, 1973) show considerable excess in the UV with U-B values
of -0.8 to -0.9 (Sharov, 1975). Therefore their absolute bolometric
magnitudes are probably even larger than -10. One may speculate therefore that the Hubble-Sandage-variables represent the later rather violent evolutionary phases of the very massive stars, observed as extreme
B supergiants a few 10^3 years earlier.

Acknowledgement: One of us (C.S.) thanks the National Foundation of
Collective Fundamental Research of Belgium for support. The research
on objects with extended atmospheres at the Landessternwarte Heidelberg
is supported by the Deutsche Forschungsgemeinschaft (SFB 132).

References:

Hearn, A.G.: 1975, Astron.Astrophys. 40, 277
Hubble, E., Sandage, A.: 1953, Astrophys.J. 118, 353
Hutchings, J.B.: 1976, Astrophys.J. 203, 438
Lamers, H.J.G.L.M., Morton, D.C.: 1976, Astrophys.J.Suppl. 32, 715
de Loore, C., De Grève, J.P., Lamers, H.J.G.L.M.: 1977, Astron.Astrophys.
    61, 251
de Loore, C., De Grève, J.P., Vanbeveren, D.: 1978, Astron.Astrophys.
    67, 373
Rosino, L., Bianchini, A.: 1973, Astron.Astrophys. 22, 453
Sharov, A.S.: 1975, IAU Coll. "Variable Stars a. Stellar Evolution"
    (Eds. Sherwood, Plaut, pg. 275)
Sterken, C., Wolf, B.: 1978a, Astron.Astrophys. (in press)
Sterken, C., Wolf, B.: 1978b, Astron.Astrophys.Suppl. (in press)

DISCUSSION FOLLOWING WOLF AND STERKEN

Lamb: The F- and G-type supergiants lie in the Hertzsprung gap.
The evolutionary timescale for crossing this region is only a few times
10^4 yrs. Thus, the interesting question concerns the absence of red
supergiants at high luminosity rather than the absence of F- and G-type
supergiants. A comparison of evolutionary timescales suggests that
there should be at least 100 times more early-type supergiants (O, B,
and A) than F- and G-type supergiants.

Chiosi: I would like to remark about the comment sometimes made
that the observed lack of very massive red supergiants can be explained
in terms of the neutrino cooling in phases beyond the core He burning.
Red supergiants are in the core He burning phase and according to the
current wisdom the fraction of lifetime spent in this area ranges from
a significant value to almost the totality. In particular the most massive objects are thought to have the whole core He burning as red supergiants no matter what criteria are adopted for intermediate semiconvec-
tive/fully convective instability. In such a case the τ_B/τ_R ratio does
not vary (increase) in the presence of the neutrino cooling as the central He burning is almost insensitive to it -- see Chiosi (1978, IAU
Symp. No. 80, in press).

ON THE FORMATION OF CONTINUOUS SPECTRUM AND EMISSION LINE PROFILES OF P CYGNI

T. Nugis, I. Kolka and L. Luud
Tartu Astrophysical Observatory
Tõravere 202444
Estonia U.S.S.R.

The attempts to interpret the values of the observed radio and infrared fluxes of P Cygni via simple mass outflow models (constant outflow velocity or usual radiative acceleration outflow) lead to a strong discrepancy between the observed and the calculated values.

Barlow and Cohen (1977) found also that constant velocity or usual radiative acceleration outflow models cannot explain the observed continuous spectra. They concluded that a more "extended" acceleration law is needed to fit the observed and the calculated radio and IR fluxes - a law by which the acceleration of matter takes place also at a comparatively large distance from the star.

In our study we analyze such type of expanding envelope models which can explain the shape of line profiles in the spectrum of P Cygni, too. Considering according to Kuan and Kuhi (1975) the hydrogen lines origin in the deceleratively moving part of an envelope and according to Oegerle and Van Blerkom (1976) the neutral helium lines form in the acceleratively moving part of an envelope, we assume that there must exist at least two differently moving regions in the envelope - an initial acceleration zone and a deceleration zone. To get correct values for IR and radio fluxes one must assume that there exists also an acceleratively moving outer zone, which explains very large wings of H_α and H_β lines as well.

We call a model consisting of three differently moving zones ADA-model (accelerating-decelerating-accelerating).

The compilation of the observed spectral energy distribution was carried out using the following data:

Wavelength or frequency	Author
4.995 and 10.68 GHz	Wendker et al. (1973)
3.3 mm	Schwarz and Spender (1977)
2.3 - 19.5 μ	Gehrz et al. (1974)
1.6 μ	Allen (1973)
U, B, V, R, I, J	Johnson et al. (1966)
Celescope: U2, U3	Davis et al. (1973)

P. S. Conti and C. W. H. de Loore (eds.), Mass Loss and Evolution of O-Type Stars, 39–42.

The interstellar absorption corrections are made using $E_{B-V}=0^m.61$ (Woolf et al. 1970) and reddening law by Nandy et al. (1975).

The observed energy distribution corrected for interstellar absorption is given in Fig. 1. There are also presented fluxes for outflow models $v = const$ and $v = v_\infty [1-(1-v_0^2/v_\infty^2) R_0/R]^{1/2}$ (lower lines) which show strong discrepancy between observations and calculations.

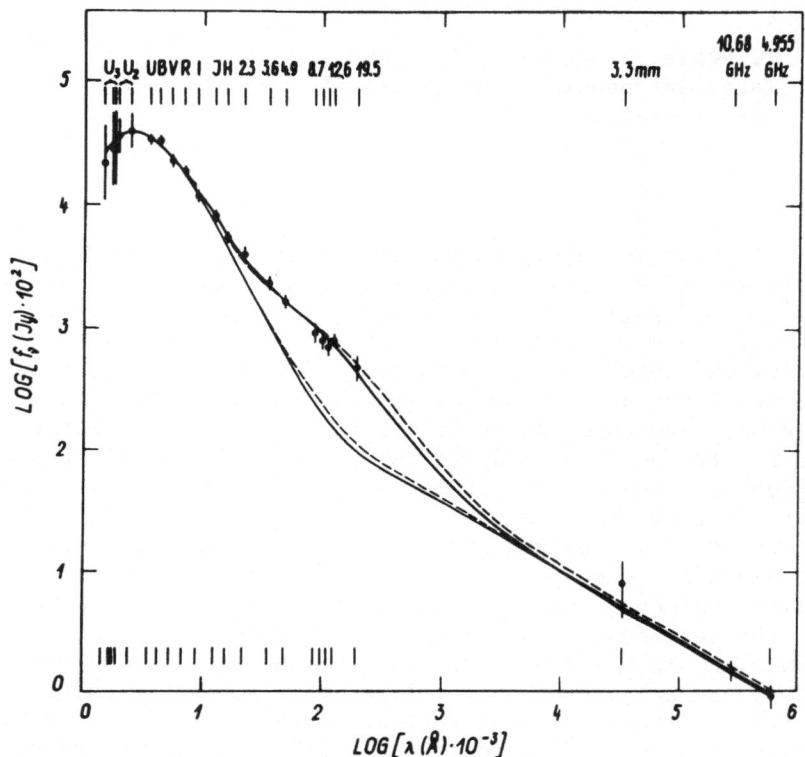

Fig. 1. The energy distribution in the continuous spectrum of P Cyg compared with model fluxes (explanations in text).

In Fig. 1 the upper lines presented fluxes for two ADA-models which well fit the observed ones. The calculations are depending on many parameters and they must be chosen on qualitative considerations of line profile shapes. Actually the good fit of observations can be reached within a large variety of them. We mention that the calculated fluxes are fitted with the observed ones supposing them to be equal at V and 10.68 GHz bands. The mass-loss rates for calculated ADA-models are in the range between $6.8 \cdot 10^{-5}$ and $9.3 \cdot 10^{-5}$ $m_\odot/yr$ according to $M_V = -8^m.4$ (Hutchings, 1976), this is in agreement with spectroscopic determinations $\sim 9 \cdot 10^{-5}$ $m_\odot/yr$ (Luud, 1967), $\sim 2 \cdot 10^{-4}$ $m_\odot/yr$ (De Groot, 1969) and $\sim 3.5 \cdot 10^{-4}$ $m_\odot/yr$ (Hutchings, 1976).

The reality of ADA-model can be checked by comparison of the observed and the calculated spectral line profiles. Here we present only semiquantitative study of hydrogen lines in which the level populations are estimated from simple outflow model solutions.

In Fig. 2 the comparison of the observed H_α profile and the calculated ones for two ADA-models is given. The observed profile is taken
according to Luud et al. (1967).

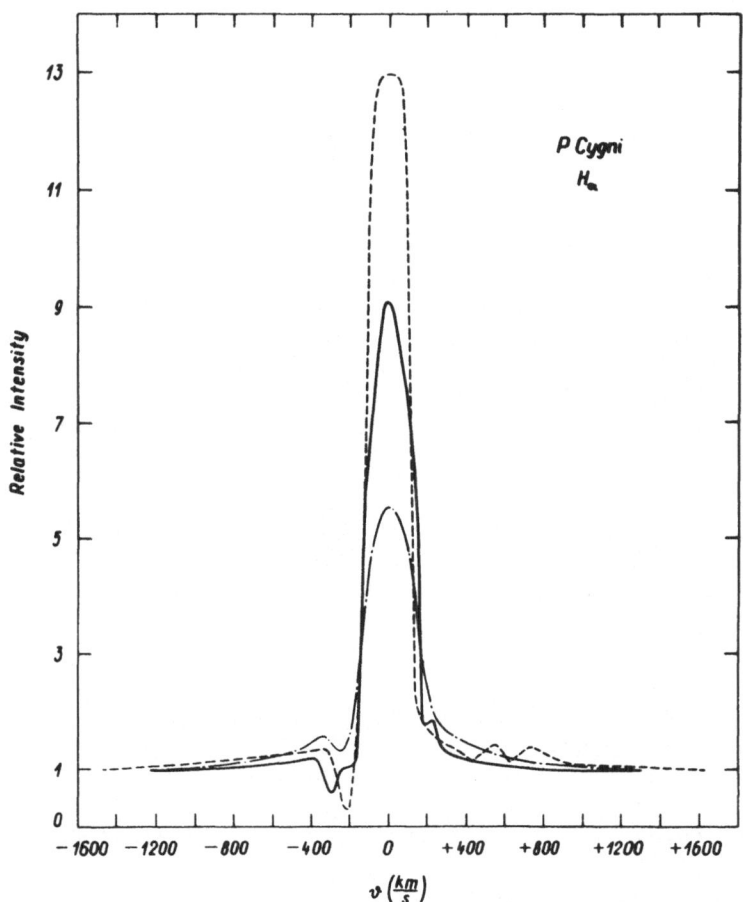

Fig. 2. H_α profile compared with the calculated ones for two
ADA-models: – – – the observed, ——— and ——— · ——— ADA-models.

The qualitative accordance of the calculated and the observed profiles is present, the intensive H_α line has also broad wings and violet absorption. The H_9 line has two absorption components as the observed one.

The origin of ADA structure must be physically founded. Here we
briefly discuss a possibility of forming three zones.

In the outer zone the matter is accelerated probably by radiation
pressure. Initial acceleration near the core is possibly caused by some
kind of mechanical dissipation processes partly assisted by radiation
pressure. According to the stellar structure calculations the massive
stars are vibrationally unstable that leads to the mass loss (Appenzeller, 1970a, 1970b).

To have the decelerating zone we must check the possibility of the

existence of an intermediate zone where gravitation force is high enough
to slowdown matter outflow accelerated in the inner part of the envelope. The analysis lead us to the conclusion that if the mass of P Cygni
is higher than $60\,m_\odot$ that deceleration may be present.

A strongly extended version of our paper will be published after
providing complementary calculations.

References

Allen, D.A. 1973, M.N.R.A.S., 161,145.
Appenzeller, I. 1970a, Astr. Ap., 5, 355.
Appenzeller, I. 1970b, Astr. Ap., 9, 216.
Barlow, M.J. and Cohen, M. 1977, Ap. J., 213, 737.
Davis, R.J., Deutchman, W.A. and Haramundanis, K.L. 1973, Celescope
          Catalog of Ultraviolet Stellar Observations, Smithsonian
          Astrophysical Observatory, Washington,
De Groot, M. 1969, B.A.N., 20, 225.
Gehrz, R.D., Hackwell, J.A., Jones, T.W. 1974, Ap. J., 191, 675.
Hutchings, J.B. 1976, Ap, J., 203, 438.
Johnson, H.L., Mitchell, R.I., Iriarte, B. and Wisnievski, W.Z. 1966,
          Comm. Lun. Plan. Lab., 4, 99.
Kuan, P. and Kuhi, L.V. 1975, Ap. J., 199, 148.
Luud, L.S. 1967, Astrofizika, 3, 379.
Luud, L.S., Põldmets, A. and Leesmäe, H. 1967, Publ. Tartu Astrophys.
          Obs., 36, 211.
Nandy, K., Thompson, G.I., Jamar, C., Monfils, A. and Wilson, R. 1975,
          Astr. Ap., 44, 195.
Oegerle, W.R. and Van Blerkom, D. 1976, Ap. J., 208, 453.
Schwartz, P.R. and Spender, J.H. 1977, M.N.R.A.S., 180, 297.
Wendeker, H.J., Baars, J.W.M. and Altenhoff, W.J. 1973, Nature Phys.
          Sci., 245, 118.
Woolf, N.J., Stein, W.A. and Strittmatter, P.A. 1970, Astr. Ap., 9, 252.

DISCUSSION FOLLOWING NUGIS, KOLKA AND LUUD

Snow: The Copernicus ultraviolet data on P Cygni show it to be
quite unlike other early B supergiants with regard to its wind. The
primary anomaly is the low ionization. No trace of N V is seen, for
example, while "normal" early B supergiants show this ion, usually with
a fully developed P Cygni profile. In the UV data for P Cygni, ions
such as Si II and C II show small shifts and possibly weak emission.
There are other anomalies about this star, and it properly should be regarded as peculiar, not representative of early B supergiants as a class.

Van Blerkom: I have recently seen a paper from Bernat and Lambert
who have observed very broad wings on some P Cygni lines which extend to
1500 km/sec, which they interpret as being due to electron scattering.
This might be an alternative explanation.

Luud: I believe the envelope temperature in P Cygni itself is not
enough to cause this effect.

EVIDENCE FOR NON-ISOTROPIC MASS LOSS FROM CENTRAL STARS OF SOME
EMISSION NEBULAE

Paris Pişmiş
Instituto de Astronomía, Universidad Nacional Autónoma de México

1. INTRODUCTION

In our general program of research on the velocity field of emission
nebulae, using the photographic Fabry-Pérot technique, we have included:
NGC 6164-5, an H II region with striking symmetry, NGC 2359 a "ring
nebula" and M1-67. The exciting stars in all three are centrally
located and are of spectral types O6f (showing P Cygni profiles), WN 5
and WN 8 respectively.

The detailed analysis of the velocity distribution of these three
regions affords evidence that these have originated from gas ejected by
the central star; that ejection has occurred not isotropically but
rather from localized regions, spots, on a rotating star. These spots
tend to be situated on opposite hemispheres on the star, approximately
at the extremities of a diameter which is oblique to the rotation axis.
It is suggested that the present mass loss from the central stars observed by their spectra may also be occurring from localized regions.
We shall discuss briefly the three regions.

2. NGC 6164-5

Both the velocity distribution and the morphology of this H II region
show a striking bi-symmetry around the exciting O6f star (HD 148937).
A confrontation of the velocity field with optical features has led to
a model, described earlier (Pişmiş 1974). In Figure 1, we give a sketch
of the main features, blobs, of emitting matter. Average radial velocities with respect to the sun are indicated for each feature. A plausible
interpretation is that the ejection from the central star which is at
present of type O6f, and presumably single, has occurred in puffs.

3. NGC 2359

This nebula has a "double ring" structure surrounding the exciting star.

P. S. Conti and C. W. H. de Loore (eds.), Mass Loss and Evolution of O-Type Stars, 43–46.
Copyright © 1979 by the IAU.

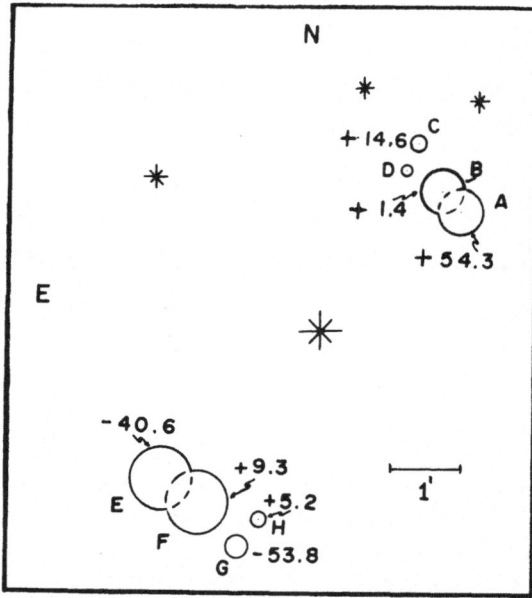

Figure 1. Sketch of the main features, blobs, of NGC 6164-5. Numbers indicate the average velocities of points within the features.

A sketch of the main features and the average radial velocities relative to sun are shown for each feature in Figure 2. The overall radial velocity of the nebula is 71 km s^{-1}. It is clear that the inner ring is approaching the observer relative to the central star while the outer ring is receding indicating that the rings denoted by f_1 and f_2, were formed by ejection from the central star, HD 56925. The detailed velocity field and a model proposed to explain it are given in a recent paper (Pişmiş et al. 1977).

4. M1-67

This object is so far considered to be a planetary nebula and is the only one to have a central WR star of the nitrogen sequence (WN 8). The structure of this nebula, dominated by blobs, shows again bi-symmetry. This object, which may well be an H II region rather than a planetary, is rather small

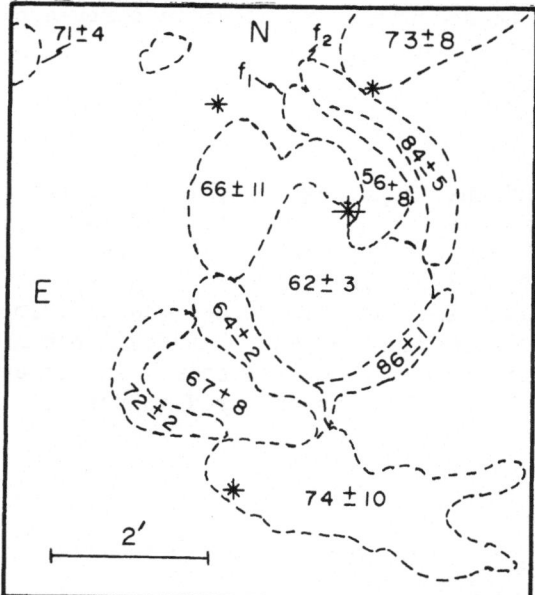

for our equipment to give as detailed a velocity field as in the former two H II regions. But clearly, the outer filaments are moving away from the central star while the regions closer to the star show expansion (Pişmiş and Recillas-Cruz, 1977). Thus, here again, we have a nebula formed probably by mass ejected from the central star.

Figure 2. A sketch of the main feature of NGC 2359 and the average heliocentric velocities within each feature.

5. A MODEL FOR THE MASS EJECTION

The following model which we propose may explain the velocity field and the main structural properties of NGC 2359 and NGC 6164-5. Ejection of matter started t years ago (t= 1-2×10^5 for NGC 2359 and 4-5×10^3 for NGC 6164-5) from active regions, spots, located nearly at the extremities of a diameter on a fast rotating star. This direction of the ejecting regions is oblique to the axis of rotation. In the case of NGC 2359 the axis of rotation is close to the line of sight whereas that of NGC 6164-5 is close to the plane of the sky.

6. CONCLUSION

If the proposed model of non-spherical ejection is correct one should expect other nebulae ejected from the parent star to be observed at varying projection angles. Objects formed in this manner in general would show bi-symmetry and sometimes ring structure. It would be interesting to determine the velocity field of H II regions with axial symmetry to check the validity of our model.

The parent stars of NGC 2359, of NGC 6164-5 and of M1-67 are all losing mass at present. It is reasonable to expect that if the gas ejected from these stars in the past has been non-isotropic the present mass loss may also be taking place in a similar fashion, that is from localized regions on the star.

In the light of this suggestion we may ask whether the line profiles of the Wolf Rayet and Of stars would not be consistent with a non-isotropic ejection of matter at the present time. It may be worthwhile to construct synthetic line profiles for rotating stars with active spots and compare them with observed profiles. Perhaps the variations of the spectral line profiles in some WR star may find an explanation by this mechanism of mass ejection.

REFERENCES

Pişmiş, P.:1974, Rev. Mexicana Astron. Astrof., 1, 45.
Pişmiş, P., Recillas-Cruz, E., Hasse, I.: 1977, Rev. Mexicana Astron. Astrof., 2, 209.
Pişmiş, P., Recillas-Cruz, E.: 1977, B.A.A.S., 9, 601.

DISCUSSION FOLLOWING PISMIS

Seggewiss: The H II-region M1-67 is listed as planetary nebula in the catalogue of Perek and Kohoutek. What are your reasons that M1-67 is not a planetary nebula but a population I object? My second question refers to the high velocity (200 km s^-1) of the nebula and the central star as well.

Pismis: According to the "classical" criteria for a planetary M1-67 will not qualify as a planetary since (1) its excitation degree is too low, (2) the object has a peculiar morphology and too large dimensions for a planetary, the internal motions show too high a dispersion and though it is a high velocity object, the velocity is too high, even for a planetary, its galactic orbit is hyperbolic. We probably have here a real runaway object!

Hutchings: Can you estimate the size and age of the nebula around HD 148937?

Pismis: In my paper on NGC 6164-5 I have given an estimate of the size of the nebula as 3 pc and an age of $4-5 \times 10^3$ yrs. These are however very rough estimates since the distance of the nebula, and star, is only approximate.

Niemela: Are there studies of any of the central stars to determine if they are binaries? If they are binaries the mass loss could be through the two external equilibrium points.

Pismis: According to Conti the central star of NGC 6164-5 is single. The central star of NGC 2359 is also believed to be single but I don't think we can be certain about this; more observations are needed. The same can be said about the central star of M1-67.

Conti: HD 148937, the central star of NGC 6164-5, has been the subject of an extensive spectroscopic study by myself, Garmany, and Hutchings [Ap. J. 215, 561 (1977)]. There is no evidence of velocity variation that could be attributed to a binary nature.

Bolton: A graduate student of mine at Toronto, Steven N. Shore, has succeeded in modeling the helium-rich spectrum variables on the upper main sequence, the model is based on the oblique rotator geometry like that used in Ap star models. A consequence of the model is that the wind streams out preferentially at the magnetic poles. A similar model could account for the phenomena you have observed.

Pismis: It is difficult to imagine that active spots on a star located, as we suggest, nearly diametrically opposite on the star may be due to anything but magnetic phenomena. In fact in my paper on NGC 6164-5 I have suggested that the agent funneling the ejecta is likely to be a magnetic dipole along the direction of ejection, that is along a diameter oblique to the rotation axis of the star.

Hummer: Two other possibilities for asymmetric ejection have been suggested by those persons dealing with planetary nebula (1) Harwit showed there was a correlation of the dipolar structure in planetary nebulae with direction of galactic disc. This suggests the IS medium may control the morphology, (2) bipolar shapes could be due to gravity darkening in the rotating central stars. Then the polar regions could tend to get blown away preferentially when the planetary forms, giving a donut structure.

MASS LOSS FROM ETA CARINAE

C. D. Andriesse
Kapteyn Astronomical Institute, Roden, The Netherlands

R. Viotti
Laboratorio di Astrofisica Spaziale, Frascati, Italy

ABSTRACT: This high luminosity (5×10^6 $L_\odot$) star since 1840 is losing mass at the rate of 7.5×10^{-2} $M_\odot$ per year. The large mass loss could be the result of vibrational instabilities produced in the CNO hydrogen burning phase of a very massive (160 $M_\odot$) star. The presence of high excitation lines in the ultraviolet spectrum of Eta Car suggests the idea of a hot zone excited by dissipation of the supersonic turbulent flow.

1. LUMINOSITY

The nature and evolutionary stage of Eta Car is still very controversial, because of the many peculiarities of the star. The infrared observations of Neugebauer and Westphal (1968) and others have disclosed that the object is one of the brightest infrared sources of the sky. The integrated energy flux is about 2.7×10^{-8} W m^{-2}, the main part of which (99 per cent) being emitted in the infrared. At a distance of 2500 pc this amounts to a luminosity of 5×10^6 $L_\odot$. The study of the light curve of Eta Car leads to the conclusion that the star has been a very luminous object since at least 150 years. In fact, during its bright phase (1820–1850) the star had a bolometric magnitude of -13^m (taking into account the interstellar extinction of $E_{B-V}= 0.4 - 0.5$), which is one magnitude brighter than the present bolometric magnitude derived from the infrared observations. Since this difference may be accounted for by the mechanical power required to push off the large mass lost by the star, the conclusion is that the star has no significantly changed its luminosity since its bright stage.

2. MASS LOSS

Eta Car is one of the few cosmic objects for which there is clear evidence

P. S. Conti and C. W. H. de Loore (eds.), Mass Loss and Evolution of O-Type Stars, 47–50.

for dust condensation in the circumstellar expanding envelope. This problem has been studied in details by Andriesse, Donn and Viotti (1978) who found that dust has been condensing initially at the rate of 10^{-4} $M_\odot$ per year, and presently at a somewhat higher rate. For cosmic element abundances in the gas envelope of Eta Car the mass flow from the star should be <u>at least</u> 100 times larger than the condensation rate. Actually it may be one order of magnitude larger, allowing for inefficiencies in the condensation process. On the other hand, the fact that the star is now intrinsically 1 mag fainter than before 1840 implies that, if the nuclear production in the star is unchanged during the last 140 years, 60 per cent of the energy production is presently employed in the expulsion of the gas. The mass flow driven by this power is 0.1 $M_\odot$ per year. Andriesse et al. concluded for a mass loss rate of 7.5×10^{-2} $M_\odot$ per year, a value which is in agreement with the estimates of the electron density in the ionized region near Eta Car as derived from the spectroscopic observations, and with theoretical considerations discussed below.

3. MASS AND EVOLUTION

It is clear that such a large luminosity and mass loss rate rise serious problems in attempting theoretical models for Eta Car. Since the star is located near the center of the young OB association Trumpler 16, which includes main sequence O3 stars, it should belong to the association. Feinstein et al. (1973) estimated for Tr 16 an age of 3×10^6 years, and this may be the age of Eta Car. Being the most luminous, hence the most massive member of the association, Eta Car should be evolved from the main sequence. According to Chiosi et al. (1978) during the core hydrogen burning, massive stars with large mass loss rates evolving from the main sequence do not join the red giant region. Their evolutionary tracks shrink towards the main sequence, and eventually cross it. The lifetime spent in the core hydrogen burning phase for a 100 $M_\odot$ star is 3×10^6 yr, with a mean mass loss rate of $1-2 \times 10^{-5}$ $M_\odot$ per year, more than three orders of magnitude smaller than that of Eta Car. The mass of Eta Car can be estimated from the binding condition that $M_* > \mu L/(4\pi cG)$, Andriesse et al. (1978), that is that the star should be stable against radiation pressure. The critical mass is 145 $M_\odot$; but in the case of considerable mass loss, the ratio of the radiative to the gravitational acceleration is probably close to unity. We thus estimate a stellar mass some 10 per cent heavier than the above limit, say 160 $M_\odot$. A star with such a large mass is not expected to be stable. According to Hoyle et al. (1973) an excess of nuclear energy is produced in these stars during the CNO hydrogen burning phase as a result of temperature variations in stellar pulsations. Thus the large mass loss could be the result of shock waves which take up the excess of nuclear energy. According to Hoyle et al. the continuous mass flow would be about 6×10^{-9} $M_\odot$ per year. Inserting the complete $L = 2.5 \times L_{rad}$ of Eta Car we obtain 7.5×10^{-2} $M_\odot$ per year.

4. IUE OBSERVATIONS

The ultraviolet spectrum of Eta Car may provide useful information about the structure of the expanding envelope and the mass loss. Viotti, Cassatella and Giangrande (1978) have recently observed the high resolution ultraviolet spectrum of Eta Car with the International Ultraviolet Explorer. They have identified many emission lines including the resonance lines of CIV, NV, MgII, AlIII, SiII, SiIV , MnII and FeII which have P Cygni profiles. The broad absorptions extend from about -230 km s^{-1} to -650 km s^{-1}, confirming the presence of a large velocity gradient in the envelope. The identification of high energy lines, namely HeII, NV, CIV and in the optical region NeIII, in the spectrum of Eta Car may be explained by the presence in the expanding envelope of a hot zone which could be produced by dissipation of the supersonic turbulent flow as suggested by Andriesse et al. (1978).

REFERENCES

Andriesse, C.D., Donn, B.D. and Viotti, R., 1978, Mon. Not. Roy. Astron. Soc., in press.

Chiosi, C., Nasi, E. and Sreenivasan, S.R., 1978, Astron. Astrophys. 63, 103.

Feinstein, A., Marraco, H.G. and Muzzio, J.C., 1973, Astron. Astrophys. Suppl. 12, 331.

Hoyle, F., Solomon, P.M. and Woolf, N.J., 1973, Astrophys. J. 185, L89.

Neugebauer, G. and Westphal, J.A., 1968, Astrophys. J. 152, L89.

Viotti, R., Cassatella, A. and Giangrande, A., 1978, 4th Colloquium on Astrophysics of the Trieste Astronomical Observatory "High Resolution Spectroscopy", Ed. M. Hack, Trieste 3-7 July 1978.

DISCUSSION FOLLOWING ANDRIESSE AND VIOTTI

<u>Conti</u>: I would like to suggest an alternative explanation for the evolutionary status of η Car. Chiosi has emphasized that the status of an early type star burning H in the core may be quite different from one burning He in the core. I think η Car might be a star undergoing this "event," changing from one kind of nuclear burning to another. I recognize this is clearly speculative.

<u>Chiosi</u>: I myself am somewhat uncomfortable with the quoted rates. The theoretical rates for mass loss by vibrational instability (Appenzeller) are $10^{-3} \sim 10^{-4}$ solar masses per year. This is two orders of magnitude discrepant.

WOLF-RAYET STARS IN THE MAGELLANIC CLOUDS

J. Breysacher
European Southern Observatory, Casilla 16317, Stgo. 9, Chile
and
M. Azzopardi
Observatoire du Pic du Midi et de Toulouse, 1 Av. C. Flamma-
rion, F-31500 Toulouse, France.

1. INTRODUCTION

Up to now 4 WR stars were known in the SMC (Breysacher and Westerlund, 1978) and 76 in the LMC (Fehrenbach et al., 1976). Because no systematic search for WR stars in the SMC had ever been made, a survey was carried out with the ESO Objective Prism Astrograph which resulted in the iden- tification of 4 new WR stars of the WN type, afterwards confirmed by slit spectroscopy. In the LMC, two fields were also observed and 13 new faint WR stars detected. The results presented here mainly concern the SMC WR stars.

2. OBSERVATIONS

The survey was done at La Silla, Chile, in October 1977, with the 40 cm Objective Prism Astrograph (Fehrenbach et al., 1964) using an interfe- rence filter centered at λ4650 (pass band 120 Å). WR stars show up strong- ly in this spectral region due to the emission mainly from either λ4650 C III (WC) or λ4686 He II (WN). This detection technique, described in detail by Azzopardi and Breysacher (1978) enables, by reducing the back- ground fog and the number of overlapping images, to study very crowded regions.

The present survey (limiting magnitude of m_{pg} 16.5) covered the Bar and the Wing of the SMC. In the LMC, the two fields observed, each 85' in diameter, were centered on stars Sk -68°82 and Sk -69°243 (Sanduleak, 1969). IIa - O nitrogen baked plates were used.

For the newly discovered WR stars, spectrogrammes at 114 Åmm^{-1} were ob- tained with the Boller and Chivens Cassegrain spectrograph equipped with a Carnegie image-tube at the ESO 3.6 m telescope.

3. RESULTS

Table 1 lists the data concerning the 8 WR stars now known in the SMC. These data are taken from the paper by Azzopardi and Breysacher (1978)

51

P. S. Conti and C. W. H. de Loore (eds.), Mass Loss and Evolution of O-Type Stars, 51-54.

in which identifications of the stars can be found. The absolute magnitudes were calculated adopting the value of 19.2 as absorption-free distance modulus of the SMC (Westerlund, 1974).

TABLE 1

Wolf-Rayet Stars in the Small Magellanic Cloud

SMC/AB	$\alpha\{1975.0\}\delta$		Spectral type	V_{pg}	E_{B-V}	M_V
1	$0^h42^m.80$	$-73°37'.1$	WN3 + OB	15.48	0.050	-3.9
2	0 47.63	-73 23.7	WN4.5 + O4	14.43	0.088	-5.0
3	0 49.12	-73 30.2	WN3-4 + O4	14.55	0.055	-4.8
4	0 49.87	-73 35.1	WN4.5 + O4-6 V-III	13.43	0.067	-6.0
5	0 58.62	-72 17.9	WN3p + OB	11.88	0.054	-7.5
6	1 02.62	-72 14.7	WN3 + O7 Ia	12.36	0.092	-7.1
7	1 02.78	-72 11.4	WN3p: + OB	13.16	0.092	-6.3
8	1 30.6	-73 33	WC4 ? + O4	12.97	-	-6.:

4. DISCUSSION

All the SMC WR stars studied show direct or indirect evidence of having an OB-type star companion. Direct meaning that absorption lines are indeed seen in the spectra and indirect that the binarity is inferred from both the strength of the continuum relatively to the emission features and the absolute magnitude. For SMC/AB 1,2,3 the absolute magnitudes determined using $(m-M)_o = 19.2$ are hardly compatible with the absolute magnitude calibrations for WR and OB stars (Smith, 1973; Walborn, 1972) and it is suggested that these 3 WR binaries, located in the same region of the SMC are, in fact, seen at a considerable depth into the Cloud.

Considering now the distribution amongst the WR subclasses, Smith (1968, 1973) noted the complete absence of subclasses WC 6-9 and probably WN 6 from the LMC. In the SMC only subclasses WN 3-4.5 are present with, in the WC sequence, one doubtfully (Breysacher and Westerlund, 1978) extreme Wolf-Rayet of type WC 4. Then, except for subclass WC 5 which possibly escaped our detection due to the employed technique (width of the λ4650 emission feature comparable in this case to the filter pass band) no WR stars belonging to subclasses WC 6-9 seem to exist in the SMC. It is also remarkable that no "transition" WN 7 stars (Conti, 1975) are found in the SMC.

After the present survey the census of the WR population in the Small Cloud can probably be considered as quite complete, at least within the survey limits.

It appears that the number of SMC WR stars is small and that their distribution amongst the WR subclasses is different from those of the LMC and the Galaxy. This is possibly related to the differences in chemical abundances, the SMC being known for its metal deficiency.

References

Azzopardi, M., and Breysacher, J., 1978, in preparation

Breysacher, J., and Westerlund, B.E., 1978, Astron. Astrophys. 67, 261

Conti, P.S., 1975, Mém. Soc. Roy. Sci. Liège, 6ieme série, t. IX, 193

Fehrenbach, Ch., Duflot, A., Duflot, M., 1964, UAI-URSI Symp. 20, 228

Fehrenbach, Ch., Duflot, M., Acker, A., 1976, Astron. Astrophys. Suppl. 24, 379

Sanduleak, N., 1969, Cerro Tololo Inter-American Obs. Contr. No. 89

Smith, L. F., 1968, Mon. Not. Roy. Astron. Soc. 140, 409

Smith, L. F., 1973, IAU Symposium N°49, 15

Walborn, N. R., 1972, Astron. J. 77, 312

Westerlund, B. E., 1974 Proceedings of the First European Astronomical Meeting, Athens, September 4-9, 1972, Vol 3, p. 39, ed. Springer-Verlag, Berlin - Heidelberg - New York

DISCUSSION FOLLOWING BREYSACHER AND AZZOPARDI

Conti: I consider this a very interesting result. The fact that the SMC WR stars are all early WNs, and furthermore they all appear to be binaries in telling us something very important. One question: Is the WR classification biased by the fact that the O star dominates the spectrum?

Breysacher: No, because in most of the cases the $\lambda\lambda4604$-4620 N V emission feature is visible above the continuum and can effectively be used for the classification. For the two "peculiar" WR stars of Table 1 the situation is slightly different. In both cases, the spectra are characterized by a broad $\lambda4686$ He II feature which is dominant, a strong continuum but no visible absorption lines. One of the stars, SMC/AB 5 shows spectral variations and the WN 3 type is inferred from the sometimes detected $\lambda\lambda4604$-4620 N V feature.

Crampton: Peter, what fraction of WN3, 4, 5 stars in the Galaxy are binaries? Since WN3, 4, 5's are less luminous than the later types, the O stars will tend to dominate the spectrum more in these cases.

Conti: The actual numbers, which I have recently collected, come from the Smith catalogue as follows (early WN:WN3, 4, 5); Galaxy 7 single, 5 double (42% binaries); LMC 8 single, 4 double (33% binaries). These are detected companions, i.e., the absorption lines are observed. This is very different, apparently, from the SMC where all are double (so far).

Parsons: The Skylab S-019 UV spectra for 6 WN and 6 WC (Galactic) stars showed almost all to be double, with C IV absorption and diluted emission lines indicating an O companion.

Underhill: Is there any "confusion" problem in the spectra of these WR stars? Could there be close multiple systems in which an O star is also observed spectroscopically? Such close systems are quite common in the Galaxy.

Breysacher: The spectral classifications given in Table 1 are based on slit spectra obtained afterwards and not on the spectra recorded on the objective prism plates.

Underhill: Yes, but the slit has a finite size. Might not optical companions be included?

Breysacher: Perhaps, but while guiding the stars onto the slit, no companions were seen.

STELLAR OUTFLOW: RELATIVE MOTIONS OF NEBULAE AND Of STARS

Beverly T. Lynds
Kitt Peak National Observatory[1]

On the basis of arguments presented by Roberts (1972) and of Shu et al. (1972), Minn and Greenberg (1973) argued that the velocity differences between newly formed hot stars and the surrounding interstellar medium are sufficiently different so that typical H II regions should consist of material which is continually being replaced by the ambient medium and which should therefore possess the velocity of the medium rather than that of the star. Obviously, the critical test of this hypothesis will be a comparison of nebular velocities with the velocities of the exciting stars.

The extensive Fabry-Perot observations of Georgelin and Georgelin (1970) and the radio measures of the hydrogen alpha recombination lines have provided a wealth of data on radial velocities of emission regions. The recent paper by Conti et al. (1977) contains a list of radial velocities of O and Of stars having known orbits or identified as single stars. Cruz-Gonzalez et al. (1974) have identified the H II regions in which the O and Of stars appear on the Palomar Schmidt prints; those stars too far south for such identification have been tentatively identified with emission nebulae by using the Rodgers et al. Atlas of the Southern Milky Way (1960).

Conti's list includes 87 O stars and 16 Of stars with well-determined radial velocities. Of this group, 42 O stars and 11 Of stars appear to be associated with H II regions of measured radial velocities. A comparison of the stellar and nebular velocities is given in Figure 1. A least squares solution fits a straight line with a slope of nearly 1 to the O stars, with a zero-point shift of +3.5 km/sec. Data for the Of stars have much greater scatter and suggest a zero-point shift of -28 km/sec.

All velocities were next reduced to the local standard of rest using the standard solar motion traditionally used by radio astronomers for the 21-cm measures, i.e., 20.0 km/sec toward 18^h right ascension and +30° declination, and then corrected for circular galactic rotation by using the distance modulus for the stars given by Cruz-Gonzalez

P. S. Conti and C. W. H. de Loore (eds.), Mass Loss and Evolution of O-Type Stars, 55-57.

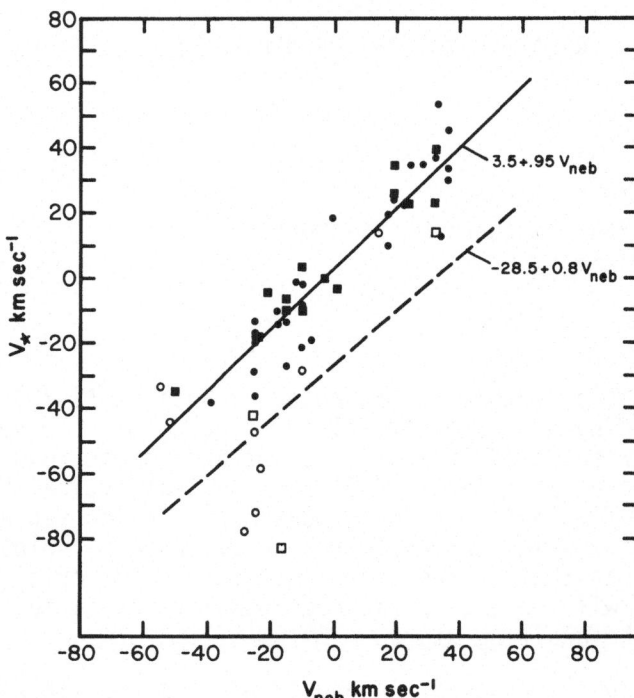

Figure 1. The filled squares and circles represent O stars of known orbital or single-star velocity, respectively; the open squares and circles refer to the Of stars in the same manner.

et al. and the galactic model of Burton and Gordon (1978) derived from CO data. The following table summarizes the results.

Comparison of Radial Component of Peculiar Velocity

	No.	V_{pec} (km/sec) *	Neb	σ (km/sec) *	Neb
Average by Nebulae					
O-stars	22	+ 5	+ 1	10	6
Of-stars	9	-24	- 4	24	16
Of-stars (alt)	7	-27	+ 2	23	9

The velocities were found by averaging nebulae, the stellar velocities for which may be the mean of several stars associated with a given nebula.

Two of the Of nebulosities appear to be anomalous. One is NGC 6164 associated with HD 148937, for which Pismis (1974) has obtained detailed Fabry-Perot interferograms which clearly show that the large negative velocity obtained by Georgelin and Georgelin represents an expansion velocity rather than a spacial kinematic velocity. The other object is NGC 7635 and may exhibit similar expansion-type velocities. Averages of the velocities after eliminating these two nebulae are shown on the Of star (Alt) line of the table.

The probable errors of all means are about 1-3 km/sec and therefore it is concluded that all groups of peculiar velocities except those of the Of stars average to about zero. The Of stars exhibit the "K-term" found by Conti (from the same data, of course) of about -25 km/sec. The fact that the nebular velocities for both O and Of stars appear to have the same velocity distribution suggests that it is the Of velocities which are anomalous.

It seems reasonable to conclude that for this sample there is no large velocity difference between the exciting stars and the associated nebulosities produced by galactic streaming motions. The large negative K term of the Of stars is best interpreted as Conti has proposed, i.e., an effect produced by the absorption-line producing layer of the star.

REFERENCES

Burton, W. B. and Gordon, M. A.: 1978, Astron. Astrophys. 63, p. 7
Conti, P. S., Leep, E. M., and Lorre, J. J.: 1977, Astrophys. J. 214,
     p. 759.
Cruz-Gonzalez, C., Recillas-Cruz, E., Costero, R., Peimbert, M., and
Torres-Peimbert, S.: 1974, Revista Mexicana Astr. Astrofisica 1, p. 211.
Georgelin, Y. P. and Georgelin, Y. M.: 1970, Astron. Astrophys. 6,
     p. 349.
Minn, Y. K. and Greenberg, J. M.: 1973, Astron. Astrophys. 24, p. 393.
Pismis, P.: 1974, Revista Mexicana Astr. Astrofisica 1, p. 45.
Roberts, W. W.: 1971, Astrophys. J. 173, p. 259.
Rodgers, A. W., Campbell, C. T., Whiteoak, J. B., Bailey, H. H., and
     Hunt, V. O.: 1960, An Atlas of H-Alpha Emission in the Southern
     Milky Way, Mount Stromlo Observatory, A.N.U., Canberra.
Shu, F. H., Millone, V., Gebel, W., Yuan, C., Goldsmith, D. W., and
     Roberts, W. W.: 1972, Astrophys. J. 173, p. 557.

[1]Operated by the Association of Universities for Research in Astronomy, Inc., under contract with the National Science Foundation.

GENERAL DISCUSSION

Van Blerkom: Quite a point was made of the binary nature of the WR stars in the SMC. I'd like to ask, Peter, what's the significance of this?

Conti: Let me give the following speculative possibility. We know the SMC is deficient in CNO elements with respect to the Galaxy. If stellar winds are driven primarily by radiation pressure from CNO lines themselves, then perhaps winds would be weaker in SMC stars of similar type compared to Galactic stars. Maybe it would then be possible to evolve massive stars to WR types only in binary systems, unlike in our Galaxy where single WR stars can evolve from massive O stars by stellar wind mass loss.

Hutchings: Our optical spectra of some SMC O and Of stars show them to be like galactic ones, hence the winds are similar, and the evolution would be similar.

Conti: It sounds like my speculation didn't make it beyond the first rejoinder, but I prefer to wait for U.V. spectroscopy where the winds are observed directly.

Crampton: I'd like to ask Dr. Breysacher again about the SMC stars. You indicated that you only inferred the presence of companions for WN3, 4, 5 stars in the SMC by their higher luminosity in some cases. Is that correct? If so, then perhaps the frequency of binaries in the SMC and the Galaxy is not too different since in the Galaxy we do not know the luminosity and hence cannot infer binary nature.

Breysacher: Yes, namely for stars SMC/AB4, 5 and 7 of Table 1 no absorption lines are seen and the binarity is inferred from the strength of the continuum and the absolute magnitudes. I am also aware of the fact that some WN stars show absorption lines themselves and would like to ask Dr. Niemela if these are hydrogen lines only.

Niemela: Hydrogen and He II absorption lines are seen in several Galactic WR stars. Many are composite spectra, but in some cases the absorptions in fact arise in the WR star itself. This is to be discussed later in this Symposium (Sessions 5 and 8). I don't know that the M_v are all that well known for WR stars, however.

Conti: I have been looking into this and will report more fully in Session 8. But for now let me say the following: The M_v calibration for WR stars is primarily based on LMC stars (Smith, M.N. 1968). The "single" early WN stars (i.e., no absorption lines) have $M_v \sim -4$ to -5. A number of early WN stars are considerably brighter but without exception all have absorption lines, presumably due to a companion. In our Galaxy, we now know of some later type WN stars to have intrinsic absorption lines, as we shall see. But then, maybe these bright early-type WN stars in the SMC (and LMC) are not composite either but single

59

P. S. Conti and C. W. H. de Loore (eds.), Mass Loss and Evolution of O-Type Stars, 59–61.
Copyright © 1979 by the IAU.

stars. That would be even more interesting. They have not been
studied in detail as yet.

Underhill: I'm still worried that many of those early WN stars
discussed by Breysacher are unresolved doubles, where the O star dominates the continuum and contributes.

Conti: Nolan Walborn would have made this same point had he been
able to attend this meeting. He has pointed out a few cases of O sub-
groups in the LMC that become resolved at the 4-meter telescope which
were previously only "fuzzy" at lower resolution.

Hutchings: I have obtained slit spectrograms of one of these O-WN
stars in the SMC -- R31. The absorption lines vary by 100 km s^{-1} so it
is clearly a binary.

Vreux: I would like to ask Ann Underhill why, if there is another
star by chance in the line of sight to SMC WR stars, it's an O star
rather than, say, an A type?

Underhill: That's easy. The luminosity you need to see absorption
lines against a WR star means M_V must be of the same order, $\sim-4^m$ to -5^m.
This corresponds to an O type; an A type would be completely invisible.

Crampton: A better argument might be that those groups usually
are made up of O types, not A types.

Van den Heuvel: 30 Doradus is perhaps a good example of this.
It's made up of five or so stars.

Leung: I would like to comment on how binary systems might end up
as contact systems. Recent calculations (Flannery, Ulrich, Burger)
taking into account the mass accreting components suggest that further
evolution of semidetached systems lead to contact systems. The following slides illustrate this point quite well. The configuration of
BF Aur represents a stage before the mass ratio reversal; μ'Sco represents a stage after the mass ratio reversal; and SX Aur represents a
stage where a system is approaching a contact configuration.

Niemela: Can you use a Roche model alone; shouldn't you take into
account the radiation pressure force the equipotentials in such luminous
stars?

Leung: I think this is all right for B stars.

R. E. Wilson: The important point is how close the luminosity is
to the Eddington limit. We will hear about this later in one case
(UW CMa). However, for most systems, I feel this isn't important.

Bidelman: I'd like to ask John Hutchings the following: What is the duplicity situation in the extreme Of stars? Also, do you consider +40°4220 an extreme Of star? It has at times strong hydrogen emission.

Hutchings: Several stars, HD 151804, HD 152408 and HD 148937, do appear to be single [Conti, Garmany and Hutchings, Ap. J. (1977)]. I am not familiar with BD+40°4220.

Crampton: In connection with mass loss I would like to call attention to one nebula, NGC 6334, described by Dr. Pismis, which contains four nebular concentrations. Three are normal H II regions, but one looks very much like the ring nebulae that contain WN stars and may in fact be the result of mass loss from the WR stars. However, this nebula, which looks like NGC 6888, has two central O type stars which appear to be completely normal.

Lamb: I'd like to ask Dr. Pismis what evidence is there that the material in the ring is ejected, rather than swept up interstellar material?

Pismis: Essentially, the symmetry of form and the structure of the velocity field. For details, see the references in the text.

Abbott: If one believes that these WN stars are evolved from massive O-stars then the wind from the progenitor star would have evacuated a large cavity around the star -- an interstellar bubble -- leaving no material to be swept up. Thus, ring nebulae are probably ejected rather than swept up.

Pismis: I still worry about the hypothesis of spherically symmetric stellar winds. Some material probably is swept up but most of what we observe in the rings was ejected non-isotropically.

Kwok: I would like to offer the following speculative argument. If these WN central stars have evolved from O stars, they may have been red supergiants in between these phases. During the red supergiant phase a low velocity wind could have carried away an appreciable fraction of stellar mass. Then when the WR star develops, it has a high velocity wind. The ring could be the shock phenomena associated with this rapidly moving wind colliding with the red supergiant ejecta.

Underhill: A really fundamental problem in physics is the ubiquitous appearance of stellar winds, and their energy input, in stars of many spectral types, not just O and WR stars.

A man who claims all ejection
Is driven below by convection
Gave a rather cold shoulder
To theorists from Boulder
Whose work he felt lacked French connection.

SESSION 2

U.V. SPECTROSCOPY

Chairman: D.G. HUMMER
Introductory Speaker: T.P. SNOW

1. H. LAMERS: Modelling of UV resonance lines.

2. D. MORTON: O VI in stellar winds.

3. S.B. PARSONS, J.D. WRAY, K.G. HENIZE, and G.F. BENEDICT:
   C IV resonance profiles in O stars.

4. S.R. HEAP: Winds in hot, subluminous stars.

5. A.B. UNDERHILL, L. DIVAN, V. DOAZAN and M.L. PREVOT-
   BURNICHON: Temperatures and radii of O stars.

6. C. MOROSSI, R. STALIO and L. CRIVELLARI: Analysis of UV
   spectrophotometric observations for O-type stars.

7. S.R. POTTASCH, P.R. WESSELIUS and R.J. VAN DUINEN (read
   by A.G. HEARN): Effective temperatures of the O stars.

AN ULTRAVIOLET VIEW OF STELLAR WINDS

Theodore P. Snow, Jr.
Laboratory for Atmospheric and Space Physics
and
Department of Physics and Astrophysics
University of Colorado

ABSTRACT

Ultraviolet observations of mass-loss effects in O stars have, over the past decade, revealed a broad picture of a phenomenon whose extent was only partially evident from earlier ground-based observations. Ultraviolet resonance lines of a variety of ionization stages of several common elements provide a comprehensive probe of the low-density, extended winds. Three general types of information have been derived from ultraviolet spectroscopy of mass-loss profiles: (1) the nature of the stars which experience mass loss via radiatively-driven winds; (2) the physical conditions in the winds; and (3) variability in the outflow, which in turn may yield clues to the origins of the winds. Observations and results in each of these areas are reviewed, and some new results are included. A good correlation of mass loss rate and luminosity is indicated by the data, in agreement with theoretical predictions. Time variations in the P Cygni profiles may be quite common, with variability on times of hours or longer. Anticipated new observations, which should be possible with existing and planned instrumentation, are described.

I. Introduction

Classical spectroscopy has revealed a great deal of information on mass-loss in hot stars, as Hutchings has shown in his review (1979, this volume). Nevertheless, to obtain a complete picture of stellar winds, it is necessary to observe their low-density outer portions, and this is best done in ultraviolet wavelengths. The numerous strong resonance lines which are accessible in this portion of the spectrum provide information on the velocity, density, and ionization conditions in the outflowing material which, when combined with visible and infrared data, can provide sufficient information to be of use in constraining possible models for the winds. Some important questions remain unanswered, however, and certainly these will be central to the discussions to follow.

P. S. Conti and C. W. H. de Loore (eds.), Mass Loss and Evolution of O-Type Stars, 65–80.
Copyright © 1979 by the IAU.

The most striking manifestation of a high-velocity stellar wind
is the characteristic P Cygni profile, consisting of an emission
component at rest in the stellar frame, and a superposed absorption
which is shifted towards shorter wavelengths. The emission results
from scattering throughout the volume filled by the expanding atmosphere, while the absorption takes place only in the observer's line
of sight, along which material flows outwards from the star. The
strength of the emission is controlled by the size of the volume,
and in stars with relatively weak winds, is absent. In these cases
the presence of the wind may be indicated only by asymmetric absorption lines, with extended short-wavelength wings. In any case, the
effects are strongest in resonance lines, which require no excitation, leading to a variety of P Cygni profiles throughout the ultraviolet. The important ions which have been accessible to current and
previous UV experiments include He II, C III, C IV, N III, N IV,
N V, O IV, O VI, S III, S IV, S VI, Si III, Si IV, and P V. In
Figure 1 are shown some example P Cygni profiles of the N V resonance doublet near 1240 A.

The first rocket ultraviolet spectra with sufficient resolution
to clearly show mass-loss effects were obtained just over a decade
ago by Morton (1967), and were followed shortly by the observations
of Carruthers (1968) and Stecher (1968). Since that time, a number
of rocket and satellite experiments have provided a large pool of
data, and at this moment significant new additions to this pool are
being contributed by the International Ultraviolet Explorer (IUE)
(see the contribution to this volume by Heap). Copernicus data have
provided high quality profiles for a number of OB stars (e.g. Morton
1975; Snow and Morton 1976).

To date, ultraviolet observations of stellar winds from O stars
have contributed information bearing primarily on three main questions:
(1) the nature of the stars which experience mass loss, and the
correlations of wind strength with stellar parameters; (2) the physical
conditions in the outflow, particularly the run of velocity, density,
and ionization degree with height; and (3) variability with time. This
paper will provide a brief survey of results in these three areas, with
some new data to be included in the discussion of variability. The
concluding section contains a description of further observations which
may be useful to carry out with present or planned ultraviolet experiments.

II. Stellar Parameters

Apparently all O stars are losing mass. Relatively low-resolution
data such as those from the early rocket experiments and from the
objective-prism surveys carried out by Skylab (Henize et al., 1975)
or the Orion series (Gurzadyan, 1975) revealed only the fully-
developed P Cygni profiles characteristic of the supergiants.
Higher-resolution data, especially those supplied by Copernicus,
have shown that even in O dwarfs there are at least extended absorption wings symptomatic of mass loss (Snow and Morton, 1976).

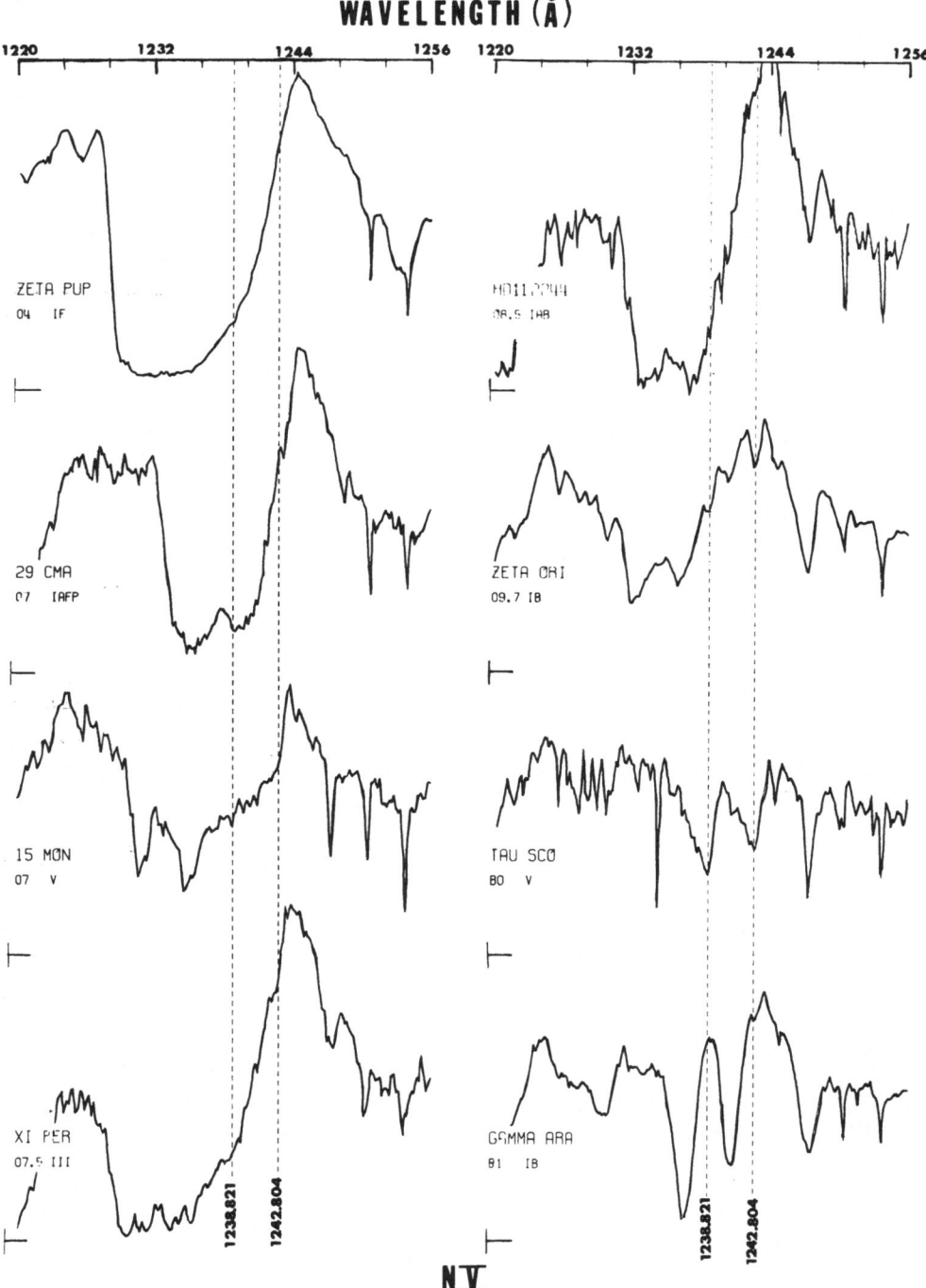

Figure 1. Example of NV P Cygni profiles. The manifestations of mass loss can consist of fully developed profiles with emission and absorption components in the case of a strong wind, or only asymmetric absorption in the case of a weaker wind.

III. Physical Conditions in the Winds

A. Ionization

One of the early discoveries made with Copernicus was the presence of asymmetric O VI absorption in the spectrum of the B0 V star τ Sco (Rogerson and Lamers, 1975). This ion is produced thermally at a temperature in excess of 10^5 K, yet it appears in the spectrum of a star with T_{eff} = 32,000 K. Rather than being a peculiarity in a single star, the presence of highly-ionized species is a general feature of the winds in OB stars (Lamers and Snow, 1978). The ion O VI would not be expected to exist in the photospheres of even the O stars, yet it is almost invariably present in the winds. Similarly, N V P Cygni profiles are found in all O stars and a number of early B stars whose photospheres are too cool to produce this ion radiatively. Evidently some form of non-radiative heating is present. Recent soft X-ray upper limits for a number of early-type stars (Cash, private communication) may provide significant constraints on the nature of this heating.

The variation of ionization with height can be deduced from ultraviolet observations by comparison of observed profiles with empirical models. Here some contrasts show up from star to star, and it is not yet known whether any systematic trends exist. In the O4If star ζ Pup, it appears that the ionization is nearly constant out to large heights in the wind (Lamers and Morton, 1976), whereas in τ Sco the degree of ionization decreases outwards (Lamers and Rogerson, 1978). Work in progress by Lamers and Snow (1979), in which profile-fitting will be done for a number of OB stars, should show how the ionization gradient varies with stellar parameters.

B. Velocities

Even from low-resolution UV spectra it is apparent that velocities as high as 2-3000 km s^{-1} or greater exist in the outer portions of the winds from hot supergiants (e.g., Morton, 1967). The regions where these extreme velocities occur are transparent to visible-wavelength photons; hence, the magnitude of the wind terminal velocities was unknown before UV observations were possible.

For strongly saturated absorption components, the wind terminal velocity is readily determined from the position of the short-wavelength edge. For the stars with weaker mass loss, however, there is no clear-cut edge, and only a lower limit can be found directly from the profile, by estimating the wavelength at which the extended absorption wing returns to the continuum level. From the UV profiles, terminal velocities ranging from about 300 km s^{-1} to over 3400 km s^{-1} were found in the Copernicus survey of mass loss (Snow and Morton, 1976). Anomalously high velocities were found from the N V profiles in several Orion supergiants; subsequent examination of Copernicus data on early B stars reveals that a blend of photospheric lines near 1234 A may contribute to the absorption in these stars, exaggerating the derived terminal velocity.

Evidently the atmosphere is not sufficiently extended in these lower-luminosity objects to produce strong emission except in the earliest types. From the Copernicus data, it is seen that mass loss generally occurs in all OB stars with $M_{bol} \leq -6$ (Snow and Morton, 1976), and is found in addition in a number of B dwarfs below this luminosity (Snow and Marlborough, 1976; Lamers and Snow, 1978).

Although mass-loss rates are not yet known for a large number of stars, some discussion of the dependence of mass loss on stellar parameters is possible at this point. For example, a crude index of mass loss rate can be formulated directly from the observational data and correlated with stellar parameters such as luminosity or temperature. Since the mass loss rate is roughly given by:

$$\dot{M} \sim 4\pi\rho v r^2 \qquad (1)$$

we can make rough approximations and write:

$$\dot{M} \propto R_* V_\infty W\lambda, \qquad (2)$$

where R_* is the stellar radius, V_∞ is the terminal velocity, and $W\lambda$ is the absorption equivalent width of an ion observed in the wind. The following assumptions are included in this simplification: (1) that all of the material in the wind is at a height of $1R_*$ and has the velocity V_∞; and (2) that the column density $N = \rho/r$ is proportional to $W\lambda$ (i.e., that saturation and photospheric absorption are not important, clearly an oversimplification in many cases). Finally, in order to compare this index from star to star, it is necessary to assume that the ion chosen is dominant throughout the spectral types covered. For this purpose N V was chosen, since it appears to be the dominant form of nitrogen in the winds of all the O stars, because it is not predominant in their photospheres, and because it is well-observed by Copernicus.

Figure 2 shows the correlation of the mass loss index $R_* V_\infty W\lambda$ with absolute bolometric magnitude for several O stars. It is clear that a good correlation exists, confirming that luminosity alone strongly governs the mass-loss rate for these stars. One interesting sidelight is that for these stars at least, other parameters such as rapid rotation (λ Cep) or the presence of a close binary companion (29 CMa) do not appear to strongly affect the mass-loss rate. There is evidence for the stars of marginal luminosity that rotation may play a decisive role in producing mass loss (Snow and Marlborough, 1976; Marlborough and Snow, 1976), but evidently once the luminosity is great enough to produce mass flow unassisted, as it is for all the O stars, other influences are insignificant.

In Figure 3 is shown the plot of $R_* V_\infty W\lambda$ versus temperature, where it is seen that no correlation exists. Similar results were found from correlations with other parameters such as vsini.

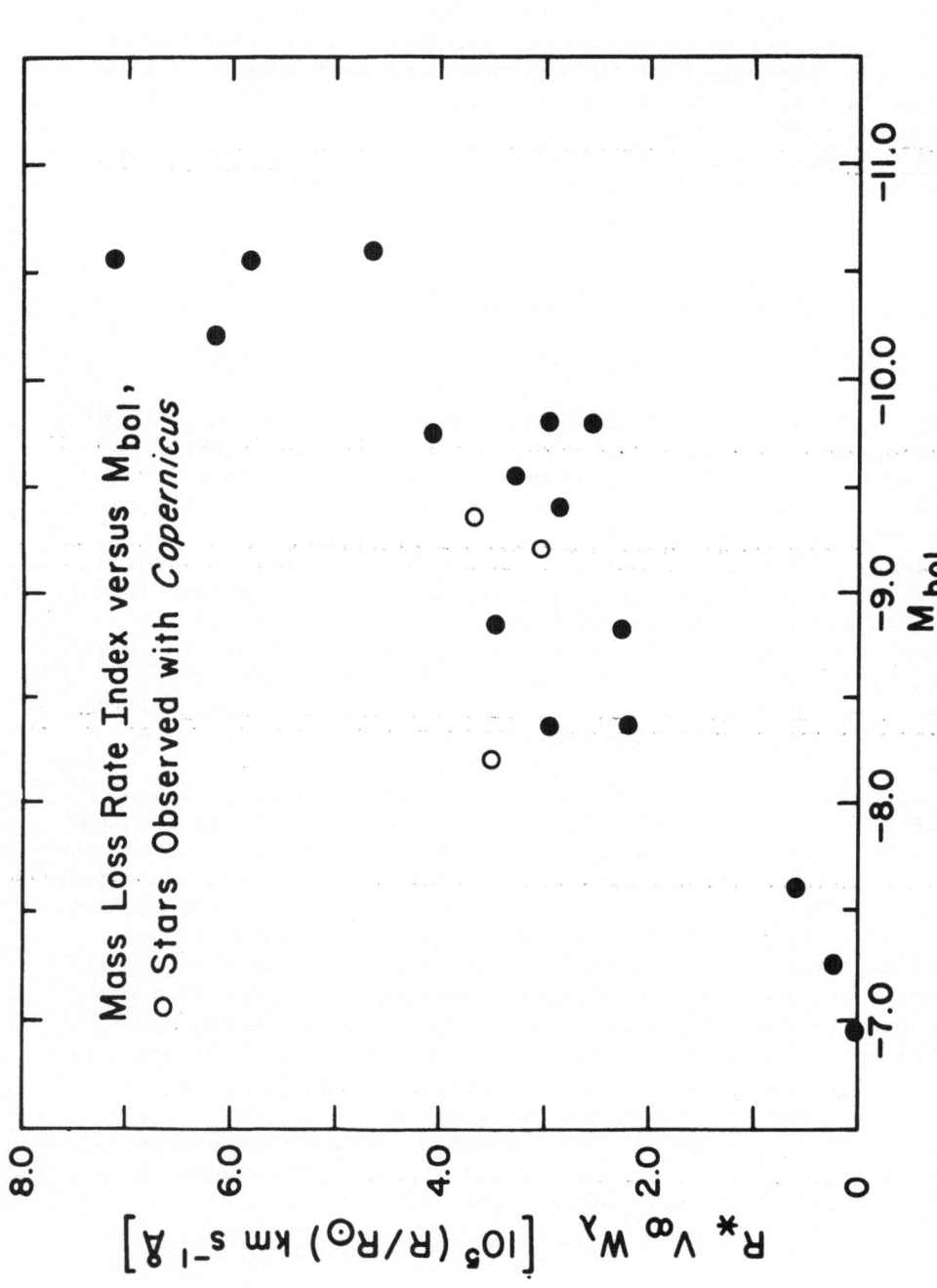

Figure 2. Correlation of the mass-loss rate parameter $R_* V_\infty W_\lambda$ with bolometric absolute magnitude. For the O stars, there appears to be a good correlation, confirming
theoretical expectations that the mass-loss rate should be strongly dependent upon
luminosity.

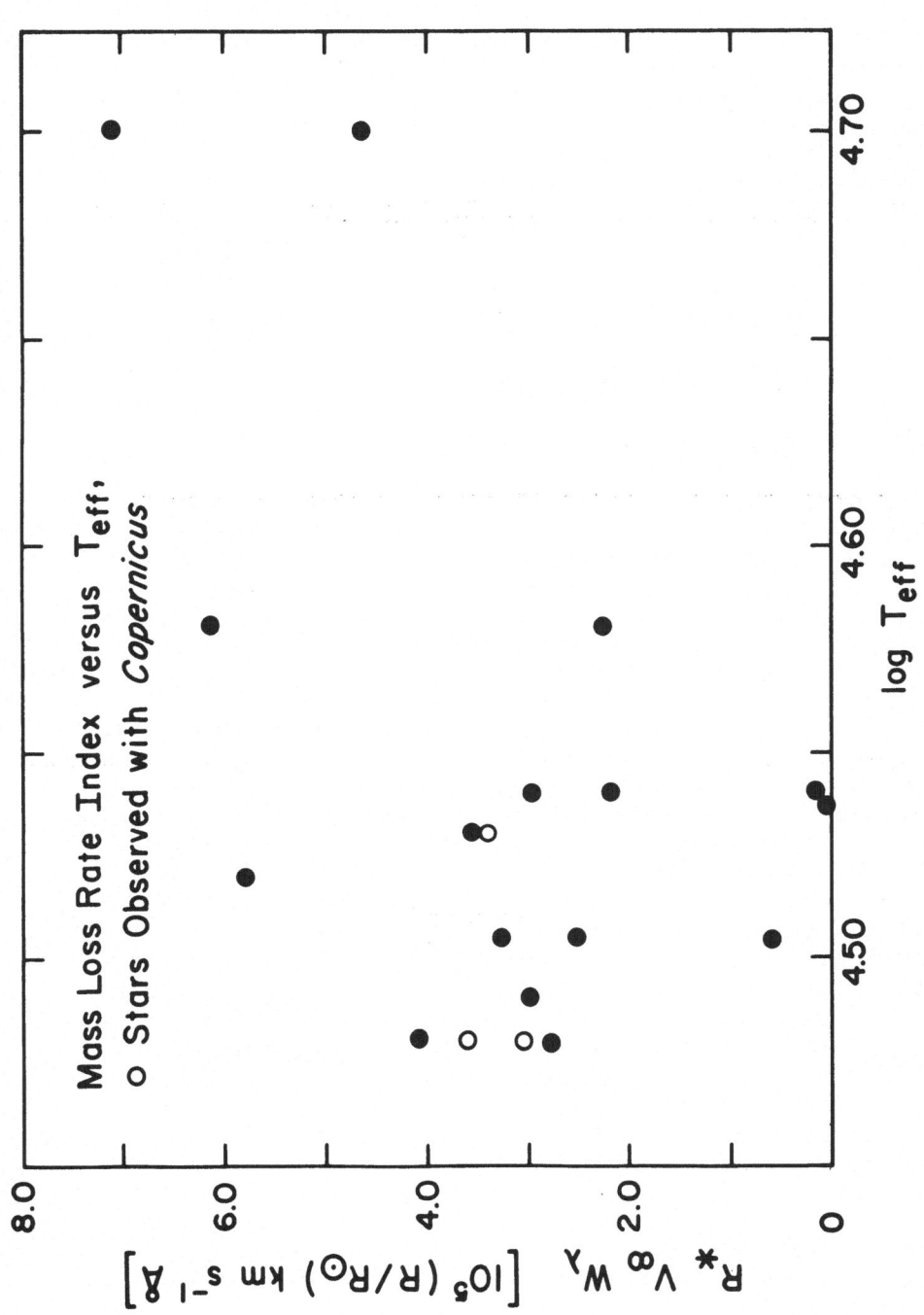

Figure 3. Correlation of the mass-loss parameter with effective temperature. No strong correlation is seen, indicating that stellar luminosity is a more important influence than temperature in governing the mass-loss rate.

While there was a general tendency for the hottest stars to have the greatest velocities, there was no strong correlation of terminal velocity with either stellar effective temperature or luminosity. Abbott (1978) has shown, however, that the terminal velocity does vary in a uniform way with the stellar escape velocity, as expected from the theory of radiatively-driven winds. In order to estimate the escape velocities, it was necessary to estimate the stellar masses from evolutionary calculations which take mass loss into account, and the gravitational binding force was modified to allow for radiative pressure. Abbott found that the terminal velocity correlates with the escape velocity according to:

$$V_\infty = (535 + 2.64 \; V_{escape}) \; \text{km s}^{-1}. \qquad (3)$$

This is consistent with the theoretical expectation that:

$$V(r) \propto V_{escape} \; (1-R_*/r)^{\frac{1}{2}} \qquad (4)$$

or

$$V_\infty \propto Vescape. \qquad (5)$$

The slope of the observed correlation allows an empirical evaluation of the net acceleration created by the radiation pressure.

From empirical profile-fitting techniques (e.g., Lamers and Morton, 1976; Lamers and Rogerson, 1978), it is possible to derive the form of the velocity law for the winds. This has been done only for two stars so far (Lamers and Morton, 1976; Lamers and Rogerson, 1978), but it is expected that similar analyses of additional stars will soon be completed. In general, smooth laws similar to equation (3) above produce adequate fits to the overall profiles, neglecting detailed structure.

The presence of relatively narrow (PWHM $\sim$100 km s^{-1}) absorption components superposed on the broad P Cygni profiles in some stars may indicate the presence of structure in the velocity law, however. Such components are quite common, and could form as a result of plateaus in the height dependence of the velocity, so that enhanced column densities occur at specific velocities. If this interpretation is correct, then the data imply the existence of several plateaus at different velocities in some stars, since the velocities of these components are seen to vary from ion to ion. This further implies that each plateau has a different characteristic temperature. Perhaps the plateaus could result from changing ionization conditions with height, since the radiative acceleration at each height depends on the abundances of the ions which absorb most strongly.

On the other hand, the narrow absorption components could simply reflect steep gradients of the ionization equilibrium in a wind with a smooth velocity law, so that within a narrow velocity range, the observed ionization stage is dominant, producing enhanced absorption at that velocity.

In either case, the apparent permanence of these components in times of years (Snow, 1977) implies that the velocity law structure or ionization gradients which cause them are stable. The optical depth of a particular absorption component may vary with time (see discussion below), but the velocities are quite invariant.

IV. Time Variations in the Winds

Since it appears likely that some form of mechanical heating is present in the outer layers of early-type stars with mass loss, it is important to gather data on the nature of this heating. If the situation is analogous to the solar chromosphere and corona, it might be expected that mass motions occur near the surface, which could result in turbulence or other forms of random or non-random variations in the material in the wind. To observe such variations would provide valuable information on the nature and stability of the lower regions of the winds, and could help answer fundamental questions about the entire phenomenon of mass loss.

Evidence of variability in mass loss from hot stars has been accumulating at a rapid pace. Visible-wavelength indications have been reported previously by several authors (e.g., Rosendhal and Wegner, 1970; Rosendhal, 1973a,b; Brucato, 1971; Conti and Frost, 1974) showing that variations on timescales as short as days occur commonly for B supergiants and the Oe star λ Cep. More recently, Conti and Niemalä (1976) found variations in the Hα profile of the Of star ζ Pup over a time of three years. Other results on variability in O star Hα profiles are being reported in this volume (Vreux and Andrillat).

In the ultraviolet, recent data obtained with Copernicus have revealed significant variability in the UV P Cygni profiles for many O stars. York et al. (1977) found O VI profile variations in three stars (ι Ori, δ Ori A, and ζ Pup) which occurred in times of hours, and Snow (1977) found upon re-observing some 15 O stars that most had undergone significant alterations in their profiles over 3-4 years. Lamers, Stalio, and Kondo (1978) found similar results for the MgII lines in B supergiants, using BUSS data. To date, no UV experiments have allowed searches for variability on timescales shorter than about one hour, and the new IUE satellite will not offer much improvement in this regard. Figure 4 shows an example of long-term variability found by Snow (1977).

While the true timescales for variability are not yet known, some remarks can be made, based on very recent results. Copernicus data have been used to study variations in several ions in the stars δ Ori A (O9.5II; studied by Snow and Hayes, 1978) and ζ Pup (O4If; analyzed by Wegner and Snow, 1978). Neither study showed changes as large in magnitude as those reported by York et al. (1977) for the same stars, but both did reveal significant fluctuations in the UV P Cygni profiles. In both cases it appeared that the variations can be characterized as sporadic events which disturb the otherwise quiescent profile. Such events apparently

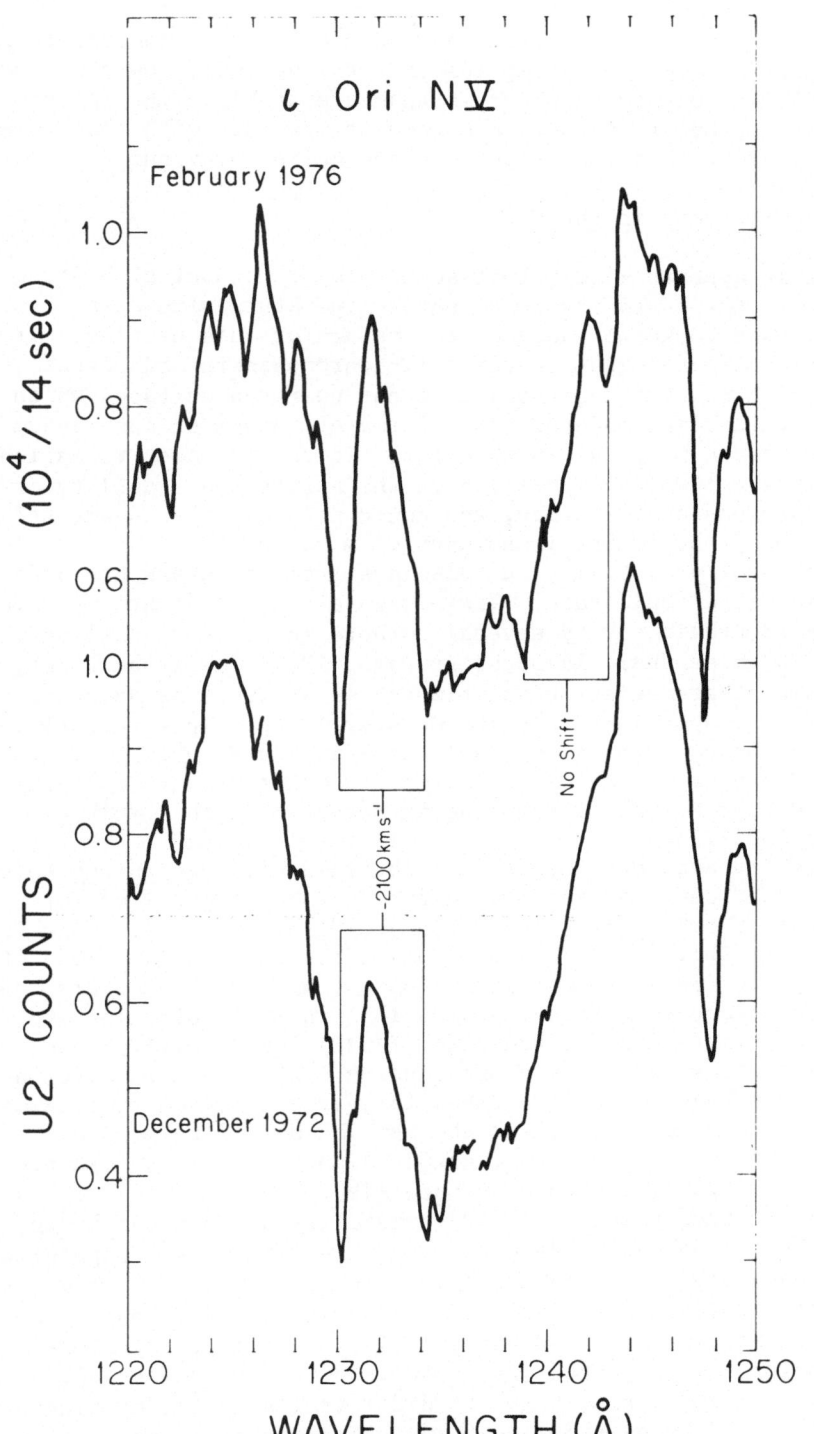

Figure 4.   Example of time variations in an ultraviolet
P Cygni profile.

can occur and diminish, at least in the portion of the wind where
the UV profiles form, in times of hours. This is suggestive of
perturbations which flow outward at the wind velocity, since the
transit time for a fluid element through variable portions of the
profile at the large observed velocities is typically of order
2-3 hours.

The work of Wegner and Snow (1978) on ζ Pup included simultaneous UV and visible-wavelength spectroscopy, with the Hα and
He II λ4686 profiles being sampled at a rate of roughly one observation every 10 minutes through several nights. The ground-
based data showed strong variations in both the Hα and λ4686 profiles,
consisting of the appearance of a secondary blue-shifted emission
peak over a time of roughly one day, with the feature persisting
for at least one day thereafter. From the good time coverage
during the nights the star was observed, it appears that this
feature developed gradually over the one-day timescale. Unfortunately, the feature did not diminish before the observing run was
interrupted, so it is impossible to determine how rapidly it died
out. In any case, it had disappeared some 27 days later, but then
began to re-appear again within two days.

If these variations are all caused by outbursts at the surface
which result in perturbations which flow out through the wind, then
it is quite reasonable to expect that longer timescales would be
found for the visible-wavelength profile variations than for the
UV, because the Hα and λ4686 lines are formed at low levels in the
wind, where the flow velocity is still quite low, of order 10^2 km
s^{-1}.

In the case of ζ Pup, as well as in a recent study of the
variable Oe star λ Cep (Leep and Conti, 1978), the magnitude of the
variations seen in the visible lines is greater than that found in
the UV. This implies that the disturbances which cause the fluctuations must diminish in strength as they flow outward; or that
they are so localized that they can flow out of our line of sight
by the time they reach high levels in the wind if they don't happen
to be ejected directly towards us (yet are so intense that they can
cause significant effects in the integrated profile when they are
at low levels in the wind). The latter possibility may be ruled
out, however, since even strong disturbances which did not flow
outward exactly along our line of sight should noticeably affect
the UV emission components, and strong variations in these emission
lines have not been seen. Hence, it may be concluded that, at
least in the two cases where simultaneous UV and visible-wavelength
spectroscopy have been carried out, the events which disturb the
profiles diminish in strength by the time they reach high levels in
the wind.

In the cases of both δ Ori A (Snow and Hayes, 1978) and ζ Pup
(Wegner and Snow, 1978), the strength and frequency of variations
found in the UV P Cygni profiles seem to change with time. Both of
these stars were evidently very active when observed by York et al.
(1977), but not so active when observed two years later in the

studies cited here. Perhaps the variability in O-star winds is
itself a variable phenomenon, analogous to the solar activity
cycle.

One of the most intriguing and potentially useful aspects of
the variability observed is the fact that the narrow velocity
components which are often seen within the broad P Cygni profiles
do not change in velocity, even though their strength may vary
drastically. This seems to be telling us that, even though ephemeral
density enhancements can occur, there is an underlying structure to
the winds which is constant.

V.   Future Observations

New instrumentation, some of it still in the planning stage,
should allow continued advances in our empirical knowledge of
stellar winds. The IUE satellite is now in successful operation,
with an echelle spectrograph capable of 0.1 A resolution throughout
the region from 1200 A to 3400 A, covering several important ions.
The Space Telescope will, in a few years, provide greater sensitivity and resolution over the same spectral region.

The first goal of these instruments with regard to mass loss
will be to determine better the limits of the region in the HR
diagram where the phenomenon occurs. Already IUE has observed
subluminous hot objects such as central stars of planetary nebulae
(Heap, this volume) and early-type stars in the Magellanic clouds
(Conti, private communication). In the future we may expect a
greater variety of objects to be sampled, including stars in a
number of nearby galaxies, where it will be interesting to see the
effects of abundance variations and to test our ideas of the relationship of mass loss and stellar evolution.

Another obvious area of improvement will occur in our understanding of time variability and hence of stability, structure, and
energy deposition in the winds. A cooperative effort is planned
for late 1978 using IUE from both sides of the Atlantic to get full
24-hour coverage of a single O supergiant (α Cam) for a few days.
This project, which will be coordinated with ground-based spectro-
scopists, will allow simultaneous sampling of a range of ions,
including N V, C IV, Si IV, H I, and He II, on a timescale of some
30-40 minutes.

Other experiments, perhaps on space shuttle payloads, should
provide coverage of shorter timescales in the UV to complement data
already being obtained from the ground. While most of the current
evidence for the O stars seems to indicate that UV variability
occurs on times of one hour or more, there have been hints of more
rapid variability in the visible-wavelength lines (e.g., Brucato,
1971 Vreux and Andrillat, this volume).

While the past decade of research in space astronomy has
revealed the broad outlines and general nature of the phenomenon of
mass loss from hot stars, the next decade should begin to fill in
the details and provide answers to many of the questions raised
during this symposium. We can look forward not only to this, but
also to the challenges of new questions which will arise.

REFERENCES

Abbott, D.C. 1978, Ap.J., in press.
Brucato, R.J. 1971, M.N.R.A.S., 153, 435.
Carruthers, G.R. 1968, Ap.J., 151, 269.
Conti, P.S. and Frost, S.A. 1974, Ap.J. (Letters), 190, L137.
Conti, P.S. and Niemalä, V.A. 1976, Ap.J. (Letters), 209, L37.
Gurzadyan, G.A. 1975, Space Sci. Rev., 18, 95.
Henize, D.G., Wray, J.D., Parsons, S.B., Benedict, G.F., Bruhweiler,
    F.W., Rybski, P.M. and O'Callaghan, F.G. 1975, Ap.J. (Letters),
    199, L119.
Lamers, H.J.G.L.M. and Morton, D.C. 1976, Ap.J. (Suppl.), 32, 715.
Lamers, H.J.G.L.M. and Rogerson, J.B. 1978, Astr. Ap., in press.
Lamers, H.J.G.L.M. and Snow, T.P. 1978, Ap.J., 219, 504.
Lamers, H.J.G.L.M. and Snow, T.P. 1979, in preparation.
Lamers, H.J.G.L.M., Stalio, R. and Kondo, Y. 1978, Ap.J., in press.
Leep, E.M. and Conti, P.S. 1978, Ap.J., submitted.
Marlborough, J.M. and Snow, T.P. 1976, IAU Symposium 70, The Merrill-
    McLaughlin Memorial Symposium on Be and Shell Stars (ed. A.
    Slettebak, Boston: D. Reidel).
Morton, D.C. 1967, Ap.J., 147, 1017.
Morton, D.C. 1976, Ap.J., 203, 386.
Rogerson, J.B. and Lamers, H.J.G.L.M. 1975, Nature, 256, 190.
Rosendhal, J.D. 1973a, Ap.J., 182, 523.
Rosendhal, J.D. 1973b, Ap.J., 186, 909.
Rosendhal, J.D. and Wegner, G.A. 1970, Ap.J., 162, 547.
Snow, T.P. 1977, Ap.J., 217, 760.
Snow, T.P. and Hayes, D.P. 1978, Ap.J., submitted.
Snow, T.P. and Marlborough, J.M. 1976, Ap.J. (Letters), 203, L87.
Snow, T.P. and Morton, D.C. 1967, Ap.J. (Suppl.), 32, 429.
Stecher, T.P. 1968, Proc. Symp. on Wolf-Rayet Stars, ed. K.B. Gebbie
    and R.N. Thomas, NBS Special Pub. No. 307, p. 65.
Wegner, G.A. and Snow, T.P. 1978, Ap.J. (Letters), submitted.
York, D.G., Vidal-Madjar, A., Laurent, C. and Bonnet, R. 1977,
    Ap.J. (Letters), 213, L61.

DISCUSSION FOLLOWING SNOW

Pismis: You emphasized the good correlation between the strength
of the stellar wind and the absolute magnitude of O stars. Now the
mass-luminosity law is expected to hold for O stars: Therefore a correlation between the wind strength and the absolute magnitude would
imply a good correlation between wind and mass of the stars. Physically
the mass is a more fundamental property of a star. I understand that
the reason why the absolute magnitude was discussed in these correlations is due to the fact that the absolute magnitude is obtained more
directly from observation than the mass.

Snow: Not only is luminosity more easily determined, but it is
also the parameter which, according to the theory of radiatively-driven
winds, directly affects the mass loss rate. Hence, in this context the
luminosity can be considered the more fundamental quantity. Furthermore, for the O stars with high mass loss rates, the masses are changing
significantly with time while the luminosities no longer follow the
usual mass-luminosity relation.

Carrasco: The correlation between mass loss rates and luminosities
for the O-type stars shown may be subject to calibration problems.
Among the stars used are several runaway stars and these might be under-
luminous by a factor of 10.

Snow: We took our luminosities from data in the literature, and
certainly it is possible that errors exist in the values of M_{bol} we
adopted.

Underhill: Some central stars of planetary nebulae have sufficiently strong winds that these winds are seen by means of the visible
spectrum. These stars have masses near 1 solar mass. What does this
observation mean in terms of your proposed significant correlation between luminosity and mass and rate of mass loss?

Snow: Other parameters such as surface gravity must play a role
which doesn't show up very clearly in this small sample. Dr. Heap is
going to discuss some recent IUE data later in this session which show
strong ultraviolet P Cygni profiles in low-luminosity, hot stars.

Castor: Bowyer's X-ray upper limits are already very interesting.
For ζ Puppis the limit on an optically thin corona is a volume emission
measure $\int n_e^2 \, dV < 10^{55 \cdot 6}$ if $T_c = 10^6$K, or $< 10^{55 \cdot 5}$ if $T_c = 5 \times 10^6$K.
These limits are 10 times less than that required in Joe Cassinelli's
optically thin coronal model (3×10^{56} for the observed O VI in ζ Pup).
The coronal model might be adjusted to come under the limit, but
clearly these limits are already useful.

[Comment added after the session: The HEAO-1 upper limits quoted
here are for soft (E < 1 keV) X-rays, and Cassinelli's and Olson's
optically thin model produces none of these.]

Sreenivasan: What is the uncertainty in the observationally estimated mass loss rates?

Snow: Substantial, even in the cases where detailed profile-fitting procedures as described by Lamers in this Symposium have been used.   The formal uncertainties are about ±50%, I think.

Sreenivasan: Effects of rotation (V sin i) or differential rotation (V sin i and macroturbulence) amount to an increase of ~20% or 30% with inclusion of centrifugal force and to about a factor 2 or 3 for centrifugal force + turbulent pressure due to differential rotation over mass loss rates for non-rotating models.   So this may not show up in a correlation analysis, if the observational uncertainty is comparable or greater.   Bodenheimer [Ap. J. (1971)] showed that the luminosity of a star in rapid (differential) rotation is much less than that of a non-rotating star, e.g., a 60 $M_\odot$ star in strong differential rotation could have the luminosity of a 30 $M_\odot$ non-rotating star!   So, this could also be responsible for the absence of correlation with rotation, etc.

Hutchings: Since you show that $\dot{m}$ is correlated with L, you should normalize to a given L value before looking for $\dot{m}$ as a function of T_{eff}.

Snow: I agree, but I'm not certain this is worth trying on the basis of this crude mass loss rate parameter; it would be preferable, of course, to have better-determined mass loss rates for any discussion of correlations such as this.   Ongoing work by Lamers and myself should produce a uniform set of mass loss rates within a few months for most of the stars in the Snow and Morton survey.

Dearborn: Is the lack of correlation shown by you between mass loss rate and T_{eff} possibly due to the fact that you consider only O stars?   Would it not be more obvious if you considered later type stars?

Snow: Certainly this is possible, and furthermore, M_{bol} and T_{eff} are not entirely independent, so as Hutchings just pointed out, the luminosity must be normalized before one looks at temperature effects. In any case I would hesitate to extend this to B stars now, because the assumption that N V dominates would certainly not hold so this mass loss rate parameter based on N V absorption would no longer be expected to correlate with the true mass loss rate.   Again, my inclination is to wait until we have actual mass loss rates for more stars.

Heap: What is the highest terminal velocity observed in young O stars?

Snow: 3400 km/s seen in 9 Sgr.

Bidelman: The high-latitude B1 supergiant Rho Leonis appears to show, from Copernicus data, substantially higher mass loss than would be expected from its spectral type and luminosity.   Perhaps this is an indication of low mass or some other stellar peculiarity.

Morton: The Copernicus UV scans show many examples of stars with similar visual spectral types but widely different P Cygni profiles. Since the P Cygni profiles originate in a region where small changes in conditions can have large effects I would believe the visual spectral type.

Conti: The night-to-night optical variability in $\lambda 4886$ and Hα observed in ζ Pup are reminiscent of similar behavior in λ Cep, another relatively rapidly rotating Of star. I would like to stress that this variability is not a common phenomenon.

Snow: Yes, in fact for both of these stars, which as far as I know are the only ones for which simultaneous UV and visible-wavelength observations of variability have yet been made, the fluctuations seen in the UV are weaker than those in the visible, as though the disturbances dissipate as they flow outwards.

MODELLING OF UV RESONANCE LINES

Henny J.G.L.M. Lamers
The Astronomical Institute at Utrecht
Space Research Laboratory

John C. Castor
Joint Institute for Laboratory Astrophysics
National Bureau of Standards and University of Colorado

ABSTRACT

The envelopes of early type stars can be studied by comparing observed and theoretical P Cygni profiles of UV resonance lines. The lines are formed by scattering in the expanding envelope and complete frequency redistribution is a reasonable assumption. The Sobolev approximation can be used for the prediction of the profiles, except within a Doppler shift of $2xv_{thermal}$ from the line center. The profiles are sensitive to the optical depth $\tau_{rad}(v) \propto n_i(r) \, (dr/dv)$, but insensitive to the velocity law. We present a fairly simple method to calculate the effects of a photospheric profile or partly overlapping doublets.

I INTRODUCTION

The study of the ultraviolet P Cygni profiles is the major source of information about the mass loss rates of early type stars and the structure of their expanding envelopes. (The radio fluxes can be used to determine the mass loss rates accurately, but the observations are limited to only a very small number of the brightest stars. The infrared fluxes up to 22 μm also give information on the mass loss rate, but, since these fluxes arise from regions very close to the star with small velocities, they do not give information on the structure of the major part of the envelope.) Profiles of ultraviolet lines in about 60 early type stars observed by the Copernicus satellite have been published by Snow and Morton (1976) and Snow and Jenkins (1977); the forthcoming observations with the International Ultraviolet Explorer will extend this number dramatically. In order to derive information about the structure of the expanding envelopes, the observed profiles have to be compared with calculated profiles.

We will discuss the process of line formation, the Sobolev approximation for the calculation of the profiles and its accuracy and the accuracy of the physical quantities that can be derived from the profiles. We will show that the optical depth $\tau(v)$ can be derived fairly

P. S. Conti and C. W. H. de Loore (eds.), Mass Loss and Evolution of O-Type Stars, 81-92.
Copyright © 1979 by the IAU.

accurate, but that the profiles give little information on the velocity
law.

II THE PROCESS OF LINE FORMATION

The P Cygni profiles of the ultraviolet resonance lines in the
spectra of early type stars are formed by resonance scattering in the
expanding envelopes. The probability of destruction of a photon and the
corresponding rate of creation of photons are small, as we can see by
computing the ratios.

$$\varepsilon = \frac{n_e C_{21}}{A_{21}} \left[1 - \exp\left(-\frac{\chi_{12}}{kT_e}\right)\right] \simeq 6.2 \, T_e^{-1/2} n_e \lambda^3 \left[1 - \exp\left(-\frac{\chi_{12}}{kT_e}\right)\right] \tag{1}$$

and

$$\eta = \frac{R_{2k}}{A_{21}} \simeq 5.11 \times 10^{19} \, W \, a \, f_{21}^{-1} \, \lambda^2 \, \chi_2^2 \, T_* \, \exp(-\chi_2/kT_*) \tag{2}$$

where $n_e C_{21}$ is the rate of collisional deexcitation and R_{2k} is the rate
of photoionization, both per excited ion. We adopted Van Regemorter's
(1962) approximation for C_{21} and a photoionizing radiation field of
$J_\nu = W \, B_\nu(T_*)$. a is the photoionization cross section of level 2 (in
cm^2), f_{21} is the emission oscillator strength, λ is the wavelength (in
cm), χ_2 is the ionization potential of level 2 (in eV) and W is the dilution factor. The electron densities in the envelope range from about
10^8 to 10^{11} cm^{-3} with the result that ε ranges from less than 10^{-8} to
10^{-5}. The value of η is less than 3×10^{-5} if $T_* \gtrsim 4 \times 10^4$ K, $\chi_2 \gtrsim 40$ eV
and $W \lesssim 0.5$. The effect of ε or η is negligible unless

$$\varepsilon\tau > \frac{WI_c}{B_\nu(T_e)} \quad \text{or} \quad \eta\tau > \frac{WI_c}{B_\nu(T_*)} \tag{3}$$

(see Castor 1970). In these expressions I_c is the intensity of the
photosphere at the line frequency. For a line at 1000 Å, with
$T_e \lesssim 2 \times 10^5$ K and $T_* \gtrsim 2.5 \times 10^4$ K, $I_c/B \gtrsim 10^{-2}$. We may suppose that
$B^* \approx I_c$. Therefore, these effects are not important unless τ is very
large, $\tau \gtrsim 10^3$.

One can assume somplete frequency redistribution for the scattering
process. Mihalas et al. (1976) showed that the assumption of complete
frequency redistribution gives an accurate approximation to the source
function of the lines in a moving atmosphere. They found that even if
the lines have a coherent scattering wing, the assumption of complete
redistribution is still to be preferred to coherent scattering for the
line as a whole provided that the optical thickness is not too large.

III THE SOBOLEV APPROXIMATION

The transfer equation for a spherically symmetric expanding envelope can be solved if the Sobolev approximation is adopted. This approximation is valid if the velocity gradient is so large that the scale
length for the density and temperature variations in the envelope is
much larger than the length in which the expansion velocity increases by
the thermal velocity. The intrinsic line width can then be neglected
relative to the Doppler displacements. There are two methods for calculating the profiles of resonance lines in a spherically symmetric expanding envelope with the Sobolev approximation. In Lucy's (1971) method
the angular dependence of the intensity is approximated by two streams
with a properly chosen boundary between them, whereas in Castor's (1970)
method the angular dependence is treated exactly using escape probabili-
ties. For the same model, the profiles calculated with the two methods
agree within one per cent.

In order to estimate the effect of the Sobolev approximation on the
profile we compared the profiles for a rapidly expanding plane-parallel
envelope calculated by Noerdlinger and Rybicki (1974) using the modified
Feautrier method including thermal broadening with the profiles for the
same envelope calculated without thermal broadening using Lucy's method.
This comparison is shown in Figure 1, where the profiles computed by
Noerdlinger and Rybicki with the parameters (ZM = 1, VM = 10) and
(ZM = 5, VM = 20) are compared with our profiles for the same case.

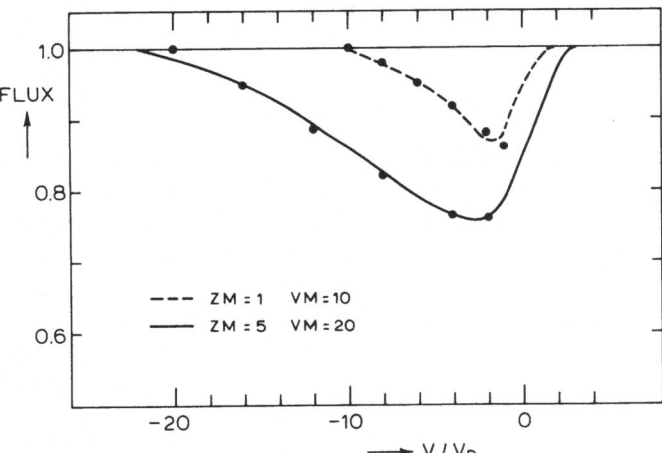

Figure 1. A comparison between theoretical profiles of scattering lines
in a plane parallel expanding atmosphere, calculated with the
Sobolev approximation (dots) and without the Sobolev approximation (full and dashed lines; from Noerdlinger and Rybicki,
1974). The agreement is very good in the region where $|V| >$
$2xV_{Doppler}$.

We note that the agreement is better than 0.5% throughout the violet
absorption wing, except within two Doppler widths of the line center. At
such small velocities the thermal motions of the ions cannot be neglected and so the Sobolev approximation breaks down. This implies that for
O stars, where the photospheric turbulent velocity can amount to about
30 km/s (Ebbets, 1978), profiles calculated with the Sobolev approximation can not be trusted within 60 km/s from the line center.

IV THE ION DENSITY AS A FUNCTION OF DISTANCE: $n_i(r)$

The information that one wants to derive from the study of the P
Cygni profiles of UV resonance lines is: the velocity law, $v(r)$, and
the density of the absorbing ions, $n_i(r)$. These two quantities enter the
calculations of the profiles in a coupled way via the radial optical
depth.

The radial optical depth in an expanding envelope is given by

$$\tau_{rad}(r) = \frac{\pi e^2}{mc} f \lambda_o n_i \left(\frac{dr}{dv}\right) , \tag{4}$$

where f is the absorption oscillator strength, λ_o (in cm) is the rest
wavelength of the transition, n_i (in cm^{-3}) is the number density of the
absorbing ions at a distance r, and dr/dv (in s) is the inverse velocity
gradient. The quantities n_i and dr/dv are evaluated at the radial distance r. We assume that the velocity increases monotonically with radius,
so that dr/dv is uniquely defined. This τ_{rad} is the total optical depth
presented by the envelope to a photon traveling radially at that frequency which would be resonantly absorbed at radius r. Most of this optical depth is contributed by a thin shell across which $\Delta v = v_{thermal}$.

The radial optical depth is expressed in equation (4) as a function
of distance from the star. If the velocity increases monotonically, τ_{rad}
can also be expressed as a function of the velocity in the envelope:
$\tau_{rad}(v)$. Calculations of line profiles show, as expected, that the absorption part of the P Cygni profiles, expressed in relative flux versus
velocity-shift, depends very strongly on $\tau_{rad}(v)$ and very little on the
velocity law $v(r)$. On the other hand, the emission component depends on
$v(r)$ but not on its detailed structure. The total amount of emission is
a measure of the extent of the envelope: a soft velocity law, (i.e.
v/v_∞ increases slowly outwards) gives a stronger emission component than
a fast velocity law.

If the line is strong and saturated, the profiles yield little information on $\tau_{rad}(v)$. In contrast to the case for static atmospheres,
for which the profiles continue to change if τ increases, the P Cygni
profiles become completely independent τ when $\tau_{rad}(v)$ is larger than
about 3 at all velocities (but smaller than 10^3, see section II). Such
saturated profiles depend on the velocity law only, but the dependence

is weak except near the line center.

In conclusion: we find that for lines which are not saturated the function $\tau_{rad}(v) \propto n_i(r).(dr/dv)$ can be derived from the profiles, except at small velocities $v < 2v_{thermal}$. For the determination of the ion density or the ionization balance as a function of distance one has to know the correct velocity law. As the P Cygni profiles are not very sensitive to the velocity law, this law may have to be derived by other methods and from other observations.

V   THE VELOCITY LAW: $v(r)$

The strongest P Cygni profiles of the UV resonance lines of luminous O and B stars have a steep violet edge at the short wavelength side of the absorption component and this edge velocity is about the same for all strong lines in the spectrum of a star. (see e.g. Snow and Morton, 1976). The emission components are strong at small velocities and decrease towards large velocities. These two facts suggest that the expansion velocity $v(r)$ increases outwards and asymptotically approaches a terminal velocity, v_∞, at large distances. Abbott(1978) found that the terminal velocity of the most luminous stars is about 3.5 times the effective escape velocity at the stellar surface.

For those stars which have weak P Cygni profiles or extended violet absorption wings at the UV resonance lines, like early-B mainsequence stars and giants, the terminal velocity cannot be determined and may in fact be considerably larger than the extent of the violet wings.

In Figure 2 we show several velocity laws. The broken lines are theoretical laws for the radiation driven wind models by Castor, Abbott and Klein (1975) and by Abbott (1977).

The steep law (CAK) includes no effects of lineblending and the slow law (A) includes a maximum of blending. These two laws can be taken as the upper and lower limit for the steepness of the velocity law predicted for the radiation driven winds. The full lines are the laws derived from the observations by Lamers and Morton (1975) from the UV lines, using the observed ionization balance, and by Barlow and Cohen (1977) from the radio and infrared flux distribution of P Cygni. We can only conclude that either there is no unique velocity law for all early type stars or the true velocity law is still unknown.

VI   THE EFFECT OF A PHOTOSPHERIC ABSORPTION LINE

There are two additional effects which have to be taken into account for an accurate analysis of the P Cygni profiles of the UV resonance lines:
a.   the presence of an underlying photospheric absorption component and

b.   the overlap of doublet lines.
In the literature one finds examples where these effects are allowed for
by simple addition or superposition of the profiles. However, calculations of profiles in which both effects are properly taken into account
show that these simple corrections can give a large error in the profiles. Therefore we have developed a simple method, based on the separate superposition and addition of flux contributions, arising from different parts of the envelope or photosphere. This simple method gives
profiles which are fairly accurate approximations of the exact calculations.

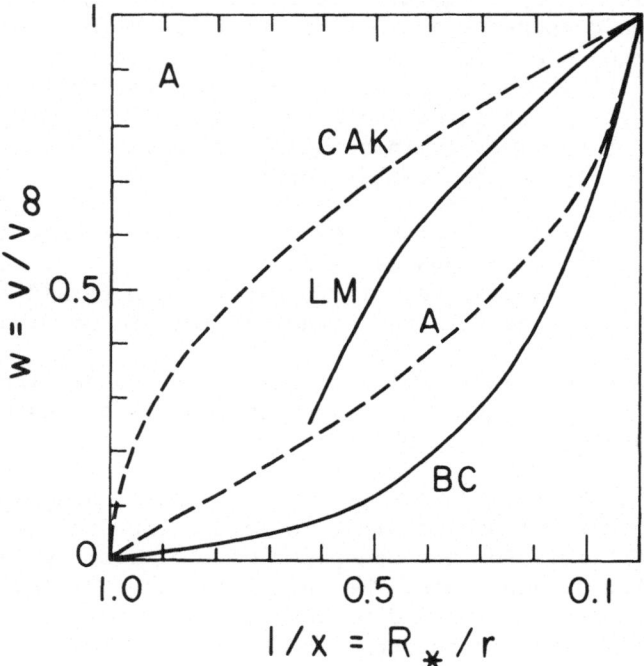

Figure 2. Theoretical (dashed lines) and observed (full lines) velocity
          laws normalized to $v_\infty = 1$ and $R_* = 1$. The theoretical laws are
          for radiation driven wind models without line blending (Cas-
          tor, Abbott and Klein, 1975; CAK) and with maximum line blend-
          ing (Abbott, 1977; A). The observed laws are derived from the
          infrared and radiofluxes of P Cygni (Barlow and Cohen, 1977;
          BC) and from the ultraviolet lines of ζ Puppis (Lamers and
          Morton, 1976; LM).

Let $F_{phot}(v)$ be the flux of the photospheric profile relative to the continuum, where v is the Doppler shift from the line center, $v = c (\lambda-\lambda_0)/\lambda_0$. Let $F_{env}(v)$ be the flux of the envelope P Cygni profile calculated for a <u>continuous</u> photospheric spectrum. The profile which results from the interaction of the two profiles can be approximated with reasonable accuracy by the following expressions:

$$F(v) = F_{phot}(v) + q[F_{env}(v) - 1] \tag{5a}$$

for the long wavelength side of the profile (v>o), and

$$F(v) = F_{phot}(v) \left[F_{env}(v) - \{F_{env}(-v) - 1\}\right] + q\left[F_{env}(-v) - 1\right] \tag{5b}$$

for the short wavelength side of the profile (v≤o),

where   v=o

$$q = \int_{v=-v_\infty} F_{phot}(v) \; d(v/v_\infty) \tag{5c}$$

The approximation can be understood as follows: The P Cygni profile consists of the sum of a shortward shifted absorption, produced in the part of the envelope that is seen projected against the stellar disc, and an emission that is symmetrical around $v = 0$ produced by photons that are scattered by the envelope into the line of sight. If the photospheric spectrum contains an absorption profile, the amount of photospheric radiation that can be scattered into the line of sight is reduced by a factor q, the mean photospheric flux in the wavelength interval corresponding to $v = -v_\infty$ to $v = o$; thus the long-wavelength emission is reduced by a factor q (second term of eqs 5a and 5b). On the short-wavelength side, the reduction factor $F_{env}(v) - F_{env}(-v) + 1$ that accounts for envelope absorption is impressed on the photospheric profile before the reduced emission is added. Figure 3 shows an example of this simple correction.

VII PARTLY OVERLAPPING DOUBLETS

Many of the ultraviolet resonance lines in the spectra of early type stars are doublets with a separation smaller than their widths, which results in partly overlapping lines. One of the authors (J.I.C.) has adapted the escape probability method to solve the transfer equations for doublets. The calculations show that the doublet profile differs strongly from the profile obtained by a simple addition or superposition of two separately calculated components. This is due to the fact that radiation scattered by the blue component in the direction of the observer can be scattered again by the red component. So the radiation scattered by both components appears as an enhancement of the emission on the long wavelength side of the red component.

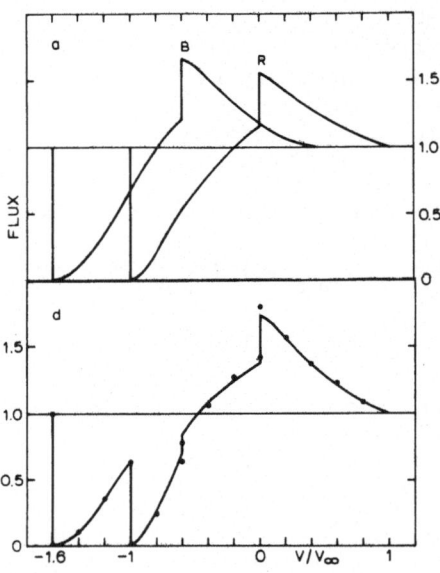

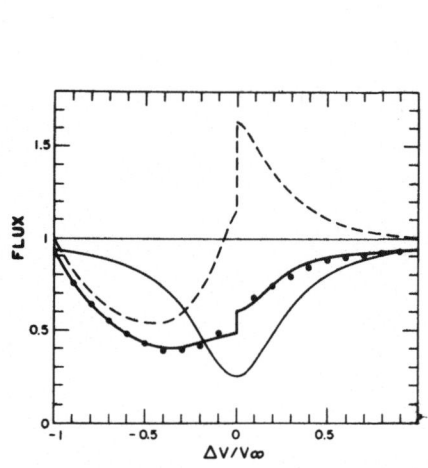

Figure 3   Figure 4

Figure 3. The effect of a photospheric absorption line. The dashed line
is the profile if the stellar spectrum were continuous. The
thin solid line is the photospheric absorption line. The thick
solid line is the profile calculated with the photospheric absorption properly taken into account as a boundary condition
to the equation of transfer in the envelope. The dots show the
approximate profile calculated with the simple method of section VI. The approximation fits the exact profile very nicely.

Figure 4. The effect of partly overlapping doublet lines. The top part
shows the profiles of the two components, if calculated separately. The blue (B) and red (R) components are calculated
for the model with $v = v_{\infty}(1-R_*/r)^{\frac{1}{2}}$ and $\tau^B_{rad} = 1.5/1-(v/v_{\infty})^2$
and $\tau^R_{rad} = 0.75/1-(v/v_{\infty})^2$. The lower part shows the exact
profile with the overlap properly taken into account in the
equation of transfer (solid line) and the approximate profile
calculated with the method of Section VII (dots). The approximation fits the exact profile nicely. Note the increased
emission of the blue component due to the effect of overlap.

    Let $F^B(v)$ and $F^R(v)$ be the flux of the blue and red components rel-
ative to the continuum if calculated separately and let $F^D(v)$ be the
flux profile of the doublet. The velocity v is the Doppler velocity corresponding to the shift relative to the wavelength λ_o^R of the undisplaced red component, $v = c(\lambda - \lambda_o^R)/\lambda_o^R$. Let $\tau^R(v)$ be the radial optical depth of the red component. The profile of the doublet can be ap-
proximated by

$$F^D(v) \simeq F^R(v) + [F^B(v)-1]\ e^{-\tau^R(|v|)} + Q \cdot R(v) \tag{6a}$$

with

$$Q = \int_{-v_\infty}^{v_\infty} \left[F^B(v)-1\right] \cdot \left[1-e^{-\tau^R(|v|)}\right] \cdot d(v/v_\infty) \tag{6b}$$

and

$$R(v) = \left[F^R(|v|)-1\right] \Big/ 2 \int_0^{v_\infty} \left[F^R(|v|)-1\right] \cdot d(v/v_\infty). \tag{6c}$$

The first term of equation (6a) is the profile of the red component,
which is not affected by the presence of a blue component. The second
term is the contribution to the blue component reduced by a factor $e^{-\tau R}$ due to
scattering in the red component. The third term is the radiation from
the blue component which is scattered and redistributed over frequency
by the red component. The factor Q is the amount of scattered radiation,
i.e., the complement of the second term integrated over the velocities
in the region where the lines overlap. The function R(v) describes the
redistribution of the radiation Q over frequency or velocity in the profile. We assume that this redistribution is the same as the distribution
of the emission, $F^R(|v|)-1$, of the red component. In Figure 4 we show
an example of this approximation.

VIII CONCLUSIONS

    Accurate observations of a P Cygni type profile supply two functions
of the velocity, v, namely the fluxes at the displacement v in either direction from line center. Therefore in principle it should be possible
to derive both the velocity law and the optical depth function from a
single line profile. In practice, the computed profiles have shown us
that the short wavelength wing of the line is primarily sensitive to
$\tau_{rad}(v)$, and not sensitive at all to v(r), while the long wavelength
wing is sensitive to both. Furthermore, the sensitivity of the long
wavelength wing to the velocity law is large only near the center of the
line, and vanishes toward the extreme. The central part of the profile
is afflicted with uncertainties due to the underlying photospheric pro-

file and the error of the Sobolev approximation. The Sobolev approximation is valid, in the case of a rapidly rising velocity law, where
$v \gtrsim 2v_{thermal}$. Thus observations readily yield only the optical depth
law $\tau_{rad}(v)$. If the central part of the profile can be corrected for the
effect of a photospheric component, then the total amount of emission on
the long wavelength side of the profile can be used to distinguish between a slowly rising $v(r)$ and a steeply rising one. Finer discrimina-
tion of the velocity law than this may not be possible.

We have presented a fairly simple method for estimating the correction due to a photospheric absorption line or the overlap of doublets.

One has to keep in mind that one of the basic assumptions used in
the analysis of UV resonance lines -- spherically symmetric outflow --
may be incorrect. Aspect effects in an asymmetric flow can cause systematic changes in the absorption and emission strengths (Cassinelli and
Rumpl 1978). Turbulence is a definite possibility, and its existence
would vitiate the Sobolev approximation if $v_{turb} \approx v$. (A small amount
of turbulence would have no effect.) The effect of stellar rotation may
also be important. The profiles for stellar envelopes with a combination
of expansion and rotation have a bewildering variety (Magnan 1970) and
such cases are better treated individually.

(This paper contains some of the results of a study of theoretical
P Cygni profiles by the authors. The details of this study and an atlas
of theoretical profiles will be published in the Astrophysical Journal
Supplement Series).

REFERENCES

Abbott, D.C. 1977, Ph.D. thesis, University of Colorado.
Abbott, D.C. 1978, Ap. J., in press.
Barlow, M. and Cohen, M. 1977, Ap. J., 213, p. 737.
Cassinelli, J. and Rumpl, W. 1978, in preparation.
Castor, J.I. 1970, M.N.R.A.S., 149, p. 111.
Castor, J.I., Abbott, D.C., and Klein, R.I. 1975, Ap. J., 195, p. 157.
Ebbets, D.C. 1977, Ph.D. thesis, University of Colorado.
Lamers, H.J.G.L.M. and Morton, D.C. 1976, Ap. J. Suppl. 32, p. 715.
Lucy, L.B. 1971, Ap. J., 163, p. 95.
Magnan, C. 1970, J.Q.S.R.T., 10, p. 1.
Mihalas, D., Kunasz, P.B. and Hummer, D.G. 1976, Ap. J., 210, p. 419.
Noerdlinger, P.D. and Rybicki, G.B. 1974, Ap. J., 193, p 651.
Snow, T.P. and Jenkins, E.B. 1977, Ap. J. Suppl., 33, p. 269.
Snow, T.P. and Morton, D.C. 1976, Ap. J. Suppl, 32, p. 429.
Van Regemorter, H. 1962, Ap. J., 136, p. 906.

DISCUSSION FOLLOWING LAMERS AND CASTOR

Underhill: Your way of estimating when it is valid to use the Sobolev approximation seems to miss the essential character of the physical assumptions lying behind this drastic simplification of radiative transfer in the presence of a velocity field. What is involved has been clearly expressed by Rybicki (1970, NBS Special Publ. 332) and by Rybicki and Hummer (1978, Ap. J.). If one follows this discussion for the case of a spherically symmetric atmosphere in which a velocity field $v(r)$ exists, then one may make the Sobolev approximation if the following inequality is valid everywhere in the atmosphere:

$$\frac{v_{th}}{v(r)} << \frac{\ell_o}{r} \left(1 - \frac{(r^2 - p^2)}{r^2} \left\{ 1 - \frac{d \ln v(r)}{d \ln r} \right\} \right) \quad .$$

Here v_{th} is the thermal velocity at the radius r in the atmosphere, that is the velocity which describes the width of the line absorption coefficient, ℓ_o is the characteristic length in the atmosphere over which the state properties of the gas are constant, r is the radius to the point under consideration and p is the perpendicular distance to the line of sight being considered. The emergent flux spectrum is obtained by integrating the specific intensity I_ν emerging along each line of sight for the range in p from 0 to R_{shell}. Usually r runs from the radius of the photosphere, R_*, to the outer radius of the atmosphere R_{shell}, which lies in the range 5 to 10 R_* for early-type stars. From a simple diagnosis of observed line widths and displacements and guesses at the electron temperature or microturbulence in the flowing atmosphere, one finds that the left side of the inequality lies in the range 0.01 to 0.3. If microturbulence is significant it lies in the range 0.1 to 0.3. The quantity ℓ_o is never defined by authors desiring to use the Sobolev approximation, but it seems unlikely that it would exceed R_*. Typically r may lie in the range 3 to 5 R_*. If $v(r)$ increases outwards, the term in braces is less than unity. It is multiplied by a factor in the range 0 to 1; typically this factor may be 0.5.

Consequently for the Sobolev approximation to be valid one must have $v_{th}/v(r) << 0.1$ to 0.2. Such conditions are only met if v_{th} is small, i.e., in a cool outer atmosphere in which there is little or no microturbulence. This is not the case for the layers in which the strong resonance lines from high ions are formed. It certainly is not so at deep layers where $v(r)$ is just beginning to increase from its small photospheric value.

Olson: In response to A. Underhill's remarks on the accuracy of the Sobolev approximation, I think that the worst case of low velocity near the stellar surface is being used as an example to claim that the approximation is poor is true everywhere. The low velocity results are inaccurate as Lamers has said, but this does not mean that an inaccuracy at low velocity affects the high velocity value of Sobolev-type calculations.

O VI IN STELLAR WINDS

Donald C. Morton
Anglo-Australian Observatory

SUMMARY

The region of the O VI resonance doublet around 1030 Å has been observed in 9 stars with the high resolution (0.051 Å FWHM) detector on the Copernicus satellite spectrometer. Although the spectra are confused with interstellar Lyman β and H_2 absorption, P Cygni profiles or shifted absorption lines indicating mass flow were found in ζ Pup (O4If), 15 Mon (O7V), HD 151804 (O8Iaf), 10 Lac (O9.4V), υ Ori (B0V) and τ Sco (B0V). However, no O VI could be seen in ρ Leo (B1Iab) or γ Ara (B1Ib). The O VI profiles display a considerable variety that is not correlated with spectral type. Zeta Pup has a broad emission and deep absorption ranging over all velocities to about −2800 km s^{-1}, while HD 151804 has only narrow absorption from −1060 to −1660 km s^{-1}, and 15 Mon has a narrower feature from −1700 to −2100 km s^{-1}. Both 10 Lac and τ Sco have broad relatively shallow absorptions extending from positive velocities to −1380 and −1000 km s^{-1} respectively. In contrast, the absorptions are narrower with steeper wings in ζ Oph and υ Ori. The latter star also has emission, while in ζ Oph the absorption is split into two components from about −1550 to −1280 and −1280 to −980 km s^{-1}.

DISCUSSION FOLLOWING MORTON

Castor: It is not true that the ionization must go <u>up</u> with radius in a photoionization model. If the wind is opaque to the ionizing radiation, the ionization goes <u>down</u> with radius as the radiation is absorbed. This occurs in the Auger-ionization model, as the softer X-rays are absorbed low in the wind.

Lamers: This depends on the mass loss rates. Joe can answer this better than I.

P. S. Conti and C. W. H. de Loore (eds.), Mass Loss and Evolution of O-Type Stars, 93-94.
Copyright © 1979 by the IAU.

Cassinelli: For optically thick winds one finds that the degree of ionization decreases in the outward direction, as I will show in my discussion tomorrow.

Morton: Since the wind of τ Sco is optically thin, the ionization should increase in the outward direction and hence I wonder about the interpretation of the profile wings.

C IV RESONANCE LINE PROFILES IN O STARS

S.B. Parsons, J.D. Wray, K.G. Henize, and G.F. Benedict
Department of Astronomy, University of Texas at Austin

I. INTRODUCTION AND OBSERVATIONS

The S-019 experiment on Skylab (cf. Henize et al. 1975; "Paper I") recorded far UV spectra in about 160 4° x 5° fields, covering 10% of the sky, on 101 film with a 15 cm aperture objective-prism telescope. Several hundred early-type stars were observed in the vicinity of 1550 Å with a resolution between 3 and 4 Å, as well as thousands of stars at longer wavelengths and correspondingly lower resolution. An atlas of spectra for types O4 to B5 is illustrated in Paper I. That figure shows that the P Cygni profile is a characteristic of all supergiants earlier than B3 and main sequence stars earlier than O8.

Subsequent to the qualitative presentation of the S-019 UV spectra, we converted intensities to absolute fluxes. The flux calibration was determined by comparison of the S-019 spectral intensities (averaged over the appropriate wavelength interval) with the fluxes derived from the S2/68 experiment on the TD-1 satellite (Jamar et al. 1976) or the WEP on OAO-2 (Code and Meade 1976). Details of the calibration will be published elsewhere. The standard deviation of resulting calculated fluxes with respect to TD-1 or scaled OAO-2 fluxes is about 30% for SL2 and SL3 and 40% for SL4. The relative accuracy in energy distributions, i.e. after allowing for a mean scale error in each spectrum, is normally better than 15%.

Two examples of reduced and averaged S-019 spectra are shown in Figure 1. The fluxes are normalized to an essentially flat continuum level and plotted against a linear wavelength scale. In the present results, one can see quantitatively the pronounced luminosity effects. The two O9.5 stars are markedly different in this region of the spectrum. The Si IV resonance lines are present but weak in the dwarf σ Ori, while in the supergiant ζ Ori there is an extreme P Cygni profile with two absorption components. The C IV λλ 1548, 1551 resonance feature makes a transition from a photospheric absorption line to a profile characteristic of a thick, rapidly expanding envelope. Because of the blends on the short wavelength side and the probable saturation

95

P. S. Conti and C. W. H. de Loore (eds.), Mass Loss and Evolution of O-Type Stars, 95–98.
Copyright © 1979 by the IAU.

of the C IV absorption, however, it has not been possible to extract quantitative information on mass loss rates.

II. C IV PROFILES AND THE H-R DIAGRAM

By studying the behavior of the C IV resonance profile with respect to T_{eff} and M_{bol}, one can learn about the conditions under which mass loss occurs without knowing precise mass loss rates. To do this, we have constructed an H-R diagram in these coordinates with the C IV profiles displayed directly for a representative sample of stars. We have used MK temperature classifications as the basis for a homogeneous set of estimates of T_{eff} and B.C., while for absolute magnitudes we have drawn from numerous sources, using cluster or association deter- minations when available, then quantitative classification methods such as Hγ equivalent width, and finally the MK luminosity class as a last resort. The adopted correlations between T_{eff} and spectral class are based primarily on the work of Conti (1973) for the O stars and Code et al. (1976) for the B stars. MK types are drawn from Morgan and Keenan (1973), Hiltner, Garrison and Schild (1969), Lesh (1968), Walborn (1973), and Conti and Leep (1974).

In Figure 2 the fiducial mark for each spectrum (zero intensity, λ = 1549 Å) is plotted at our best estimate for the bolometric absolute magnitude and effective temperature, except where a slight shift was required to avoid crowding. For many O stars, we give both Walborn's and Conti and Leep's spectral classifications. The star ξ Per is an anomaly and we have chosen to plot it using a quantitative determination of M_V = -6.5 from Borgman and Blaauw (1964).

III. DISCUSSION

There is a clear trend in the C IV feature toward more fully developed P Cygni profiles at higher luminosities. We have published before (Paper I) a bolometric limit M_{bol} = -8.4 above which we see definite P Cygni emission at the C IV and Si IV lines. We can now add that there is at least a significant velocity shift in the C IV absorp- tion down to $M_{bol} \simeq$ -7.7. At lower luminosities, given our moderate resolution and slight uncertainty in fixing the wavelength scale, we do not see evidence in this ultraviolet region for significant mass flow, although Snow and Morton (1976) detected velocity shifts with Copernicus for stars as faint as M_{bol} = -6.0. Our results and Copernicus data both show, however, that major mass loss effects, substantial envelopes and "stellar hurricanes" (Weymann 1977), occur for all stars more luminous than -8.4 or about $1.5 \times 10^5 L_{\odot}$.

This work is supported at the University of Texas under NASA contract NAS 8-31459 and NASA grant NSG 7371.

Figure 1. Normalized fluxes derived from objective-prism spectra. The zero point of the ζ Ori spectrum is indicated by the horizontal bar. Many differences in line strengths and profiles are evident.

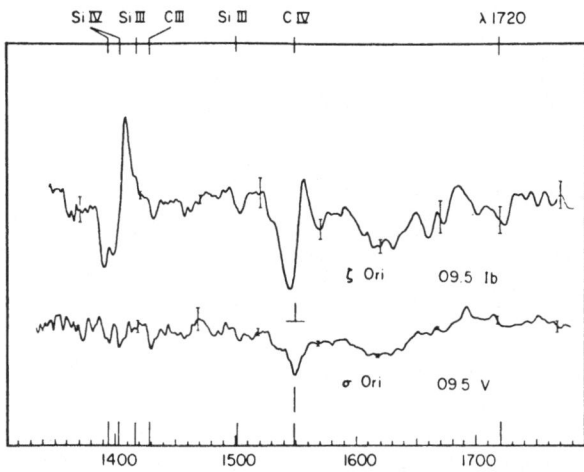

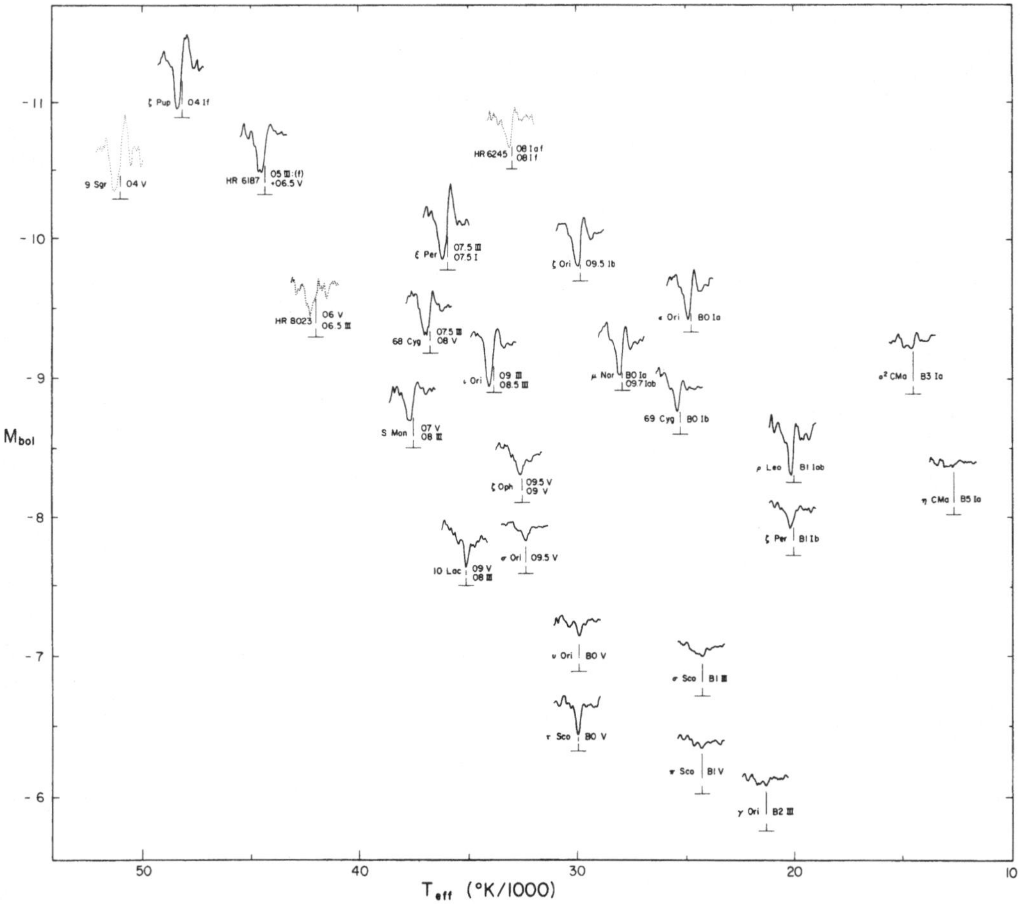

Figure 2. C IV profiles displayed with respect to estimated bolometric luminosities and effective temperatures of the stars. The interval 1507-1584 Å is extracted from plots of the type shown in Figure 1. The vertical bars indicate the rest position for the C IV resonance doublet.

REFERENCES

Borgman, J., Blaauw, A. 1964, Bull. Astr. Inst. Netherl. 17, 358.
Code, A.D., Davis, J., Bless, R.C., Hanbury Brown, R. 1976, Astrophys. J. 203, 417.
Code, A.D., Meade, M.R. 1976, Wisconsin Astrophysics, No. 30.
Conti, P.S. 1973, Astrophys. J. 179, 181.
Conti, P.S., Leep, E.M. 1974, Astrophys. J. 193, 113.
Henize, K.G., Wray, J.D., Parsons, S.B., Benedict, G.F., Bruhweiler, F.C., Rybski, P.M., O'Callaghan, F.G. 1975, Astrophys. J. 199, L119.
Hiltner, W.A., Garrison, R.F., Schild, R.E. 1969, Astrophys. J. 157, 313.
Jamar, C., Macau-Hercot, D., Monfils, A., Thompson, G.I., Houziaux, L., Wilson, R. 1976, "Ultraviolet Bright-Star Spectrophotometric Catalogue", ESA SR-27.
Lesh, J.R. 1968, Astrophys. J. Suppl. 17, 371.
Morgan, W.W., Keenan, P.C. 1973, Ann. Rev. Astr. Astrophys. 11, 29.
Snow, T.P. Jr., Morton, D.C. 1976, Astrophys. J. Suppl. 32, 429.
Walborn, N.R. 1973, Astr. J. 78, 1067.
Weymann, R.J. 1977, quoted in Science News 112, 107.

DISCUSSION FOLLOWING PARSONS et al.

Cassinelli: Are you aware of any observations of C IV in β Ori B8 Ia?

Parsons: We do have observations of β Ori but I don't recall any C IV. There is a definite C IV profile in o^2 CMa (B3 Ia) and just a hint of the profile in η CMa (B5 Ia).

Snow: What is your spectral resolution at the wavelength of the C IV doublet?

Parsons: It is in between 3 and 4 Å, depending slightly on the stability of the spacecraft.

WINDS IN HOT, SUBLUMINOUS STARS

Sara R. Heap
Laboratory for Astronomy and Solar Physics
Goddard Space Flight Center

Recent observations with the International Ultraviolet Explorer (IUE) satellite show that two very different types of hot stars have stellar winds: not only do the young, massive OB stars (the subjects of our discussion in this symposium, so far) undergo high-velocity mass-loss, but so also do hot evolved, solar mass stars, among them the central star of planetary nebulae. In this talk, I would like to show ultraviolet spectra of two central stars, the nuclei of NGC 6826 and Abell 78, and to describe how the characteristics of these spectra may be used to derive information concerning winds in these stars and in hot stars in general.

Figure 1 shows the ultraviolet spectra of the two central stars over the wavelength range, 1150-1750 Å. The nominal spectral resolution is about 5 Å (FWHM). The nucleus of NGC 6826 is a tenth-magnitude star having an O3f-type spectrum (Heap 1977) very much like that of the young star, ζ Puppis. You will note that the ultraviolet-spectrum indicates that, like ζ Puppis, the central star has a wind: the resonance lines of C IV $\lambda 1550$ and N V $\lambda 1240$ are strong with P-Cygni profiles; and lines from excited levels, such as N IV $\lambda 1720$, O IV $\lambda 1340$, and O V $\lambda 1370$ have P-Cygni profiles or appear as shortward-displaced absorption lines. The apparent terminal velocity of the wind, about 2000 km/s, is somewhat lower than that of ζ Pup, which Lamers and Morton (1976) estimate at 2600 km/s. (I use the term, "apparent" terminal velocity, because I have not taken instrumental line-broadening into account.) The general level of ionization in the wind is similar to that in the wind of ζ Pup. (I should qualify this statement by saying that the OVI resonance lines are not observable with the IUE because of a short-wavelength cutoff in sensitivity at 1150 Å. The inaccessibility of OVI may amount to a serious limitation of the IUE in persuing the problem of stellar winds).

The nucleus of Abell 78 is a thirteenth-magnitude star having an OVI-type spectrum (Greenstein and Minkowski 1964). Such a high-level of ionization is unknown among young hot stars. From Abell's (1966) derivation of the Zanstra temperatures and from Pottasch et al.'s

99

P. S. Conti and C. W. H. de Loore (eds.), Mass Loss and Evolution of O-Type Stars, 99-102.
Copyright © 1979 by the IAU.

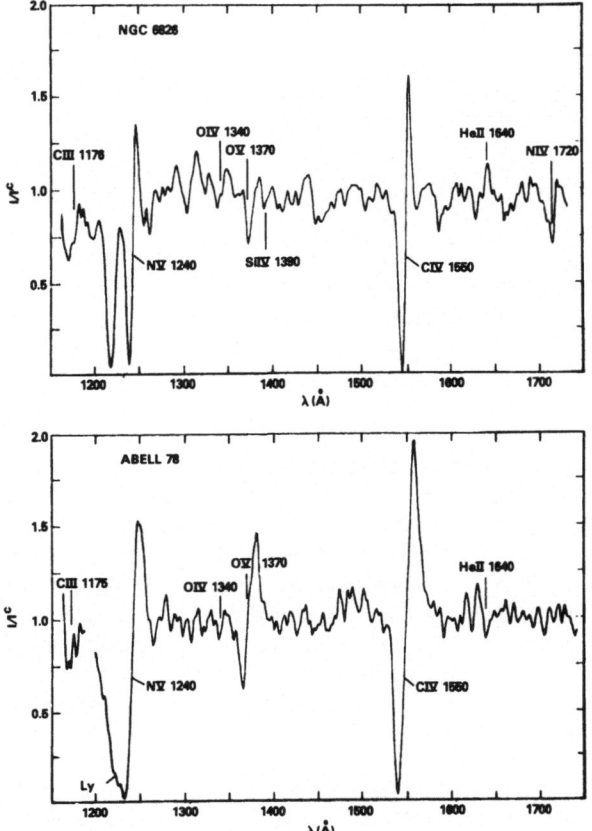

Figure 1.   Ultraviolet Spectra of Two Central Stars

(1978) derivation of the ultraviolet color temperatures of Abell objects, I estimate that the central star has an effective temperature in the range, 100,000 - 150,000 °K, and a gravity greater than 10^5 cm/s². The ultraviolet spectrum of the central star, as shown in Figure 1, indicates that this star also has a wind, whose apparent terminal velocity is large (about 3400 km/s) and whose general level of ionization is very high.

How is it that these subluminous stars, especially the nucleus of Abell 78 which has such a high-gravity, have a wind? And why is the terminal velocity of the nucleus of Abell 78 significantly larger than that of the nucleus of NGC 6826? These questions can best be answered upon consideration of the stellar luminosity-to-mass ratios ($\mathcal{L}/m$) and mass-to-radius ($m/\mathcal{R}$) ratios. We heard earlier this afternoon from Ted Snow that among young hot stars, mass loss appears to set in at a bolometric magnitude, M_{bol} = -6. An almost equivalent statement would be to say that mass-loss is characteristic of stars whose $\mathcal{L}/m$ (in solar units) is greater than about 1600. This statement assumes that mass-losing stars evolve off the main sequence at a constant $\mathcal{L}/m$ -- an assumption that is only roughly correct, as de Loore et al. (1977) have shown. The reason for restating this

empirical mass-loss criterion in this way is that it is now applicable
to all hot stars including planetary nuclei. Or more strictly speaking,
we may now test whether the parameter, $\mathcal{L}/m$, is the essential criterion
for both young massive stars and evolved, solar-mass stars. Referring
to the data in the table, we see that the presence of winds in the
central stars is consistent with our $\mathcal{L}/m$ criterion. Furthermore, we
see that the apparent terminal wind velocities of the two central stars
are at least qualitatively consistent with the view that the terminal
velocity is correlated with the escape velocity, which goes as $(m/R)^{\frac{1}{2}}$.

Star	$\mathcal{L}/m$	m/R
ζ Pup	14000	6.3
NGC 6826	24000	0.48
Abell 78	>11000	>2.8

A more general conclusion is that both the presence of stellar
winds in hot stars of high $\mathcal{L}/m$ and the correlation of terminal wind
velocity to escape velocity of hot stars are consistent with the
radiation-driven theory of stellar winds in hot stars (Castor, Abbott,
and Klein 1975).

Abell, G.O.: 1966, Astrophys. J. 144, p.259.
Castor, J.I., Abbott, D.C., and Klein, R.I.: 1975, Astrophys. J. 195,
    p. 157.
de Loore, C., DeGreve, J.P., and Lamers, H.J.G.L.M.: 1977, Astron.
    Astrophys. 61, p.251.
Greenstein, J. and Minkowski, R.: 1964, Astrophys. J. 140, p.1601.
Heap, S.R.: 1977, Astrophys. J. 215, p.864.
Lamers, H. and Morton, D.C.: 1976, Astrophys. J. Suppl. 32, p.715.
Pottasch, S.R., Wesselius, P.R., Wu C.C., Fieten, H., and
    van Duinen, R.J.: 1978, Astron. Astrophys. 62, p.95.

DISCUSSION FOLLOWING HEAP

Snow: Could the presence of a strong C IV P Cygni profile while
Si IV is absent be attributed to an abundance anomaly, rather than un-
usually high ionization, in view of the fact that the required ionization energies to form these two species are not very different?

Heap: I'm not sure, but I don't think so: In ζ Pup for example,
the C IV absorption is certainly more strongly saturated than is the
Si IV absorption, and this star has high ionization in its wind.

Niemela: R. Mendez and I have just finished a study of three central stars of planetary nebulae; and two of them, namely those of He 2-
131 of type 08f and He 2-138 of type BOI show blueshifted lines with
respect to the nebular velocity, thus suggesting they have expanding
atmospheres. So the mass loss effects are seen also in the optical
spectra of some central stars of planetary nebulae.

Heap: The spectrum of He 2-131 is very interesting indeed. Last summer, I published tracings of portions of its optical spectrum. As I remember, lines of He I, N III, H I, etc. all had P Cygni profiles, and the unidentified Of lines at 4486 and 4503 Å are stronger in the spectrum of this star than in any young Of-star spectrum that I've ever seen.

Abbott: The theory of radiatively-driven winds predicts $\dot{M} \propto L(\overline{\Gamma/1-\Gamma})^{1-\alpha/\alpha}$, where Γ = Stellar Luminosity/Eddington Luminosity and α is a parameter expected to obey $0.5 < \alpha < 0.9$. Thus, planetary nebulae, where Γ approaches 1, will be predicted to scale to higher mass loss rates for a given luminosity.

Heap: Two stars having the spectral type, say a young Of star and a Of-type planetary nucleus, will have the same L/m ratio and since $\Gamma = (\sigma_c/4\pi Gc)(L/M)$ they will have the same Γ. From this it follows that $\dot{M}/M \propto L/M$ will also be the same according to the radiatively-driven wind theory.

Lamers: According to the radiation-driven wind models, two stars with the same spectral type but different gravities should have different mass loss rates and different terminal velocities: Smaller gravities give larger mass loss rates but smaller terminal velocities. Did you observe these trends?

Heap: Two stars of the same spectral type (including luminosity class) should have the same effective temperature and gravity. Since $g \propto M/R^2$, $v_{esc} \propto (M/R)^{1/2}$ and $v_{term} \propto v_{esc}$, then $v_{esc} \propto R^{1/2}$ for two stars of the same spectral type. Since the nucleus of NGC 6826 must have a considerably smaller radius than ζ Pup, I had expected that the P Cyg profiles in the central star would be sharper than observed. Maybe the central star has a higher temperature and gravity than ζ Puppis.

Bisiacchi: You have shown the spectra of two nuclei of planetary nebulae; do you know if their spectra in the visible region resemble Population I stars?

Heap: The visual spectrum of the central star of NGC 6826 is nearly a ringer for that of ζ Pup. (The main differences between the two spectra is that the line widths of the central star are sharper than those of ζ Pup and the N IV spectrum is different.) The visual spectrum of the central star of Abell 78 has no counterpart among young, hot stars.

TEMPERATURES AND RADII OF O STARS

A.B. Underhill[*], L. Divan[**], V. Doazan[***], M.L. Prévot-
Burnichon[**]
*Goddard Space Flight Center, Greenbelt, MD; temporarily at[**]
**Institut d'Astrophysique, Paris, France
***Observatoire de Paris, Paris, France

ABSTRACT

Angular diameters have been estimated for 18 O and 142 B stars
using absolute intermediate-band photometry in the near infrared and
they have been combined with integrated fluxes to yield effective temperatures. The effective temperatures of the O stars lie in the range
30000 K to about 47000 K. For a given subtype, the luminosity class I
stars have lower effective temperatures than the main-sequence stars by
about 1000 K. The absorption-line spectral types of the supergiants of
types O and B reflect electron temperatures which are higher than can be
maintained by the integrated flux which flows through the stellar atmosphere. Distances have been estimated for all the stars and linear
diameters found. The average radius for an O8 to O9.5 supergiant is
about 23.3 $R_\odot$; the radii for luminosity class III and Class V O stars
lie in the range 6.8 to 10.7 $R_\odot$.

METHOD

The flux effective temperature of a star is defined by the relation

$$T_{eff}(\text{flux}) = (4f/\sigma_R)^{\frac{1}{4}} \Theta^{-\frac{1}{4}}, \tag{1}$$

where f is the integrated flux from the star over the full extent of the
spectrum, σ_R is the Stefan-Boltzmann constant and Θ is the angular
diameter. We find f over the range 1380 to 11084 Å using the S2/68
absolute ultraviolet spectrophotometry (Jamar et al. 1976) and the
absolute-energy 13-colour photometry of Johnson and Mitchell (1975). We
correct for interstellar extinction using the reddening law of Nandy et
al. (1975, 1976). The parts of f shortward of 1380 Å and longward of
11084 Å are estimated using the predicted continuum fluxes from the
adopted model atmospheres with the derived angular diameters.

The angular diameters for our stars are found from the relation

103

P. S. Conti and C. W. H. de Loore (eds.), Mass Loss and Evolution of O-Type Stars, 103–108.

$$\theta_\lambda = 2(f_\lambda / \mathcal{F}_{S,\lambda})^{\frac{1}{2}}, \tag{2}$$

where f_λ is the absolute flux in a Johnson and Mitchell intermediate-width passband having an effective wavelength in the range 6356 to 11084 Å. We usually found a mean value from the results from the 6 longest passbands. If an infrared excess seemed to be present, we omitted the results from the longest wavelength band. Here $\mathcal{F}_{S,\lambda}$ is the monochromatic flux emitted in wavelength λ at the star. It is approximated by the monochromatic flux from a model atmosphere. We adopted the LTE, classical model atmospheres of Kurucz (1977) and their continuum fluxes in 25-angstrom-wide bands. We found that the predicted continuous spectra from these models at the wavelengths we needed were essentially the same as those of the NLTE models of Mihalas (1972), and that they were more consistent and given in greater detail than the Mihalas results. To select a representative model for each star, we adopted the effective temperature scale of Conti (1975) for the O stars, which is based on an interpretation of the strengths of the He I and He II lines using the NLTE calculations of Auer and Mihalas (1972). We assigned log g equal 4.5 for main-sequence stars and 4.0 or 3.5 for giants and supergiants.

If the angular diameters are to be correct, the monochromatic fluxes used in eqt. (2) must refer to single stars. For those of our stars for which the spectrophotometry refers to two or more stars, we made a correction, representing the light from the companion(s) by photometry from a single B star of the same spectral type scaled to allow for the difference in magnitude and interstellar extinction between it and the companion.

STARS STUDIED

The O stars which we have studied are α Cam O9.5Ia, δ Ori A O9.5II, ν Ori O9V, λ Ori A O8III((f)), σ Ori AB O9.5V, ζ Ori A O9.5Ib, S Mon A O8III((f)), HD 48099 O6.5V, UW CMa O8.5If, τ CMa O9I, ζ Pup O4ef, ζ Oph O9Ve, HD 151804 O8Ifp, 9 Sge O8If, HD 188209 O9.5Ib, HD 193322A O8.5III, λ Cep O6ef and 10 Lac O8III. We also studied 142 B stars.

RESULTS

We have 14 stars in common with Hanbury Brown et al. (1974). Fig. 1 shows that we can systematically reproduce the angular diameters measured by the intensity interferometer provided that for the B-type supergiants and giants we use a representative model atmosphere with an effective temperature corresponding to the $(B-V)_0$ colour. The one widely divergent point, at $\theta_{LD} = 19.2 \times 10^{-4}$ arc sec, is α Eri. We do not know the reason for this discrepancy. Usually the agreement between our angular diameters and those of Hanbury Brown et al. is excellent.

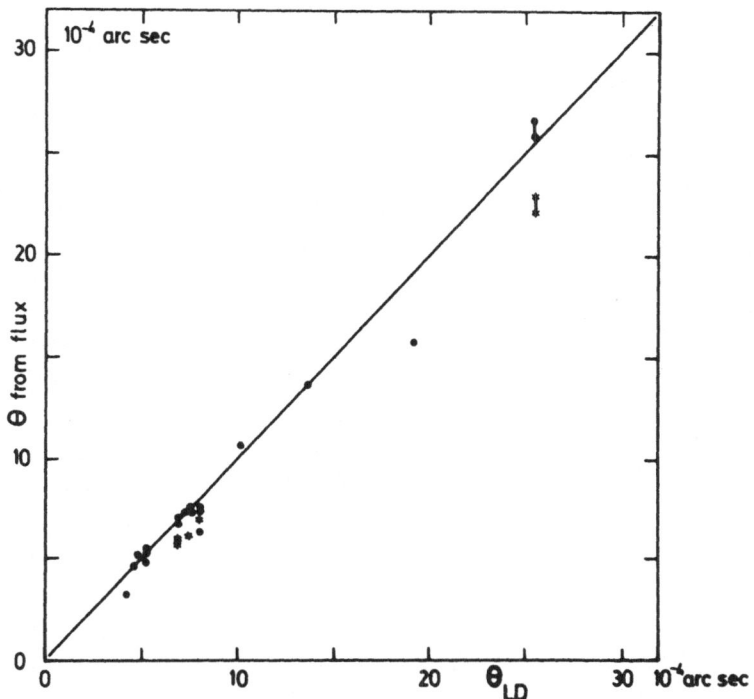

Fig. 1. The relation between angular diameters found from
long-wavelength fluxes and the angular diameters measured by
Hanbury Brown et al. (1974) for 14 stars. Points using the
same model fluxes but different values of E(B-V) are joined
by a vertical line. The angular diameters of ε Ori, BOIa,
η CMa, B5Ia, β Ori, B8Ia and ε CMa, B2II are plotted with
filled circles when a model effective temperature correspon-
ing to the (B-V)$_0$ is used, and by a star when a model effective
temperature like that suggested by the line spectrum is used.
The other stars, α Eri, γ Ori, ζ Ori A, β CMa, ζ Pup,
α Leo, δ Sco A, ζ Oph, α Pav and α Gru are represented
by filled circles.

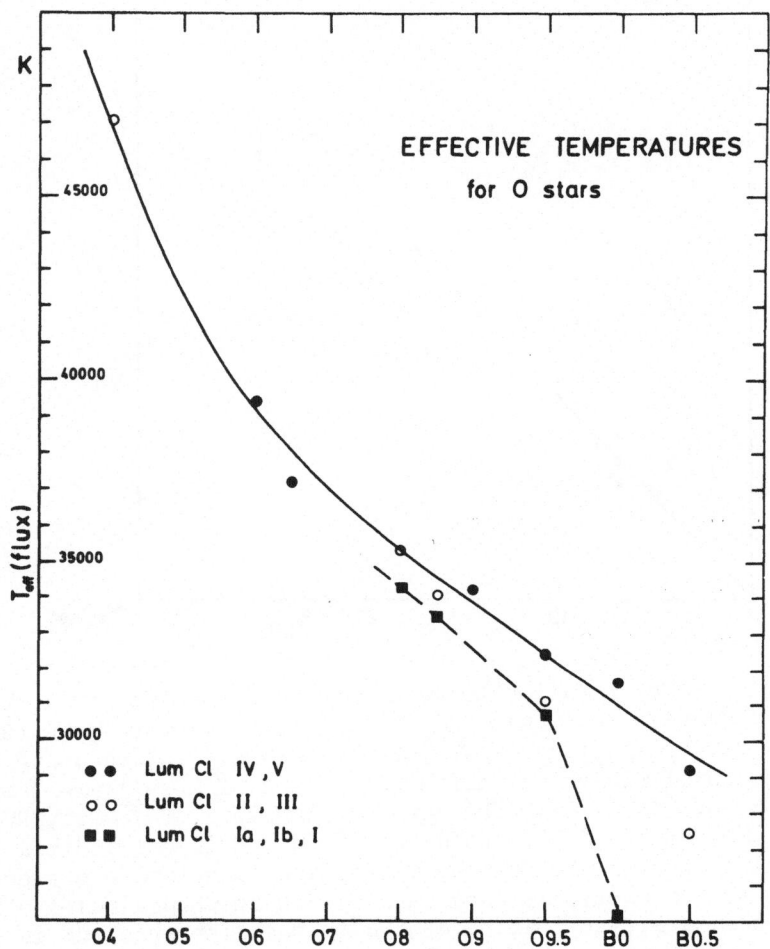

Fig. 2. Flux effective temperatures for the O stars as a
function of spectral type.

The average flux effective temperatures for each O subtype represented in our data are shown in Fig. 2 as a function of spectral type. The effective temperatures of the main-sequence stars are the largest at each subtype. Those of the giants and the supergiants are smaller. The result for ζ Pup, 47120 K, is significantly higher than that of Code et al. (1976), 32510 K, and it is reasonable for the assigned spectral type. We find a high effective temperature for ζ Pup because we find $\Theta = 3.27 \times 10^{-4}$ arc sec in place of 4.2×10^{-4} arc sec as found by Hanbury Brown et al. No reasonable change in our choice of representative model atmosphere or of reddening correction permits us to reproduce the result of Hanbury Brown et al. It is unlikely that the discrepancy is due to our use of planar geometry to calculate the continuous spectrum.

The flux effective temperatures of the O-type supergiants are about 1000 K lower than those of main-sequence stars of the same subtype. This systematic trend becomes marked in the B supergiants. The discrepancy is more than 6000 K at type B0. We think that the assigned absorption-line spectral types of O and B supergiants correspond to higher temperatures than the integrated flux can support because the strengths of the lines selected as spectral-type criteria reflect conditions in a superheated chromospheric layer rather than conditions in the photosphere.

The linear radii of the class I stars on our list average 23.3 $R_\odot$. The linear radius for δ Ori A is 15.8 $R_\odot$; for ζ Pup it is 15.1 $R_\odot$ while for λ Cep it is 13.9 $R_\odot$. The linear radii for the other stars of luminosity classes V and III range from 6.8 to 10.7 $R_\odot$.

REFERENCES

Auer, L.H. and Mihalas, D. 1972, Astrophys. J. Suppl., 24, pp.193-246.
Code, A.D., Davis, J., Bless, R.C. and Hanbury Brown, R. 1976, Astrophys. J., 203, pp.417-434.
Conti, P.S. 1975, in "H II Regions and Related Topics", eds. T.L. Wilson and D. Downes, (Berlin: Springer Verlag), pp.207-221.
Hanbury Brown, R., Davis, J. and Allen, L.R. 1974, Mon. Not. R. astr. Soc., 167, pp.121-136.
Jamar, C., Macau-Hercot, D., Monfils, A., Thompson, G.I., Houziaux, L. and Wilson, R. 1976, "Ultraviolet Bright-Star Spectrophotometric Catalogue", ESA SR-27.
Johnson, H.L. and Mitchell, R.I. 1975, Rev. Mexicana Astron. Astrofys., 1, pp.299-324.
Kurucz, R.L. 1977, magnetic tape.
Mihalas, D. 1972, "Non-LTE Model Atmospheres for B and O Stars", NCAR Technical Note, NCAR-TN/STR-76.
Nandy, K., Thompson, G.I., Jamar, C., Monfils, A. and Wilson, R. 1975, Astron. Astrophys., 44, pp.195-203.
Nandy, K., Thompson, G.I., Jamar, C., Monfils, A. and Wilson, R 1976, Astron. Astrophys., 51, pp.63-69.

DISCUSSION FOLLOWING UNDERHILL <u>et al</u>.

<u>Morton</u>: The problem with ζ Pup is not likely an underestimate in the error on the measured angular diameter; it seems that for ζ Pup, which has a large mass outflow, the intensity interferometer is measuring a larger diameter than the photosphere.

<u>Underhill</u>: The 13-color photometry yields the same angular diameter from 4000 to 8600 Å, so I have some confidence in our result.

<u>R. E. Wilson</u>: A check on radii and luminosities such as those you have found can be provided by eclipsing binaries with good velocity data. The problem is that suitable examples are hard to find, especially on the ZAMS. I (Wilson and Rafert, in preparation) have some results of this kind for δ Pictoris, which you may want to compare with your work. Both components are on the ZAMS within the observational uncertainties.

<u>Lamers</u>: The situation on the effective temperatures of B-supergiants is very confusing. The detailed studies of the visual spectra by model atmosphere techniques (e.g., Lamers, 1974, <u>Astron. Astrophys.</u> 37, 237) clearly indicate much larger temperatures than those derived from absolute integrated fluxes. This might indicate that the model atmospheres are not adequate. Would the choice of wrong models for deriving the angular diameter and for extrapolating the flux to wavelength shortward of 1200 Å affect the resulting values of T_{eff} for early-B supergiants?

<u>Underhill</u>: The choice of T_{eff} for the model to predict the continuous spectrum is closely constrained by the requirement that the <u>shape</u> be correct from 6356 to 11084 Å and that the Hanbury-Brown <u>et al</u>. angular diameters be found for the supergiants which he observed (types B0 Ia, B5 Ia, B8 Ia). We tried low and high effective temperatures for the supergiants. Only low effective temperatures corresponding to $(B-V)_0$ reproduce the measured angular diameters and have the correct shape. The "absorption-line" effective temperatures are clearly too high. This point will be discussed fully in the paper which Divan, Doazan, Prévot-Burnichon and I will be submitting to Astron. Astrophys.

ANALYSIS OF UV SPECTROPHOTOMETRIC OBSERVATIONS FOR O-TYPE
STARS.

Carlo Morossi, Roberto Stalio and Lucio Crivellari.
Astronomical Observatory of Trieste.

The ESA "Ultraviolet Bright Star Spectrophotometric Catalogue"
(Jamar et al. 1976) and the "Catalogue of 0.2 A Resolution Far-ultraviolet Stellar Spectra Measured with Copernicus" (Snow and Jenkins 1976)
have been used to try to separate, by means of simple diagrams, O-type
stars belonging to different luminosity classes and having different
temperatures. The ESA catalogue gives absolute fluxes for 43 O-type
stars, mostly reddened, with a spectral resolution of 35–40 A, depending
on the channel, in the wavelength range 1350–2550 A. The Copernicus
catalogue gives spectra between 1000 and 1450 A for 17 O-type stars,
with low reddening, at 0.2 A resolution.
We have selected 32 stars from the ESA catalogue with known
spectral class and reddening and calculated the spectral indices Δm_{2100}
and Δm_{1500}, after having corrected the observed fluxes for the color
excess with the mean interstellar law of Code et al. (1976). The
spectral indices, defined by the following relation

$$\Delta m_\lambda = (m_\lambda - m_{5500})_{star} - (m_\lambda - m_{5500})_{Vega}$$

are the difference of flux, expressed in magnitudes, at λ, between the
star and Vega, when the observed spectra are normalized to 5500 A. From
the RMS deviation of the observations used to compute the mean fluxes,
we have estimated the error in the spectral indices to be on the order
of 10% or less. Additional errors of about 15% are due to the calibration of the instrument (Jamar et al. 1976). The dependence on the
assumed mean interstellar extinction law has been tested by comparing
our spectral indices with the Δm_λ calculated using the law of Nandy et
al. (1976). The difference is unappreciable at 2100 A and very small
(about 3% for star with E(B-V)=0.6) at 1500 A. A table with all the
data is available on request.
Theoretical indices have also been calculated from the grid
of LTE line blanketed model atmospheres of Kurucz, Peytremann and
Avrett (1974) by taking different values of T_{eff} and log g appropriate
to O stars, and using a "theoretical" model of Vega with fluxes at
1500, 2100 and 5500 A given by the observations. Figure 1 reports the
curves for Δm_{2100} and Δm_{1500} versus the effective temperature, and the
position of the observed indices on a temperature scale given by

109

P. S. Conti and C. W. H. de Loore (eds.), Mass Loss and Evolution of O-Type Stars, 109–112.
Copyright © 1979 by the IAU.

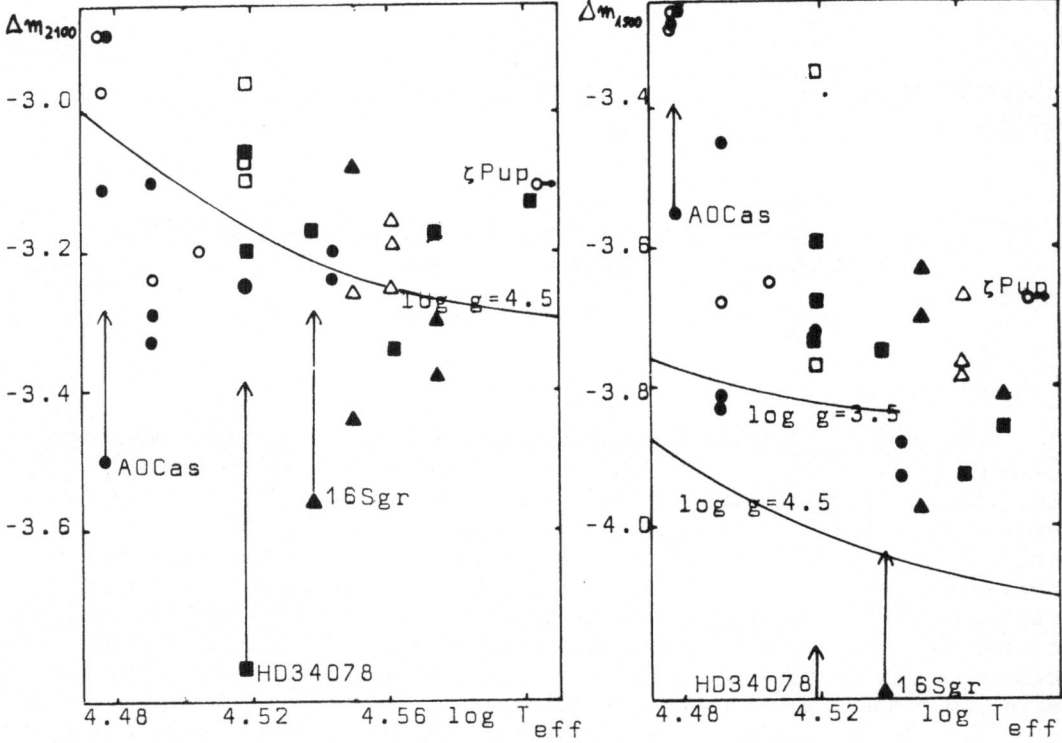

Figure 1:   The spectral indices versus log T_{eff}.   Circles are super-
            giants, triangles are giants, squares are dwarfs.   Black
            symbols refer to stars with E(B–V) 0.2.

Conti's (1973) spectral type – effective temperature relation.   The
fluxes are not very affected by line blanketing in this region, and are
still far from the maximum of the Plank curve.   Thus one expects that
the Δm_{2100} indices are almost independent of the luminosity.   This is
confirmed by the negligible separation (not shown in the figure) of the
theoretical curves for the two values of log g=3.5 and 4.5 .   The m_{1500}
curves are both temperature and gravity dependent.   We are measuring
a region which is rich in lines of highly ionized elements; thus more
flux is absorbed by lines at lower gravities.   The observed points are
systematically above the theoretical curves, probabily because the
models both do not include all sources of line opacity, and underestimate the flux absorbed by the available lines.   This last argument is
particularly significant for explaining the large separation from the
theoretical curves of most of the supergiants of our sample: the models
have been calculated with a 2 km/s velocity of microturbulence, which
is much less than the generally adopted values for hot supergiants.   In
addition the observed fluxes may be influenced by the displaced envelope
component of the strong resonance line doublet of C IV at 1548–50 A.

There is evidence, in figure 1, of a well established relation between observed spectral indices and empirical temperatures. A satisfactory agreement with the models is also achieved for the 2100 indices. The stars which show greater deviation from the mean relations have probably an incorrect value of the reddening parameter, or an incorrect spectral class assignation. To illustrate the effect of a bad choice of $E(B-V)$ we have calculated the Δm_λ indices for AO Cas, HD 34078 and 16 Sgr with 10% less reedening. The indices go up to the values indicated by the arrows. For ζ Pup the situation is unclear: the spectral indices require a lower temperature, but this might be a spurious effect due to the presence of numerous emission lines throughout the spectrum.

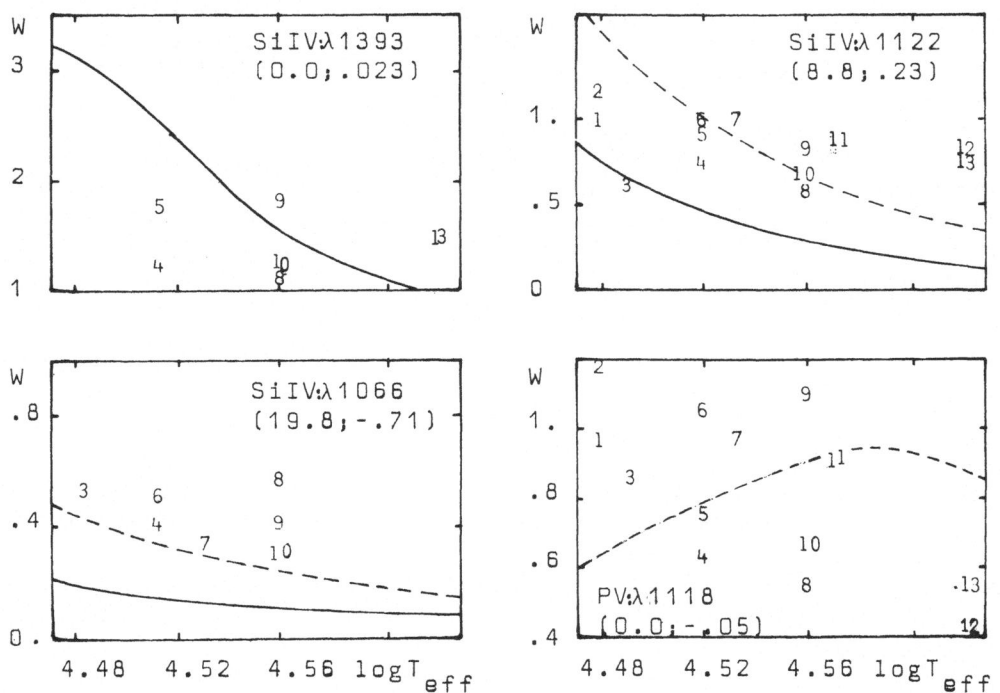

Figure 2:   Theoretical and observed equivalent widths (in A); dashed curves are computations with $\Gamma = 10 \times \Gamma_{cl}$; numbers in parentheses are the excitation potential (in eV) and log gf respectively. 1-ζOri, 2-μNor, 3-τCMa, 4-μCol, 5-ζOph, 6-δOri, 7-^{1}Ori, 8-10Lac, 9-λOri, 10-15Mon, 11-ξPer, 12-HD199579, 13-9Sgr.

        Ultraviolet spectroscopic criteria for the separation of O stars into their different spectral classes have been investigated by comparing equivalent widths of selected lines with LTE model atmosphere computations. The equivalent widths have been measured from the

Copernicus spectra with the usual method of drawing a local continuum
and integrating the line profile over an appropriate wavelength interval.
After a careful inspection of the tracings and of the available identification lists, we have measured the photospheric lines presented in
figure 2: their equivalent widths are given versus the temperature determined from the spectral type.

Theoretical line strengths, computed with a code that solves
the LTE line transfer problem from the model atmosphere input of Kurucz,
Peytremann and Avrett at log g=4.5 and different values of the temperature, are also given in figure 2. We have assumed solar element abundances and a constant value of the microturbulence velocity of 5 km/s.
Although the number of observations is too small and the theory used is
not fully appropriate (we should have used, for example, more realistic
values for log gf and non-LTE theories), nonetheless the general behaviour of the silicon lines looks consistent with the computations. It is
not so for the P V resonance line, whose equivalent widths appear
scattered on the W-log T_{eff} plane, as if some unidentified blended
component contribute to the line. The separation by luminosity
classes is not seen in the diagrams of figure 2.

We are presently working on improving the equivalent width
obsevations over a larger sample of stars and taking other lines into
account. In the near future, we hope to be able to drawn a more consistent picture of the relations between O star spectral classes and UV
spectrophotometric observations.

REFERENCES

Conti,P.S. 1973, Ap. J. 179, 181.
Kurucz,R.L., Peytremann,E. and Avrett,E.H. 1974, "Blanketed Model
         Atmospheres for Early-type Stars", Smithsonian Institution,
         Washington D.C.
Jamar,C. et al. 1976, "Ultraviolet Bright Star Spectrophotometric
         Catalogue", ESA SR-27.
Nandy,K. et al. 1975, Astr. Ap. 44, 195.
Snow,T.P. and Jenkins,E.B. 1976, Ap. J. Suppl. 33, 269.

THE EFFECTIVE TEMPERATURES OF THE O STARS

S. R. Pottasch, P. R. Wesselius and R. J. van Duinen
Kapteyn Astronomical Institute and Space Research Laboratory,
University of Groningen

ABSTRACT

Far ultraviolet photometric observations (in 5 bands between λ = 1550 Å and 3300 Å) are presented for ten O stars which are also the exciting stars of diffuse nebulae. Since the number of photons shortwards of the Lyman limit is known for these stars, sufficient information on the total flux is available to determine the effective temperature without making any assumptions concerning a model atmosphere. Conversely, the distribution of flux with wavelength can be used to determine the applicability of a given model. A detailed discussion of this is presented.

A comparison of these 'normal' O stars and stars with O type spectra which excite planetary nebulae is given. It is concluded that the atmospheres of these two types of O stars are very similar; probably they have the same effective gravity. Further, a discussion of the Zanstra He II temperature is presented.

Editor's Note: The contents of this paper are in press in <u>Astronomy and Astrophysics</u>.

DISCUSSION FOLLOWING POTTASCH, WESSELIUS AND VAN DUINEN

<u>Underhill</u>: The photometry in the 5400 Å band is a critical parameter in Pottasch's logic for obtaining T_{eff} and unfortunately Pottasch does not give the source or reliability of the absolute photometry which he uses in the preliminary paper he has circulated for comment. His effective temperatures for λ Ori and δ Mon are about 4000 K lower than I find. My scale is in agreement with Conti's scale.

<u>Conti</u>: Right-on, Anne! Pottasch's temperature scale is about 10% cooler than mine, a little outside of what I feel is reasonable. Even

P. S. Conti and C. W. H. de Loore (eds.), Mass Loss and Evolution of O-Type Stars, 113–114.
Copyright © 1979 by the IAU.

though Stuart Pottasch is not here to defend himself, and Tony Hearn
probably can't reply either I must describe one worry. The fundamental
magnitude calibration Pottasch used was taken from a paper of Kwan and
Kuhi (1976) who did scanner observations for another purpose. This
paper had some serious difficulties with wavelength dependence, particularly since the implied E(B-V) was completely inconsistent with the observed excesses in some cases. This tie into the Kwan and Kuhi paper
might be a problem of this temperature scale. Kuhi tells me a student
of his is redoing the observations.

Bolton: Could anyone give a quantitative estimate of how sensitive
the temperature determination is to some perturbation of the λ5400 photometer?

Underhill: Yes. Basically Pottasch gets the angular diameter
from this measurement. It's fairly sensitive.

GENERAL DISCUSSION

<u>Plavec</u>: Could someone say something about the status of some of these stars with very large mass loss rates? P Cygni, for example.

<u>Hummer</u>: Is there any expert on some of these pathological cases?

<u>Bidelman</u>: I don't have anything to say in defense of P Cygni, but I think that one should always be suspicious of a star that just suddenly appears! I have often suspected that P Cygni might be star nearing the end of its initial gravitational contraction stage, though I have no idea of how this can be proved or disproved. It is certainly very different from normal supergiants. It also appears rather red for a star of its class.

<u>Snow</u>: From <u>Copernicus</u> data we have derived the H I column density towards P Cygni, and from the literature we have found the measured diffuse interstellar band strenghts. Both of these quantities correlate well with interstellar reddening, and both indicate that E(B-V) for P Cygni is between $0^{m}.3$ and $0^{m}.4$, much less than the value E(B-V) = 0.6-0.7 which is derived from the UBV photometry and the assumption that the star's intrinsic colors are those of a normal early B supergiant. The lower value of E(B-V) leads to a reduced distance estimate and hence a lower luminosity for the star than has commonly been assumed. The UBV photometry may be influenced by the star's infrared excess, which may contribute to the V band more than to B, introducing a spuriously high value of E(B-V).

<u>Van Blerkom</u>: P Cygni differs from the O stars in that: (1) there is no evidence for outflow velocities in excess of 300 km s^{-1}; the emission wings on Hα and He I can be attributed to thermal electron scattering; (2) the excitation of the wind is quite low -- no He II is detected, for example; (3) H line profiles have been interpreted by three different models -- decelerating flow, a monotonically increasing velocity with radius in which a slow acceleration occurs, and an accelerating-decelerating-accelerating envelope. Thus, there is an obvious non-uniqueness in the models which does not seem to be as severe for the O stars.

<u>Underhill</u>: Some years ago Mart de Groot studied the spectrograms of P Cygni obtained over the years at Mt. Wilson. He showed that most H lines often appear to have three absorption components. Two remain stationary; the third appears to oscillate in about 114 days. This suggests standing waves of density at some places in the very extensive atmosphere. P Cygni is not at all typical of normal B-type supergiants.

<u>van den Heuvel</u>: As to P Cygni one can make some speculative theories on the origin of the mass outflow (van den Heuvel 1976). If one looks at a mass-exchange close binary with a large initial mass ratio, one expects such a system to go through a common envelope stage during which much of the transferred matter is expelled from the system,

P. S. Conti and C. W. H. de Loore (eds.), Mass Loss and Evolution of O-Type Stars, 115-116.
Copyright © 1979 by the IAU.

as the low-mass component cannot accept it (Flannery and Ulrich 1976).
Now, one can go one step further, and presume that the companion star
is a compact object, i.e., that P Cygni is a later stage of evolution
of a massive X-ray binary. In such systems one expects the compact star
to be swallowed by the envelope of its supergiant companion. The accretion luminosity will come out in the optical region, as the envelope is
optically thick to X-rays. A spiral-in binary of this type can be quite
long lived (Bodenheimer, Taam and Ostriker 1978), and, as we see quite
a number of X-ray binaries in the sky, one also expects to see a couple
of these spiral-in binaries in the sky. A possible support for this
idea in the case of P Cygni may be the photometric period of 0.5 days,
claimed by Magalasvili and Kharadze, some years ago. This seemed quite
a reasonable period for a spiral-in binary descending from an X-ray
binary within a period of a few days. It seems like an exciting idea
that P Cygni would be a descendant of a massive X-ray binary like
Cygnus X-1 or Cen X-3.

    Ludd: I want to make two remarks about P Cygni: (1) Using ex-
tended series of spectrograms it was found that the absorption components
of H9 and H10 have 47^d period; that is two times shorter than obtained
by de Groot; (2) using all observational data kindly presented by Prof.
Kharadze and period-searching computer routine the 0^d5 photometric
period by Kharadze and Magalasvili was not confirmed.

    Thomas: I note the repetition of the "belief" that luminosity --
or luminosity/escape energy -- is what describes mass loss. I think
you are being too serious and religious. You are _assuming_ that (T_{eff},
g) suffices to model an atmosphere -- but _none_ of these "theories"
have proved this. It is in no way clear that just because two stars
lie in the same (T_{eff},g) box that they will have the same mass loss:
Or that even [luminosity, spectrum] suffice to define mass loss, emission lines, etc. These problems are to be investigated -- _not_ assumed
as so many of you are doing. Some of you say P Cyg is an unusual,
highly individual star. Agreed -- and so what? We must show that _all_
stars are not highly individual before we _assume_ that they are not.

A theorist who puts thin coronae
In O stars (though some think it phoney)
Is ready to swear
That they really are there
On the gospel according to Tony.

SESSION 3

INFRARED AND RADIO DATA

Chairman: A. UNDERHILL
Introductory Speaker: M.J. BARLOW

1. J.M. VREUX and Y. ANDRILLAT: Near infrared spectra of
   O stars and related objects.

2. P. PERSI, M. FERRARI-TONIOLO and G. SPADA: Infrared observations of HDE 226868/Cyg X-1 and HDE 245770/
   AO535+26.

3. A.R. HYLAND: Infrared excesses and mass loss: Implications
   for OB stars in Carina and the LMC.

4. P.R. SCHWARTZ: Radio emission from flows.

5. S. KWOK and C.R. PURTON: Radio observations of stellar
   mass loss.

6. D.C. MORTON and A.E. WRIGHT: Radio emission from hot stars
   at two centimeters.

117

RADIO AND INFRARED EMISSION BY O-TYPE AND RELATED STARS

M. J. Barlow
Anglo-Australian Observatory, P.O. Box 296,
Epping, N.S.W. 2121, Australia.

1. INTRODUCTION

In addition to O-type stars, this review will discuss the radio and infrared properties of B supergiants and Wolf-Rayet stars, since it is generally accepted that these objects represent later stages in the evolution of O stars. The radio properties of X-ray binaries which exhibit non-thermal emission will not be discussed. Hjellming (1977) gives an excellent review of the radio properties of close binaries, including X-ray binaries.

2. RADIO OBSERVATIONS OF MASS LOSS

Johnson (1971) recognised that Wolf-Rayet stars, due to their very large mass loss rates, offered the best chance of a radio detection of a stellar wind. Consequently, he observed nine Wolf-Rayet stars at 2.7 GHz, but due to the low receiver sensitivities and large beamwidths then current, no detections were obtained. Balick (1972) set upper limits of between 20 and 50 mJy at 2.7 and 8.1 GHz for 32 O, B and Wolf-Rayet stars. The first definite detection of mass loss from an early-type star was by Wendker, Baars and Altenhoff (1973), who detected P Cygni at 5 and 10.7 GHz and found a spectral index of 0.7 ± 0.3 (if the flux S_ν at a frequency ν is proportional to ν^α, then α is defined as the spectral index). Wendker et al. (1975) found a source of 3.5 ± 1.5 mJy at 5 GHz coincident with the WN6 Wolf-Rayet star HD 192163. Seaquist (1976) detected the Wolf-Rayet binary γ Vel at three frequencies between 5 and 8.87 GHz and found a spectral index of $\alpha = 0.64 \pm 0.32$. The lists of 10.7 GHz upper limits given by Altenhoff et al. (1976) and by Woodsworth and Hughes (1977) include several O and B stars. Florkowski and Gottesman (1977) detected the WC7+O5 Wolf-Rayet system HD 193793 at 2.7 and 8.1 GHz. This star will be discussed in more detail in Section 4. Schwartz and Spencer (1977) obtained a 2σ detection of P Cygni at 90 GHz, confirming that the spectral index of 0.7 between 5 and 10 GHz extended to higher frequencies. Morton and Wright (1978, and this

119

P. S. Conti and C. W. H. de Loore (eds.), Mass Loss and Evolution of O-Type Stars, 119–130.
Copyright © 1979 by the IAU.

Symposium) detected Zeta Pup at 14.7 GHz and also measured the flux from γ Vel at the same frequency. Schwartz (this Symposium) has detected the Oe star HD 60848 at 2.7 GHz.

Seaquist and Gregory (1973) have shown that a spherically symmetric plasma density distribution, which is radially decreasing outward, leads to a spectral index intermediate between α = -0.1 and +2. The latter values correspond to optically thin and thick free-free emission. Panagia and Felli (1975) and Wright and Barlow (1975) showed that a density distribution $N \propto r^{-\beta}$ gives rise to a radio spectral index

$$\alpha = \frac{4\beta - 6.2}{2\beta - 1} \tag{1}$$

They solved the radiative transfer equation for the case of β = 2 (constant expansion velocity) in order to derive expressions for the total flux at a given frequency and the mass loss rate corresponding to an observed flux (Thorne (1975) independently rederived the same results not long afterwards). Thus, for an isothermal wind it was found that

$$S_\nu \propto \left(\frac{\dot{M}}{v_\infty}\right)^{4/3} \frac{\nu^{2/3} g^{2/3}}{D^2} \tag{2}$$

i.e. $\dfrac{\dot{M}}{v_\infty} \propto S_\nu^{3/4} D^{3/2} \nu^{-\frac{1}{2}} g^{-\frac{1}{2}}$ \hfill (3)

$\dot{M}$ is the mass loss rate, v_∞ is the terminal velocity of the wind and D is the distance of the star. S_ν is the flux received at frequency ν and g is the Gaunt factor. It is the weak frequency dependence of the Gaunt factor in the radio which flattens the spectral index from α = $2/3$ to α = 0.6. Relationship (2) shows that for a given mass loss rate a lower terminal velocity will lead to a larger radio flux and consequently a better chance of detection.

Relationships (2) and (3) are valid for the radio spectral region only if the wind has reached its terminal velocity in the radio-emitting volume. This can easily be shown to hold, since the characteristic radio-emitting radius, obtained from equation (11) of Wright and Barlow (1975) is typically between several tens to hundreds of stellar radii, at which point the wind will have reached terminal velocity for any plausible acceleration law. If $n \propto r^{-2}$ then S_ν is independent of the wind temperature, except weakly through the Gaunt factor. However, if the wind temperature varies systematically with radius, then this Gaunt factor dependence on temperature can cause the spectral index to be modified from the value of α = 0.6 appropriate for an isothermal wind to values up to α = 0.83, as shown by Thorne (1975) and by Chiuderi and Torricelli Ciamponi (1978). Both these treatments give formulae for the spectral index for the case of a wind undergoing cooling by adiabatic expansion. The expression for α from the latter paper is

$$\alpha = \frac{1.5 - 0.6\ \gamma}{2.85 - 1.35\ \gamma} \qquad\qquad\qquad (4)$$

Where γ is the adiabatic exponent of the gas. Thus, α can range from the standard isothermal value of $\alpha = 0.6$ for $\gamma = 1$, up to $\alpha = 0.83$ for $\gamma = {}^5/_3$.

Thorne (1975) has discussed the criteria for adiabatic cooling to be important in a wind, pointing out that until adiabatic cooling is significant, the electron temperature and ionization state in the flow will remain constant, since they depend only on the Tarter-Tucker-Salpeter parameter $L/\rho r^2$, which is a constant for regions where $v = v_\infty$. The gas will start to cool when the adiabatic cooling rate $\dot{E}_{ad} = -2kT_e v/r$ is comparable to the radiative cooling rate due to recombination, bremsstrahlung and collisional excitation, $\dot{E}_{rad} = Ln_e n_H$. From the work of Cox and Caltabuit (1971) and Raymond, Cox and Smith (1976) I can crudely but adequately set $L = 5 \times 10^{-27}\ T_e$ erg cm^3 s^{-1}, for electron temperatures T_e in the range 10^4 K $-$ 2×10^5 K (this is strictly only valid for collisionally ionized plasmas at a temperature T_e, but a glance at Figure 3 of Cox and Daltabuit (1971) shows that the most important permitted coolant ions are those known to be present in O star winds). Setting $\dot{E}_{ad} = \dot{E}_{rad}$, we obtain the radius r_{cool} at which adiabatic cooling becomes important

$$r_{cool} = 4.3 \times 10^{15}\ \frac{\dot{M}_{-6}}{v_8^2}\ cm \qquad\qquad\qquad (5)$$

where $\dot{M}_{-6}$ is the mass loss rate in units of $10^{-6}\ M_\odot$ yr^{-1} and v_8 is the terminal velocity in units of 10^8 cm s^{-1}. Adopting $\dot{M}_{-6} = 7$ and $v_8 = 2.66$ from Lamers and Morton (1976) and Morton and Wright (1978) we obtain $r_{cool} = 4.3 \times 10^{15}$ cm for Zeta Pup. This can be compared with the characteristic radius for 5 GHz radio emission, equal to 7.6×10^{13} cm for Zeta Pup, using equation (11) of Wright and Barlow (1975). For all likely OB star mass loss rates, the adiabatic cooling radius is always much larger than the characteristic radio emitting radius. For Wolf-Rayet stars the radiative cooling rate will be slightly different from the cosmic abundances rate used above, but the conclusions remain the same. Since no mechanical energy deposition is expected at the large characteristic radio emitting radii of OB and WR stars ($\sim$ 30-150 stellar radii), the isothermal wind temperatures in these regions should be $\sim 0.9\ T_{eff}$ (Klein and Castor 1978), where T_{eff} is the effective temperature of the star.

Thus it is predicted that early-type stars undergoing mass loss should always have isothermal winds in their radio emitting regions, giving rise to a spectral index of 0.6. The only early-type star with radio detections of high enough signal-to-noise for detailed checking of this prediction is γ Vel. Seaquist (1976) found 36 ± 5 mJy at 5 GHz and 52 ± 6 mJy at 8.87 GHz and Morton and Wright (this Symposium) found 69 ± 5 mJy at 14.7 GHz. Thus, spectral indices of 0.55 ± 0.25 and 0.64 ± 0.26 are found between 14.7 and 8.87 GHz and 8.87 and 5 GHz,

respectively, and over the largest frequency baseline of 14.7 to 5 GHz a
spectral index of 0.60 ± 0.14 is found, confirming the isothermal model
prdiction.

Olnon (1975) has derived the radio spectral indices corresponding
to power-law, Caussian and inner-truncated density distributions, while
Marsh (1975) has considered the case of an outer-truncated density
distribution. Panagia and Felli (1975) demonstrated that the radius at
which a stellar wind interacts with the interstellar medium is much
larger than the characteristic radio-emitting radius. Therefore, in the
absence of variability, a simple r^{-2} density distribution and thus a 0.6
spectral index should always apply to the radio emission from OB and WR
stars.

3. INFRARED OBSERVATIONS OF MASS LOSS

3.1 OB stars

Infrared observations of early-type stars are scattered throughout
the literature and are thus less easy to summarise than the radio data.
I will accordingly concentrate mainly on the work which has been concerned with mass loss aspects.

Woolf, Stein and Strittmatter (1970) interpreted the 2-10μm excesses
of a number of early-type stars (mainly classical Be stars, apart from P
Cyg) as due to free-free emission from hot ionised shells surrounding
the stars. Gehrz, Hackwell and Jones (1974) obtained 2-20μ photometry
of a much larger sample of Be stars, and in addition included a number
of early-type Supergiants such as P Cyg. The analysis used in the above
papers invoked a constant density shell of finite radius surrounding
each star, giving optically thin and thick free-free radiation shortward and longward of a critical wavelength λ_c. This wavelength λ_c was
deduced for each star as the intersection of $\nu^{0.1}$ and ν^{-2} power laws
fitted to the excess flux distributions. While perhaps valid for 'shell'
Be stars, this approach would clearly not be applicable to stars undergoing continuous mass loss, since then a range of densities is present
around each star, giving an infrared spectral index intermediate between
-0.1 and +2, as discussed in the previous section.

The 2-20μm free-free emission from mass loss outflows originates
from regions close to the stellar surface (typically between 1 and 5
stellar radii). Consequently we may expect that the density distribution in this region will be steeper than r^{-2} since the wind will be
undergoing acceleration up to its terminal velocity. Additionally, the
temperature T of the wind may vary with radius, due to mechanical energy
input or other reasons. Hartmann and Cassinelli (1977) have shown that
if $N \propto r^{-\beta}$ and $T \propto r^{-m}$, the (frequency) spectral index α is given.

$$\alpha = \frac{4\beta - m - 6}{2\beta - \frac{3}{2}m - 1} \tag{5}$$

For m = 0 (isothermal) this reduces to the standard case (1) (ignoring the Gaunt factor), while for β = 2 (constant expansion velocity) $\alpha = {}^2/_3$, independent of temperature variations (again ignoring the Gaunt factor effects discussed in the previous section).

Thus, if temperature gradients are present in the region of acceleration of a wind, the interpretation of the infrared data will be somewhat complicated. However, judicious model fitting can be used to set some limits on temperature gradients in cases where the mass loss rate is known from radio data.

The infrared spectrum of P Cyg (B1I$^+$) has been discussed in detail by Barlow and Cohen (1977). The optical and ultraviolet spectra of this star show only low-excitation emission lines and so the assumption of an isothermal wind of temperature T $\sim$ T$_{eff}$ would appear to be justifiable. Barlow and Cohen showed that the observed infrared fluxes were too large and the spectral index between the infrared and radio regions too steep, to be consistent with a constant outflow velocity in the infrared-emitting region. An extended acceleration zone was required in order to increase these quantities to their observed values. The decelerating wind model of Kuan and Kuhi (1975) was shown to be inconsistent with the observed infrared flux distribution. The derived velocity law was significantly less steep than the original theoretical velocity law of Castor, Abbott and Klein (1975). Barlow and Cohen were also able to show that, for a given velocity law, density clumping in the wind would lead to larger mass loss rates being derived from Hα fluxes than from infrared fluxes. Since the observed fluxes from a number of stars do not give this result, it would appear that significant density clumping does not occur in mass outflows from O stars.

The application by Barlow and Cohen of the empirical P Cyg velocity law to the infrared excesses of other B supergiants and a few O stars allowed their mass loss rates to be determined. The run of ϵ (ϵ = mass loss rate $\dot{M}$, divided by L/v$_\infty$c, the upper limit for mass loss by single scattering radiation pressure) versus effective temperature, and the dependence of $\dot{M}$ on stellar luminosity L ($\dot{M} \propto L^{1.15}$) were both consistent with the radiation pressure mechanism of Lucy and Solomon (1970), as developed by Castor et al. (1975).

It is obvious that additional velocity law determinations are needed for stars more 'normal' than the extreme supergiant P Cyg. Work underway on the infrared flux distribution of Zeta Pup indicates that application of a more appropriate 'faster' velocity law to O star infra-red excesses should not increase the derived mass loss rates by more than about a factor of two over the values determined using the P Cyg velocity law.

Cassinelli and Hartmann (1977) have shown that an extended coronal region ($\Delta R/R$ = 0.3, T $\sim$ 2 x 10^6 K) at the base of an O star wind would give rise to a rather flat infrared flux distribution with a broad bump at $\lambda \sim$ 30 μm. The observed 2.2-20μm spectrum of Zeta Pup rules out this

model since the excess flux distribution is rather steep ($\alpha \sim 1$). The
infrared observational data do not however rule out the thin corona
model ($\Delta R/R < 0.1$) of Cassinelli, Olson and Stalio (1978) and Cassinelli
and Olson (1978).

Barnes, Lambert and Potter (1974) obtained a high resolution 0.9-
1.7µm spectrum of Zeta Pup which showed an absorption line spectrum
apart from very weak He II emission at 10124 Å (Pickering α). Emission
lines are expected to make a negligible contribution to the infrared
excesses of OB stars. No evidence has ever been found for infrared
thermal dust emission from OB star winds. Some O stars are of course
embedded in H II regions which are infrared sources.

3.2 Wolf-Rayet stars

Allen, Harvey and Swings (1972) observed a large sample of Wolf-
Rayet stars at 1.6 and 2.2µm. They found that the infrared excesses
from WN and early WC stars were consistent with free-free emission,
whereas the WC9 stars had excesses which were too large to be explained
by this mechanism. By obtaining photometry to longer wavelengths (3.5-
10µm), Allen et al. showed that the WC9 excesses could be fitted almost
perfectly by single-temperature black-body curves of temperature T $\sim$
1000 K, attributable to dust emission. Further long-wavelength photometry and more detailed dust shell models for the WC9 stars can be found
in Gehrz and Hackwell (1974) and Cohen, Barlow and Kuhi (1975). The
lack of strong spectral emission features together with the known carbon-
richness of WC stars strongly suggests that the dust is a condensate of
carbon, such as graphite. Thomas, Robinson and Hyland (1976) obtained
intermediate bandwidth photometry ($\Delta\lambda \sim 2$µm) in the 8-13µm region for
the WC9 star Ve2-45. They found evidence for a weak emission feature at
$\lambda \sim 10$-12µm, possibly due to emission by SiC particles. The SiC emission
feature is known to be present in the infrared spectra of many carbon
stars whose infrared excesses are thought to be dominated by carbon
particle emission.

In addition to the above WC9 stars, CRL 2104 (WC8) and CRL 2179
(WC9), have been found to show strong circumstellar dust emission (Cohen
and Kuhi 1976, Allen et al. 1977). Another WC8 star, CV Ser, probably
has a dust shell, the infrared excess being rather large to be explained
by free-free emission (Cohen et al. 1975). The same probably holds for
the 3µm excess of the WC7+Be star HD 192641 (Cohen and Vogel 1978). The
WC8+O9I star γ Vel shows no dust emission in the infrared (Allen and
Porter 1973; Seaquist 1976). Thus all northern (Allen et al. 1972,
Gehrz and Hackwell 1974, Cohen et al 1975) and southern (Allen and
Porter 1973; Cohen 1975) WC9 stars show circumstellar dust emission,
whereas some WC7 and WC8 stars show dust emission and some do not.

The analysis by Hackwell, Gehrz and Smith (1974) and Cohen et al.
(1975) of the free-free emission from WN and early-WC stars was done in
the framework of a constant density, finite radius, shell model. A
more detailed analysis in terms of an expanding wind model is therefore

needed.   The 10μm-(5-15) GHz spectral indices of γ Vel (WC8+09I) and
HD 192163 (WN6) are 0.71 and 0.67, respectively, whereas the spectral
indices of P Cyg (B1I$^+$) and Zeta Pup (O4f) between the same spectral
regions are 0.76 and 0.74 respectively. The less steep spectral indices
of the Wolf-Rayet stars can be interpreted as evidence that accelerative
effects are a little less important in distorting the spectral index
from the constant velocity value. Thus the 10μm radiation from Wolf-
Rayets probably originates from a region closer to the wind terminal
velocity than is the case for the OB stars.

Although detailed models are needed for the infrared emitting
regions, one may use 10μm fluxes, combined with the mean infrared-radio
spectral index from γ Vel and HD 192163, to derive approximate mass loss
rates for a variety of Wolf-Rayet stars. The mass loss rates obtained
range from 2×10^{-5} M$_\odot$/yr for some WN5 stars up to 1×10^{-4} M$_\odot$/yr for
WN7, WN8 and WC8 stars. These mass loss rates are larger by factors of
5 to 40 than the upper limit for mass loss by single-scattering radiation
pressure, $L/v_\infty c$, in contrast to the OB stars, which have mass loss rates
less than this upper limit. Thus the mass loss phenomenon in Wolf-Rayet
stars is orders of magnitude more extreme than O star mass loss.

Hartmann and Cassinelli (1977) obtained a fit to the 1-10μm energy
distribution of HD 50896 (WN5) by using a density distribution which
incorporated an extended deceleration zone out to several stellar radii,
followed by a zone of rapid acceleration to terminal velocity. Hartmann
(1978) modeled the density structure of the wind of the WN5 component of
V444 Cyg by using 2.2 and 3.5μm photometry of the eclipse of the system
and obtained results consistent with the decelerating wind hypothesis.
Since a decelerating wind is a surprising result physically and is of
intrinsic interest if confirmed, more work should be carried out on the
structure of Wolf-Rayet winds, using data from all spectral regions.

Barnes et al. (1974) obtained high resolution spectra of γ Vel
between 0.9 and 1.7μm, at a resolution of $\sim 4 \times 10^3$. Their spectra show
several strong emission lines of He I, He II, C III and C IV. Cohen and
Vogel (1978) obtained 2-4μm spectrophotometry at a resolution of ~ 70
for several late WC stars and one WN6 star. Their data show mainly
emission lines of He I and He II in this spectral region, although an
unidentified emission feature at 3.29μm is often present. Those
stars known to have dust shells exhibit diluted gaseous emission lines,
some having dust emission so strong that the lines appear to be completely washed out, since no trace of them is apparent. The spectrophotometry of Cohen and Vogel (1978) shows that line emission is not a
significant contaminant of WR free-free fluxes deduced using standard
broad-band 2.2 and 3.5μm filters.

4.   INFRARED AND RADIO VARIABILITY

Hackwell et al. (1976) presented evidence for a systematic decrease
between 1970 and 1975 of the infrared fluxes of the Wolf-Rayet stars

HD 193793 (WC7+O5) and HD 192641 (WC7+Be), the changes being most pronounced in the former star. No other Wolf-Rayet star which they monitored showed evidence for variability. They interpreted their results
for these two stars in terms of a steadily decreasing mass loss rate
with time. Florkowski and Gottesman (1977) detected HD 193793 in
October 1975 at 2.7 and 8.1 GHz. The spectral index of -0.19 which they
found was inconsistent with an r^{-2} density distribution, corresponding
instead to optically thin free-free emission. Florkowski has since
reported (Wood 1977) that observations of HD 193793 in March 1977 revealed that the radio flux had declined by at least a factor of five
since the 1975 measurements and that the spectral index was at least
0.6. Hackwell has also reported (Wood 1977) that infrared observations
in May 1977 showed the infrared flux of HD 193793 to have increased and
reached almost its 1970 value. The infrared magnitudes of HD 193793
obtained by Williams et al. (1977) in September 1977 were brighter by a
full magnitude at K and L than the 1970 data of Hackwell et al. (1976).
Williams et al. suggested that a dust shell had formed. Thus HD 193793
has exhibited complex variability in both the infrared and radio regions.
In this regard it is interesting that Davies et al. (1967) claimed a
detection of HD 193793 at a 2.7 GHz flux level of 320 ± 90 mJy, although
the 15 arcmin beamwidth used by these authors makes confusion more of a
problem. The flat radio spectral index found by Florkowski and Gottesman
(1977) in October 1975 might be interpretable as a shell ejection event,
the shell having become optically thin by the time it had travelled out
to the radio emitting region (travel time ∿ 1 month). Continued monitoring of HD 193793 at all wavelengths is essential. Morton and Wright
(this Symposium) have found no evidence for any radio variability of the
WC8+O9I star γ Vel.

Infrared photometry which I obtained for Xi Per (O7.5III) in August
1976 showed this star to be systematically fainter at all wavelengths
than the 1972-3 photometry of Gehrz et al. (1974), the difference being
greater than 1 magnitude at 10μm. Since the infrared excess in 1976 was
similar to that of other O stars of the same spectral type, it would
appear that Xi Per was undergoing an outburst when observed by Gehrz et
al. Some evidence for this exists also in the radio data of Bohnenstengel
and Wendker (1976), who found a radio component at 2.7 GHz coincident
with Xi Per, in observations made during the last quarter of 1972.
Later observations which they made apparently failed to show the same
radio component. Thus a supposedly ordinary O7.5III star shows evidence
for large variations in its mass loss rate. The importance and frequency of such mass loss events needs to be fully explored for a range
of luminosity and spectral types.

5.   FUTURE PROSPECTS

The radio region is probably the best spectral domain for determining accurate mass loss rates for individual stars. Once the radio
flux at a given frequency has been measured, only the terminal velocity

of the wind (obtainable from ultraviolet spectra) and the distance of
the star are required in order to obtain the mass loss rate. Once $\dot{M}$ is
known, model fitting to the infrared flux distribution, in combination
with Balmer line profile fitting, probably provides the best method of
determining the velocity law for an individual stellar wind. When $\dot{M}$ and
the velocity law have been determined using these methods, the detailed
physics of the wind can be explored using ultraviolet spectra.

Only a small number of early-type stars are detectable by the
present generation of radio-telescopes. However, this situation will be
completely changed with the completion of the Very Large Array (VLA) and
the proposed Australian Synthesis Telescope (AST). When complete, the
VLA is expected to be able to detect (3σ) fluxes as small as 10 micro-
Janskys (μJy) at 5 GHz. With this sensitivity a WC8 star like γ Vel
will be detectable to a distance of 30 kpc, an extreme B supergiant such
as P Cyg will be detectable to 54 kpc, an O4f star such as Zeta Pup will
be detectable to 10 kpc and a main-sequence O8 star would be detectable
at distances up to 1.6 kpc. The proposed AST is projected to be able to
detect 80 μJy at 10 GHz, which will allow detections of the objects
listed above to distances up to 60% as far as those appropriate for the
VLA. Thus the AST should be able to detect S Doradus and a few other
extreme early-type supergiants in the Magellanic Clouds. Between them
the VLA and AST will be able to detect every Wolf-Rayet star in the
Galaxy (although some on the far side of the Galaxy may be optically
invisible).

The proposed cryogenically cooled 1-m telescope to be flown in the
Space Shuttle (Gillett 1977) should be able to detect, at the 3σ level
is one hour's integration, fluxes as small as 4 mJy at 100μm, allowing a
star such as Zeta Pup to be detected out to 3.3 kpc. This sytem should
be factors of 7 and 40 more sensitive at 30μm and 10μm compared to
100μm, although at the shorter wavelengths acceleration of the wind
close to the star will affect the spectral index.

Another possible application of radio interferometers is to the
determination of wind temperatures, since the characteristic radio-
emitting radius is proportional to $T^{-\frac{1}{2}}$. However, even the 35 km baseline of the VLA will probably be inadequate to resolve the radio emission
from most winds, e.g. the 5 GHz angular diameter of P Cyg is expected to
be $\sim$ 0.16 arcsec, compared to a VLA resolution of $\sim$ 0.3 arcsec at that
frequency. Going to higher frequencies does not help much, since although the angular resolution is proportional to ν^{-1}, the radio-emitting
radius of a wind is proportional to $\nu^{-2/3}$, giving only a factor of $\nu^{1/3}$
advantage. Thus much longer baselines will be required for wind temperature determinations. Systems such as the Jodrell Bank Multi-Telescope
Radio-linked interferometer with baselines of hundreds of km appear best
suited to this application.

REFERENCES

Allen, D.A., Harvey, P.M. & Swings, J.P. 1972, Astr. Astrophys.,
     20, 333.
Allen, D.A., Hyland, A.R., Longmore, A.J., Caswell, J.L., Goss, W.M.,
     & Haynes, R.F. 1977, Astrophys. J., 217, 108.
Allen, D.A. & Porter, F.C. 1973, Astr. Astrophys., 22, 159.
Altenhoff, W.J., Braes, L.L.E., Olnon, F.M. & Wendker, H.J. 1976,
     Astr. Astrophys., 46, 11.
Balick, B. 1972, Astrophys. Lett., 12, 21.
Barlow, M.J. & Cohen, M. 1977, Astrophys. J., 213, 737.
Barnes, T.G., Lambert, D.L. & Potter, A.E. 1974, Astrophys. J.,
     187, 73.
Bohnenstengel, H.-D. & Wendker, H.J. 1976, Astr. Astrophys., 52, 23.
Cassinelli, J.P. & Hartmann, L. 1977, Astrophys. J., 212, 488.
Cassinelli, J.P. & Olson, G.L. 1978, Astrophys. J., in press.
Cassinelli, J.P., Olson, G.L. & Stalio, R. 1978, Astrophys. J., 220, 573.
Castor, J.I., Abbott, D.C. & Klein, R.I. 1975, Astrophys. J., 195, 157.
Chiuderi, C. & Torricelli Ciamponi, G. 1978, preprint.
Cohen, M. 1975, Mon. Not. R. astr. Soc., 173, 489.
Cohen, M., Barlow, M.J. & Kuhi, L.V. 1975, Astr. Astrophys., 40, 291.
Cohen, M. & Kuhi, L.V. 1976, Publ. astr. Soc. Pacific, 88, 535.
Cohen, M. & Vogel, S.N. 1978, Mon. Not. R. astr. Soc., in press.
Cox, D.P. & Daltabuit, E. 1971, Astrophys. J., 167, 113.
Davies, J.G. Ferriday, R.J., Haslam, C.G.T., Moran, M. & Thomasson, P.
     1967, Mon. Not. R. astr. Soc., 135, 139.
Florkowski, D.R. & Gottesman, S.T. 1977, Mon. Not. R. astr. Soc.,
     179, 105.
Gehrz, R.D. & Hackwell, J.A. 1974, Astrophys. J., 194, 619.
Gehrz, R.D., Hackwell, J.A. & Jones, T.W. 1974, Astrophys. J.,
     191, 675.
Gillett, F.C. 1977, in 'Infrared and Submillimetre Astronomy',
     ed., G.G. Fazio (D. Reidel), p.195.
Hackwell, J.A., Gehrz, R.D. & Smith, J.R. 1974, Astrophys. J., 192, 383.
Hackwell, J.A., Gehrz, R.D., Smith, J.R. & Strecker, D.W. 1976,
     Astrophys. J., 210, 137.
Hartmann, L. 1978, Astrophys. J., 221, 193.
Hartmann, L. & Cassinelli, J.P. 1977, Astrophys. J., 215, 155.
Hjellming, R.M. 1977, Proc. IAU Colloquium No.42, eds. R. Kippenhahn,
     J. Rahe, W. Strohmeier, p.279.
Johnson, H.M., Proc. IAU Symp. No.49, eds., M.K.V. Bappu, J. Sahade
     (D. Reidel), p.42.
Klein, R.I. & Castor, J.I. 1978, Astrophys. J., 220, 902.
Kuan, P. & Kuhi, L.V. 1975, Astrophys, J., 199, 148.
Lamers, H.G.J.L.M. & Morton, D.C. 1976, Astrophys. J. Suppl., 32, 715.
Lucy, L.B. & Solomon, P.M. 1970, Astrophys. J., 159, 879.
Marsh, K.A. 1975, Astrophys. J., 201, 190.
Morton, D.C. & Wright, A.E. 1978, Mon. Not. R. astr. Soc., 182, 47P.
Olnon, F.M. 1975, Astr. Astrophys., 39, 217.
Panagia, N. & Felli, M. 1975, Astr. Astrophys., 39, 41.

Raymond, J.C., Cox, D.P. & Smith, B.W. 1976, Astrophys. J., 204, 290.
Schwartz, P.R. & Spencer, J.H. 1977, Mon. Not. R. astr. Soc., 180, 297.
Seaquist, E.R. 1976, Astrophys. J. (Lett)., 203, L35.
Seaquist, E.R. & Gregory, P.C. 1973, Nature Phys. Sci., 245, 85.
Thomas, J.A., Robinson, G. & Hyland, A.R. 1976, Mon. Not. R. astr. Soc.,
     174, 711.
Thorne, K.S. 1975, Orange Aid Preprint No. 421, California Institute
     of Technology.
Wendker, H.J., Baars, J.W.M. & Altenhoff, W.J. 1973, Nature Phys. Sci.,
     245, 118.
Wendker, H.J., Smith, L.F., Israel, F.P., Habing, H.J. & Dickel, H.R.
     1975, Astr. Astrophys., 42, 173.
Williams, P.M., Stewart, M.J., Beattie, D.H. & Lee, T.J. 1977, IAU Circ.
     No. 3107.
Wood, F.B. 1977, Proc. IAU Colloquium No. 42, eds. R. Kippenhahn,
     J. Rahe, W. Strohmeier, p.639.
Woodsworth, A.W. & Hughes, V.A. 1977, Astr. Astrophys., 58, 105.
Woolf, N.J., Stein, W.A. & Strittmatter, P.A. 1970, Astr. Astrophys.,
     9, 252.
Wright, A.E. & Barlow, M.J. 1975, Mon. Not. R. astr. Soc., 170, 41.

DISCUSSION FOLLOWING BARLOW

Hearn: Cassinelli and Hartmann have shown that the infrared
intensity of a star is given by $4\pi R_\lambda^2 B_\lambda$. At different wavelengths the
radius of the star is determined by where the optical depth $\zeta_\lambda \simeq 1$ occurs.
With ordinary hydrostatic models the density decreases so quickly that
the radius is not significantly dependent on the wavelength and the infrared intensity is determined only by the variation of the Planck function.

The interpretation of the infrared data gives the density temperature structure. The interpretation of mass loss involves a further interpretation of the density structure.

Conversely the IR region, particularly in the range 10 μ to 500 μ,
will give information on the density-temperature of the stars in the
region between the photosphere and corona.

Underhill: What are the temperatures in your IR emitting regions?

Barlow: For P Cygni itself, the temperature can safely be taken
to be 20,000°K. For some other stars we have taken somewhat higher
values, of order 40,000°K as given by the example of ζ Pup.

Underhill: They do not seem to be as high as a few hundred
thousand degrees?

Barlow: We cannot use the IR data to make a statement, one way
or the other. Admittedly, the temperature adopted is important. The
ratio data give the mass loss rates without much temperature dependence.

Castor: Tony is right that from the modeling you find $\rho(r)$, not
$v(r)$. However, even with inhomogeneities it is true that $\overline{v(r)}$ = con-
stant/$[r^2\overline{\rho(r)}]$ so as long as we realize that ρ and v are means, we can
use the usual relationship.

Hearn: If the extended density structure obtained from the IR observations results from some other physical process, such as the wave
pressure on turbulent pressure, the mass loss interpretation could be
quite wrong. With the right sort of turbulent pressure, presumably one
could explain the IR observations without any mass loss at all.

D. Van Blerkom: HD 50896 (WN5) was studied by Hartmann and
Cassinelli, who suggested a region of decelerated flow in the envelope
was necessary to account for the observed infrared flux distribution.
Has any subsequent work been done on this star to confirm this unexpected velocity distribution?

Cassinelli: We could get a fit to the IR flux with a density
falling as $1/r^2$, if the radius of the star is small ~2.5 $R_\odot$. One can
also get information about the density structure by looking at an
eclipsing binary at IR wavelengths. Recently Lee Hartmann published
results of his IR observations of V 444 Cyg, an eclipsing WR system.
He found that the extended density distribution that we derived for
HD 50896 gave an excellent fit to the IR light curve.

NEAR INFRARED SPECTRA OF O STARS AND RELATED OBJECTS

VREUX, J.M.
Institut d'Astrophysique, Liège, Belgium
ANDRILLAT, Y.
Observatoire de Haute Provence, France

ABSTRACT

The λ 10124 He II line has been found to be a measurable emission in three stars. Our new data are compared to theoretical predictions. The λ 10830 He I line is observed in emission in 74 % of the O5-O8 supergiants but only seen in 29 % of the dwarfs, all of the latter exhibiting some "pecularities" i.e. classified as Oef, Oe, ON or On. An envelope with a sufficient amount of material seems to be a favorable condition to get the λ 10830 line in emission. However the mechanism leading to the observed emission is temperature dependent as well.

INTRODUCTION

The results given here have been obtained with the "Roucas" grating spectrograph attached to the Cassegrain focus of the Haute Provence Observatory 1.93 m (77 inch) telescope. The characteristics of this instrument have been given elsewhere (Andrillat, et al., 1973). We only recall here that the dispersion is 230 Å mm^{-1} and that the 200 μ slit corresponds to 7 Å in the plane of the receiver. The latter is a cooled (around -60°C) ITT F-4718 two stage image tube; 103aD film is used behind the fiber optics output. The accuracy of equivalent widths larger than 2 Å is estimated to be of the order of ± 25 %. It is worse for fainter features, the detection threshold being of the order of 1 Å. The noise is not the only limiting factor : strong telluric absorption and contamination by OH night sky emission combined with the sharp sensitivity dropoff at λ > 1 μ make sometimes the definition of the continuum rather imprecise.

P. S. Conti and C. W. H. de Loore (eds.), Mass Loss and Evolution of O-Type Stars, 131-137.
Copyright © 1979 by the IAU.

THE He II λ 10124 LINE

The most recent theoretical predictions are due to
Klein and Castor (1978). According to their model this line
should be a strong emission in some Of stars, the ratio of
λ 10124 to λ 4686 ranging from 2.16 to 3.64. Using previous
observations of two Of stars by Mihalas and Lockwood (1972),
Klein and Castor (1978) derived an observed ratio of about
1.3. Out of a sample of 67 O stars we do observe emission at
λ 10124 in three objects : HD 16691 (04If[+]), HD 190429(04If[+])
and HD 228766(05.5f), the last one being a transition object
according to Massey and Conti (1978). Following Klein and
Castor (1978) the equivalent width of the emissions have
been corrected for the underlying absorption on the basis of
non LTE/L calculations (Auer and Mihalas, 1972) and then
compared to the available values of He II λ 4686 (Conti and
Alschuler, 1971; Conti and Leep, 1974; Massey and Conti,
1978). For the first two stars we get a maximum value of the
ratio less than 1.4 i.e. again smaller than the ratio
predicted by the model. But HD 228766, the transition object,
gives a value of the order of 2.1, i.e. close to the
prediction. We have used the observed ratio for "normal" Of
stars and the calculations of Auer and Mihalas quoted above
to predict the intensity of the λ 10124 line in all the
stars for which we had data both at λ 4686 and at λ 10124.
The result is given in table 1.

Table 1 - Comparison between observed and
predicted He II λ 10124 intensities in Of stars

Star	Spectral type (Conti)	Observed λ 4686 log W(mA) (emission)	Predicted λ 10124 W ($\overset{\circ}{A}$)	Observed λ 10124 W ($\overset{\circ}{A}$)
108	07If	2.11	1.3	4.0
14442	06ef	2.59	1.2	2.5
14947	05.5f	3.36	E 1.1	0
16691	05f	3.78	E 6.1	E 3.0
				E 6.0
57060	08.5If	2.91	0	0
166734	07.5If	2.56	0.9	A ou 0?
167971	07.5If	2.57	0.9	0?
188001	08If	2.26	1.0	A?
190429	04f	3.57	E 3.0	E 3.5
				E 4.0
				E 3.5
210839	06ef	2.86	0.7	A ou 0?
+60°2522	06.5IIIef	2.88	0.6	0

Except for HD 108 (which has a P Cygni profile at λ 4686)
and in a lesser extend for HD 14442 and HD 14947, the
agreement is good, the difference between prediction and
observation being of the order of the estimated error. It
may be worthwhile stressing that this result is obtained
with the observed ratio, not the theoretical one and that
in addition to HD 108 and HD 14442 we have a few other O
star with an unexpectedly strong absorption.

THE He I λ 10830 LINE

Since many years the He I λ 10830 line (2^3s-2^3P) is
known to be a strong emission in most of the WR stars (Kuhi,
1968) and to be a strong absorption in B stars with extended
atmosphere (Underhill, 1970). The present results deal with
the stars between these two classes. The two O stars spectra
exhibiting the strongest emission at λ 10830 Å are
illustrated in fig. 1.

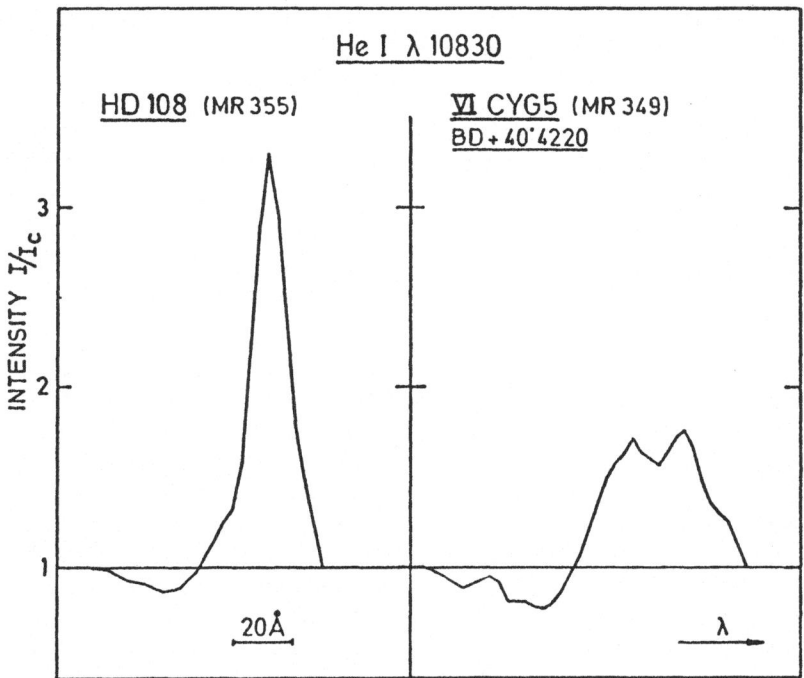

Figure 1. The two strongest He I λ 10830
emissions recorded (uncorrected for
instrumental profile).

The broadest profile is observed in VI Cyg 5 (BD +40°4220)
which, according to Bohannan and Conti (1976), is on the
way to become a Wolf Rayet. In order to try to find which

physical conditions are requested to produce λ 10830
emission we have plotted in fig. 2 the observed behavior

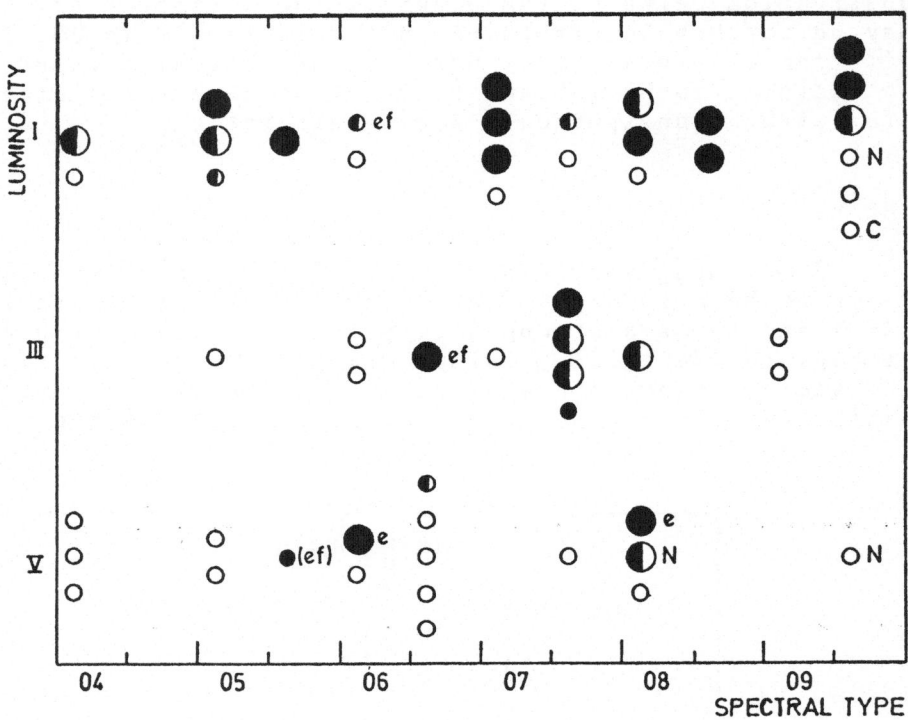

Figure 2. Distribution of He I λ 10830 emission
        (black filled circles) as a function of
        spectral type and luminosity class.

of this line as a function of spectral type and luminosity
class. Empty circles are used for stars with well exposed
spectra where no emission is observed. Black filled circles
mean that λ 10830 has been observed in emission, half
circles being used when the emission is variable and has
not been found on all the spectra of the object. A larger
circle has been used when an equivalent width has been
obtained. From this figure it appears that an emission is
observed at λ 10830 (with or without a P Cygni profile) in
74 % of the 04-08 supergiants but only seen in 29 % of the
dwarfs all of the latter exhibiting some peculiarities in
their spectra (i.e. classified as Oef, Oe, ON or On),
peculiarities which are believed to be characteristic of
the presence of an envelope or of a companion. But an
envelope or a companion is not a sufficient condition. To
try to go further we have searched for a correlation
between the mass loss rate and the intensity of the λ 10830
emission. The results are given in table 2. The available
data show a nice correlation between the two parameters in

Table 2 - O Star Mass Loss Rates and λ 10830 Intensity

Star	Spectral type (Conti)	$\dot{M}$ $(10^{-7} M_o yr^{-1})$	W(λ 10830) (Unit : Å) Emission
A. Mass Loss from Hutchings (1976)			
108	07If	1000	35.5
29CMa 57060	08.5If	85	9
9 Sge 188001	08If	75	6 (var)
14947	05.5f	60	8
λ Cep 210839	06ef	30	< 1
Cyg X-1 226868	09I	25	4.5 (var)
46150	05.5((f))	15	0
48099	06.5V	1	0
B. Mass Loss from Barlow and Cohen (1977)			
108	07If	85	35.5
14947	05.5f	24	8
60848	08Ve	24	∿ 15 (var)
39680	06Ve	19	18
45314	0Be	4	E

05-09 stars. This effect is temperature dependent : it is strongly weakened among the B stars (He I λ 10830 being only observed in emission in three objects out of a sample of 22) and looks weakened among the 04 stars too, but we only have observations for four 04 stars with known mass loss rate. In conclusion our sample indicates that the most favorable conditions to "push" He I 10830 in emission are that the envelope has a sufficient amount of material and that the central star has a temperature between 30 000°K and 45 000°K. These He I observations undoubtedly put constraints on the ionization balance of Helium in the theoretical models of the stellar wind.

REFERENCES

Auer, L.H., Mihalas, D. : 1972, Astrophys. J. Suppl. 24, pp. 193-246
Andrillat, Y., Baranne, A., Duchesne, M. : 1973, Mém. Soc. Roy. Sci. Liège V, pp. 51-55
Barlow, M.J., Cohen, M. : 1977, Astrophys. J. 213, pp. 737-755

Bohannan, B., Conti, P.S. : 1976, Astrophys. J. 204,
   pp. 797-803
Conti, P.S., Alschuler, W.R. : 1971, Astrophys. J. 170,
   pp. 325-344
Conti, P.S., Leep, E.M. : 1974, Astrophys. J. 193, pp. 113-
   124
Hutchings, J.B. : 1976, Astrophys. J. 203, pp. 438-447
Klein, R., Castor, J. : 1978, Astrophys. J. 220, pp. 902-
   923
Kuhi, L.V. : 1968, in "Wolf Rayet Stars", NBS Special
   Publication 307, pp. 101-144
Massey, P., Conti, P.S. : 1977, Astrophys. J. 218, pp. 431-
   437
Mihalas, D., Lockwood, G.W. : 1972, Astrophys. J. 175, pp.
   757-764
Underhill, A. : 1970, Dilution Effects in Extended
   Atmospheres, in "Spectroscopic Astrophysics", ed. G.H.
   Herbig, pp. 159-172

DISCUSSION FOLLOWING VREUX AND ANDRILLAT

Lamers: Can you express the extent of the He I emission lines in km/sec, in order to get an impression about the part of the envelope where the line is formed?

Vreux: The full width of HD 108 emission is about 1000 km/sec -- the absorption component is marginally present: The sensibility of the instrument is varying rapidly around one micron and the definition of the continuum is not easy -- the accuracy of the equivalent width of HD 108 is better than 20% but to know the extension of the wings we need the new detector under development at the Haute Provence Observatory.

Castor: From the figure of HD 108 it appears that the absorption on the blue side extends ~35 or 40 Å from line center, which is about 1000 km/sec. That's probably rough, but it is of the order we should think about.

Bidelman: This is slightly off the topic but I would like to make a point here about nomenclature. Is VI Cyg #5 the same star as BD+40° 4220?

Bohannan: Yes, it is also known as V729 Cyg.

Bidelman: Well, I'd like to say, as a member of an IAU Commission concerned with such things, that one should adopt a consistent labeling for a star.

Underhill: I'd like to second that.

Bidelman: We don't all have encyclopedic memories.

Conti: My feeling is that when talking about the spectrum one should use the HD (HDE,BD) name, and when talking about the photometry, the variable star name. However, our practice has always been to quote both initially.

Bidelman: We really should come to some formal agreement about this. Another point: we should stop talking about γ "two" Vel. Just say γ Vel.

Morton: What do you call the other star then?

Bidelman: You don't care about the other star.

Underhill: It's a nice B2IV.

Bidelman: The number 1 and 2 business is a mistake from a long time ago. If you really want to be fussy it's γ Vel A and B. When one star is so much brighter than its companion, and so close, there is no point in using the superscripts.

Underhill: Getting back to the IR spectral data: There probably is an eventual dropoff in temperature far out in the stellar wind. He I $\lambda 10830$ is a metastable transition of a neutral atom. Could it be that this profile is formed far from the star, rather than close to the photosphere?

INFRARED OBSERVATIONS OF HDE 226868/CYG X-1 AND HDE 245770/A0535+26

P. Persi, M. Ferrari Toniolo and G. Spada
Laboratorio Astrofisica Spaziale, CNR, Frascati, ITALY

ABSTRACT: We know from Copernicus ultraviolet observations that all O-type stars are losing mass by stellar wind. The ionized expanding circumstellar envelope formed by the stellar wind is emitting through free-free and bound-free radiation processes. This radiation is detectable at the infrared wavelengths where the stellar continuum is negligible. The measurement of the IR excess (defined as the difference between the total flux and the stellar continuum at a given wavelength) and the knowledge of the terminal velocity of the envelope, allow us to derive for OB stars the mass loss rate. From the analysis of our IR observations of two O stars, HDE 226868 and HDE 245770, identified as optical counterpart of X-ray sources, we give an estimate of their mass loss rate. The IR observations were carried out with the Jungfraujoch 76 cm telescope using a GE bolometer with a focal plane chopping system and with the Merate 132 cm telescope using an InSb detector.

HDE 226868/CYG X-1

The O9.7IIab star HDE 226868 is known to be the optical counterpart of the X-ray source 3U 1956+35. Our IR observations in the J,H,K,L and M bands were obtained in May 1976 and September 1976. During these periods no significant variations were observed.

Figure 1 displays the unreddened optical-infrared energy distribution of HDE 226868. We used a colour excess $E_{B-V} = 1.08$. The adopted U, B and V magnitudes are mean values obtained by Bregman et al. (1973) and Liutyi et al. (1975).

We compared the observed spectral points with a blackbody distribution at the same effective temperature of an O9.7 supergiant star ($T_{eff} = 32,000°K$). Looking at the figure, we see that J,H,K and L fluxes lie on the blackbody curve, while the magnitude corresponding to 5μ (M-band), is 4σ above the stellar continuum. This evidence of an IR excess at 5μ is in agreement with a mass loss rate greater than 10^{-6} $M_\odot$ y^{-1}. The decrease of the statistical error at 5μ, and the IR observations at

139

P. S. Conti and C. W. H. de Loore (eds.), Mass Loss and Evolution of O-Type Stars, 139-142.
Copyright © 1979 by the IAU.

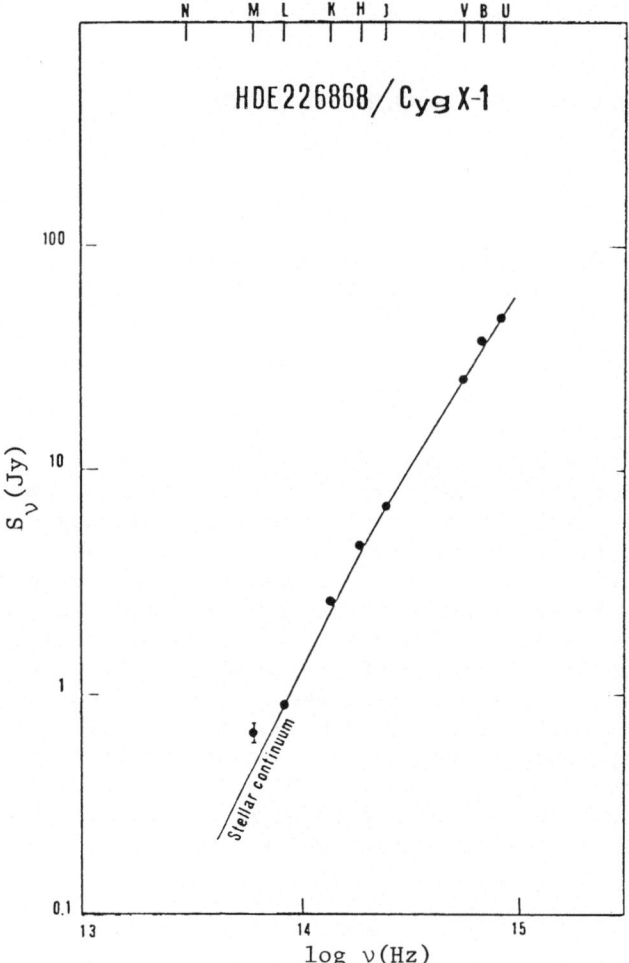

Figure 1. The optical and near IR energy distribution of HDE 226868.

longer wavelengths (10μ) would allow us to derive a better value of $\dot{M}$ for HDE 226868.

HDE 245770/A0535+36

This star, recently classified by Giangrande et al. (1977) as 09.7IIe, has been suggested as the possible optical counterpart of the transient X-ray source A0535+26. A model proposed by Rappaport et al. (1976) consists in a long-period orbit neutron star (>17^d) around the OBe star with a variable stellar wind. The presence of a stellar wind is suggested by the broadening of the Hα emission line (~300 km s^{-1}) observed by Soderblom (1976).

In order to derive the mass loss rate from HDE 245770, observations were carried out in the J,H,K and L photometric bands on Nov.-Dec. 1976 and March 1977. In the same period the star was observed in U,B,V by Giangrande et al. (1978). No variations were found in optical and IR wavelengths.

Figure 2 shows the unreddened ($E_{B-V} = 0.80$) energy distribution for HDE 245770. By comparison we also show (crosses in figure) the narrowband spectrophotometric observations of Wade and Oke (1977). The stellar continuum is represented by a blackbody with an effective temperature of 32,000°K. The IR excess (filled squares) shows a distribution of the type $S_\nu \propto \nu^\alpha$ with a spectral index $\alpha \sim 0.6$ which is consistent with a thermal free-free + bound-free emission from a circumstellar envelope with an electron density $n_e \propto r^{-2}$. Taking as a model an expanding

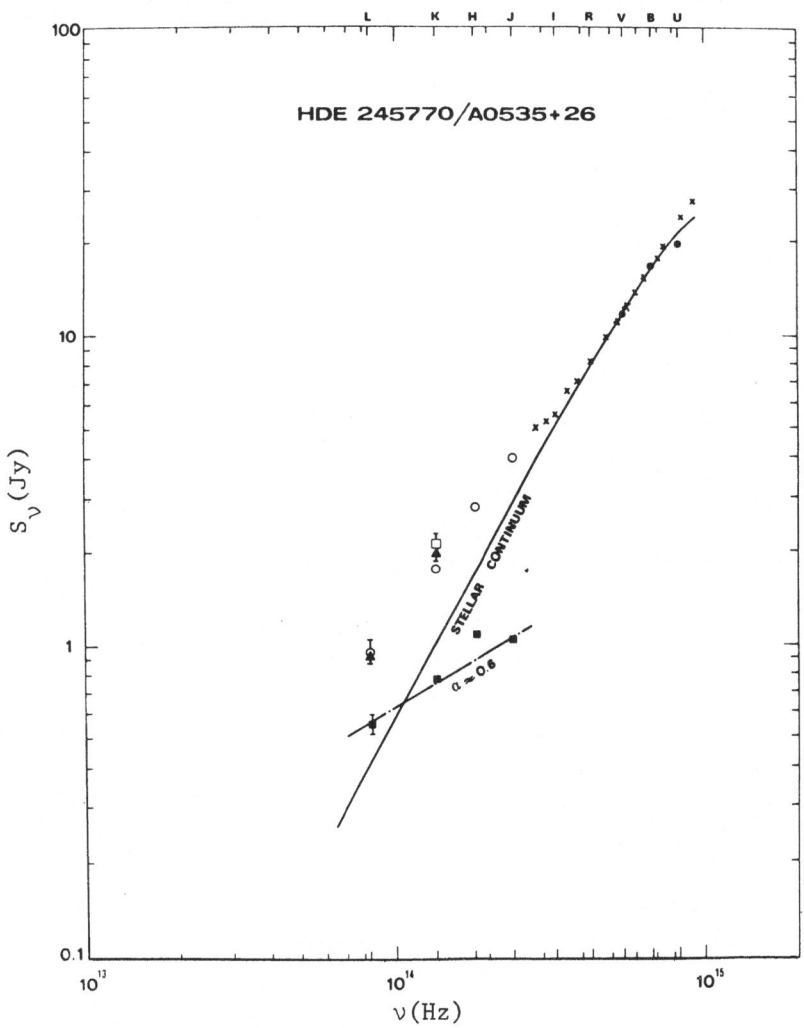

Figure 2. The optical and near IR energy distribution of HDE 245770; ○: JHKL observations on 1976 Dec. 11, 13; □: K observations on 1976 Nov. 16; ▲: KL observations on 1977 March 16.

spherical envelope with a constant velocity described by Wright and
Barlow (1975) and Panagia and Felli (1975), we can estimate the value
of $\dot{M}$ from the observed IR excess.

Adopting for HDE 245770 a distance D $\sim$ 1.8 kpc, an envelope temperature T_e = 20,000°K, and the measured IR excess at 3.6μ S_ν = 0.56 ±
0.04 Jy, we obtain a ratio $\dot{M}/v_{exp}$ ≅ 1.7×10^{-8} $M_\odot$ y^{-1}/km s^{-1}. For a
v_{exp} ≅ 300 km s^{-1} we derive for the mass loss $\dot{M}$ ≅ 5×10^{-6} $M_\odot$ y^{-1}. This
value of $\dot{M}$ could be overestimated because of the breakdown of the simple
uniform-flow model.

Finally, from the evolutionary tracks computed by Chiosi et al.
(1978) for a massive star with mass loss, we derive the mass of HDE
245770. Adopting a luminosity of L^* = 2.9×10^5 $L_\odot$ and the above mass
loss rate we obtain M $\sim$ 35–40 $M_\odot$.

REFERENCES

Bregman, J., Butler, D., Kemper, E., Koski, A., Kraft, R.P. and Stone,
    R.P.S.: 1973, Astrophys. J. Lett.185, p. L117.
Chiosi, S., Nasi, E. and Sreenivasan, S.R.: 1978, Astron. Astrophys.
    63, p. 103.
Giangrande, A., Giovannelli, A., Bartolini, C., Guarnieri, A. and
    Piccioni, A.: 1977, IAU Circ. 3129.
Giangrande, A., Giovannelli, F., Bartolini, C., Guarnieri, A. and
    Piccioni, A.: 1978, submitted to Astrophys. J.
Liutyi, V.M., Sunyaev, R.A. and Cherepashchuk, A.M.: 1975, Sov. Astron.
    18, p. 684.
Panagia, N. and Felli, M.: 1975, Astron. Astrophys. 39, p. 1.
Rappaport, S., Joss, P.C., Bradt, H., Clark, C.W. and Jernigan, J.G.:
    1976, Astrophys. J. Lett. 208, p. L119.
Soderblom, D.R.: 1976, IAU Circ. 2971.
Wade, R.A. and Oke, J.B.: 1977, Astrophys. J. 215, p. 568.
Wright, A.E. and Barlow, M.J.: 1975, Mon. Not. R. Astr. Soc. 170, p. 41.

DISCUSSION FOLLOWING PERSI et al.

Cowley: Because A0535+26 is a transient X-ray source, it would be
especially interesting to redetermine the mass loss rate from the IR
data at a time when the X-rays are strong, as it may be an indication
of an enhanced mass flow.

INFRARED EXCESSES AND MASS LOSS: IMPLICATIONS FOR OB STARS IN CARINA
AND THE LMC

A. R. Hyland
Mount Stromlo and Siding Spring Observatories,
The Australian National University,
Canberra, Australia

Abstract: A method of deriving mass loss rates using near IR excesses
is established and applied to OB stars in Carina and the LMC.

1.   IR OBSERVATIONS AS A DIAGNOSTIC TOOL FOR MASS LOSS

Methods to derive precise values of the radiation pressure driven
mass loss rate are crucial to the development of a broader understanding
of the consequences of mass loss from hot luminous stars. One of the most
successful methods appears to be the analysis of IR excesses by Barlow
and Cohen (1977), (hereinafter referred to as BC). For reasons of observational difficulty all previous analyses have dealt with bright stars.
Wright and Barlow (1975) developed a theory for predicting the IR and
radio fluxes from a uniform velocity ionized mass loss outflow. Free-
free emission from the stellar wind produces an IR excess, the wavelength
dependence of which depends upon the rate of mass loss and the form of
the velocity law. For a uniform velocity outflow $S_\nu \propto \nu^{2/3}$ while for an
accelerating flow BC showed that the exponent of ν is greater than 2/3.
In BC's extensive study of mass loss from O stars and B supergiants using
observations of IR excesses in the 3-10 μm region, they found that an
accelerating flow is in better accord with the data for P Cyg, and
applied such a law to derive mass loss rates for their program stars.
Surprisingly they found excesses at 3.6 μm to be small or zero.
Predicted excesses for 3.6-10 μm and 1.25-2.2 μm colors under two
assumptions for the flux distribution in the stellar wind are shown in
Figure 1 and indicate that a one to one correlation should exist between
them. This is confirmed by comparison of the 1.25-2.2 μm excess (which
we denote as $E_f(J-K)$) with BC's 10 μm excesses. Thus, values of $\dot{M}$ should
be obtainable from observations of $E_f(J-K)$, and this method has several
advantages over longer wavelength measurements e.g., (a) the magnitudes
and excesses can be derived more precisely, (b) much fainter objects
can be measured, and (c) the 1.25-2.2 μm region is less affected by
emission from cool dust.

P. S. Conti and C. W. H. de Loore (eds.), Mass Loss and Evolution of O-Type Stars, 143-146.

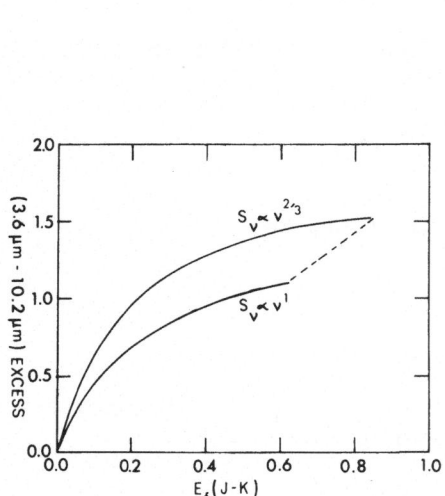

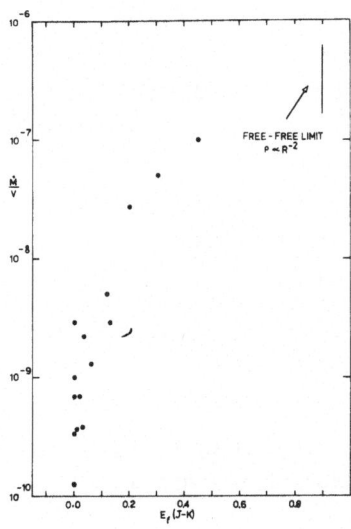

Figure 1. Predicted excesses

Figure 2. Empirical calibration of $E_f(J-K)$ against $\dot{M}/v_\infty M_\odot yr^{-1}/(km/sec)$.

2. CALIBRATION OF $E_f(J-K)$ IN TERMS OF MASS LOSS

We have calibrated $E_f(J-K)$ in terms of $\dot{M}/v_\infty$ using the results of BC. Values of $E_f(J-K)$ were determined from published colors (see Schultz and Wiemer 1975, and references therein) using the normal van der Hulst reddening curve and intrinsic (B-V) colors (Johnson 1966). This procedure is prone to an uncertainty ~0.03, due to the fact that the intrinsic colors derived by Johnson include any component due to free-free emission in the colors of normal stars, our values being too small by this amount. This empirical calibration is shown in Figure 2. Although for $E_f(J-K) < 0.10$ the result provides little quantitative information, for $E_f(J-K) \geqslant 0.10$ the method holds promise for clear estimates of mass loss rates for OB stars.

3. RESULTS AND DISCUSSION

Infrared J, H, K magnitudes have been obtained for a series of OB and WR stars in Carina, the LMC, and in interstellar bubbles in the LMC. The observations were made with the Mount Stromlo IR photometer attached to the 3.9m AAT and the 1m telescope at Siding Spring Observatory. The results are given in Figure 3, from which these points can be made:
(a)   The Carina O stars (HD93250, 93204, 93205, F100 and F104) (Feinstein, Marraco and Muzzio 1973) have excesses typical of the calibration O stars, where $0.05 < E_f(J-K) > 0.12$, depending upon the value of R.
(b)   The excesses of the LMC OB emission line stars range up to the limit predicted for radiation pressure induced mass loss. Contamination of

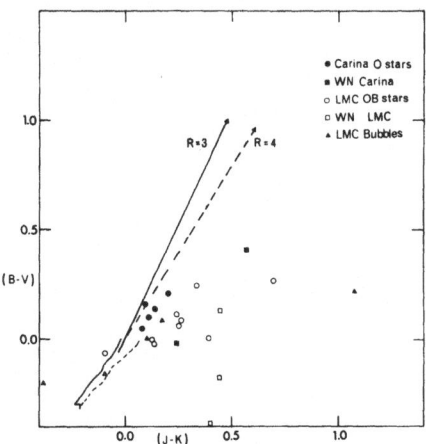

Figure 3. Observations of OB stars in the (B-V) vs. (J-K) plane. Loci of the intrinsic colors of main sequence and supergiant OB stars are shown as solid and dashed lines respectively. Arrowed lines are reddening lines.

the colors by dust emission was excluded by consideration of the fluxes from 1.25-3.6 μm.
(c) The Carina WN stars (HD93162 and HD93131) and those in the LMC (Hyland, Thomas and Robinson 1978) have E_f(J-K) $\gtrsim$ 0.50, implying extreme rates of mass loss. The LMC stars appear more extreme, but their (B-V) colors may be contaminated by surrounding nebulosity.
(d) The establishment of a method for obtaining mass loss rates of the faint central stars in interstellar bubbles in the LMC was the prime motivation for this study. Our aim was to relate these rates with those inferred from Hα intensities and dynamical measurements of the bubbles. Observations were obtained for stars in the bubbles N70, N9, N185, and N186 (two stars). Apart from N186 #1, whose extreme excess is due to its composite nature, all other bubble stars have small excesses. The central star of the best studied bubble, N70, lies to the extreme left of Figure 3 and has an unexplained IR deficiency. Hence the observed mass loss rates are insufficient to produce and support the bubbles, and continuous mass loss is unacceptable as a mechanism for their formation. An intermittent or eruptive process which produces high mass loss for short periods of time, appears to be required.

REFERENCES

Barlow, M.J., and Cohen, M.: 1977, Astrophys.J., 213,pp.737-755.
Feinstein, A., Marraco, H.G., and Muzzio, J.C.: 1973, Astron.Astrophys. Supplement, 12,pp.331-350.
Hyland, A.R., Thomas, J.A., and Robinson, G.: 1978, Astron.J., 83,pp.20-25.
Johnson, H.L.: 1966, Ann.Rev.Astron.Astrophys., 4,pp.193-206.
Schultz, G.V., and Wiemer, W.: 1975, Astron.Astrophys.,43,pp.133-139.
Wright, A.E., and Barlow, M.J.: 1975, Mon.Not.R.astr.Soc., 170,pp.41-51.

DISCUSSION FOLLOWING HYLAND

Conti: If one has trouble interpreting the nebular shell as coming from the present star it might be that a previous "event" caused the ejection. The central star now may have a completely different spectrum than previous to this episode.

Hyland: That is certainly true, and the existence of such events may have some bearing on the unusual colors of the central star in N 70. Unfortunately, the hypothesis of continuing steady mass outflow as the cause of the "bubble" appears to be untenable.

Cowley: A number of the early OB supergiants in the LMC, which show "normal" blue spectra, in fact turn out to be composites in a fainter red supergiant. Can you distinguish these from other kinds of IR excesses?

Hyland: Yes. The three common types of infrared excesses (i.e., the presence of a companion, free-free emission or dust emission) are easily distinguished in a (J–H) vs. (H–K) color-color diagram. Several emission line OB stars in the LMC have been identified as being composite on the basis of their positions in such a diagram.

Underhill: Could you remind us of the wavelengths of HJK?

Hyland: H is 1.65 μm, J is 1.25 μm and K is 2.20 μm.

RADIO EMISSION FROM FLOWS

P. R. Schwartz
E. O. Hulburt Center for Space Research
Naval Research Laboratory, Washington, D. C. 20375

Many emission line stars are radio sources including early type stars such as ζ Pup and P Cyg, peculiar objects such as MWC 349 (a possible proto-planetary system) and V1016 Cyg, nebular variables such as T Tau, and at least one late M giant. These "radio stars" are characterized by (1) a rising spectrum with spectral index α = 0.6 to 1.1, (2) positional coincidence between star and radio source, and (3) no rapid (less than 10^7 sec) time variability. These characteristics are in contrast with the non-thermal spectrum and rapid variability of Algol type "radio stars" and the radio emission associated with x-ray sources. The radio component of the spectrum of emission line stars originates in the substantial mass loss flows which are a feature of their outer atmospheres and which, when ionized by stellar radiation, radiate by free-free electron thermal bremsstrahlung. These circumstellar H II regions are, in general, detectable with present radiotelescopes if the characteristic size of the optically thick region of the flow is greater than about 10^{-7} of the distance to the star.

In the radio emission region (typically at radii greater than a few A.U.), the flow has reached terminal velocity, V_∞, and thus the electron density varies as the inverse square of the radius. Wright and Barlow (1975) and Panagia and Felli (1975) have discussed emission from such flows and shown that the resulting spectral index is α = 0.6. The mass loss rate may be calculated from the observed radio flux, S_ν, at frequency ν, the distance, d, and the terminal velocity:

$$\frac{dM}{dt} = 2.9 \times 10^{-9} \ [\frac{S_\nu}{mJy}]^{0.75} \ [\frac{d}{kpc}]^{1.5} \ [\frac{\nu}{10 \ GHz}]^{-0.45} \ [\frac{V_\infty}{km/s}] \ M_\odot/yr$$

where it is assumed that the electron temperature is 10^4 K and understood that dM/dt refers only to the ionized component of the flow.

The mass loss rates derived from radio observations for both early type and other emission line stars for which reasonable distance and terminal velocity measurements or estimates are available are tabulated below. With the exception of BN Gem, for which there is only one λ = 13 cm observation from an NRL "radio star" survey conducted with the

147

P. S. Conti and C. W. H. de Loore (eds.), Mass Loss and Evolution of O-Type Stars, 147–149.
Copyright © 1979 by the IAU.

300 m antenna at Arecibo, Puerto Rico, all of these stars have constant, thermal (α = 0.6 to 1.1) spectra and thus the estimate of the mass loss rate is reliable to about ± 50%. If the BN Gem result is confirmed and its spectrum verified, it represents an example of a main sequence O star with a very large mass loss rate which may be contrasted with a negative result from the same survey for S Mon. Although several nearby emission line stars in NGC 2264 are radio sources (indicating mass loss rates of the order of 10^{-6} M$_\odot$/yr) S Mon (07n) is undetected at a flux level indicating a mass loss rate of less than 3 x 10^{-6} M$_\odot$/yr.

Stellar Mass Loss Rates Derived from Radio Observations

	Sp	d(kpc)	V_∞(km/s)	$\frac{dM}{dt}$ (M$_\odot$/yr)	
I. Early Type Stars					
BN Gem	08 V pec	1.61	500	13	1
γ Vel	WC8 + 09I	.45	2900	39	2
MWC 349	Be pec	2.1	100	67	3
P Cyg	B1e I	1.8	240	20	3
RY Sct	Be pec	2.0	30	18	4
ζ Pup	04 If	.45	2660	6.3	2
II. Other Stars					
α Ori	M2I	.2	10	.03	3
LkHα 101	Fe	.8	100	4.3	3
T Tau	K1e	.16	225	.37	3
V1016 Cyg	Me pec	1.4	105	6.8	3

References:
1. d: Barlow and Cohen (1977), V_∞ assumed
2. Morton and Wright (1978)
3. Schwartz and Spencer (1977)
4. d, V_∞, flux: Hughes and Woodsworth (1973)

Although the stars listed above undoubtedly represent the upper bound mass loss rates for their respective types, the general statement can be made that mass loss rates are as high as 10^{-5} M$_\odot$/yr in some OB stars (peculiar objects such as MWC 349 have rates an order of magnitude greater) and mass loss rates are as high as 10^{-6} M$_\odot$/yr in some nebular variables. The latter result almost certainly reflects the fact that radiation pressure alone does not drive all flows and emphasizes the fact that high mass loss rates are a feature of the evolution of low to intermediate mass stars as well as of very massive stars. Despite the result for α Ori, which only relates to the unionized portion of the flow around a low temperature star, mass loss rates of the order of 10^{-6} M$_\odot$/yr are characteristic of most late M giants where the gas is coupled to the very intense infrared radiation field by dust (Gehrz and Woolf, 1971). In some late type stars, particularly those with circumstellar maser emission, even higher mass loss rates than those indicated for OB stars probably exist. Thus mass loss is also an important feature of post-main-sequence evolution. As new and more powerful observational instruments in the radio and infrared become available, mass loss and interaction between stars and the interstellar medium will probably be seen to be an even more important factor in evolution at many stellar masses and ages.

REFERENCES

-Barlow,M.J. and Cohen,M.:1977, Astrophys. J. 213, pp. 737-755.
-Gehrz,R.D. and Woolf,N.J.:1971, Astrophys. J. 165, pp. 285-294.
-Hughes,V.A. and Woodsworth,A.:1973, Nature Phys. Sci. 242, pp. 116-117.
-Morton,D.C. and Wright,A.E.:1978, Mon. Not. R. Astr. Soc. 182, pp. 47P-51P.
-Panagia,N. and Felli,M.:1975, Astr. & Astrophys. 39, pp. 1-5.
-Schwartz,P.R. and Spencer,J.H.:1977, Mon. Not. R. Astr. Soc. 180, pp. 297-303.
-Wright,A.E. and Barlow,M.J.:1975, Mon. Not. R. Astr. Soc. 170, pp. 41-51.

DISCUSSION FOLLOWING SCHWARTZ

Conti: In reference to Billy Bidelman's comments earlier, we may have here another example of the nomenclature obscuring the physics; one should use both the HD name and variable star name together. I know about this star, BN Gem, discussed by Schwartz because I have spectra of it as HD 60848. The high mass loss rate may be related to its Oe type (analogous to Be classification), where rotation probably plays a role in the outflow.

Underhill: It's interesting that S Mon (=15 Mon) is not detected. It's a "well-behaved" near main sequence star.

RADIO OBSERVATIONS OF STELLAR MASS LOSS

Sun Kwok and C. R. Purton
C.R.E.S.S., York University,
Downsview, Ontario, Canada.

A program to search for steady-state thermal emissions from stars has been in progress for several years in Canada (Purton 1976). In this program we have specifically excluded flaring objects (such as β Lyr or HR1099) where non-thermal emission is probably responsible. Out of the 30 or so detections, about 1/4 of them show a ~+1 spectral index, suggesting a $1/r^2$ density profile in the emitting region. This implies the existence of a non-static circumstellar envelope which is mostly likely the result of continuous mass outflow.

In order for thermal emission from a stellar wind to be detectable, two conditions must be satisfied: a high emission measure and a source of ionization. In a stellar wind situation, the radio flux F_ν is proportional to $(\dot{M}/V)^{4/3}$ (Wright and Barlow 1975) where $\dot{M}$ is the mass loss rate and V the ejection velocity. The feasibility of studying stellar mass loss is therefore controlled by the mass loss mechanism.

Table 1
Mass Loss Mechanisms for Stars of Different Spectral Types

Mechanism	Spectral Type	$\dot{M}(M_\odot\,yr^{-1})$	$V(Kms^{-1})$
1. Radiation Pressure on Grains	Late Type Giants & Supergiants	$< L/Vc$ $\sim 10^{-6}-10^{-5}$	5-30
2. Radiation Pressure on Gas	O,B Supergiants, WR stars	$< NL/c^2$	500-1500
3. Mass Motions a. Coronal Evaporation	Lower MS stars	$\sim 10^{-14}$	~500
b. Macro-turbulence	G & K Supergiants e.g. HR8752 (G0Ia)	4×10^{-7}	~16
4. Radiation Pressure on Grains + Mass Motions	G, K, early M Supergiants e.g. HR5171 (G8Ia) μ Cep (M2Ia)	3×10^{-5} 10^{-5}	~14

P. S. Conti and C. W. H. de Loore (eds.), Mass Loss and Evolution of O-Type Stars, 151-154.

In table 1 we have outlined the major mass loss mechanisms for
stars of different spectral types. Dust grains, having a continuum
opacity to stellar radiation, can eject the atmosphere of cool luminous
stars on a large scale and at a low velocity (Gilman 1972, Kwok 1975).
In comparison, radiation pressure on ionized gas has a mass loss rate
dependent on the number of resonance lines (N) and the ejection velocity
is much higher. Mass loss processes which rely on energy input (e.g.
the solar wind) rather than momentum input generally have a much lower
mass loss rate. Macro-turbulence in the chromospheres of G and K
supergiants are capable of increasing the atmospheric scale height to
bring the gas to a point where the escape velocity is low, but this
process is not very efficient. However if dust grains are condensed,
they can greatly increase the atmospheric opacity and enhance the mass
loss rate (Gilman and Woolf, unpublished manuscript).

We note that stellar winds ejected under radiation pressure on
grains have a large $\dot{M}$ and small V and therefore should be the easiest
to detect whereas stellar winds ejected by radiation pressure on gas
have ejection velocities so high that the radio flux is expected to be
very low unless the star is nearby.

As for sources of ionization, photo-ionization is easily achieved
by the presence of a hot star. In some cases, a strong shock wave may
be responsible.

By combining the respective mass-loss and ionization mechanisms
we have attempted to classify the detected mass outflow radio stars
into 5 types.

Class I	OB Supergiants, Wolf-Rayet stars: radiation pressure on ionized gas. Examples: P Cygni, γ^2 Vel, ζ Pup.
Class II	Binary (M giant + hot star): stellar wind from M giant ionized by hot companion. Example: AG Peg (M3III + WN6)
Class III	Mass loss and ionization by supersonic mass motions. Examples: HR8752 (GO Ia), the sun.
Class IV	Young Stars: re-expansion of co-coon, dust and gas of the proto-stellar nebula ejected by radiation from the new born star. Examples: LHα 101, MWC 349 (?)
Class V	Very young Planetary Nebulae: Ionization of the remnant red-giant envelope through exposure of hot core. Examples: V1016 Cygni, HM Sge.

Although the number of stars belonging to Class I is large, yet the expected radio flux is low for reasons given above. Class II represents a subset of symbiotic stars which have been extensively searched and resulted in a number of detections (Wright and Allen 1978). Although late-type supergiants have large $\dot{M}/V$ but normally the outflowing material is neutral, therefore explaining the low search success rate (Smolinski et al. 1977). Post-shock heating due to large scale mechanical motion in the outer-atmosphere may ionize part of the stellar wind and lead to radio emission. The large turbulent velocity (38 km s^{-1}) observed in the only detected cool supergiant (HR8752) is consistent with this suggestion. HR5171 having a turbulent velocity of 50-100 km s^{-1} may also be a good candidate for future observations.

The most interesting category resulting from our study of radio stars is Class V. Continuous mass loss in the red-giant stage is expected to eventually deplete the hydrogen atmosphere and expose the hot core. The change in stellar surface temperature leads to a change in mass loss mechanism from radiation pressure on grains to radiation pressure on gas, resulting in a large increase in ejection velocity. The collision of the new high velocity wind with the remnant red giant wind will create a high density shell which in time will develop into a planetary nebula (Kwok, Purton and FitzGerald 1978, Kwok and Purton 1978). In very young planetary nebulae where the old wind has not yet been totally swept up by the shell, thermal emission from the old wind will be produced as a result of its ionization by the exposed hot core.

There is a group of radio stars (H1-36, VY2-2, Hb12, Hen1044, He2-90) whose nature we are uncertain of. Their spectra are similar to V1016 Cygni and HM Sge, having a +1 spectral index at low frequency and becoming optically thin between 10-100 GHz. This implies that the base of the outflow has a radius $10^{14} - 10^{15}$ cm, much larger than the size of a star. The existence of such a large bubble is difficult to understand although it is consistent with the properties of very young planetary nebulae where the base of the remnant red-giant wind is detached from the star.

In conclusion the nature of radio stars undergoing mass loss is very diverse and they do not form an homogeneous group. Mass loss from OB supergiants, having a high ejection velocity, only emit weakly in the cm region and are not particularly suitable for study in the radio.

References

Gilman, R. C.: 1972. *Ap. J.* 178, 423.
Kwok, S.: 1975, *Ap. J.* 198, 583.
Kwok, S., Purton, C. R.: 1978, submitted to *Ap. J.*
Kwok, S., Purton, C. R., FitzGerald, P. M.: 1978, *Ap. J. (Let.)* 219, L125.
Purton, C. R.: 1976, in IAU Symp. 70, ed. A. Slettebak.
Smolinski, J., Feldman, P. A., Higgs, L. A.: 1977, *Astron. Astrophys.* 60, 277.
Wright, A. E., Allen, D. A.: 1978, *M.N.R.A.S.* in press.
Wright, A. E., Barlow, M. J.: 1975, *M.N.R.A.S.* 170, 41.

DISCUSSION FOLLOWING KWOK AND PURTON

Noerdlinger: Dust is observed in planetary nebulae, often fairly
deep inside. There are theories for formation of dust grains in such
hot winds, but I wonder if an instability could develop at the inter-
face of the two winds leaving inclusions behind so the existing dust
grains could act as nuclei for new ones?

Kwok: In our examples of very young planetary nebulae V 1016
Cygni and HM Sge, both show two dust components. One is the silicate
component which is probably the remnant of the progenitor red giant.
Another is a featureless 1000 K blackbody similar to dust grains observed in say Nova Vul. We interpret this hot dust component (possibly
graphite) to be recently formed after the exposure of the hot core. In
several hundred years, the silicate feature will disappear and the new
grains will dominate the IR spectrum as we observe in planetary nebulae.
For instability we are currently investigating this possibility for it
may explain the presence of condensations in planetary nebulae. Dust
grains are seen to be forming near the nuclei of planetary nebulae A 30
and A 78 (Cohen and Barlow, Ap. J. 193, 401). If this is true, then it
is unlikely that old grains serve as nuclei for new grains. Also dust
grains seem to be able to condense around novae with no problem.

RADIO EMISSION FROM HOT STARS AT TWO CENTIMETERS

Donald C. Morton and Alan E. Wright
Anglo-Australian Observatory

Radio fluxes of 7.2 ± 1.1 and 69 ± 5 mJy at 14.7 GHz have been detected from the stars ζ Pup (O4If) and γ Vel (WC8+O9I) respectively with the 64-m telescope at Parkes, and upper limits have been determined for 9 more hot stars. The interpretation of these fluxes as free-free emission from a shell of ionized gas resulting from a stellar wind gives mass-loss rates $\dot{M}$ = (6.5 ± 1.4) x 10^{-6} $M_\odot$ yr^{-1} for ζ Pup and (3.9 ± 0.7) x 10^{-5} $M_\odot$ yr^{-1} for γ Vel if H/He = 10 by number and the He is fully ionized. If the gas around γ Vel originates mainly from the WC8 star, the helium predominates and $\dot{M}$ = (5.2 ± 0.9) x 10^{-5} or (17 ± 3) x 10^{-5} $M_\odot$ yr^{-1} depending on whether the helium is doubly or singly ionized.

1. INTRODUCTION

The mass ejected by a hot star will form an expanding shell of ionized gas emitting bremsstrahlung which may be observable at radio and infrared wavelengths. Thus, the radio fluxes detected from P Cyg by Wendker, Baars and Altenhoff (1973) and from the Wolf-Rayet binary γ Vel by Seaquist (1976) probably originate from stellar winds. Similarly, the infrared excesses that Barlow and Cohen (1977) found at 10 μm in several hot supergiants can be explained by the same model.

In a previous paper (Morton and Wright 1978, Paper I) we described the first measurement of radio emission from ζ Pup and additional radio data on γ Vel. In this paper we report further observations of these stars, upper limits for 9 more hot stars, and a re-analysis of γ Vel for the case of a helium-rich atmosphere. The observations were made with the Parkes 64-m radio telescope. A high frequency is very useful for such a programme because the radio flux density S is approximately proportional to $\nu^{0.6}$ in the mass-loss model, while the background confusion is less than at lower frequencies due to the narrower telescope beam and the spectra of the contributing sources.

In several ways the radio flux provides valuable information about stellar mass ejection. First, the radio emission, which arises from all

P. S. Conti and C. W. H. de Loore (eds.), Mass Loss and Evolution of O-Type Stars, 155-163.
Copyright © 1979 by the IAU.

Table 1.   Radio Flux Densities at 14.7 GHz.

Star	b	Date	No. of runs	Flux S_ν (mJy) Observed	Adopted
ζ Pup	$-4° \; 42'$	1977 Jun	7	6.7 ± 1.5	
		1977 Nov	9	7.5 ± 1.2 }	7.2 ± 1.1
γ Vel	$-7° \; 41'$	1977 Jun	4	67 ± 5	
		1977 Jun	scan	70 ± 10	
		1977 Nov	3	77 ± 7	69 ± 5
		1978 May	3	67 ± 7	
		1978 May	scan	65 ± 8	
θ Mus	$-2° \; 29'$	1977 Jun	1	-0.9 ± 3.0	<7
HD 151804	$+1° \; 57'$	1977 Jun	1	-0.4 ± 3.4	
		1978 May	1	0.8 ± 1.7 }	<5
HD 152408	$+1° \; 30'$	1977 Jun	3	2 ± 14	
		1977 Nov	1	-3 ± 3	<5
		1978 May	1	2.2 ± 1.7	
HD 112244	$+6° \; 02'$	1977 Jun	1	0.4 ± 2.6	<7
δ Ori	$-17° \; 45'$	1977 Jun	1	-2.5 ± 3.5	<7
ε Ori	$-17° \; 15'$	1977 Jun	1	-1.0 ± 3.2	<7
ρ Leo	$52° \; 46'$	1977 Jun	1	-0.5 ± 6.0	<12
$ζ^1$ Sco	$+0° \; 53'$	1977 Jun	3	14 ± 5	
		1977 Nov	1	10 ± 4	<10
		1978 May	1	4 ± 6	
		1978 May	scan	10	
η Cen	$+16° \; 41'$	1977 Jun	1	0.8 ± 3.4	<7

ionized species including hydrogen and helium, can be a useful check on
the mass-loss rate obtained from ultraviolet or visual line profiles,
where the results may be highly sensitive to estimates of the ionization
equilibrium.   Secondly, it may be possible to combine the data with
information at other wavelengths to obtain constraints on the temperature and velocity law in the outer parts of the stars.   Thirdly, since
the radio emission originates over a larger radial extent, time variations are not as important as at infrared and shorter wavelengths.
Also, as discussed by Wright and Allen (1978), when an ejection event
does occur, it is expected to appear first at optical frequencies and
later in the infrared and radio regimes as the material moves outwards.

2.   OBSERVATIONS

Three separate observing runs were made with a new 14.7 GHz cryo-
genic, parametric receiver on the Parkes 64-m radio telescope (see Paper
I). At this frequency, only the 37-m centre section of the dish is
illuminated. The receiver had two beams of 2.3 arcmin half power width,
one on axis and the other 6.2 arcmin off axis. Usually the telescope
was operated in the wagging mode (Wright 1974) in which the feed was
rotated to align the two beams to the same zenith angle, and then the
telescope was moved so that alternate beams were pointed at the source,
with typical cycle times of 30 sec. Table 1 lists the sources, galactic
latitude b, dates, and flux densities. In some cases a star was chosen
more for zenith angle than being a candidate for strong radio emission.
The periods of observation were 1977 Jun 20-24, 1977 Nov 22-29, and 1978
May 5-8 when the receiver was tuned to 14.7, 14.45 and 14.4 GHz respectively. For simplicity we have adopted 14.7 GHz for all the analysis. Sources were observed one or more times with runs of wags ranging
from 8 to 40, depending on the stability of the signal. Cirrus cloud in
the far-field of the telescope can add considerable noise at 2 cm. The
number of such separate runs is noted in column 4 of Table 1. Scans in
α and δ also were made on γ Vel and ζ^1 Sco with scan lengths of about
± 10 arcmin. Finally, the tables list our best judgement of the flux
density and rms error, or a conservative upper limit.

ζ Pup: Resonable signals above the noise were obtained from ζ Pup on
two separate occasions. Observations at positions 15 and 2 arcmin off
the source gave signals comparable with the noise, showing that the
region is not seriously confused by other sources. We consider that the
detection of ζ Pup is highly probable and adopt S_ν = 7.2 ± 1.1 mJy at
14.7 GHz. This is the weakest source yet measured by the Parkes telescope at any frequency.

γ Vel: A strong signal was observed from γ Vel, as expected from the
measurements by Seaquist (1976) at lower frequencies. Scans across the
source gave a peak in good positional agreement with the Wolf-Rayet
binary. The three separate observing sessions give consistent results
so that we adopt S_ν = 69 ± 5 mJy. The suggestion of variation in
Paper I was due to one measurement in May 1977 which we now know to be
incorrect.

ρ Leo: The run suffered from thin, cirro-stratus cloud near sunset,
making the limit abnormally high.

ζ^1 Sco: On the first two occasions, ζ^1 Sco seemed close to detection,
but the results were not confirmed in May 1978. Observations a few
arcmin off the source imply the presence of other emission of similar
strength within 3 arcmin. We suspect that ζ^1 Sco is indeed a source of
about 9 mJy, but the detection above the background is not clear, unlike ζ
Pup which is farther from the galactic equator. Thus we adopt the limit
S_ν < 10 mJy.

3. ANALYSIS

Wright and Barlow (1975) and Panagia and Felli (1975) have derived simple relations between the rate of mass loss $\dot{M}$ and radio flux density S_ν at frequency ν when the flow is uniform and spherically symmetric with a velocity V_∞ independent of radius and time. The majority of the radio emission should originate far from the star so that the velocity can be approximated by the terminal value. The Wright-Barlow formulation gives

$$\dot{M} = \frac{0.095\ \mu V_\infty\ S_\nu^{3/4}\ D^{3/2}}{Z\ \gamma^{\frac{1}{2}}\ g^{\frac{1}{2}}\ \nu^{\frac{1}{2}}}\ M_\odot\ yr^{-1}$$

with V_∞ in km s^{-1}, S_ν in Jy, ν in Hz and D the distance in kpc. The remaining parameters depend on the composition, kinetic temperature T, and ionization equilibrium in the radiating gas. Specifically, μ is the mean atomic weight per nucleon, γ is the number of electrons per ion, and Z is the rms average charge of the ions. In each case it is necessary to sum over all elements weighing each by its abundance, though in practice only hydrogen and helium are important unless both are depleted relative to the heavies. Finally, g is the Gaunt factor for which we use Spitzer's (1962) approximation for radio waves

$$g = \frac{3^{\frac{1}{2}}}{\pi}\left[\ln \frac{(2kT)^{3/2}}{\pi Ze^2\nu m_e^{\frac{1}{2}}} - 1.443\right]$$

There is considerable debate whether the temperature in the wind from a hot star is close to the effective temperature or near 2×10^5 K. This high value was found in ζ Pup by Lamers and Morton (1976) and in τ Sco (B0V) by Lamers and Rogerson (1978), due to the presence of N V and particularly O VI P-Cygni profiles. Both N V and O VI also occur in ζ Pup, γ Vel, δ Ori, ε Ori and HD 152408 and at least N V is present in ρ Leo, HD 112244, θ Mus and HD 151804 according to the observations of Morton (1976, 1978), Snow and Morton (1976), Hutchings (1976b) and Johnson (1978). Both ions seem to be absent in η Cen and the status of ζ^1 Sco is not yet known. Thus we have adopted T = 2.5×10^4 for η Cen, and 2×10^5 for all others. In the case of ζ Pup, if we had taken T = 5×10^4 there would be only a 20% reduction in g, but γ and Z would change by larger amounts due to the helium not being fully ionized. A range of possibilities for the parameters μ, γ, Z and g is listed in Table 2. In normal stars, H/He $\approx$ 10 by number, but H/He = 1 or 0.1 is probably more appropriate for a Wolf-Rayet star (Smith 1973). The small effect of the cosmic abundances of the heavier elements has been included. Either decreasing T from 2×10^5 to 2.5×10^4 K or making helium the dominant element increases $\dot{M}$ by at most 40%, but changing both so that the main source of electrons is singly ionized helium can increase $\dot{M}$ by more than a factor 4.

Table 2. Effects of Changing the Adopted Temperature and Helium Abundance.

T	2×10^5	2×10^5	5×10^4	2.5×10^4	2.5×10^4
H/He	10	0.1	10	10	0.1
γ(He)	2	2	2	1	1
μ	1.3	4.1	1.3	1.3	4.1
γ	1.1	2.0	1.1	1.0	1.0
Z	1.13	2.0	1.13	1.0	1.02
$g(\nu = 14.7 \times 10^9)$	6.9	6.6	5.7	5.2	5.2
$\dfrac{\mu}{Z\,\gamma^{\frac{1}{2}}\,g^{\frac{1}{2}}}$	0.42	0.56	0.46	0.57	1.76
ratio	1	1.3	1.1	1.4	4.2

Table 3 lists the adopted parameters for each star and the derived mass-loss rate or its upper limit. In most cases the spectral types, distances and terminal velocities are the same as used by Snow and Morton (1976). For γ Vel we adopted the spectrum derived by Conti and Smith (1972), the distance of ζ Pup, and the terminal velocity from the C IV line plotted by Johnson (1978). The spectrum and distance of θ Mus are from Moffat and Seggewiss (1977), and the terminal velocity from N V as quoted by Hutchings (1976b). Since ζ^1 Sco appears to be associated with the cluster containing HD 151804 and 152408, the same distance was used for all three stars. The spectral type of ζ^1 Sco is from Lesh (1972) and the terminal velocity from the C IV P-Cygni profile observed with the International Ultraviolet Explorer. We have revised the velocity for HD 151804 slightly to include new data on O VI and we have adopted values for δ and ε Ori from Abbott (1978) who ignored the N V extreme short wavelength wing which may be due to a blend.

The mass-loss rates are listed in the last row of Table 3. For the two Wolf-Rayet systems it was assumed that the WR star is the prime contributor to the wind and hence the helium is the main constituent. Since the nitrogen is ionized 4 times in the wind, the helium should be fully ionized. However, we cannot rule out the possibility of helium recombination farther out where the radio emission originates or Auger ionization of the heavier elements by a corona at $7 \approx 10^6$ K leaving the helium singly ionized. If so, a considerable correction must be applied as shown by Table 2. The quoted error on $\dot{M}$ does not include this possibility, so that the distance is the major source of uncertainty.

Table 3. Mass-Loss Rates Based on Observations at ν = 14.7 GHz.

Star	ζ Pup	γ Vel	θ Mus		
HD	66811	68273	113904	151804	152408
Spectrum	04If	WC8 +O9I	WC6 +O9.5I	O8Iaf	O8:Iafpe
D (kpc)	0.45	0.45	2.0	2.2	2.2
S_ν (mJy)	7.2±1.1	69±5	< 7	< 5	< 5
V_∞ (km s^{-1})	2660 ±150	2900 ±150	1650	1700	1430
T	2×10^5	2×10^5	2×10^5	2×10^5	2×10^5
H/He	10	0.1	0.1	10	10
γ(He)	2	2	2	2	2
$\dfrac{\mu}{Z\,\gamma^{\frac{1}{2}}\,g^{\frac{1}{2}}}$	0.42	0.56	0.56	0.42	0.42
$\dot{M}(10^{-6}$ M$_\odot$ yr$^{-1})$	6.5 ±1.4	52 ±9	<50	<34	<29

Star		δ Ori	ε Ori	ρ Leo	ζ^1 Sco	η Cen
HD	112244	36486	37128	91316	152236	127972
Spectrum	O8.5Iab	O9.5II	B0Ia	B1Iab	B1.5Ia+p	B1.5V
D (kpc)	1.5	0.46	0.46	1.0	2.2	0.10
S_ν (mJy)	<7	<7	<7	<12	<10	<7
V_∞ (km s^{-1})	1900	2410	2010	1580	950	810
T	2×10^5	2×10^5	2×10^5	2×10^5	2×10^5	2.5×10^4
H/He	10	10	10	10	10	10
γ(He)	2	2	2	2	2	1
$\dfrac{\mu}{Z\,\gamma^{\frac{1}{2}}\,g^{\frac{1}{2}}}$	0.42	0.42	0.42	0.42	0.42	0.57
$\dot{M}(10^{-6}$ M$_\odot$ yr$^{-1})$	<28	<6.0	<5.0	<19	<32	<0.28

4. DISCUSSION

The rate of mass loss for ζ Pup is in excellent agreement with $\dot{M} = (7 \pm 3) \times 10^{-6}$ M$_{\odot}$ yr^{-1} which Lamers and Morton (1976) derived from the far-ultraviolet line profiles and 8×10^{-6} M$_{\odot}$ yr^{-1} which Cassinelli, Olson and Stalio (1978) obtained from the Hα profile even though the UV analysis led to $T = 2 \times 10^{5}$ K and the Hα gave best agreement at $T \approx 3 \times 10^{4}$ K. To confirm the radio result it would be good to measure ζ Pup at another frequency to check that $S_{\nu} \propto \nu^{0.6}$.

In the case of the Wolf-Rayet binary γ Vel, our best estimate of 5.2×10^{-5} M$_{\odot}$ yr^{-1} is significantly below 9×10^{-5} M$_{\odot}$ yr^{-1} that Willis and Wilson (1978) derived from the C III] $\lambda 1909$ absorption line, or 2.7×10^{-4} that Barlow and Cohen (1978) obtained from the IR excess. If we had adopted their assumption that helium is singly ionized, our $\dot{M}$ would increase to 1.7×10^{-4} M$_{\odot}$ yr^{-1}. Presumably, the rest of the discrepancy is due to the acceleration that is occurring in the region of infrared emission, as discussed by Barlow and Cohen (1978).

For the extremely luminous supergiants 151804, 152408 and ζ^{1} Sco, Hutchings (1968) estimated mass-loss rates $\dot{M}$ of 1×10^{-5}, 1×10^{-4} and 1×10^{-5} M$_{\odot}$ yr^{-1} respectively from the visible P-Cygni lines. Our upper limits are consistent for 151804 and ζ^{1} Sco, but show that Hutchings' value is too high for 152408. Barlow and Cohen (1977) obtained 1.4×10^{-5} M$_{\odot}$ yr^{-1} from the IR excess in ζ^{1} Sco, but they assumed $V_{\infty} = 500$ km s^{-1}. Increasing the terminal velocity to 950 km s^{-1} determined from the C IV resonance absorption gives $\dot{M} = 2.7 \times 10^{-5}$ consistent with our suspicion that $S_{\nu} \approx 9$ mJy.

In δ Ori and ϵ Ori our upper limits on $\dot{M}$ are consistent with 1.0×10^{-6} and 1.7×10^{-6} M$_{\odot}$ yr^{-1} respectively, estimated by Morton (1967) from the UV P-Cygni absorptions and 0.9 and 1.7×10^{-6} found by Barlow and Cohen (1977) using the IR excess. Our limit on ρ Leo is well above 8.6×10^{-7} M$_{\odot}$ yr^{-1} that these authors obtained and our limit for η Cen is consistent with 7×10^{-9} M$_{\odot}$ yr^{-1} that Lamers and Rogerson (1978) derived from the UV P-Cygni lines in τ Sco (B0V).

For the remaining stars HD 112244 and θ Mus our upper limits agree with the rough estimates of 5×10^{-6} and 8×10^{-6} M$_{\odot}$ yr^{-1} respectively by Hutchings (1976a, b), who compared the mass-loss features in their visible and UV spectra with stars for which $\dot{M}$ had been determined. His standards included HD 152408, which we believe has been overestimated, and P Cyg itself, which now seems to be highly abnormal.

Our observations imply that for most OB stars, including the extreme supergiants 151804, 152408 and ζ^{1} Sco, $\dot{M} < 5 \times 10^{-5}$, and that WC stars have the same limit (or possibly a factor 3 larger if the helium is mainly singly ionized). Therefore, if the radio flux is to be detectable easily with an antennae the size of Parkes, the star will have to be closer than $\sim$ 500 pc. Thus, stars worth observing include ξ Per

(07.5III), λ Ori (08III), ι Ori (09III), ζ Oph (09.5V), ζ Ori (09.7Ib) and ζ Per (BlIb). Otherwise it will be necessary to use interferometric techniques to distinguish the weak point sources from the background.

We wish to thank the CSIRO Division of Radiophysics for observing time at Parkes, and Dr. Michael Barlow for valuable discussions and permitting us to include data he had obtained on HD 151804 and 152408.

REFERENCES

Abbott, D.C.: 1978 (preprint).
Barlow, M.J. and Cohen, M.: 1977, Astrophys. J. 213, 737.
Barlow, M.J. and Cohen, M.: 1978 (in preparation).
Cassinelli, J.P., Olson, G.L. and Stalio, R.: 1978, Astrophys. J. 220, 573.
Conti, P.S. and Smith, L.F.: 1972, Astrophys. J. 172, 623.
Hutchings, J.B.: 1968. Monthly Notices Roy. Astron. Soc. 141, 329.
Hutchings, J.B.: 1976a, Astrophys. J. 203, 438.
Hutchings, J.B.: 1976b, Astrophys. J. Letters. 204, L99.
Johnson, H.M.: 1978, Astrophys. J. Suppl. 36, 217.
Lamers, H.J.G.L.M. and Morton, D.C.: 1976, Astrophys. J. Suppl. 32, 715.
Lamers, H.J.G.L.M. and Rogerson, J.B.: 1978, Astron. Astrophys. 66, 417.
Lesh, J.R.: 1972, Astron. Astrophys. Suppl. 5, 129.
Moffat, A.F.J. and Seggewiss, W.: 1977, Astron. Astrophys. 54, 607.
Morton, D.C.: 1967, Astrophys. J. 150, 535.
Morton, D.C.: 1976, Astrophys. J. 203, 386.
Morton, D.C.: 1978 (paper presented at this Symposium).
Morton, D.C. and Wright, A.E.: 1978, Monthly Notices Roy. Astron. Soc. 182, 47 (Paper I).
Panagia, N. and Felli, M.: 1975. Astron. Astrophys. 39, 1.
Seaquist, E.R.: 1976, Astrophys. J. Letters. 203, L35.
Smith, L.F.: 1973, IAU Symposium No.49, Wolf-Rayet and High-Temperature Stars, ed. M.K.V. Bappu and J. Sahade.
Snow, T.P. and Morton, D.C.: 1976, Astrophys. J. Suppl. 32, 429.
Spitzer, L.: 1962, Physics of Fully Ionized Gases, 2nd Edition, Interscience Publ., New York.
Wendker, H.J., Baars, J.W.M. and Altenhoff, W.J.: 1973, Nature, Physical Science. 245, 118.
Willis, A.J. and Wilson, R.: 1978 (preprint).
Wright, A.E.: 1974, Monthly Notices Roy. Astron. Soc. 167, 251.
Wright, A.E. and Allen, D.A.: 1978 (preprint).
Wright, A.E. and Barlow, M.J.: 1975, Monthly Notices Roy. Astron. Soc. 170, 41.

DISCUSSION FOLLOWING MORTON AND WRIGHT

Cassinelli: What temperature did you assume in your derivation of $\dot{M}$ and what would be the effect of changing T from 2×10^5 to 3×10^4 or vice versa?

Morton: Since O VI is present, we used 2×10^{5}°K. Decreasing to 3×10^{4}°K has little effect on $\dot{M}$, provided the helium remains fully ionized.

Underhill: How do you know that the radio emission comes from the gas that produces the O VI?

Morton: The radio emission probably originates further out than the O VI, but I doubt that any recombination has occurred.

Niemela: I'd like to ask whether γ Vel B may contribute to the radio emission from γ Vel A? γ Vel B is also a binary star which Dr. Sahade is working on. There are radio observations from binaries ... .

Barlow: The known radio binaries usually don't have a $\gamma^{2/3}$ spectrum. Their radio spectra look nonthermal and so can be distinguished that way.

Hutchings: The revised $\dot{M}$ rates I gave yesterday, which agree with Barlow and Cohen's values, are ~2 times lower in the range occupied by ζ^1Sco, 152408, 151804 and they also fit in well with your upper limits.

Kwok: What velocities did you use for γ Vel?

Morton: I used 2900 km s^{-1} from the C IV line plotted by H. Johnson (Ap. J. 1978). The other values were from the UV mass loss survey (Snow and Morton 1977).

GENERAL DISCUSSION

Underhill: We have heard that a few O-type stars have been detected as radio sources, but many more O stars are known to have infrared excesses. We have also heard that a considerable number of late-type stars and some B stars with extended atmospheres are detected in radio wavelengths. I wonder if the relative intensities of infrared and radio emissions are similar for all stars or whether there is a systematic difference that can be related to spectral type. Information on this point should yield information about the structure, temperature and density of the winds around various types of stars. Such data would be a constraint on the theories of stellar einds, a topic which we shall discuss this afternoon. Can anyone comment on what infrared and radio observations of stars tell us about the physics of stellar winds in various parts of the HR diagram and in particular about the winds from O stars?

Kwok: When stellar radio emission is detected a hot star is usually present. The question remains whether the gas is actually ejected from the star. Mass loss from late type giants have velocities of ~10 km/sec compared to ejection velocities of ~1000 km/sec for early giants. Since the radio flux density is proportional to $(m/v)^{4/3}$, there exists a two-order of magnitude difference in favor of detecting a red giant wind. This accounts for the small number of detections of O star winds. As for winds from late-type stars, we have to consider the possibility of the free-free spectrum becoming optically thin before reaching the infrared and the presence of dust. Therefore I do not consider the comparison of the radio/IR intensity ratio between early and late type stars to be particularly useful.

Cassinelli: Free-free opacity varies as λ^2 so as one goes from the visible to infrared wavelengths one is "seeing" to points farther out in the flow. Since the effective radius increases, you get an increasingly large infrared excess.

Conti: I should like to emphasize two points of these data on mass loss rates presented in this session and in the previous two. There is good agreement in the mass loss rate of ζ Pup among the U.V., optical, IR and radio data of $\sim 6 \times 10^{-6}$ yr^{-1}. This rate is significant for the evolution of this star and emphasizes the applicability of the consequences of mass loss to be discussed in sessions later on. As we have seen earlier in this Symposium the mass loss rates for WR stars are even larger, typically factors of 10. This is also significant on an evolutionary time scale for these stars.

Underhill: Yes, it's clear that the mass loss rates in WR stars are much larger than in O types. Another point is that Wolf-Rayet stars don't vary in spectral appearance all that much. The initial descriptions by Wolf and Rayet in the mid 1800's are nearly identical with the present spectra. By contrast, Be stars have large spectral variations. The evolutionary state of these two objects is also very different.

P. S. Conti and C. W. H. de Loore (eds.), Mass Loss and Evolution of O-Type Stars, 165-166.
Copyright © 1979 by the IAU.

de Loore: I would like to make a remark concerning evolution in connection with what Anne was saying. I think there are two different things. Before nuclear changes in the interior can really affect the atmosphere it takes a large amount of time.

I have a question concerning the WC star 193793 mentioned by Barlow, quoted as WC binary (+O5 star) and a remark that it could possibly be a single star. How strong is the evidence for the fact that it could be single? Peter Conti probably can answer this.

Conti: The evidence is as follows. There are very nice absorption lines (e.g., Balmer lines) observed in the spectrum of HD 193793; Ms. McDonald at Victoria many years ago obtained extensive spectroscopic coverage. The radial velocities have a relatively large range but no period could be found. About 10 years ago I too began to obtain coude plates of this object. I thought all one needed was better resolution. Well, the upshot of a large accumulation of data from that time is that one still finds a variation but no periodicity. It does seem to be two stars since you see O type absorption lines and WC emission lines. You could have the situation where two stars are in the line of sight or else it's an unfavorable inclination. These are the most likely possibilities but cannot be proven for certain. A less likely possibility is that there is just one star. We have some examples of WN stars in which absorption and emission lines are observed in the same object. Dr. Niemela will talk about this in Session 5. It's possible in a WN star to have hydrogen absorption lines since nitrogen enhancement can come about before all the hydrogen is used up. However, this does not seem to be possible in a WC star as Eggleton has emphasized to me (e.g., Dearborn and Eggleton 1977). If WC stars have enhanced carbon, all the hydrogen is probably consumed, and helium also dominates the surface. From the absorption line spectrum of HD 193793 we still have evidence of hydrogen. So my feeling is that it's probably two stars. Hα is in emission, by the way.

Castor: Is there any Pickering absorption?

Conti: Yes.

Underhill: Yes, I have worked on it myself.

Conti: This star does hold the record for outflow velocity in a WR star from Anne's work on the narrow λ 3889 He I. Also, the emission lines are not as strong as in some other WC stars so one infers the O-type companion contributes to the continuum and somewhat drowns out the WC star. But this has not been investigated quantitatively as yet.

When the argument got very heated
And things, which should not be repeated,
Were shouted out loud
Came a voice from the crowd
"Its all bull (expletive deleted)".

SESSION 4

PANEL DISCUSSION ON STELLAR WIND THEORIES

Chairman and Introductory Speaker: A.G. HEARN

Panel Members:

1. J. CASTOR: Radiatively driven stellar winds: Model improvements, ionization balance and the infrared
   spectrum.

2. H. LAMERS: The warm wind model.

3. J. CASSINELLI: The coronal plus cool wind model for Of
   stars and OB supergiants.

4. R. THOMAS: The thermodynamic requirements on atmospheric
   models imposed by observed stellar nonthermal mass-
   fluxes and by those observed nonthermal features
   enhanced in Xe stars.

Editors' note: The Scientific Organizing Committee agreed that the following questions were to be presented to the panelists in advance. The panelists were to use these as guidelines in their presentation at the Symposium and for their written contributions. The precise formulation of the questions was made by Dr. A. Hearn.

1) How does the temperature structure and stellar wind velocity in the theory vary with height? A geometrical distance in units of stellar radii or kilometers would seem to be the best representation.

2) How does the theory account for the observations of:
   a) the U.V. region and in particular the simultaneous presence of high and low ionization stages;
   b) the hydrogen and helium lines;
   c) the observed upper limits of X-ray emission;
   d) the I.R. emission?

3) How does the theory explain the observed differences in ionization and mass loss rate between a supergiant such as ζ Orionis and a main sequence star such as τ Scorpii?

4) How does the theory account for the variability of the wind with time?

5) What specific observations are needed to prove or disprove the theory?

6) What does the theory predict for the variation of mass loss rates with stellar characteristics?

THE THEORY OF MASS LOSS FROM HOT STARS

A.G. Hearn
Sterrekundig Instituut, Utrecht, The Netherlands

Although some evidence had been available earlier, it was the work of Morton (1967) that proved that the OB Supergiants are losing mass. He obtained spectra with a rocket borne spectrograph of the three stars in Orion's belt, δ, ϵ and ζ Orionis. These stars are all supergiants of spectral type 09.5 or BO. The spectra showed resonance lines of ions such as C III, C IV, Si IV, N V with P-Cygni type profiles. The deepest part of the absorption profile corresponded to a velocity away from the star of 1400 km s^{-1}, and the absorption profile extended out to wavelengths corresponding to velocities of 2000 km s^{-1} away from the star. Since these velocities are much greater than the escape velocities from the surface of these stars, it is clear that matter is escaping from them. With a number of simplifying assumptions Morton deduced that the rate of mass loss is about 10^{-6} M$_{\odot}$ yr^{-1}. This order of magnitude has remained unchanged by the many high quality observations made since that time.

On the basis of Morton's observations, Lucy and Solomon (1970) suggested that the mass loss could be explained by the radiative forces associated with the observed resonance lines. Radiation carries momentum, and if radiation coming from one direction is absorbed by a resonance line and then re-emitted isotropically, the result is that the absorbing matter experiences a force. Just one of the typical resonance lines absorbing the radiation from the photosphere of an OB supergiant contributes, if it is unsaturated, an outward force which is 300 to 1000 times the inward force due to gravity. Therefore it can explain the mass loss.

A radiation pressure explanation of mass loss from hot stars was attractive for several reasons. At the time there was no observational evidence of coronae round hot stars and further model atmosphere calculations show quite insignificant convection zones which are incapable of heating a corona in the way that they do for the Sun. If the high terminal velocity of the mass loss is explained by a simple Parker type stellar wind then a coronal temperature of at least 10^{8} K is required and that is quite inconsistent with the observations of ions

P. S. Conti and C. W. H. de Loore (eds.), Mass Loss and Evolution of O-Type Stars, 169–173.
Copyright © 1979 by the IAU.

such as C III, C IV, etc.

An objection to this simple explanation was raised by Marlborough and Roy (1970) and later by other authors. They studied the effect of an outward radiative force on the equations which give the Parker solar wind solution with an acceleration of the atmosphere from subsonic velocities near the star to supersonic velocities at greater distances, and they found that if the outward radiative force is greater than the inward force due to gravity the subsonic-supersonic solution disappears. The result of a large radiative force is not an acceleration of the atmosphere, but a compression of the flow resulting in a deceleration. This result seems to be an excellent example of the sheer perversity of nature: the harder you push it, the slower it goes. But the conclusion is clear, it is not possible to accelerate a steady flow to supersonic velocities by large radiative forces. However Cassinelli and Castor (1973) showed that when the flow is supersonic, a large radiative force will cause a further rapid acceleration of the flow.

It seems to be universally agreed that the radiative forces associated with the resonance lines are responsible for accelerating the flow from supersonic velocities to the very large observed final velocities. What is being argued about, sometimes rather hotly, is what causes the mass loss, or at least what accelerates the flow to supersonic velocities so that the radiative forces may finish the work.

In the session which follows, the four main theories of mass loss from hot stars will be discussed. The physical basis of these theories is really rather different, and the conclusions that they draw about the physical conditions of the expanding envelope is very different indeed.

Since the original work of Morton, many high quality observations have been made with the Copernicus satellite, and now the I.U.E. satellite is producing more. One of the most important observations with Copernicus was of O VI in τ Scorpii by Rogerson and Lamers (1975). τ Scorpii is a BO V star. Later O VI was observed in many OB stars. O VI cannot be explained by pure radiative equilibrium models for these stars and its observation is the first clear evidence that the outer layers of hot star atmospheres have some sort of mechanical heating. Three of the four theories for mass loss explain the formation of O VI by different physical processes and this is the main cause of the very different temperatures deduced by the theories from the ultraviolet observations.

The first theory is a development of the radiative equilibrium, radiation pressure driven theory of mass loss of Lucy and Solomon (1970).

Castor, Abbott and Klein (1975) found a way round the objection of Marlborough and Roy (1970).

They pointed out that these resonance lines are also found in the photosphere and they are saturated which substantially reduces the radiative force associated with them. As the atmosphere expands and accelerates the lines desaturate and the radiative force becomes greater. They produced a model wherein the velocity gradients in the subsonic part of the flow are so small that the lines are sufficiently saturated to reduce the outward radiative force to less than the inward force due to gravity. The objection of Marlborough and Roy is thereby overcome and the flow can accelerate to supersonic velocities. Once supersonic, the velocity gradient becomes large giving a strong acceleration from the desaturated resonance lines. They used the Sobolev approximation to relate the radiative force to the velocity gradient. Their model has a rapid acceleration, so that the critical point, the point where the flow attains supersonic velocities, is right at the surface of the star.

The observation of O VI has caused some difficulty for this theory. It shows that the assumption of radiative equilibrium is not valid. More recent work by Castor, which will be discussed in the following session, has shown that the O VI observations can be explained if the atmosphere is mechanically heated up to 60 000 K and the expanding atmosphere is optically thick in the continuum.

The second theory is that the mass loss is a hot coronal stellar wind of the Parker type, which was proposed by Hearn (1975). To explain the observed mass loss rates, the critical point of the Parker solution must be near the surface of the star, and this means coronal temperatures from 3 to 9 million K. The corona is small in extent, and beyond the corona the expanding atmosphere is in radiative equilibrium. It is in this radiative equilibrium region that the observed resonance lines are formed and where the flow is accelerated by radiative forces up to the final expansion velocities. But the mass loss is determined by the hot coronal wind mechanism which accelerates the flow to supersonic velocities. The extent of the corona must be very small indeed. The lower limit for the X-ray measurements of ζ Puppis by the ANS satellite means that its corona cannot be greater than about 40 000 km thick. Cassinelli, Olson and Stalio (1978) came to a similar conclusion from an analysis of the Hα line profiles. There is also a physical reason why the corona should be small in extent. In the solar corona the high coronal temperature is maintained to great distances from the Sun by thermal conduction of heat from the base of the corona where it is heated. The corona round an OB supergiant will have an electron density which is 10^2 or 10^3 times greater than the solar corona. This means that the radiated energy losses, which are proportional to the electron density squared, are 10^4 or 10^6 times greater than in the solar corona. These losses are just too great for thermal conduction to maintain the corona beyond the region where it is directly heated. Olson (1978) and Cassinelli and Olson (1978) have shown that the O VI observations can be explained by Auger ionization of O IV in the radiative equilibrium region outside the corona by X-rays emitted by the corona.

The third theory is a combined coronal, radiation pressure model proposed by Rogerson and Lamers (1975). The result of Marlborough and Roy (1970) that the subsonic-supersonic solution disappears if the outward radiative force is greater than the inward force due to gravity is true for an isothermal corona. Lamers showed in his thesis that in the presence of a large positive temperature gradient, such as in the transition region between the corona and the photosphere a subsonic-supersonic flow can occur. In this case the critical point lies in the transition region. The consequences of this theory have not been worked out. Lamers and Morton (1976) have shown that the O VI observations of ζ Puppis can be explained by a collisional ionization model. This requires an electron temperature of about $2 \ 10^5$ K throughout the expanding atmosphere out to about 10 stellar radii. In τ Scorpii only the inner region of the atmosphere is at $2 \ 10^5$ K (Lamers and Rogerson).

The fourth theory of mass loss was proposed by Cannon and Thomas (1977). They use the de Laval Nozzle analogy of a stellar wind. The shape of a nozzle has to be designed to fit the input and output conditions of the nozzle. If the shape of the nozzle is not perfect then shocks will form in the flow. Now the nozzle equivalent to a radiative equilibrium stellar atmosphere is a very long thin nozzle with the throat far away from the star. This nozzle does not give the perfect match for the observed mass loss rates from OB stars. Therefore shocks will form in the flow just above the photosphere and it is these shocks that heat the corona. This is quite different from the Parker theory. The Parker theory shows that mass loss is a consequence of an extended hot corona. Cannon and Thomas say that the corona is heated as a consequence of the mass loss. To some extent their work is more a criticism of the other theories than a theory of mass loss. They take the mass loss rate as an observed quantity and investigate the effects of this mass loss on the stellar atmosphere. They do not calculate a mass loss rate. They believe that the mass loss rate is determined by the dynamical processes below the photosphere, the subphotospheric non-thermal storage modes in their language, and that until the dynamics is completely understood it is impossible to calculate a mass loss rate.

These are the four theories which will be discussed in the following session. During this session it was clear that everyone now agrees that the outer layers of hot star atmospheres have mechanical heating. How they might be heated was not discussed. This is clearly a field which needs more work.

References:

Cannon, C.J. and Thomas, R.N.: 1977, *Astrophys. J.* 211, pp. 910-925.
Cassinelli, J.P. and Castor, J.I.: 1973, *Astrophys. J.* 179, pp. 189-207.
Cassinelli, J.P. and Olson, G.L.: 1978, *Astrophys. J.* in press.
Cassinelli, J.P., Olson, G.L. and Stalio, R.: 1978, *Astrophys. J.* 220, pp. 573-581.
Castor, J.I., Abbott, D.C. and Klein, R.I.: 1975, *Astrophys. J.* 195,

pp. 157-174.
Hearn, A.G.: 1975, *Astron. Astrophys.* 40, pp. 277-283.
Lamers, H.J.G.L.M. and Morton, D.C.: 1976, *Astrophys. J. Suppl.* 32, pp. 715-736.
Lamers, H.J.G.L.M. and Rogerson, J.B.: 1978, *Astron. Astrophys.* 66, pp. 417-430.
Lucy, L.B. and Solomon, P.M.: 1970, *Astrophys. J.* 159, pp. 879-893.
Marlborough, J.M. and Roy, J.-R.: 1970, *Astrophys. J.* 160, pp. 221-224.
Morton, D.C.: 1967, *Astrophys. J.* 150, pp. 535-542.
Olson, G.L.: 1978, *Astrophys. J.* in press.
Rogerson, J.B. and Lamers, H.J.G.L.M.: 1975, *Nature* 256, pp. 190.

DISCUSSION FOLLOWING HEARN

Lamers: In your original version of the coronal model, the corona was very extended. After the revisions by Cassinelli the corona is much thinner. This means that the escape-velocity in the new coronal models is much larger than in your original model, and consequently the temperatures have to be higher in order to reach the critical point close to the star. What is the coronal temperature that you require to explain the observed mass loss rates by gas pressure in a thin corona?

Hearn: For a star such as ξ Orionis a temperature of about 3.5×10^6 K, and as high as 9×10^6 K for ξ Pup. The model given by Cassinelli et al. uses a corona of 5×10^6 K to explain the observations. This is not a hydrodynamically consistent model. But the interpretation of the observations is not sensitive to the coronal temperature, and a coronal temperature of 9×10^6 K in the model of Cassinelli et al. would not make a great difference.

RADIATIVELY-DRIVEN WINDS: MODEL IMPROVEMENTS, IONIZATION BALANCE
AND THE INFRARED SPECTRUM

John I. Castor
Joint Institute for Laboratory Astrophysics, University of
Colorado and National Bureau of Standards, Boulder, CO 80309

ABSTRACT

Recent improvements to theoretical stellar wind models and the
results of empirical modelling of the ionization balance and the
infrared continuum are discussed. The model of a wind driven by radiation pressure in spectral lines is improved by accounting for overlap
of the driving lines, dependence of ionization balance on density, and
stellar rotation. These effects produce a softer velocity law than
that given by Castor, Abbott and Klein (1975). The ionization balance
in ζ Puppis is shown to agree with that estimated for an optically
thick wind at a gas temperature of 60,000 K. The ionization model is
not unique. The infrared continuum of ζ Pup measured by Barlow and
Cohen is fitted to a cool model with a linear rise of velocity with
radius; this fit is also not unique. It is concluded we should try to
find a model that fits several kinds of evidence simultaneously.

I. INTRODUCTION

As I understand the charge to the members of this panel, each of
us is to present his favorite model of the temperature structure of the
wind of an O star, and show how it fares with all the various observational requirements; thereby will the correct theory of stellar wind
dynamics be found. Interesting as this may be, I am afraid it will not
lead to the hoped-for result, for two reasons. The first is that present observations are not sufficient to select a single model; there
are simply not enough observables that can be related to the temperature of the gas in a simple and sensitive way. The second reason is
that the temperature may not be closely coupled to the processes that
make the stellar wind go. Other forces than gas pressure may be more
important in the overall dynamical scheme. Another problem that arises
in comparing the alternative models is that some are what I would call
"theoretical models," in which there are no arbitrary functions but
everything is derived (approximately!) from first principles, and
others are "empirical models" in which temperature and velocity are

P. S. Conti and C. W. H. de Loore (eds.), Mass Loss and Evolution of O-Type Stars, 175-190.

arbitrary functions. As always happens, "theoretical" models never fit the data, and "empirical" models always do.

My own work on stellar wind models (all of it in collaboration with David Abbott, now at U. Wisconsin, and Richard Klein, now at Kitt Peak) began with a "theoretical" model -- outflow driven by radiation pressure in a large number of spectral lines á la Lucy and Solomon. This model was quite simplified in terms of its physical assumptions, such as radiative equilibrium, but it did make quite definite predictions of the velocity and temperature structure of the wind as well as of the rate of mass loss. These predictions were quickly discredited by observations of the ionization structure of the stellar wind and indirectly of the velocity field. However, no other "theoretical" model has been advanced. Since I, too, want to agree with observation as much as possible, I have dabbled with "empirical" models. This has taken the form of investigating how the ionization balance in the wind relates to the gas temperature in the perilous regime of large EUV optical depth, and also of trying to analyze the infrared observations of ζ Puppis.

My talk today is going to range over a wider area than just the temperature of the wind. On the "theoretical" side, I would like to talk about improvements in the basic model which I hope make it more realistic. These improvements address the complications of (1) overlap in frequency of the wind-driving resonance lines, (2) the effect on the driving force of the variation of ionization balance with density, and (3) stellar rotation. With regard to "empirical" models I offer comments and cautions regarding the interpretation of ionization balance data in terms of a temperature. To underscore the caution I present a model of ζ Pup that explains the ionization balance in terms of photoionization in an optically thick stellar wind (a realistic case). In this model there is a moderate elevation of the wind temperature from radiative equilibrium. My second "empirical" comment is that the infrared fluxes of ζ Puppis, observed by Barlow and Cohen, are fitted exceedingly well by a cool isothermal wind model having the observed rate of mass loss and a slowly rising velocity law, as indicated by line-profile studies.

II. IMPROVEMENTS TO THE RADIATION-DRIVEN WIND MODEL

This model is developed in its simplest form by Castor, Abbott and Klein (1975), based on the idea of Milne (1926) refined by Lucy and Solomon (1970). The identity of the driving spectral lines was investigated more carefully by Abbott (1977); see also Castor, Abbott and Klein (1976). For this model the following assumptions were made: (1) spherical symmetry; (2) steady radial flow; (3) only the forces of gravity, radiation scattered by free electrons and radiation scattered by ions are significant; (4) energy balance with no mechanical heating. This model is not advanced as a perfect representation of reality, but rather as a calculable one that accounts for what appear to be the dominant processes. Assumption (3) could be modified to include the

thermal pressure of the gas (in the model we made, it was), but this
has very little effect on the dynamics for any reasonable temperature
since the wind is so highly supersonic. (This is supported by the
calculations of MacGregor [1978], for a model with a temperature rise.)

The quantities that can be predicted with this model are the rate
of mass loss, and the dependences on radius of the velocity and the
gas temperature. The mass loss agrees within a factor 2 with observation for the supergiants for which the rate of mass loss can be found
from the radio flux or from the infrared excess. The predicted velocity law gives quite reasonable terminal velocities and a qualitatively
correct shape, but seems too steep at small radii. The gas temperature
turns out to be about 0.9 times the effective temperature of the star,
consistent with radiative equilibrium. Sadly, this is inconsistent
with the observed presence of O VI and perhaps also with N V.

The problem with the gas temperature points to the necessity of
some additional heating mechanism, the nature of which is certainly not
known at the present. The problem with the steepness of the velocity
law suggests that we have oversimplified the dynamics. Two aspects of
the dynamics that are suspect are, first, that we have assumed that
each spectral line acts independently of the others, whereas in reality
they overlap in frequency and therefore modify one another; and, second,
we have ignored the effect of ionization changes with radius on the
radiation force. Furthermore, if the star is rotating there are additional large terms in the equations of motion that we have not considered.

A. Overlapping Lines. The difficulty we face when the rest
frequencies of the strong lines are spaced by less than the Doppler
shift corresponding to the terminal velocity of the wind is that the
photons originating in the stellar continuum that a particular line
would normally scatter may have been, in some sense, "used up" by
another line. Of course, the photons are not actually "used up."
Collisional destruction of the photons is very unlikely, so interaction
with one line can only alter the frequency and direction of a photon,
after which it is still present to scatter in another line. Perhaps it
is helpful to focus our attention on a single photon from its origin at
the photosphere until it leaves the stellar envelope going outward, or
flies back into the photosphere again. In the meanwhile it may have
interacted with several different lines. Each such interaction consists
of several, perhaps a very large number, of scatterings, but owing to
the large velocity gradient in the gas, all the scatterings occur in a
small volume. Therefore we can find the impulse given to the material
at that point from the photon's initial and final directions; it is as
if the photon scattered only once.

Now comes a tricky part. Suppose, as an idealization, that the
strong spectral lines are distributed at random in frequency, within
some fairly large band. Then our photon, having scattered many times
in one line, will fly across the envelope a random distance until its

frequency matches that of the next strong line. (Distances convert to
frequencies owing to the Doppler effect in the expanding envelope.)
Then the photon scatters in this line some number of times before proceeding in a new direction, and so on. The whole process is a kind of
diffusion, in which the number of particles is conserved. The mean
free path for this diffusion is the distance in space that corresponds
to a Doppler shift equal to the mean interval in frequency between
strong lines. This is not a hard diffusion problem to solve, and the
boundary condition is the known net flux of photons at the photosphere.
The result is the angular distribution of the radiation averaged over
the large frequency band in question. How does this help us compute
the total force on the material? In order to find the force due to a
particular line we need to know the intensity of the radiation available
for scattering in this line, which therefore includes the effects of
scattering in all the lines at higher frequency. In general this is a
nasty quantity, but if the lines are randomly distributed in frequency
then the average of this intensity over several lines is the same as
the frequency average we find from our diffusion problem.

The picture outlined in the preceding paragraph leads to a
quantitative result for the line force in the following way. The expression for the force on a unit mass of material due to an individual
line is (Castor 1974)

$$f_L = \frac{2\pi\nu}{\rho c^2} \int_{-1}^{1} \mu I(\mu) \left[\mu^2 \frac{dv}{dr} + (1-\mu^2) \frac{v}{r} \right] \left\{ 1-\exp[-\tau(\mu)] \right\} d\mu \quad . \quad (1)$$

In this formula all quantities are evaluated at the distance r from
the center, μ is the direction cosine of a ray with respect to the
radial direction, $I(\mu)$ is the intensity of the radiation available
for scattering in the line, $v(r)$ is the outward flow speed, and $\tau(\mu)$
is the Sobolev optical depth of the line for the position and direction
in question. It depends on direction as
$\tau(\mu) = \tau_{rad}[\mu^2+(1-\mu^2)d\ell nr/d\ell nv]^{-1}$. According to the argument given
above, we can identify I with the average intensity in a frequency
band surrounding the line. In that case the average net flux F_ν in
the band is related to I by

$$F_\nu = 2\pi \int_{-1}^{1} I(\mu)\mu d\mu \quad .$$ (2)

Since the net flux is conserved, we can calculate F_ν from the properties of the photosphere -- it is the continuum flux. We can now recast
equation (1) in this way:

$$f_L = \frac{\nu F_\nu}{\rho c^2} \frac{dv}{dr} \left[1-\exp(-\tau_{rad}) \right] F_a \quad ,$$ (3)

in which F_a is the ratio

$$F_a = \frac{\int_{-1}^{1} \mu I(\mu) \left[\mu^2 + (1-\mu^2) d\ell nr/d\ell nv \right] \left\{ 1 - \exp[-\tau(\mu)] \right\} d\mu}{\left[1 - \exp(-\tau_{rad}) \right] \int_{-1}^{1} \mu I(\mu) d\mu} \quad . \tag{4}$$

Apart from the factor F_a, equation (3) is the formula for the force we have used up to the present. Thus F_a is the factor required to account for overlapping lines and the correct angular distribution of the radiation. In the likely event that the outward intensity is larger than the inward intensity for every ray, we can see that F_a is smaller than unity where $d\ell nv/d\ell nr$ is larger than unity, namely the inner part of the envelope, and larger than unity in the outer region where $d\ell nv/d\ell nr$ is smaller than unity. This is illustrated in Figure 1, in which the overall F_a is shown for a distribution of lines with different strengths and for which overlap is a large effect. It may seem counterintuitive that in the high velocity region, where line overlap should be more serious, the force turns out to be larger than in the inner region. The explanation of the paradox lies, first of all, in noting that the photons are not "used up." The effect is in fact due to saturation of the line. Increasing optical depth of the line reduces the force. The assumption that the photons all travel radially, our earlier model, underestimates the optical depth in the inner region and overestimates it in the outer region. Interaction with several lines broadens the angular distribution and increases the effect.

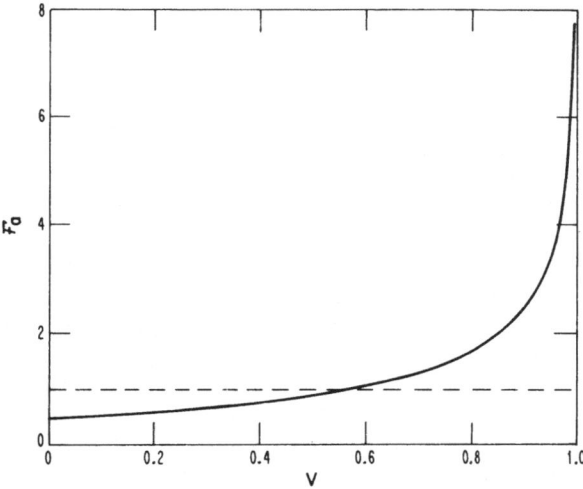

Figure 1. Angular-distribution factor in the radiation force as a function of outflow velocity. The model has $v = v_\infty (1-R/r)^{1/2}$ and three strong lines per frequency interval $v_0 v_\infty / c$.

Before investigating the dynamical effect of the new factor in the radiation force, we must note that there is a competing factor: the effect of ionization changes on the total force.

B. Ionization Changes. Except possibly for the highly ionized species like O VI and N V, the ions in the stellar wind seem to be produced by photoionization. In the simplest photoionization model the degree of ionization depends on the ratio of the diluted stellar flux to the electron density, F/N_e. Since the flux varies as $1/r^2$ and the density varies according to mass conservation as $1/vr^2$, the degree of ionization depends on the velocity. The stage of ionization that is one step lower than the most abundant one will vary in abundance as v^{-1}, while the one that is one step higher varies as v. How the force will vary with v on account of these ionization changes depends on which stage of ionization produces the most force. It happens that the higher stages of ionization have their resonance lines shifted away from the peak of the flux distribution toward the EUV, and thus contribute less. This is not an invariable rule. For example, the Li sequence ions behave in the opposite way. Nonetheless, most of the force is produced by the lower ions. The total force due to the lines of a particular ion varies with the ion abundance f_{ion} as $f_{ion}^{0.1-0.2}$. The final result is that the total force varies as $v^{-0.1}$. This result, as well as support for the general statements above, comes from the detailed force calculations of Abbott (1977).

When the effect of ionization changes is included with the line-overlap factor, discussed above, in the expression for the line radiation force, the momentum equation for steady flow (neglecting gas pressure) takes the form

$$r^2v\frac{dv}{dr} = -GM(1-\Gamma) + Ck(v)\left[r^2v\frac{dv}{dr}\right]^\alpha \quad , \quad . \tag{5}$$

where the force constant k now depends on velocity according to

$$k(v) \propto v^{-0.1}F_a \quad . \tag{6}$$

The function k(v) has a minimum at some value of v, and becomes large both when v is small and when v approaches the terminal velocity. The value of v at the minimum is about $0.1v_\infty$ for the function F_a shown in Figure 1. The correct velocity law and rate of mass loss are determined by finding the solution of equation (5) that passes through a regular singular point. That point is in fact the place where k attains its minimum. The rate of mass loss is given by the formula in CAK, except that the minimum k must be used. (Gas pressure is a 1-4% effect at the singular point, so it should not modify these results very much.)

The effect of the variation in k is that r^2vdv/dr is smaller interior to the singular point, and larger exterior to it, compared

with the constant value for the CAK model. This produces a more slowly
rising velocity law, which is the direction indicated by observation.
In order to find the new velocity law quantitatively we have to find
the angle factor F_a self-consistently with the velocity law, which
has not yet been done. (Recall that F_a depends on the solution to a
diffusion problem, one of the ingredients of which is the velocity law.)

    C. <u>Stellar Rotation</u>. The effect of stellar rotation on a
radiatively driven stellar wind is a particularly unpleasant problem.
The three dimensional aspects of even the simpler problem of a gas-
pressure-driven wind have not been fully explored (see Nerney and Suess
1975), and the tensorial character and explicit dependence on velocity
gradients of the radiation force are not likely to make the problem
easier. Marlborough and Zamir (1975) have investigated the effects of
radiation force exerted in the continuum, but I think the effect of line
force is crucial. I have had a crude stab at including rotation in the
framework of the CAK model. I assumed a flow that had zero component
of velocity in latitude, for which the angular momentum per unit mass
was conserved. For this simple model the effect of rotation is only to
add a known centrifugal term to the equation of motion, which can be
solved in the usual way. (I have <u>not</u> added the complications discussed
in the preceding sections.)

    The results for this model are (1) the mass loss is nearly
unchanged by rotation, and (2) the rise of velocity with radius is quite
appreciably slower when rotation is included, and the terminal velocity
is less. The lack of change in the rate of mass loss reflects the fact
that for this model the singular point, where the mass loss is determined, falls at a very large radius where the centrifugal force has

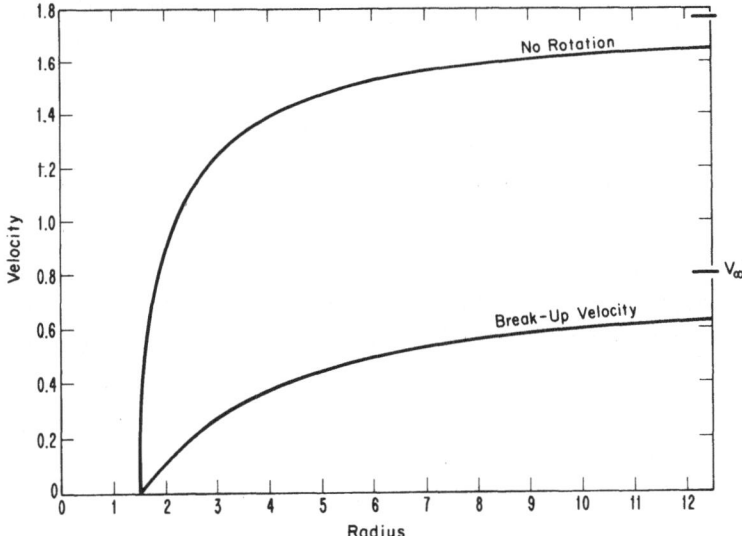

Figure 2. Velocity law for a rotating stellar wind. The
upper curve is for no rotation, the lower curve is the equatorial run of velocity for rotation at the break-up speed.

become negligible. This conclusion could change when the overlapping-
line and ionization effects are put in, since then the singular point
will be much closer in to the center. The effect on the velocity-radius
relation is shown in Figure 2 for a star rotating at the break-up velocity. (By the way, nothing dramatic happens to the stellar wind as
this limit is approached.) The fact that the velocities at a given
radius are smaller when rotation is included means that the densities
are larger, which will produce stronger emission in the Balmer and
other lines, just as if the rate of mass loss had been increased. This
model could be tested by looking for an anti-correlation between Vsini
and the limiting velocity for the UV P Cygni lines.

III. EMPIRICAL STELLAR WIND MODELS

A. Ionization Balance. What we are most interested in for today's
discussion is the temperature of the wind. Most of the efforts that
have been made toward determining the temperature directly from observations have focused on the observed ionization balance, and in particular on the presence of the highly ionized species O VI and N V in
the ultraviolet spectra of the O stars. These efforts are courageous,
because we normally expect the ionization balance to be nearly independent of the local temperature at the outside of the star. The temperature affects the ionization directly through collisional ionization by
electrons, but this process is very weak. There is also an indirect
effect through thermalization of the EUV radiation field; however, the
stellar wind is not sufficiently optically thick for this to occur.
That would be the end of the story had not the O VI absorption lines
shown up in the Copernicus spectra. One can calculate the rate of
photoionization of O V by the emergent flux from an appropriate model
atmosphere, and it is orders of magnitude too small to explain the
observed line strengths; hence we discard photoionization. Collisional
ionization being weak, we crank the gas temperature way up to give the
desired ionization rate. This model works fine, and Henny Lamers will
describe it in more detail. However, it may not be the only possible
model, even among those with a roughly uniform temperature in the wind.
Perhaps photoionization was discarded a little too hastily.

Let us first compare the rates of collisional ionization and
photoionization for an ion, when the ionizing radiation is optically
thick. Let R_{1K} be the rate coefficient for photoionizing the ground
state of the ion, given by

$$R_{1K} = \int_{\chi/h}^{\infty} \frac{4\pi}{h\nu} J_{\nu} a_{\nu} d\nu \quad , \tag{7}$$

where χ is the ionization potential, J_{ν} is the local continuum
intensity, and a_{ν} is the photoionization cross section. The quantity
R_{K1} is defined by a similar integral, with J_{ν} replaced by the local
Planck function B_{ν}. The significance of R_{K1} is that $N_1 * R_{K1}$ is

the rate of radiative recombinations to the ground state, also written
as $N_e N_+ \alpha_1$ in terms of the recombination coefficient α_1. Here N_1^*
is the ground-state population given by LTE. The departure coefficient
b_1 is defined as N_1/N_1^*. Now, if it happens that the stellar envelope
is optically thick for the radiation that can ionize this ion, then to
a reasonable approximation J_ν can be replaced by the continuum source
function, which is given by $B_\nu/\bar{b}_1$. Here $\bar{b}_1$ is the mean coefficient
for all ions that absorb this radiation. The result is that
$R_{1K} = R_{K1}/\bar{b}_1$. The rate coefficient for ionization by electron impact
is defined to be $N_e C_{1K}$, where C_{1K} is the result of folding the impact ionization cross section with the Maxwellian distribution of electron energies. If we look in Allen (1973) for reasonable estimates of
these cross sections, we come up with the formula

$$\frac{N_e C_{1K}}{R_{K1}} = 4 \times 10^{-15} \frac{n_*^{\,7}}{z^6} T_e^{-1/2} N_e \quad , \tag{8}$$

in which n_* is the effective principal quantum number for the target
valence electron and z is the charge of the higher ion. If we put
R_{K1} in terms of R_{1K} and take reasonable numbers for n_*, z, and T_e,
we get

$$\frac{N_e C_{1K}}{R_{1K}} \approx \frac{b_1 N_e}{2 \times 10^{18}} \quad . \tag{9}$$

We see that $b_1 N_e$ is the governing parameter. If it is larger
than 10^{18} or so, then collisional ionization dominates (coronal
approximation); if it is less, then photoionization is more important.
10^{18} may seem a little large for an electron density, but remember that
if the electron temperature is large but the ionization is moderate then
b_1 must be very large.

There is an upper limit to $b_1 N_e$, and hence to the importance
of collision ionization, set by the condition that the rate of collisional ionization can not exceed the total rate of radiative recombination (three-body recombinations being negligible). The requirement is
that

$$N_1 N_e C_{1K} < N_e N_+ \alpha \quad , \tag{10}$$

where α is the total recombination coefficient, including excited
states and dielectronic recombination. Since $N_1^* R_{K1} = N_1 R_{1K} = N_e N_+ \alpha_1$
the condition is

$$\frac{N_e C_{1K}}{R_{1K}} \approx \frac{b_1 N_e}{2 \times 10^{18}} < \frac{\alpha}{\alpha_1} \quad . \tag{11}$$

Equality in equations (10) and (11) defines the coronal approximation.
It appears from equation (11) that collisional ionization can never be

very large compared with photoionization, but we must remember that
dielectronic recombination can cause α to be much larger than α_1.
Ratios of 100 or more are not unusual, if T_e is over 10^5 K. Thus
the typical value of $b_1 N_e$ for the coronal approximation is about
10^{20} - 10^{21} and photoionization, even for the optically thick gas, is
small.

The very large ratio of α to α_1 indicates that recombination to
the ground state is negligible -- excited states and particularly auto-
ionizing states are more important. This raises the question: what
about the inverse of those processes? Photoexcitation can create
appreciable populations in the excited states, from which photoionization (or autoionization) can occur. In addition, the bulk of dielectronic recombination occurs through states of very large principal
quantum number for which, at electron densities of order 10^{11} cm^{-3},
collisional effects become important, quenching the process. A detailed
statistical equilibrium calculation for the stellar wind ions, which
could answer these questions, has not yet been done. This is a high
priority item for future work, but until it is done we can only worry
that α/α_1 has been overestimated.

What is the ionization balance we predict if optically thick
photoionization and collisional ionization are both taken into account?
We add $N_1 R_{1K} = N_1 R_{K1}/\overline{b_1}$ to the left side of inequality (10), making
it an equality. Now we must distinguish between the departure coefficient b_1 of the ion in question and the mean $\overline{b_1}$ of the absorbing
ions. We find

$$N_1 \left(\frac{R_{K1}}{\overline{b_1}} + N_e C_{1K} \right) = N_1 {}^* R_{K1} \frac{\alpha}{\alpha_1} \tag{12}$$

hence

$$\frac{1}{N_e b_1} = \frac{\alpha_1}{\alpha} \left(\frac{1}{N_e \overline{b_1}} + \frac{C_{1K}}{R_{K1}} \right) \approx \frac{\alpha_1}{\alpha} \left(\frac{1}{N_e \overline{b_1}} + \frac{1}{2 \times 10^{18}} \right) \quad . \tag{13}$$

If $N_e \overline{b_1}$ is larger than 2×10^{18}, the coronal approximation applies,
and $N_e b_1$ is around 10^{21}. Otherwise $N_e b_1$ is larger than $N_e \overline{b_1}$ by
a factor α/α_1. How do we decide on a value for $N_e \overline{b_1}$? One way is to
use equation (13) for all ions, in each case using the appropriate
average for $\overline{b_1}$; the result is a set of coupled nonlinear equations for
the departure coefficients. However, equation (13) may not be adequate
for the dominant absorbing ions in the gas; for these, effects like
excited-state photoionization must be included. The difficulty that
arises in treating these ions was discussed, in the case of He II, by
Klein and Castor (1978). Instead of solving for all the ions, we might
treat $\overline{b_1}$ for the dominant absorber as a free parameter. If we make
the further (naive!) assumption that α/α_1 is the same for every ion,
we find that all ions with an ionization potential as large or larger
than the dominant one are described by the same departure coefficient,

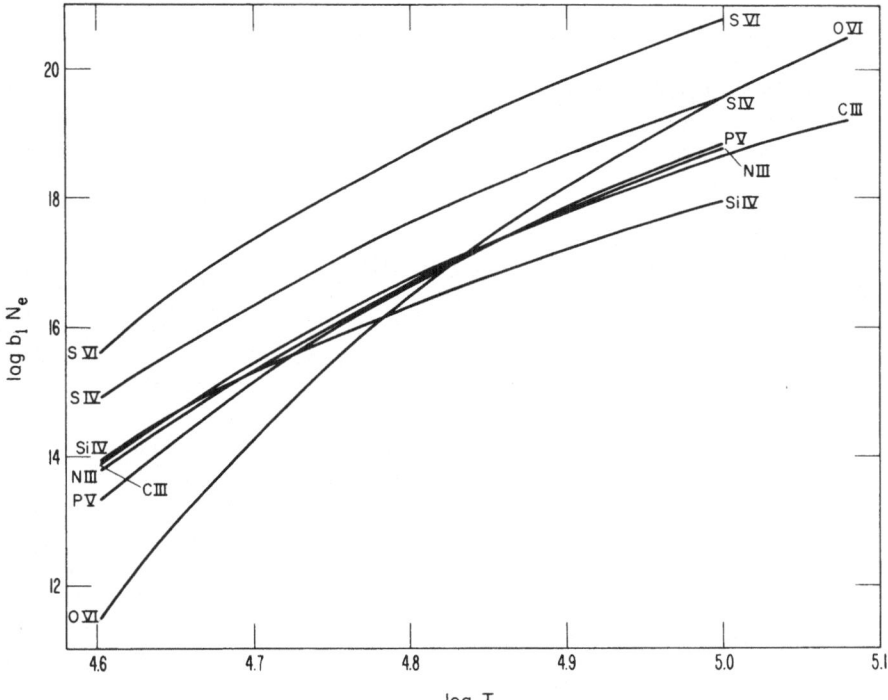

Figure 3.   Loci in $b_1 N_e$ versus T_e on which the ions attain
the abundances observed in ζ Puppis. b_1 is the departure
coefficient from the Saha equation at the temperature T_e.

with the exception of the dominant one itself. We now have a two-
parameter ionization balance: T_e and $b_1 N_e$.

One does not really expect a single departure coefficient to fit
every ion, but it is interesting to compare the ionization computed on
this hypothesis with the observed ionization fractions for ζ Pup, the
only star with a fairly complete set of data.   In Figure 3 the observed
ionization fractions for several ions have been converted to equal-
ionization contours in the parameter diagram of $b_1 N_e$ versus T_e.   Surprisingly, all the ions except sulfur agree within a factor 3 in $b_1 N_e$
provided the temperature is somewhere around 60,000 to 70,000 K.   The
best value $b_1 N_e$ is around 3×10^{16}, low enough for collisional ionization to be neglected. This is about 10 times larger than N_e times the
departure coefficient of He II at this temperature, if in fact He II
is ionized by dilute photospheric radiation at 35,000 K; this would be
consistent with He II being the dominant ion and α/α_1 being 10.

What do we conclude from this? It does appear that it is possible
to explain the observed ionization in ζ Pup with a model at a roughly
uniform temperature of order 60,000 K by taking due account of the
large optical depth in the far UV.   This model is still a crude one
at present because we have not found an accurate way of including the
excited states in the ionization balance problem; this is a difficulty

that is faced by all the ionization models. The present state of ionization modelling is a sad one. There are at least three models, with quite different temperature structures, that fit the observed ionization fairly well, and which are equally reasonable from a theoretical point of view. Clearly other kinds of data are needed if a choice is to be made.

    B. <u>Infrared Spectrum</u>. The infrared spectrum provides data that can be translated into temperatures of the wind in a much more straightforward way than can the observed ionization balance. This is because the dominant absorption in the infrared is free-free, with a simple dependence on temperature and no complications due to departures from LTE. The price that is paid is that the dependence on temperature, while unambiguous, is not very strong. The infrared spectrum also depends on the density distribution, that is, on the velocity law, so the velocity and temperature must be found at the same time. The infrared spectrum alone is not sufficient for this, but in conjunction with other data, such as the Hα profile, it should be possible to determine both variables. The best star for analysis is again ζ Pup. Morton and Wright (1978) have determined the rate of mass loss from the radio flux independently (nearly) of the velocity law and temperature. The infrared observations by Barlow and Cohen, which Mike Barlow has described earlier, are the most complete for ζ Pup of all the O stars. David Van Blerkom will discuss the analysis he has been doing on the Hα line; I would like to show the results of some model fitting I have done using the infrared data obtained by Barlow and Cohen.

    All the models are computed by solving the spherical transfer equation allowing free-free absorption and electron scattering as opacity sources. The region interior to the sonic point was assumed to radiate a Planck spectrum, and I used a constant infrared Gaunt factor for simplicity. The first attempt at fitting the spectrum was made with the standard velocity law $v(r) = v_\infty(1-R_*/r)^{1/2}$. One temperature was assumed for the stellar photosphere, and another constant temperature was taken for the wind. The rate of mass loss was fixed at the radio determination. The result is shown in Figure 4. The dots are the observations, and it is clear that in the wavelength range they span, the wind is simply too transparent to affect the emergent flux, and what is seen is the photospheric Planck spectrum. However, the observed fluxes define a spectrum slope that is definitely <u>redder</u> than a Planck function at the stellar temperature. (Lowering the assumed stellar temperature would not improve the fit very much.) We also notice in Figure 4 that varying the wind temperature does not make any difference to the emergent flux in the region that is observed.

    For a model close to hydrostatic equilibrium, the 11.67 μm radiation comes from a layer with electron density 2×10^{13} cm^{-3}, which implies an outflow velocity of about 10 km s^{-1}. The flow velocity for the 2 μm emitting layer would be corresponding less, about 2 km s^{-1}. Thus one explanation of the redness of the spectrum is that there is a rise in temperature by about a factor two between these two layers.

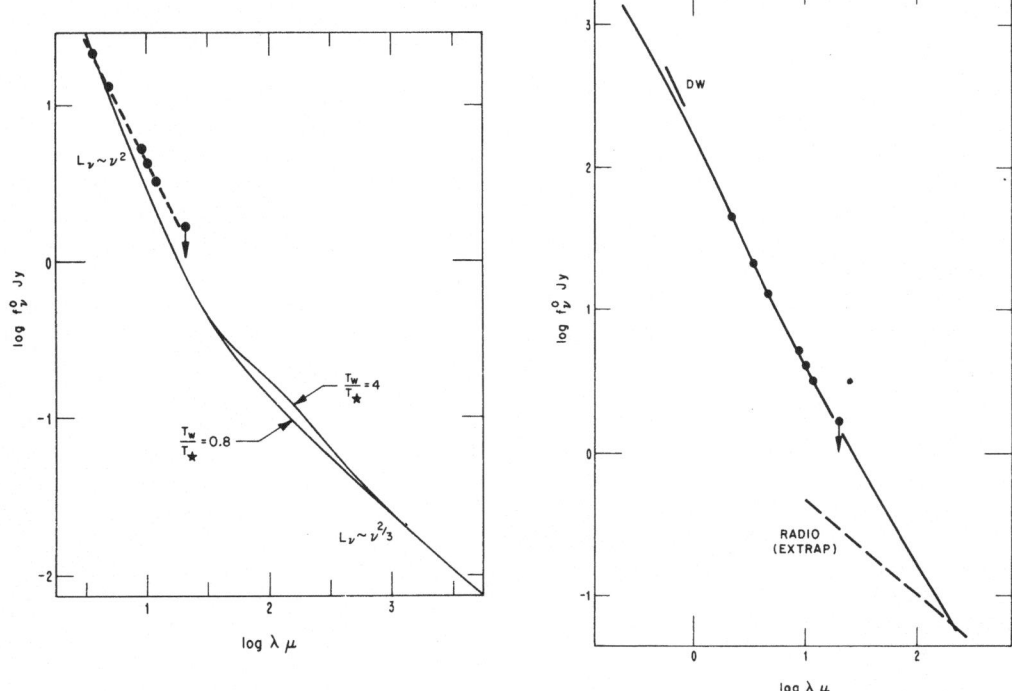

Figure 4. The solid lines are computed infrared fluxes (in Janskys) for a model with $v = v_\infty(1-R/r)^{1/2}$ and the indicated ratios of wind temperature to photosphere temperature. The dots and the dashed line show the Barlow and Cohen observations of ζ Pup.

Figure 5. Same as Figure 4 except the model has $v = v_1(r-R)$ and $T_w \approx T_*$. The short line marked DW is the visual photometry of Davis and Webb (1974). The dashed line is the radio flux extended to shorter wavelengths with a $\nu^{2/3}$ law, corrected for the Gaunt factor.

A relation between temperature and velocity of the form $T \propto v^{0.5}$ would give the observed spectral index, if the density distribution were close to hydrostatic equilibrium. A model of this kind could be tested by investigating the consequences for the stellar absorption lines, which are formed in about the same region as we know from the Balmer progression of radial velocities. I have not pursued this model, but have instead investigated a model in which the infrared excess is due entirely to geometrical extension.

In order to have the wind affect the 2-10 micron infrared, its emission measure $(\int n_e^2 dr)$ must be increased over that for the standard velocity law; this means that the velocity should rise more slowly with radius than for the standard law. I have tried the law $v = v_1(r/R_* - 1)$. This linear law guarantees a much larger emission measure than that for the square-root law. (The law was used only in the region where $v \geq 25$ km s^{-1}; the Planck boundary condition was applied at $v = 25$.) Of course, the velocity does not increase indefinitely with radius, but turns over and approaches the terminal velocity. This turn-over will have very little effect on the region shortward of 20 μm, as we see in Figure 4, so we need not try to model it. The parameters that can be adjusted in the model are the temperature T_* and the angular size of the photosphere, the velocity coefficient v_1, and the temperature T_w of the wind. Of these, T_* is assumed to be fixed at 50,000 K, and the visual flux then implies an angular diameter equivalent to $R_* = 15.9$ $R_\odot$ at a distance of 450 pc (Lamers and Morton 1976). The two remaining parameters can then be used to fit the slope and the magnitude of the emergent flux in the near infrared.

The results of a not-quite-perfect fit are shown in Figure 5. The long solid line is the model, and it fits the IR data very well. It does miss the visual data by 0.1 in log $f_\nu{}^o$, which is the not-quite-perfect aspect. In fact, the fitting is done in non-dimensional variables, and the model shown turned out to correspond to $T_* = 46,000$ K rather than the desired 50,000 K. In addition, my treatment of the photosphere as a Planck boundary condition is not very accurate, so it is not surprising there is some discrepancy. The computational aspects are better for the infrared itself, so we can have some confidence in the values of v_1 and T_w. These are $v_1 = 700$ km s^{-1} and $T_w = 42,000$ K.

These are interesting results, but what do they actually imply? Apparently a cool model with a slowly-rising velocity can fit the infrared to within the accuracy of the observations. But by no means is this the only model that will fit. I have described above a model with a temperature rise in the subsonic region that will also work. No doubt there are others. Additional data must be analyzed in conjunction with the infrared. For example, we should analyze the Hα profile, the Balmer velocity progression, and the intensities of He I lines. These are all sensitive to the velocity and temperature distributions in different ways.

Some other aspects of the model derived above may be of interest. The velocity range for the layers in which the observed infrared originates is from the sonic point out to 250 km s^{-1}. These are more than ten times higher than the velocities mentioned earlier in connection with the nearly hydrostatic atmosphere, and the reason is that in an extended atmosphere less particle density is needed to produce a given emission measure. The optical depth in electron scattering at the sonic point is about 0.4. This is large enough to explain the large angular diameter found by Hanbury Brown et al., but not so large that the treatment of the photosphere is nonsense or so that extensive wings would be expected on every line profile.

IV. CONCLUSIONS

The radiatively-driven wind models have been fairly successful in explaining the rates of mass loss and terminal velocities of the winds in O and B stars. I feel we have correctly identified the mechanism of the wind. The effects of line overlap and shifting ionization balance, as well as rotation, will improve the agreement between the predicted and observed shapes of the velocity law, as I have sketched above. I am hopeful that the models including stellar rotation will give a satisfactory account of the Be stars.

The present uncertainty about the temperature indicates that our understanding of the stellar wind is by no means comprehensive. It is likely that the models that have been made do correspond fairly well with the real star in some spherically-averaged way, but there may be significant inhomogeneities, hot and cold regions and so forth, that we have no idea of at present. This is a frustrating situation for the model maker -- imagine trying to study prominences and other features of the solar corona from a distance of 500 parsecs! We will have to study the inhomogeneities by measuring as many different integrals of the velocity and temperature structure as we can, but at some point we will simply accept our incomplete knowledge and go on to study other things.

This work was supported by National Science Foundation Grant AST77-23183 to the University of Colorado.

REFERENCES

Abbott, D. C.: 1977, unpublished Ph.D. thesis, University of Colorado.
Allen, C. W.: 1973, Astrophysical Quantities (London: The Athlone Press).
Castor, J. I.: 1974, Monthly Notices Roy. Astron. Soc. 169, p. 279.
Castor, J. I., Abbott, D. C. and Klein, R. I.: 1975, Astrophys. J. 195, p. 157.
Castor, J. I., Abbott, D. C. and Klein, R. I.: 1976, in Physique des Mouvements dans les Atmosphères Stellaires, eds. R. Cayrel and M. Steinberg (Paris: CNRS).
Davis, J. and Webb, R. J.: 1974, Monthly Notices Roy. Astron. Soc. 168, p. 163.
Klein, R. I. and Castor, J. I.: 1978, Astrophys. J. 220, p. 902.
Lamers, H. J. G. L. M. and Morton, D. C.: 1976, Astrophys. J. Suppl. 32, p. 715.
Lucy, L. B. and Solomon, P. M.: 1970, Astrophys. J. 159, p. 879.
MacGregor, K. B.: 1978, Bull. Amer. Astr. Soc. 10, p. 412.
Marlborough, J. M. and Zamir, M.: 1975, Astrophys. J. 195, p. 145.
Milne, E. A.: 1926, Monthly Notices Roy. Astron. Soc. 86, p. 459.
Morton, D. C. and Wright, A. E.: 1978, Monthly Notices Roy. Astron. Soc. 182, p. 47P.
Nerney, S. F. and Suess, S. T.: 1975, Astrophys. J. 196, p. 837.

DISCUSSION FOLLOWING CASTOR

Hearn: How well does this theory predict the mass loss rates?

Castor: I think it does quite well.

Lamers: The mass loss rate of τ Sco, as derived from the UV lines
by Lamers and Rogerson is about 25 times smaller than the mass loss rate
predicted if it is proportional to the luminosity. This indicates that
the mass loss rates drop very drastically along the main sequence near
types B0, i.e., close to the limit $M_{bol} \simeq -6$ found by Snow and Morton.

Castor: The radiatively-driven stellar wind theory was used by
Rogerson and Lamers to derive the rate of mass loss in τ Sco. So presumably the theory should give the same answer if it manages to give
the observed degree of ionization, and therefore the observed line
strengths. The rate of mass loss is proportional to luminosity to a
power larger than one, since the number of absorbing lines goes down
as the wind becomes more tenuous.

Hearn: What about variability in the wind?

Castor: You get variations out if you put variations in. Presently, I have assumed it steady. However, I do test for instabilities
which often appear to be present, so I assume they are there. The overall wind is some average, with fluctuations superimposed.

Hearn: Well, there are instabilities and instabilities ... Do
yours get large enough to destroy the over-all equilibrium?

Castor: I don't know for sure.

Thomas: What is the physical cause of these instabilities?

Castor: The instability analysis has not yet been done. I suspect
instabilities could come from the random force. This could occur in a
symmetric or asymmetric fashion. I believe it safe to say that these
will not grow and disrupt the envelope.

Cassinelli: Does the concept of overlapping lines driving the wind
lead to an increase often quoted "maximum mass loss rate" $L/v_\infty c$?

Castor: Yes, if you put in many, many lines, suppose one per
Angstrom, and the expansion velocity is, say, six Angstroms, then you
could end up with three times $L/v_\infty c$ for the mass loss rate. $L/v_\infty c$ is
not an absolute upper limit on the rate, if you have overlapping lines.

Morton: Is there enough flux shortward of the Lyman limit in a
star as cool as ε Ori (B0Ia) or ζ Ori (09.7Ib) to drive the observed
wind by radiation pressure?

Castor: I have not checked that. I think so.

THE WARM WIND MODEL

Henny J.G.L.M. Lamers
The Astronomical Institute at Utrecht
Space Research Laboratory

ABSTRACT

The analysis of the UV spectra of τ Sco (B0 V) and ζ Pup (O4 f) resulted in the empirical warm wind model. In this model the presence of O^{5+} and N^{4+} ions in the envelopes of early type stars is explained by collisional ionization in a 'warm' envelope of $T \simeq 2 \ 10^5$ K.

I   INTRODUCTION

The ultraviolet spectrum of a number of early type stars was observed by the Copernicus satellite. The spectra contain P Cygni profiles of the resonance lines of high ionization stages of abundant elements. One of the startling discoveries was the presence of wide O VI resonance lines in the spectrum of the standard B0 mainsequence star τ Sco (Rogerson and Lamers, 1975) which suggested an expanding envelope hotter than the stars' effective temperature of $3 \ 10^4$ K. These lines were later found in the spectra of supergiants of types O to B1 (Snow and Morton, 1976). Lamers and Snow (1978) have shown that an anomalously high degree of ionization is found in the envelopes of all O stars and B type supergiants up to B8. Lamers et al. (1978) found evidence for an anomalously high Fe III/Fe II ratio in the envelope of α Cyg (A2 Ia).

In order to understand the proces of mass loss and the anomalously high degree of ionization in the envelopes, two stars were selected for a detailed study: ζ Pup (O4 f) by Lamers and Morton (1976) and τ Sco (B0 V) by Lamers and Rogerson (1978). The first star has a high mass loss rate of $7 \ 10^{-6}$ $M_\odot$/yr and strong P Cygni profiles whereas the second one has a small mass loss rate of $7 \ 10^{-9}$ $M_\odot$/yr and extended violet absorption wings. The profiles of the resonance lines of O VI, N V, Si IV and C IV in the spectrum of τ Sco are shown in Figure 1. The ultraviolet lines which show P Cygni profiles or extended violet wings are listed in Table 1. The extend of the violet absorption, V_{edge}, is also indicated.

P. S. Conti and C. W. H. de Loore (eds.), Mass Loss and Evolution of O-Type Stars, 191–199.

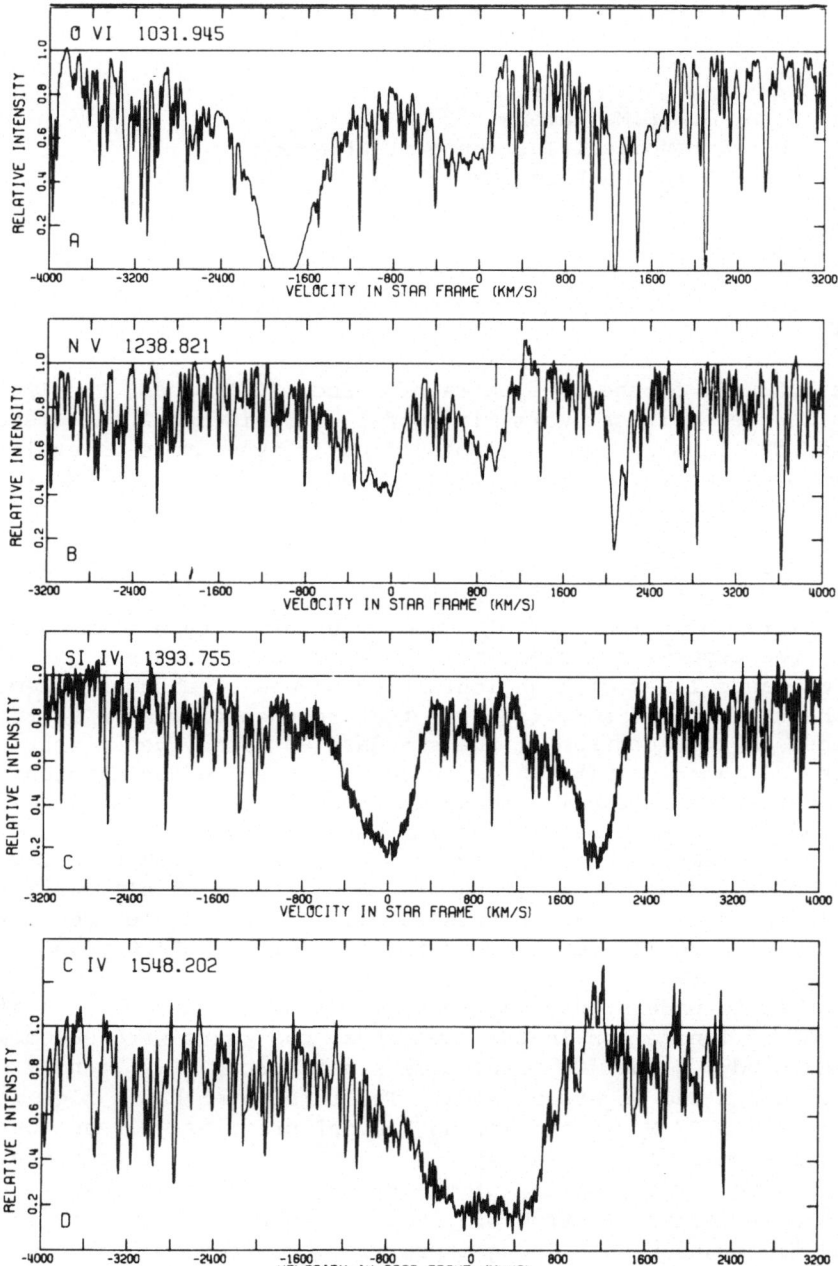

Figure 1. Parts of the tracing of the spectrum of τ Sco near the resonance doublets of O VI, N V, Si IV and C IV. The wavelength of the
strongest component is written. The abscissa shows the velocity in the
frame of the star. The laboratory wavelength of both components is indicated by vertical marks. The adopted continuum was determined from an
inspection of a wavelength region of about 50 to 100 Å wide. The continuum and the zero level near the C IV lines is uncertain.

TABLE 1. ULTRAVIOLET ENVELOPE LINES IN THE SPECTRA OF ζ PUP and τ SCO

ION	LINES λ (Å)	EXC. POT (eV)	ζ PUP profile	V_{edge} (km/s)	τ SCO profile	V_{edge} (km/s)
C III	977	0	p^+	-2660	a	-2000?
	1176	6.50	p^+	-2660		
C IV	1548+1550	0	p^+	-2660	p^+	-2000
N III	990+ 992	0	a	-2660	a	-2000?
N IV	995+1719	16.20	p	-2000		
N V	1238+1243	0	p^+	-2660	a	-1400
O IV	1338+1343	22.40	a	- 800		
O V	1371	19.69	a	- 800		
O VI	1032+1038	0	p^+	-2660	a	-1000
Si IV	1394+1403	0	p^+	-2660	a	-1600
P V	1118+1128	0	p	-2660	a	- 700
S IV	1063+1073	0	p	-2660		
S VI	933+ 944	0	p^+	-2660		

p = P Cygni profile

p^+ = strong P Cygni profile, deeper than 50 per cent

a = extended absorption wing.

Since the warm wind model was derived from the observations of the UV lines in τ Sco and ζ Pup, we will briefly review the analysis of these two stars.

II THE IONIZATION BALANCE IN THE ENVELOPES OF ζ PUP AND τ SCO

The profiles were compared with theoretical profiles for a spherically symmetric expanding envelope, using the Sobolev approximation. By fitting the observed profiles to the predicted ones the radial optical depth could be derived:

$$\tau_{rad}(v) = \frac{\pi e^2}{mc} f \lambda n_i(r) (dr/dv) \qquad (1)$$

where $n_i(r)$ is the ion density and dr/dv is the inverse velocity gradient. At any velocity v the velocity gradient is the same (but unknown)
for all ions. Consequently we compared at any velocity the ratios of
the densities of all observed ions with the following results:

ζ Pup

i. The ratios between the different ions does not change with velocity.
This implies that the ionization balance in the envelope of ζ Pup is
constant from $v = 0.25 \, v_\infty$ to $v = 0.9 \, v_\infty$, i.e. from about 1.5 R_* to
about 10 R_*.
ii. The degree of ionization in ζ Pup is higher than can be explained
by radiative ionization by photospheric radiation, even if ζ Pup has
an effective temperature of $5 \, 10^4$ K. Especially the amount of O VI is
orders of magnitudes larger than predicted.
iii. The ionization balance of all ions fits reasonably well (i.e.
within a factor 3) with the predictions for collisional ionization in
an optically thin plasma of $T_e = 2 \, 10^5$ K throughout the envelope.

Recently, Olson (1978) has reanalyzed the UV lines in the spectrum
of ζ Pup and noted that several of the lines studied by Lamers and Morton are optically thick, especially the O VI lines. These lines can
also be fitted by adopting an ionization balance which changes with
distance in a way predicted by the thin coronal model.

τ Sco

i. The violet lines of the O VI lines extend to -1000 km/s, the N V
lines to -1400 km/s, the Si IV lines to -1600 km/s and the C IV lines
to about -2000 km/s. This suggests that the degree of ionization decreases with increasing velocity in the envelope.
ii. The degree of ionization can be explained by collisional ionization
in an optically thin envelope of $T_e \sim 2 \, 10^5$ K at $v \sim 250$ km/s and $T_e \lesssim$
$1 \, 10^5$ K at $v \gtrsim 1400$ km/s.
iii. If the high degree of ionization were due to radiative ionization
from a large UV or X-ray flux one would expect the degree of ionization
to *increase* with velocity since the flux decreases less rapidly with
distance ($F \propto r^{-2}$) than the electron density ($n_e \propto v^{-1}r^{-2}$). The observations point to *decreasing* degree of ionization.

Lamers and Rogerson also noticed that the absorption of the O VI
and N V lines extend at the long wavelength side to about +250 km/s
(Fig. 1). These red absorption wings resemble Gaussian profiles with a
Doppler velocity of 100 km/s. The presence of these absorptions show
that O VI and N V ions are already present in the low layers of the en-

velope where the velocity is still small (v $\sim$ 0 km/s) and that these
layers are very 'turbulent'. One should remember that τ Sco is a slow
rotator (v sin i $\lesssim$ 5 km/s) and that the photospheric lines have a typical full width at half maximum of 23 km/s.

III   THE WARM WIND MODEL

The warm wind model, proposed by Rogerson and Lamers (1975) is
based on the analysis of the τ Sco. It is an empirical model and as
such it fits the observations of τ Sco perfectly but it lacks a solid
theoretical basis. This is in contrast to the radiation driven wind
model which is theoretically solid but cannot explain the presence of
O VI ions.

In the warm wind model the photosphere has some type of mechanical
energy which is propagated upwards and dissipates above the photosphere.
Due to this dissipation there is a chromosphere in which the temperature increases outward. In the layers where T $\sim$ 2 10^5 K about 10 per
cent of the O and N ions are ionized to O^{5+} and N^{4+}. These ions have
their resonance transitions at 1030 and 1240 A where the stellar flux
reaches its maximum. As there are no strong *photospheric* absorptions of
O VI and N IV the ions in the chromosphere absorb continuum radiation.
The radiation pressure due to the O VI and N V lines alone gives a force which exceeds the gravitational force by a factor four. Consequently,
the chromosphere will be accelerated outwards to a large velocity. As
the envelope expands the temperature decreases outwards by radiative
cooling. As a result of this the relative concentration of O VI and N V
decreases, but the relative concentration of C III, C IV, N III and
Si IV increases. These latter ions, whose resonance transitions are
also in the ultraviolet produce a large radiation pressure as the flow
velocity is large enough to shift those transitions outside the wavelength range of their photospheric counterparts. Consequently the acceleration continues up to large distances and large velocities of v $\gtrsim$
2000 km/s.

This scenario explains the presence of O VI and N V at small velocities. The large 'turbulent' motions observed in their resonance lines
represent the motions in the dissipation layer. The asymmetric wings
found in the high resolution profiles of the photospheric lines by
Smith and Karp (1978) may indicate convection whose flux can provide
the energy required for the heating of the chromosphere.

In this model the origin of the stellar wind is the radiation
pressure by the UV resonance transitions. The mass loss is triggered by
the heating as it produces the right type of ions which do not have
strong photospheric lines. The warm wind model has a strong similarity
to the imperfect flow model of Cannon and Thomas (1977) which also has
a warm trans-sonic region heated by shocks. But in the imperfect flow
model the origin of the mass loss is due to the presence of non thermal

energy in the star which leaks out and produces a mass flux.

The warm wind model applied to ζ Pup has to be modified in two aspects. Firstly, the O VI and N V lines are not the dominant contribu-
tors to the radiation pressure of ζ Pup. In this star, which has a mass
loss rate 10^3 times as large as τ Sco, the radiation pressure is provided by a large number of resonance lines in the Lyman continuum between 230 and 912 A (Lamers and Morton, 1976; Castor et al. 1976).
Secondly, the presence of O VI up to the terminal velocity indicates
the envelope is warm, T ∿ 2 10^5 K, up to large distances. Since radiative cooling is very efficient at this temperature an additional heating process throughout the flow is required to keep its temperature
high. So the warm wind model does not fit the observations of ζ Pup as
accurate as τ Sco.

IV   COMPARISON WITH THE OBSERVATIONS

The anomalous degree of ionization is explained in the warm wind
model by collisional ionization at T ∿ 2.10^5 K. Klein and Castor (1978)
have pointed out that the observed Hα/He II 4686 ratio in O stars fa-
vors a cool wind, rather than a warm wind (see also III,b). However the
dominant contribution to profiles of these lines comes from the region
of small velocities v ∿≾ 0.3 v_∞ where the temperature of ζ Pup is unde-
termined. Moreover, at these layers the profiles are sensitive to the
velocity law, since ρ ∝ v^{-1}, and the velocity law adopted by Klein and
Castor is much steeper than the UV observations indicate.

The infrared excess of ζ Pup at 2.2 μm can be explained by the
warm wind model, but this depends critically on the velocity law at
v < 0.25 v_∞ for which there is no observational information.

The ¼ keV X-ray upperlimit for ζ Pup, measured by Mewe et al.
(1975) is in agreement with the warm wind model. The warm wind is optically thick at ¼ keV. For an optically thick wind the emergent luminosity may be estimated from the Eddington-Barbier relation to be

$$L_\nu = 4\pi \, R_\nu^2 \, (2/3) \, \pi \, S_\nu \, (2/3)$$

where R_ν is the radius at which the monochromatic optical depth, τ_ν,
is 2/3, and S_ν is the source function at that radius. The source function is the ratio of the total emission to the total absorption. At ¼
keV the emission in a gas at T_e ∿ 2×10^5 K is primarily due to recombination of helium and hydrogen. The opacity is bound-free absorption by
metals, e.g. Si^{+4}, O^{+3}, Mg^{+3} and of He^+. The source function therefore
tends to decrease with radius in proportion to N_e. The effective radius
is at about 10 stellar radii. When these effects are taken into account,
the predicted X-ray flux is well below the ANS upperlimit for ζ Pup. It
should be noted that even though the warm wind may be rather optically
thick at some frequencies, the calculated ionization balance is not in-

consistent if the collisional ionization rates are larger than the radiative ionization rates.

REFERENCES

Cannon, C.J., Thomas, R.N.: 1977, Astrophys. J. 210, p. 910.
Castor, J.I., Abbott, D.C., Klein, R.I.: 1976, in *Physique des Mouvements dans les Atmosphères Stellaires*, ed. R. Cayrel and M. Steinberg, CNRS, Paris, p. 453.
Klein, R.I., Castor, J.I.: 1978, Astrophys. J.
Lamers, H.J.G.L.M., Morton, D.C.: 1976, Astrophys. J. Suppl. 32, p. 715.
Lamers, H.J.G.L.M., Rogerson, J.B.: 1978, Astron. Astrophys. 66, p. 417.
Lamers, H.J.G.L.M., Snow, T.P.: 1978, Astrophys. J. 219, p. 504.
Lamers, H.J.G.L.M., Stalio, R., Kondo, Y.: 1978, Astrophys. J. 223, p. 207.
Mewe, R., Heise, J., Gronenschild, E.H.B.M., Brinkman, A.C., Schrijver, J., den Boggende, A.J.F.: 1975, Astrophys. J. (Letters), 202, L67.
Olson, G.: 1978, Astrophys. J. (in press).
Rogerson, J.B., Lamers. H.J.G.L.M.: 1975, Nature, 256, p. 19.
Smith, M.A., Karp, A.H.: 1978, Astrophys. J. 219, p. 522.
Snow, T.P., Morton, D.C.: 1976, Astrophys. J. Suppl., 32, p. 429.

DISCUSSION FOLLOWING LAMERS

Hearn: Could you say something about the optical line observations
and IR? Do they agree with your model?

Lamers: There may be troubles with the helium lines. Our model
does not specify the temperature or velocity inside the region of about
one-fourth the terminal velocity. The He I lines are probably formed
here, and this would mean that this region is cool. However, this might
not agree with the O VI lines. For the IR flux observed by Barlow and
Cohen, we are not in difficulty. There are too many free parameters
since the velocity law is not specified. One can't make a conclusive
argument as yet.

Hearn: How about variability?

Lamers: One does observe photospheric motions in the spectrum of
τ Sco [Smith and Karp, Ap. J. 219, 522 (1978)]. Presumably these will
propagate outwards in some manner. Nobody has looked for variability
in τ Sco.

Abbott: Would you care to put an uncertainty on your mass loss
rate for τ Sco?

Lamers: The rate is $7.0 \pm 1.6 \times 10^{-9} \ M_\odot \ yr^{-1}$.

Underhill: I have heard it said that your ζ Pup model has a high
temperature out to 10 stellar radii. How do you get this result? Why
can't you have this hot region which produces the O VI close to the
star with a very rapid velocity rise?

Lamers: The ionization balance is more-or-less constant in the
region from $0.25 \ v_\infty$ to v_∞. The distance argument comes from the huge
emission components to the P Cygni profiles. The more extensive the
emission the more extended the envelope contribution.

Underhill: This assumes that the same type of emission mechanism
operates for all ions.

Lamers: Yes, we are assuming a scattering process for the emission.
This assumption is justified for the UV resonance lines [e.g., Castor
and Lamers, An atlas of theoretical P Cygni profiles, Ap. J. Suppl.
(1979) in press]. We also assumed one velocity law for all ions.

Sreenivasan: What is the source of the turbulence in these stars?
In τ Sco, for example? Does it have an appreciable rotation (V sin i)?

Lamers: No, the projected rotational velocity is smaller than
5 km/s. Smith and Karp [Ap. J. 219, 522 (1978)] proposed a He II convection zone.

de Loore and I have tried to find an explanation for the observed motions in the atmospheres of supergiants (Lamers and de Loore, 1976, in Physique des Mouvements dans les Atmosphères Stellaires, eds. Cayrel and Steinberg, CNRS, Paris). We compared the observed kinetic fluxes in 10 supergiants of types O5 to G2 with predicted fluxes.

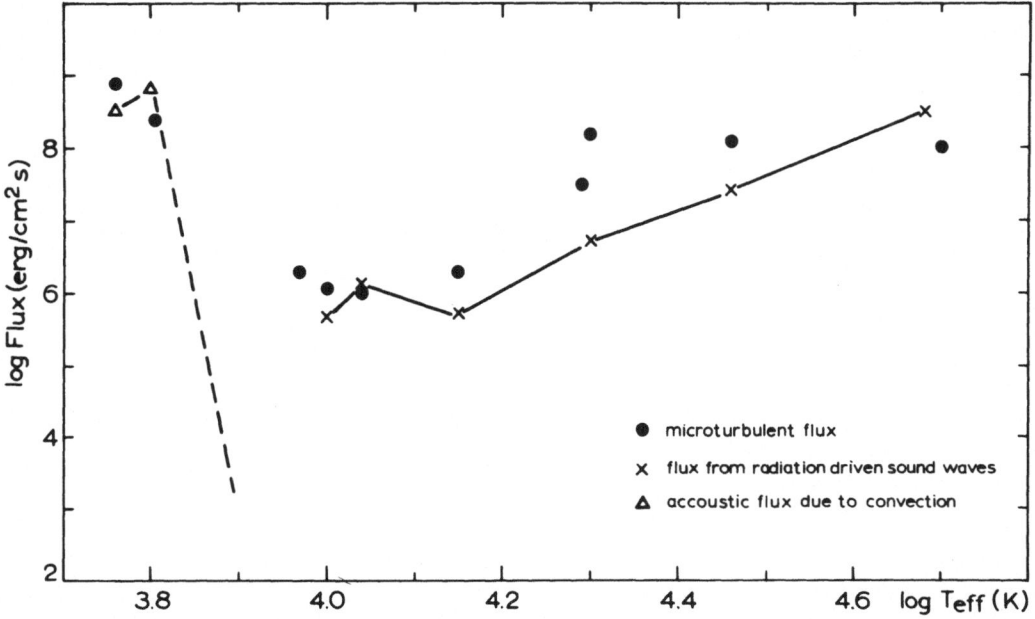

The dots show the "observed fluxes" derived from the microturbulent velocities and assuming that these motions are in fact sound-waves. The triangles are the predicted fluxes of acoustic waves, generated by convection in the models by de Loore [Astroph. Space Sci. 6, 60 (1970)]. The crosses are the predicted fluxes of radiation-driven sound waves, predicted by Hearn [Astron. Astrophys. 23, 97 (1973)]. We conclude that: 1) Although the microturbulent velocity in early type supergiants does not change very much with spectral type, the associated flux shows a minimum around type A. This suggests that microturbulence in stars earlier than A is created by a different mechanism than in stars later than A. 2) The similarity of the observed and predicted variation of flux with T_{eff} suggests that the microturbulence in O and B stars is created by interaction of matter with radiation, and in supergiants of types F, G and possibly later by convection.

THE CORONA PLUS COOL WIND MODEL FOR O$\int$ STARS AND OB SUPERGIANTS

Joseph P. Cassinelli
Washburn Observatory
University of Wisconsin - Madison
Madison, Wisconsin 53706

ABSTRACT

The anomalously strong OVI and NV lines in O stars and the CIV and Si IV lines in B supergiants may be due to Auger ionization by x-rays from a thin coronal zone at the base of their cool stellar winds. This paper summarizes the results of several studies to determine constraints on the size and temperature of coronal zones from calculations of the effects of coronae on continuum and line spectra. The model that has resulted can be tested by observations from HEAO-B of the predicted 2 kev emergent fluxes. The model explains very well the persistence of the OVI and NV in the ultraviolet spectra of supergiants to class B0.5 and B2 respectively, and it predicts that CIV should be observable in the ultraviolet spectra at and beyond B5I.

INTRODUCTION

The high stages of ionization that are seen in the Copernicus spectra of the luminous O and B stars can be explained either by assuming that the winds are at an elevated temperature and that the ions are produced by electron collisions and photoionization by the ambient diffuse field or by assuming that the winds are subjected to a radiation field which is harder than what is expected from photospheric models.

Lamers and Morton found that the overall degree of ionization in the wind of ζ Pup, O4f, is about the same as one would expect in an optically thin plasma at a temperature of 2×10^5 K. So they postulated that the winds have elevated or "warm" temperatures throughout the extended region in which OVI must exist.

One serious objection to the "warm wind" models is that an extremely large rate of mechanical energy deposition is required to maintain the elevated temperatures, because such temperatures are near the peak of the radiative cooling curves of, for example, Cox and Tucker (1969). If we assume that the temperature is maintained by the deposition of acoustic or mechanical energy that emanates from the surface of the star, the

201

P. S. Conti and C. W. H. de Loore (eds.), Mass Loss and Evolution of O-Type Stars, 201-213.

the mechanical luminosity of ζ Pup would be about 2×10^{38} ergs/sec or approximately 5 percent of the radiative luminosity of the star. Furthermore, it must be deposited over a spatially extended region of several radii.

The corona plus cool wind model is an alternate semi-empirical picture of the stellar winds of the O and B supergiants, which, as we shall see, can explain the anomalously high ionization stages and which requires a much smaller rate of mechanical deposition. In this model, we postulate that the mechanical energy is deposited only near the base of the stellar wind and it produces a thin coronal zone with a temperature near 5×10^6 K. Beyond the coronal zone, the winds are assumed to have the relatively cool temperatures of about $0.8 \, T_{eff}$ that would be appropriate for a gas in radiative equilibrium. Figure 1 shows the temperature distribution that has evolved from several studies that are reported below.

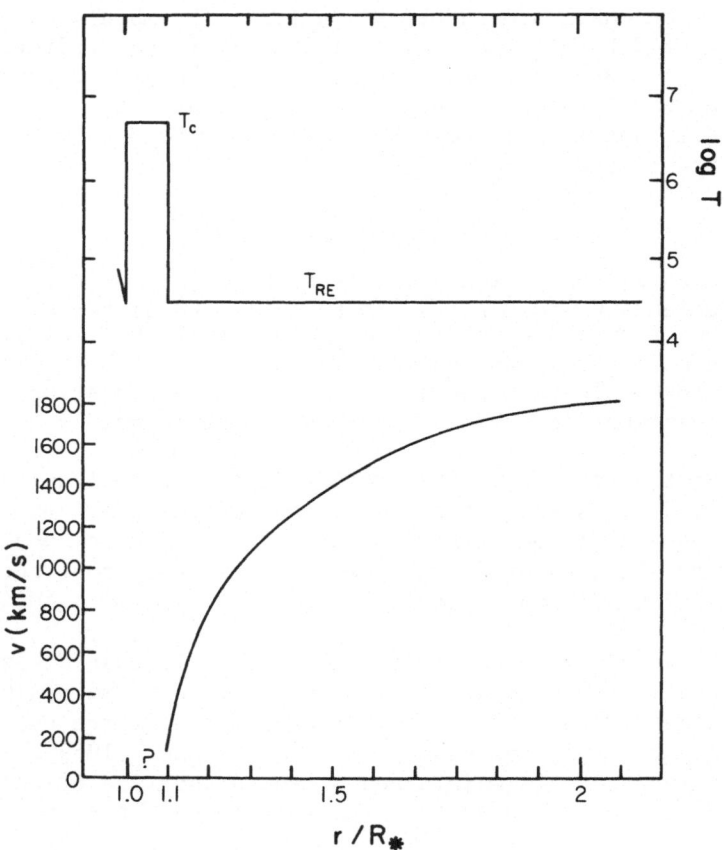

(Figure 1)

A two component temperature structure was first suggested for O and
B supergiants by Hearn (1975). He proposed that the flows are driven to
near escape speed by thermal pressure gradients in the coronal zone, and
that the final acceleration to the high terminal velocities is due to
radiation force on the line opacity in the cool wind. The transition
from coronal to cool wind temperatures should occur, rather abruptly,
where the mechanical deposition ceases, because the densities are so
large ($\sim 10^{10} cm^{-3}$) that cooling by radiative recombination is very effective. Hearn deduced from a simplified analysis of the Hα P Cygni line
in ζ Ori O9.5Ia that the coronal zone extends to 2 stellar radii.

Lee Hartmann, Gordon Olson, Roberto Stalio and I have investigated
the observational consequences of hybrid corona-cool wind models in attempt to deduce whether the coronae exist. Although definitive proof
of the existence of coronae has not resulted we have, in the process,
developed a semi-empirical model which rather nicely explains the ionization anomalies, and which is useful for suggesting new observational
tests.

In the next section is summarized the constraints that have been
derived concerning the coronal structure in several papers which are already in print. In the last two sections the explanation of the high
ionization stages for ζ Pup and for other O and B supergiants is presented.

CONSTRAINTS ON CORONAL STRUCTURE

The first models that were considered had the rather extended coronal regions that were suggested by Hearn (1975) on the basis of his Hα
analysis.

Cassinelli and Hartmann (1977) calculated the effect of extended
coronal zones on infrared continuum distributions. The infrared continuum can be used to probe the temperature and velocity structure at the
base of the massive stellar winds of O and B stars because the opacity
is rather high and optical depth unity occurs in these expanding layers.
The free-free opacity varies as λ^2, and thus as one looks to longer wavelengths one "sees" an increasingly larger star. Because of their winds,
the early type supergiants are expected to have infrared excesses.
Cassinelli and Hartmann found that coronal zones should produce an additional excess or broad bump in the continuum between 20 μm and 100 μm,
and some evidence for the bump should be seen between 10 and 20 μm.
Barlow and Cohen (1977), however, found no significant extra rise in the
continuum between 10 and 20 μm. Its absence can be explained in the context of the coronal plus cool wind model by assuming that the coronal zone
is much thinner than originally postulated and that the temperature rise
at the base of the flow is more abrupt.

Cassinelli, Olson and Stalio (1978) reinvestigated the evidence from
Hα that there may be extended coronal zones, by calculating theoretical
profiles using the Sobolev technique. It was found that models with extended coronal zones in which the outflow was raised to speeds of order

500 km/sec give rise to broad flat topped emission components of the P
Cygni profiles. This is because the corona is a zone of null contribution
to the profile and the extra emission that would have come from the denser
lower wind layers is absent. Such broad flat topped Hα profiles are not
seen in the spectra of O and B supergiants. It was concluded that the
coronal zones must be very thin ($\sim$ 0.1 R_*) and that the flow velocity at
the start of the cool wind zone is small, about 1/20 v_∞. This is a radical
departure from the original Hearn model and it suggests that the coronae
do not play the dominant role in driving the mass outflow, but may nevertheless, affect the ionization balance in the wind. Olson (1978) pursued
the studies of line profiles and developed a method somewhat like that of
Lamers and Morton (1976) for determining the ionization and velocity
structure of winds from fits to ultraviolet line profiles. He derived
the degree of ionization of several elements in the wind of ζ Pup and
found that they compared favorably with the predictions of the corona
plus cool wind model. The study showed that the fractional abundance
of NV and OVI must be at least as large as 10^{-4} to account for the strong
resonance lines in O and B supergiants. Olson found that a velocity
distribution like that shown in Figure 1b gives rise to a good fit to the
resonance lines in ζ Pup. Again, the initial velocity in the cool wind
was found to be low ($\lesssim 0.1 v_\infty$).

The Hα and ultraviolet resonance lines have no contribution from the
gas inside the coronal zone and hence the studies of these lines has led
to no information concerning the velocity structure inside the zone.
However, it is possible to derive some useful information concerning integrated properties of the corona, such as the coronal emission measure,
$EM_c = \int N_e^2$ dvol, and the average coronal temperature T_c by considering
the production of the anomalous ionization stages in the winds.
Cassinelli and Olson (1978) carried out such an analysis for ζ Pup.

EFFECTS OF CORONAL RADIATION ON THE WIND OF ζ PUP

ζ Pup is commonly chosen for the most detailed analyses because good
observational constraints on the wind structure are available. The mass
loss rate is known to be near 7 x 10^{-6} M⊙/yr from studies of Hα and
ultraviolet line profiles and from infrared and radio continuum observations. (Lamers and Morton, 1976; Cassinelli, Olson and Stalio, 1977;
Barlow and Cohen, 1977; Morton and Wright, 1978). The degree of ionization, of say OVI and NV, is know from the resonance line analyses of Lamers
and Morton (1976) and Olson (1978).

The assumed radiative flux from the photosphere and corona of ζ
Pup is shown in Figure 2. The star is assumed to have an effective temperature of 42000 K. There should be a drop in the emergent flux at the
HeII, n=1, edge at 228Å, and in absence of the coronal zone there would
be little flux shortward of the O^{+4} to O^{+5} ionization edge at 108Å. The
energy emergent at the high energies is due to the corona, which is assumed to have a temperature of 5 x 10^6 K. The underlying coronal continuum is due to bremstrahlung and there is a large contribution from
emission lines, which is shown as if smoothed into 5 to 10Å bands.

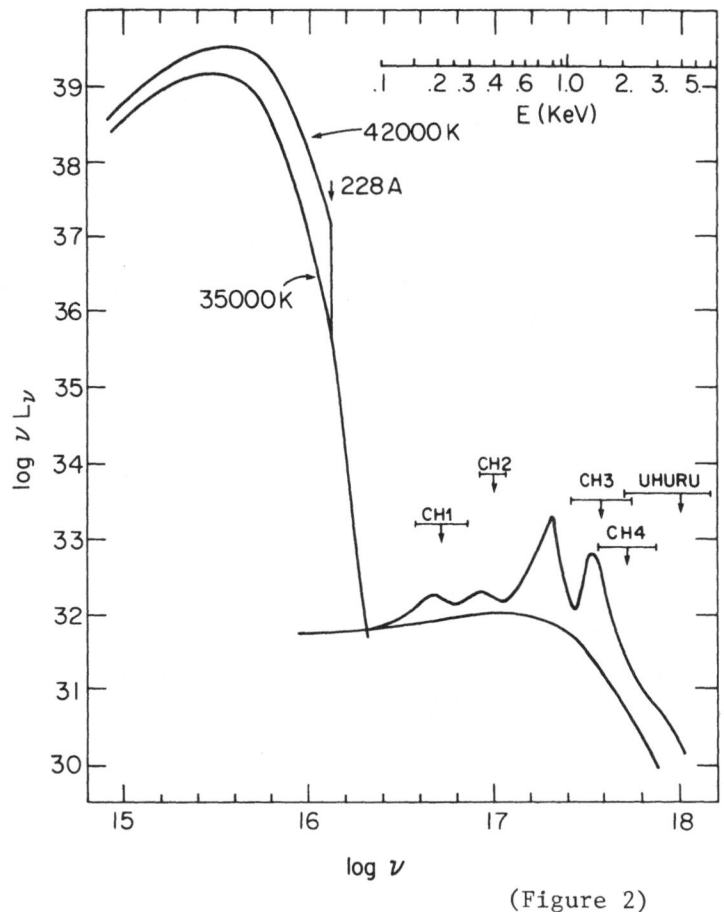

(Figure 2)

The knee in the coronal flux occurs near $h\nu/kT_c = 1$, thus, if the coronal temperature were increased, the knee would be shifted to higher energies. The coronal emission measure, EM_c, determines the magnitude of the coronal flux. The distribution shown of Figure 2 corresponds to $EM_c = 10^{56} cm^{-3}$. If the wind of ζ Pup were optically thin, that value for the emission measure would be sufficient to produce the required amount of OVI in the wind to explain the observed profile.

Also shown in the figure are x-ray upper limits for ζ Pup as were derived from four channels of the ANS detector and from Uhuru satellite observations. Thus, if the wind were thin the corona flux distribution would suffice to explain the ionization observations and to satisfy the x-ray constraints.

However, the winds of O and B supergiants and Of stars are not opti-cally thin in the frequency range which contains many important ionization edges. Figure 3 shows the run of total optical depth versus frequency for a model of ζ Pup in which the wind is assumed to have an electron temp-erature of 35000 K. The optical depth is greater than 10 at the HeII n = 1 edge and it rises to even larger values as other opacity edges are

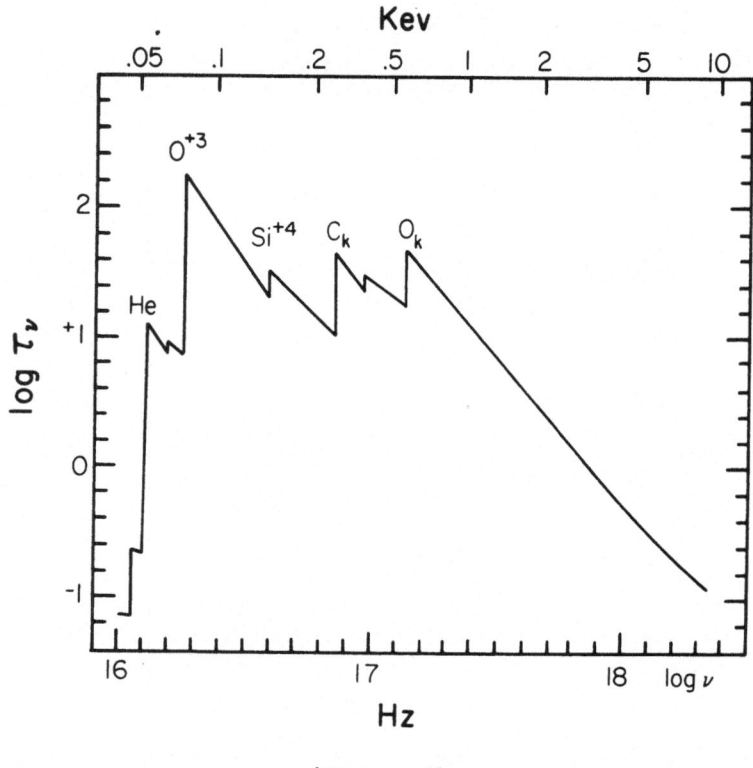

(Figure 3)

crossed. The last important opacity edges are due to K shell ionization
of carbon, nitrogen and oxygen (two of which are noted as C_k and O_k in
the figure). Beyond the oxygen K shell edge the opacity, and hence also
the optical depth, decreases as $\nu^{-2.6}$, and not until the relatively high
energy of 2 keV does the optical depth decrease to values less than unity. Therefore, only the harder x-rays penetrate far out into the wind
where the high stages of ionization are known to exist.

Since the wind is thick for a broad range in frequencies it is
necessary to account both for the attenuation of the coronal radiation
field, and for the diffuse radiation that originates in the wind. The
diffuse field plays an important but indirect role in explaining the
anomalous ionization. It is important in determining the general ionization balance of the gas and in particular, determining the dominant
stages of ionization because several important ionization edges occur
at frequencies for which the wind is very thick. To a fairly good first
approximation, in the cool wind model, one can assume that the diffuse
radiation field is equal to $B_\nu(T_e)$ at these frequencies.

What is of particular interest to us is the production of the high
stages of ionization by absorption of x-rays with energies beyond 1 keV.
For C, N and O, an absorption of an x-ray K shell ionization essentially

always leads to the net ejection of two electrons instead of one because of the Auger effect (Weisheit, 1974).

The process of determining the abundance of an interesting high stage of ionization can be carried out in two steps. The fractional abundance of the dominant stages of ionization can be determined as if there were no coronal radiation, because only a small fraction of a percent of the dominant stages are ionized to high stages by the Auger process. The second step is to determine the abundance of the "high stage," which is two ion states above. This can be determined rather easily as follows.

The coronal x-rays will be attenuated by the K shell opacities of C, N and O, and these opacities are nearly independent of the number of electrons in the outer shells of the ion that is absorbing the photon (Daltabuit and Cox, 1972). So one needs to know the column density of the nuclei of C, N and O, which depends on the mass loss rate and velocity law in the wind, but <u>not</u> on the assumed temperature of the wind. Letting $\theta_\nu(r)$ be the attenuation optical depth measured outward from the corona at r, then $\theta_\nu = \theta_o(\nu_o/\nu)^{2.6}$, where $\theta_o(r)$ is the value at the K shell edge of oxygen at ν_o(0.6 keV). The ratio of the number of high ions of an element El^{+p+2} to the number in the parent stage El^{+p} is given by

$$\frac{N(El^{p+2})}{N(El^p)} = \frac{1}{N_e \alpha} \int_{\nu_{El}}^{\infty} \frac{4\pi}{h\nu} a_\nu W F_\nu^c e^{-\theta_\nu} d\nu \quad , \tag{1}$$

where a_ν is the K shell absorption cross section $\approx a_o(\nu_o/\nu)^{2.6}$, $g W F_\nu^c e^{-\theta_\nu}$ is the spatially diluted and attenuated coronal flux, $N_e(r)$ is the electron density and $\alpha(T_e)$ is the recombination coefficient from El^{p+2} to El^{p+1}. Expressing the results in terms of the fractional abundance, $g(El^p) = N(El^p)/N_{El}$ and noting that the coronal flux distribution has the general shape of a bremmstrahlung spectrum, $C EM_c \exp(-h\nu/kT_c)$ we get

$$\frac{g(El^{p+2})}{g(El^p)} \alpha(T_e) = \frac{4\pi}{h} \frac{a_o C W EM_c}{N_e(r)} \int_{\nu_{El}}^{\infty} \frac{\exp[-(h\nu/kT_c) - \theta_\nu]}{\nu^{3.6}} d\nu \quad . \tag{2}$$

This equation nicely separates the quantities which depend on the wind temperature $g(El^p)$ and $\alpha(T_e)$ from quantities which depend only on the local radius, r, and the coronal parameters EM_c and T_c. Thus, we see immediately that the abundance of the "anomalously high stage" of ionization is directly proportional to EM_c and to the abundance of the parent stage. The dependence on coronal temperature is slightly more complicated, but we see that as T_c increases a greater fraction of the flux lies at the higher energies which can penetrate far into the wind. Because of the $\theta_\nu(r)$ term it is possible for the abundance of a high ion state to decrease in the outward direction, in contrast to the thin case in which the ionization increases in the outward direction.

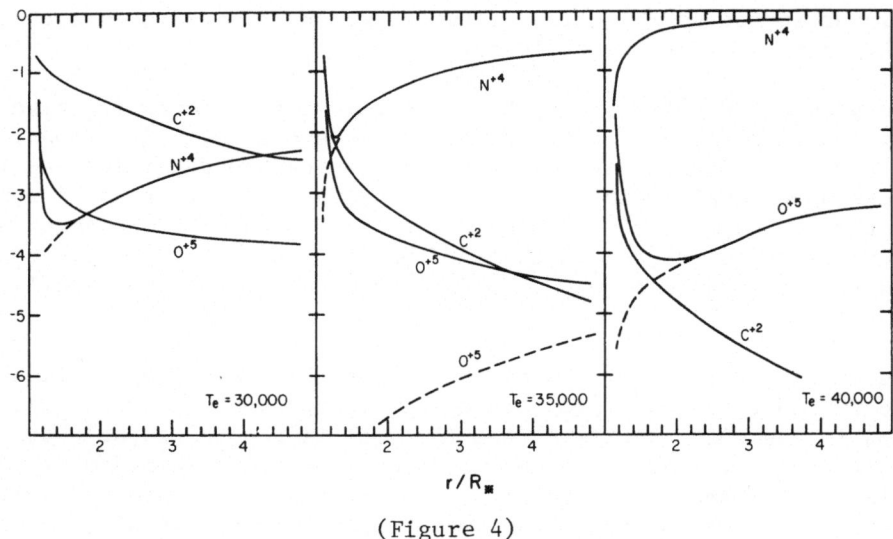

(Figure 4)

Figure 4 shows the calculated spatial structure of the ionization for several ionization stages observed in the wind of ζ Pup. Note that the O^{+5} abundance decreases, because of optical depth effects for some cases. Results are shown for three assumed wind temperatures. The dashed line labeled O^{+5} shown in the second and third panel of the figure, are results for the case in which there is no coronal radiation, i.e., $EM_C = 0$. Thus, the diffuse radiation field can produce a significant amount of O^{+5} if the wind is at temperatures larger than about 40,000 K. Castor has investigated in much greater detail the effect of the diffuse field in producing the anomalous ionization that is seen in O and B supergiants, and the model has been referred to as a "tepid wind" model to distinguish it from the Lamers "warm wind" models. The warm, tepid, and corona plus cool wind models are discussed in a review paper by Cassinelli, Castor and Lamers (1978).

Figure 5 shows the most important result of the calculations; the x-ray flux that is predicted to penetrate through the wind and hence is potentially observable. There is negligible flux transmitted between 0.1 to 1 keV because the wind is so thick at these frequencies, and hence soft x-ray detectors are not likely to find evidence of hot coronae. The soft x-ray flux expected for a warm wind model is indicated by the WW in the figure. The flux expected from the coronal plus cool wind model for ζ Pup is fairly large at 2 keV, and is just below the existing ANS upper limit. The predicted flux is above the estimated detection limit for HEAO-B which is scheduled for launch in late 1978. We can easily see that the crucial test of the corona plus cool wind model will be provided by the HEAO-B observations.

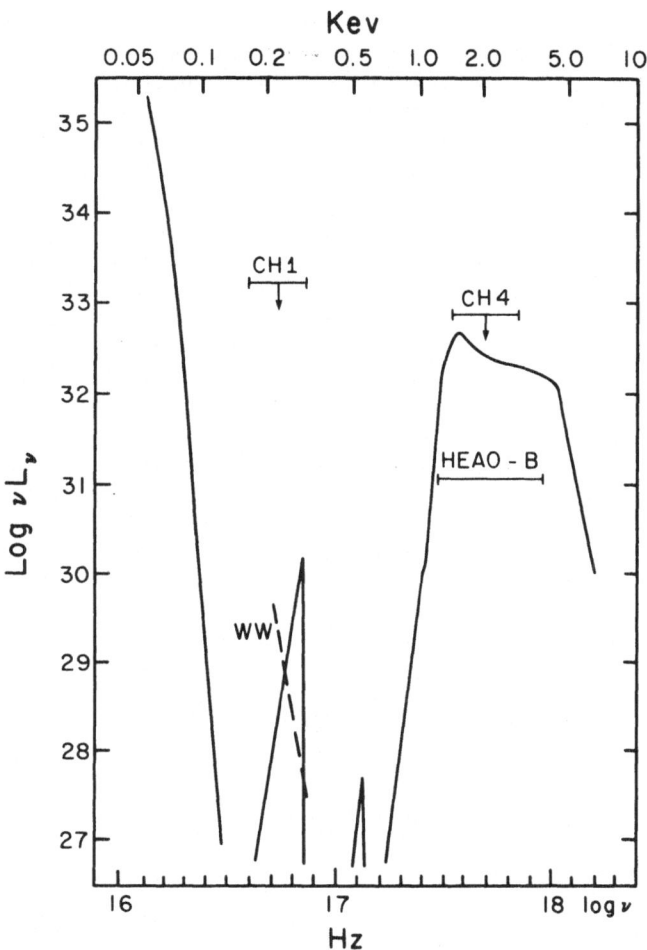

(Figure 5)   Emergent x-ray flux from ζ Pup

THE ANOMALOUS IONIZATION IN B SUPERGIANTS

There are other predictions of the corona plus cool wind model that can be tested with existing satellite data. The most interesting concerns the persistence of the anomalous ionization stages into the B supergiant spectral range.

The Auger process produces enhanced abundance of ions which are two stages above the dominant stages of ionization. Thus, if oxygen is primarily in the ionization stage OIV in the stellar wind then we should expect to find a greatly enhanced OVI abundance. As we look at stars of later spectral types, we should expect that OVI will weaken and disappear at the spectral range for which the dominant stage of ionization is shifting from OIV to OIII. A similar situation should hold for NV as produced from NIII, and CIV as produced from CII.

    The first step in the analysis is to estimate the ranges in effective
temperature for which CII, NIII and OIV are expected to be very abundant.
I have carried out some simplified calculation for this purpose, and assumed that Ne = 10^{10}cm^{-3} and used a dilution factor appropriate for two
stellar radii. The electron temperature in the cool wind is determined
by the effective temperature of the star, $T_e \approx 0.8\ T_{eff}$. From calculations of relative abundances versus effective temperature two useful
temperature ranges can be derived: 1) The range in T_{eff} over which the
ion has an abundance of at least 10 percent and is, therefore, a likely
candidate for being a parent of an anomalously high ion stage; 2) The
range in T_{eff} over which the ion has an abundance of at least 10^{-4} and
is, therefore, likely to be seen even in absence of the Auger process.

    Table 1 summarizes the results. Listed are resonance lines from
relatively high stages of ionization whose appearance in the spectrum may
indicate the presence of x-rays. The two ranges of effective temperatures
discussed above are shown, as is the parent ion and the K shell ionization
edge of that ion. For simplicity, it is assumed in this analysis that
the Auger ionization of Silicon, Sulpher and Phosphorus will also lead
primarily to enhancement of ions two stages higher. This is not entirely
correct and better calculations are in progress.

    Note from Table 1 that NV and OVI are to be expected, via the
Auger process, in the spectra of supergiants with affective temperature
as low as 20000 and 30000 respectively. (More detailed calculations
lower these estimates even further) In absence of the x-ray ionization
OVI should not be seen in the cool winds of any stars and NV should only be
seen in the very hottest. (See also, Lamers and Snow, 1978).

COMPARISION WITH OBSERVATIONS

    Table 2 again lists the lines from high ion stages and indicates
whether or not the line is present in the spectra of 10 supergiants,
with a wide range in effective temperature. A plus sign indicates that
the line is seen, an exclamation mark (+!) indicates that the line is
"anomalous", i.e., not expected to be produced by the photospheric radiation field or by collisions or diffuse radiation in a cool stellar
wind. Finally, the range in effective temperatures above which the line
could be produced by the Auger mechanism in the corona plus cool wind
model is indicated by the series of asterisks (***).

- OVI is seen in the spectra of stars of spectral type as late as BO.5.
Morton discussed earlier in this meeting that the line is not present
at B1 and later.

- NV is present in the Copernicus spectra in the catalogue of Snow and
Morton (1976) in stars as late as B1. To help me fill in the blank at
B2 in the original catalogue, Walter Upson at Princeton and Blair Savage
at Wisconsin have kindly allowed me to look at Copernicus spectra of θ
Ara, and NV is definitely present. It disappears at B3, as we would
expect from the Auger model.

Table 1

ULTRAVIOLET RESONANCE LINES LIKELY TO BE ENHANCED BY AUGER IONIZATION

LINE	(Å)	f	Abundance N_{El}/N_H	T_{eff} range normally expected (10^3 °K)	T_{eff} range ion expected via Auger process (10^3 °K)	Parent ion	K-shell edge (kev)
C IV	1548.2	0.194	3.7×10^{-4}	25 – 65	9 – 20	C^{+1}	.296
	50.8	0.097					
N V	1238.8	0.152	1.1×10^{-4}	$T_{eff} > 39$	20 – 37	N^{+2}	.432
	1242.8	0.076					
O VI	1031.9	0.130	6.8×10^{-4}	$T_{eff} > 54$	30 – 50	O^{+3}	.595
	1037.6	0.065					
Si IV	1393.8	0.528	3.5×10^{-5}	20 – 45	8 – 16	Si^{+1}	1.88
	1402.8	0.262					
P V	1118.0	0.495	2.7×10^{-7}	33 – 70	14 – 29	P^{+2}	2.24
	1280.0	0.245					
S IV	1062.7	0.038	1.6×10^{-5}	19 – 50	8 – 19	S^{+1}	2.54
	1073.0	0.037					
S VI	933.4	0.426	1.6×10^{-5}	$T_{eff} > 37$	25 – 44	S^{+3}	2.58
	944.5	0.210					

Table 2

RESONANCE LINES SEEN FROM HIGH STAGES OF IONIZATION IN OB SUPERGIANTS

STAR		T_{eff}	$\dot{M}$ 10^{-6} $M_\odot$/yr	S IV λ1065	Si IV λ1400	C IV λ1550	P V λ1120	N V λ1240	O VI λ1035	S VI λ940
ζ Pup	O4f	42000	7.0	+	+	+	+	+	+!	+
α Cam ζ Ori	O9.5 Ia	30000	2.5	+	+	+	+	+!	+!	
ε Ori	B0 Ia	24800 28800	4.3	+	+	+		+!	+!	
κ Ori	B0.5 Ia	26400	2.5	+	+			+!	+! ***	***
ρ Leo	B1 Iab	21000	1.0		+			+!	–	
θ Ara	B2 Ib	18000						+! ***	–	
o² CMa	B3 Ia	15500	2.0	–	+!	+!	***	–	–	
η CMa	B5 Ia	13300	.4	–	(+!)	(–) ***		–	–	
β Ori	B8 Ia	11600	1.0	– ***	+! ***					

+   indicates that the line shows mass loss effects; P Cygni profile or displaced absorption.

–   indicates that the line shows no mass loss effects.

     blank indicates no information.

( ) indicates an uncertainty.

+!   indicates that this line is not expected in a cool radiative equilibrium wind, yet is seen.

*** above which is the region where Auger enhancement is expected.

- CIV is not easily detected on the Copernicus scans because the spec-
trometers are not sensitive at 1550Å. However, some useful data is a-
vailable from the Skylab survey of Parsons et al (1977). The line is
fairly strong at B3 but is much weaker or absent at B5. IUE data is
now becoming available, and I think the line should be seen in the
spectra of B5 stars and later, especially since Si IV is seen at B8I.

- Sulpher seems to disagree with the predictions but this may be due to
the large K shell ionization energy (see Table 1) or due to our oversimplified analysis of the Auger process.

The predictions of the corona plus cool wind model agree very well
with the observed persistence of OVI, NV and Si IV, and there is hint of
agreement with CIV. However, this agreement of predictions and theory
cannot be considered sufficient proof of the validity of the model because Lamers and Snow (1978) have shown that the persistence of OVI to
B0.5 and Si IV to B8 could also be explained by assuming the temperature
of the warm winds are lower ($\sim$ 80000 K) for stars showing no OVI.

CONCLUSIONS

Several lines of evidence based on IR continuum, x-ray upper limits
and profiles of Hα and ultraviolet resonance lines, indicate that if there
are coronal zones at the base of the winds of luminous early-type stars,
they must be very thin ($\lesssim 0.1 R_*$). There is interesting positive evidence
for the existence of the coronae in that the model explains, in a very
natural way, the persistence of the anomalous ionization into the B
spectral class. The span of effective temperatures over which the high
ion states exists is explained by the Auger ionization process whereby
two electrons are removed from the dominant stage of ionization, and thus
the anomalously high ions cease to appear at temperatures at which the
parent ion ceases to be the dominant stage. Conclusive proof for the existence of hot coronal zones in O and B supergiants will have to await
the observations from the HEAO-B satellite.

REFERENCES

Barlow, M.J., and Cohen, M.: 1977, Ap.J. 213, p. 737.
Cassinelli, J.P., Castor, J.I., and Lamers, H.J.G.L.M.: 1978, Publ. Ast.
  Soc. Pac., October 1.
Cassinelli, J.P., and Hartmann, L.: 1977, Ap.J. 212, p. 488.
Cassinelli, J.P., Olson, G.L., and Stalio, R.: 1978, Ap.J. 220, p. 573.
Cox, D.P., and Tucker, W.: 1969, Ap.J. 157, p. 1157.
Daltabuit, E., and Cox, D.E.: 1972, Ap.J. 177, p. 855.
Hearn, A.G.: 1975, Astr. & Ap. 40, p. 277.
Lamers, H.J.G.L.M., and Morton, D.C.: 1976, Ap. J. Supp. 32, p. 715.
Lamers, H.J.G.L.M., and Snow, T.P.: 1978, Ap.J. 219, p. 504.
Morton, D.C., and Wright, A.E.: 1978, Mon. Not. Roy. Ast. Soc. 182, p. 47P.
Olson, G.L.: 1978, to appear in Ap.J., November 1.
Parsons, S.B., Wrays, J.D., Henize, K.G., and Benedict, G.F.: 1978, preprint.
Snow, T.P., and Morton, D.C.: 1976, Ap. J. Supp. 32, p. 429.
Weisheit, J.C.: 1974, Ap. J. 190, p. 735.

DISCUSSION FOLLOWING CASSINELLI

Hearn: How well does your model fit the IR observations?

Cassinelli: At one time we had a temperature maximum that extended to 2 stellar radii. This provided an IR "bump" which is not observed. Now, however, our coronal model is sufficiently thin in extent that there is no conflict with the IR data.

Hearn: Can you account for a wind in τ Sco in your model?

Cassinelli: I haven't considered this star.

Hearn: What about variability?

Cassinelli: One could imagine fluctuations in this high temperature corona of order of hours. This would drastically affect the ionization in the outer regions, and so could produce the kinds of variability observed in ultraviolet spectra by York et al. [Ap. J. 213, 261 (1977)].

Morton: Do C III and O VI resonance lines have same terminal velocity in supergiants?

Cassinelli: Yes, at least for the stars with large mass loss rates.

THE THERMODYNAMIC REQUIREMENTS ON ATMOSPHERIC MODELS IMPOSED BY OBSERVED
STELLAR NONTHERMAL MASS-FLUXES AND BY THOSE OBSERVED NONTHERMAL FEATURES
ENHANCED IN Xe STARS

Richard N. Thomas
Institut d'Astrophysique, Paris and Department of Astro-
Geophysics, University of Colorado, Boulder, CO 80309

INTRODUCTION

If I understand correctly, this session of the Symposium on Mass-
Loss and Evolution of O Stars aims at clarifying the merits and demerits
of four "theories" for the observed, nonthermal mass-loss from these
stars. Hearn has summarized what he considers to be the essential characteristics of each of the four, especially relative to a set of questions which, he considers, put the observational requirements on the
"theories" in focus. Representatives of each of the "radiation-pressure
initiated, radiative-equilibrium controlled", "hot corona", and "warm
corona" alternatives have elaborated on Hearn's summary, to stress what
they consider essential. So, now, I would do the same for what we cari-
cature as the "imperfect wind-tunnel" model --- not theory, which I
assert does not yet exist --- both for nonthermal mass-flux, and for
other observed nonthermal phenomena,[1] in stellar atmospheres generally,
not just in O stars particularly. I assert, that in studying nonthermal
mass-flux from O stars, if you limit your attention only to nonthermal
mass-flux, and only to O stars, you handicap, a priori, your chance to
understand what is required for the general model of a stellar atmosphere,
in order to produce this variety of nonthermal phenomena[2] observed, alike
in all varieties of stars.

I think the merit of my assertion is well-illustrated by: (1) con-
trasting it to another assertion, at the 1972 Goddard conference on
stellar chromospheres, by speculative-theoreticians, without considering
observations, that no star hotter than Fo could have a chromosphere; so
that any possible winds from these stars could have no relation to
chromospheres; (2) by the modeling of hot-star mass-loss as radiation-
pressure produced, cold winds, from RE-controlled, spherically-symmetric,
time-independent thermal atmospheres, again by speculative-theory, without considering observations; (3) by observations of super-ionization
(relative to RE, thermal models) in a number of these hot stars, culmi-
nating in OVI observed in Tau Scorp; (4) by historical and current observations of both rapid (2-15 min) and longer term (days to months)
variations in the emission-line spectra of hot, cool, and intermediate

215

P. S. Conti and C. W. H. de Loore (eds.), Mass Loss and Evolution of O-Type Stars, 215-226.
Copyright © 1979 by the IAU.

stars alike (eg Andrillat and Fehrenbach, 1978; Doazan, 1976; Herbig, 1960; Hutchings, 1968; Kolotilov and Zaitseva, 1975; Rosendhal, 1973). I do not see how a homogeneous, radiation-pressure and radiative-equilibrium controlled, hot atmosphere can produce the OB star variations; nor how a cool star, plus only an extended atmosphere or shell, can produce these cool-star observed variations. So, our approach focuses on empirical-theoretical modeling of that combined entity of atmosphere+ subatmosphere which can produce the observed data --- thermal and non-thermal, and, hopefully, lead to inference of those nonthermal energy storage modes required to produce these observations. Such nonthermal modes are, a priori, unknown, as witness their ignoration in these speculative-theoretical models. So, only observations, not further speculation, can remove "unknown", and substitute "empirical-theoretical" kinds and amplitudes of nonthermal modes.

I aim especially for clarity, re the characteristics of our model, in the minds of you who are basically observers, and you who are concerned with the state of the subatmosphere. I stress such aim, because I believe the importance of the mass-loss problem, together with that of the interpretation of other, observed, correlated, nonthermal phenomena,[2] lie in what these data tell us --- how they guide us --- in establishing, empirically, the general thermodynamic structure of the star, and the particular values of the amplitudes of nonthermal storage modes, within the star, of matter and energy. I think it is quite evident that, a priori, or speculatively-theoretically, we do not know either this general thermodynamic structure, nor the kinds and amplitudes of any nonthermal, subatmospheric and atmospheric, storage modes. So, the general problems we face --- of which the mass-loss is but one aspect, but apparently strongly linked to the other nonthermal aspects[2] --- can only be resolved and solved by collaboration between observer, subatmospheric gas-dynamicist, and empirical-theoretician.

In the above, I question "theory", and stress observed and nonthermal, because, before such nonthermal mass-loss was observed/inferred, there existed no theory, atmospheric or interior, which predicted/required a nonthermal mass-flux. One can say precisely the same, re those other "nonthermal" phenomena that are enhanced in "abnormal" --- which we call Xe --- stars[3]: emission lines, abnormally-broad spectral lines, superionization and excitation, "symbiotic" properties --- and of course direct evidence for mass-flux in the form of line-displacements. At those epochs, and even considerably later, when such mass-loss was first observed-inferred, all physically-consistent stellar atmospheric models were thermal, imposed radiative-equilibrium (RE), and gave negligible thermal mass-loss. In effect, the star was modeled as a closed (energy fluxes only) thermodynamic system: indeed, as a limited closed system --- only radiative energy fluxes, whose appearance varied only in evolutionary time-intervals: Observationally: "normal" stars and stellar atmospheres were classified into (assumed-) homogeneous-population boxes, each characterized by only two observed quantities (luminosity-spectral appearance) assumed to give (total energy flux, surface-temperature).

Theoretically: only thermal models, as a priori, speculative-theoretical, constructions. No observations were applied, before axiomatizing the theory to establish the basic thermodynamic characteristics of the star, according to the following alternative possibilities:

1. Is the star an isolated (no fluxes), closed (only energy fluxes), or open (both mass and energy fluxes) thermodynamic system?
2. Are the energy and matter storage modes thermal only (only random microscopic velocities), or also nonthermal (macroscopic mass-motions, systematic and possibly quasi-random; macroscopic magnetic fields)?
3. What is the degree of nonEquilibrium permitted: Equilibrium, Linear NonEquilibrium (all angle-averaged microscopic distribution functions having TE forms; all fluxes given by gradients of some thermodynamic potentials, which are expressed in terms of Equilibrium thermodynamic state parameters), or Nonlinear nonEquilibrium (no restrictions on distribution functions or fluxes)?

All attempts at modeling the "abnormal" stars were observationally-inspired, but wholly ad hoc in construction, seeking to preserve the "essential" axioms of "normal atmospheric theory", while adjoining certain features capable of superficially explaining the "abnormal" (what I call here nonthermal) observations. I would stress that an essential characteristic of the "imperfect wind-tunnel" model is that it regards "normal" and "abnormal" atmospheres as being built on the same nonthermal model, differing only in amplitudes of the subatmospheric storage modes. The observed differences between "normal and "abnormal" stars are not predicted by any existing theory: our empirical model uses the observations to infer differences in fluxes of mechanical energy and mass, between "normal (X)" and "abnormal (Xe)" stars lying in the same "classical" population-box (spectral class). Then, we try to develope diagnostic methods capable of giving at least a range of possible storage modes corresponding to these nonthermal fluxes.

The modern era of stellar "winds" began with Biermann's demonstration that the observed behavior of comet tails required a particle, not just radiative, solar flux: a "wind". The wind was modeled following Parker's demonstration that stars having coronas could not have atmospheres that are both static and smoothly merging into the interstellar medium; hence the corona must be nonstatic, and a wind must be an integral part of its structure. The existence of the (nonthermal) corona was linked to the existence of a particular kind of nonthermal energy storage in the solar subatmosphere (convection), but there still exists no theory for a convective storage mode, which both predicts the amplitude of the convection and links it to the amplitudes of nonthermal kinetic energy flux (popularly assumed to produce the chromosphere-corona) and of mass-flux. Indeed, current observations of the solar atmospheric velocity fields, and of the (space, time, amplitude) properties of the solar wind, have put into question whether a convective storage mode suffices to explain --- hence ultimately to predict --- the solar chromosphere, corona, and wind. So there does not yet exist a nonempirical theory for even the solar wind. The extension across the HR diagram of speculative-theories for possible

subatmospheric nonthermal modes --- from solar-type to the very hot stars
considered in this Colloquium --- has lagged; even having been opposed
as unnecessary and unrealistic; even in the light of the existence of the
above "abnormal" hot stars (WR,Of,Oe,Be) and the resemblance of certain
of their spectral features to the solar chromosphere-corona; until observations, not theory, have shown the presence of chromospheres in even
these hot stars. No speculative-theory has predicted them; empirical-
theories are trying to model them.

In brief, the range of possibilities for speculative-theoretical
modeling of stellar atmospheres --- even of subatmospheres and interiors
--- is so great that the usual procedure of choosing the "simplest" speculative alternative is inefficient, unproductive --- and even psycholog-
ically inhibiting, when one holds to some "speculative" universal characteristic such as only thermal storage modes, and radiative-equilibrium,
or closed system, or linear nonEquilibrium, because they are "easire" to
compute, instead of asking, observationally, what the "real star" demands.

II. IMPERFECT WIND-TUNNEL MODEL:

A. Broad Approach:

For the reasons given in the Introduction, the basis of our model
is naively simple: it is not a "theory" of why "hot" --- or any other
type --- stars must have mass-fluxes. Rather, it is an algorithm --- a
diagnostic approcah to deriving an empirical-theoretical model from as
complete a set as possible of observations --- embedded in the framework
of a self-consistent, a priori unrestricted, nonEquilibrium thermodynamic characteristics listed above: type of thermodynamic system; kind
of storage modes; degree of nonEquilibrium. Then, for each star ---
once it has been so characterized --- we attempt to obtain, empirically/
observationally, the fluxes of mechanical energy and matter --- together
with their scales --- to supplement the conventional empirical parameters
of radiative flux and gravity. Then, we ask what such fluxes require
in the way of storage modes to produce them.

Let me make clear the meaning of "imperfect" in the title of the
model. As shown some time ago (Clauser, Germain, 1965) the gas-flow
from a spherically symmetric stellar atmosphere can be considered to
have an analogy with a wind-tunnel, or converging-diverging nozzle connecting some storage-pot of gas to its environment. A "perfect" nozzle
is one where the flow passes smoothly from subsonic to supersonic flow,
without aerodynamic heating or shocks either before or after the "throat"
of the nozzle. Such a nozzle must be carefully designed, in terms of the
"storage-reservoir" conditions, and the "environmental" conditions (in
the star, the subatmosphere and the interstellar medium, respectively).
All nozzles are not perfect, expecially if they are given a priori, as
is the star. One must observe the flow, if he doesn't design the nozzle,
to diagnose whether it is perfect or imperfect. Gas-dynamic history is
full of examples where a priori, untested, speculation led to disaster:

wind-tunnels which either lost most of their energy in pre-nozzle shocks
and heating, or never became supersonic; irrigation flumes pounded apart
by the equivalent of shock-waves in the equibalent of too-rapidly con-
verging nozzles for "smooth" flow to occur --- in astronomy, "chromospheres" in Tau Scor, when an "imposed perfect-nozzle" was observed to
be imperfect, and to heat, mechanically, by the equivalent of a pre-
nozzle shock = chromosphere. In analogy, note that the nonLTE "theory"
was really only an algorithm, for diagnosing observations, whose maxim
was "be sure you have LTE, by computing reaction rates in detail, before
you assume it: allow the possibility of nonLTE". Similarly, here, the
"imperfect" model is simply an algorithm, again for diagnosing data: ask,
observationally, whether the particular star is an "imperfect" or "perfect" wind-tunnel, before imposing its "perfection": allow the possibility of imperfection. Then ask what other characteristics the star must
have, to reach such an observed state.

     In brief, that is our approach. First, establish "global" thermo-
dynamic characteristics, empirically. Next, establish particular characteristics of the storage modes for the particular star observed: thermal or nonthermal, as the star tells us, not as we tell the star.

B. Specific Approach:

     1. Global Thermodynamic Characteristics:
        a. Type of thermodynamic system:

     We see the star: so it cannot be an isolated system. Historically,
stars have been modeled as closed systems: only radiative (energy)fluxes.
Hack's (1969) summary of stars with observed/inferred mass-fluxes shows
that at least some stars --- whose types range across the HR diagram ---
must be open systems. While exact generalizations can never be established, the fact that such mass-fluxes exist so generally, even in the
Sun, suggests that we must permit any given star to show that it does
not have a mass-flux, before we "forbid" it to have one. So, we model
stars as open thermodynamic systems, observationally; keeping open the
possibility that somewhere, sometime, we may be able to establish that
some star has no mass-flux, hence is closed.

        b. Kind of subatmospheric and atmospheric storage modes:

     Using any of a variety of existing thermal models for stellar
atmospheres, we readily show that the predicted mass-fluxes by thermal
evaporation are many orders of magnitude lower than the observed mass-
fluxes, even admitting several orders of magnitude uncertainty in these
latter. Also, most of the inferred mass-fluxes come from shifts, not
just broadening, of spectral lines; indicating systematic, hence nonthermal, massflows. Moreover, larger-amplitude mass-fluxes are linked
with the presence of emission lines; and, as mentioned, emission-line
presence and absence is variable in times too short to be explained
simply by extended atmospheres, so must be associated with mechanical
heating. Thus we conclude that the stars having mass-fluxes have non-

thermal storage modes. Material is pushed out of the star from below.
IF it could be shown that (thermal) radiation-pressure alone suffices
to initiate mass-fluxes, the preceding conclusion might be disputed.
Material would be pulled out of the star from above. However, the only
stars for which such radiation-pressure initiation has been suggested are
the hottest early stars: and, these "pulling-out" theories depend upon
imposing the perfect wind-tunnel condition: smooth transition, from sub-
thermal to superthermal flow, sans shocks, mechanical heating, chromospheres. However, even for these hot stars, the cited "associated nonthermal phenomena" are observed, so we reject these theories, and interest in them as representing a possibility to preserve thermal models in
the atmosphere.

    Sometimes, in discussing "coronal" origins of stellar winds, there
is implicit implication that such models are thermal; that the winds
have their origin in a "hot" (or warm, no matter) corona. This thermal
implication is illusory; one must consider the origin of the corona
itself. Even if the major part of the mechanical heating, which produces
the corona, comes from a mechanical energy flux that does not transport
mass --- such as a system of acoustic waves --- both this mechanical
energy flux, and the mass-flux corresponding to the "coronal" wind, have
nonzero values in some subatmospheric or atmospheric regions, where there
is a nonthermal energy storage mode (convection, rotation, pulsation,
magnetic field, whatever). The boundary conditions on these nonthermal
storage modes, and the conditions that help fix their amplitudes in their
nonlinear description, must include these values of nonthermal kinetic
energy and mass fluxes. In the wind-tunnel analogy, the star --- or its
subatmospheric regions --- is a nonstatic reservior, driven by whatever
drives the (convection, rotation, pulsation, etc). So, we cannot regard
a corona as the "origin" of a wind. Rather, the corona, the wind, chromospheric phenomena, emission lines, superionization and excitation, symbiotic phenomena, etc, are all associated phenomena, whose origin lies
in the existence (for whatever reason) of subatmospheric and atmospheric
nonthermal storage modes.

    From this viewpoint, there is really very little evidence in any
spectral class across the HR diagram for stars with only thermal storage
modes, thus susceptible to thermal modeling. All main sequence stars
cooler than A are thought to have at least subatmospheric convection
zones, which produce acoustic fluxes; existing data confirm this thinking. Giants and supergiants in this region are characterized by variability and pulsational instability. Giant and supergiant stars of A
and hotter are, observationally, also characterized by variability. The
"last refuge" for thermal modeling was thought, for some years, to be
the O-B main-sequence stars, where, speculatively, there were no subatmospheric motions that coupled to the atmosphere: if one neglected Be,
Oe and Of, etc "abnormal" (because they exhibit nonthermal character)
stars. Some of us have long-argued that the so-called hydrogen deficiency, but rather a chromosphere-corona. Now, OVI observed in Tau Scor
removes the thermal possibility for O-B stars, and re-emphasizes the WR
question.

So, we think the evidence is clear that most, if not all, stars have
nonthermal storage modes in their atmosphere and subatmosphere. The
question is what they are, where they are located, and what is their
effect on atmospheric structure. For this, we must consider each star
individually, because there is not a priori reason why such modes should
be specified wholly by radiative flux and gravity, as many authors seem
to assume. To begin with some kind of categoric simplification, we have
simply chosen the (X,Xe) categorization. We conjecture --- on the basis
of the observations --- that all nonthermal phenomena are stronger in Xe
stars than in their X counterparts. An example is the T Tauri stars relative to the Sun, because it is thought these are all 1 solar mass stars.
I cite this, because we are on the way to having the best data for a detailed (X,Xe) comparison. We hope that we will soon have the same kinds
of results on O-B stars, from observational programs in progress, some
typical results of which are being reported here by MMe Andrillat. Preliminary statistical studies of low-amplitude, short-periods B-star variability have been given by Bijaoui and Doazan, Hutchings, Walker, and by
others. We hope that with these new TV scanner techniques, larger-amplitude variations will be found in Be stars.

c. Degree of nonEquilibrium:

I do not believe that, in 1978, I need to re-hash all the early
nonLTE work --- observational and theoretical --- which led us to conclude that the nonlinear nonEquilibrium approach was the only acceptable
one (Thomas and Athay, 1961; Thomas, 1965; Jefferies, 1968). The approach
is routine, by now (Mihalas, 1970). The only point which I would emphasize is that we have not yet established where, in the atmosphere, significant deviations from a Maxwellian thermal velocity distribution must
be introduced. It is not clear that these conditions are the same in
static and moving gases. The same can, of course, be said for distinc-
tions between radiation and collisional control in moving atmospheres.
But these kinds of problems can be approached parametrically; it is only
in defining the actual conditions in a stellar atmosphere, to which these
parametric results can be applied, that problems arise. And these conditions can only be delineated by nonthermal studies of actual atmospheres.

2. Particular Applications to Particular Stars:

We are interested in the observable difference between thermal and
nonthermal models: not only in terms of a wind, or a height variation of
the systematic velocity associated with the wind, but in the change in
atmospheric structure coming from the presence of these nonthermal modes,
and the resultant change in the radiation by which we diagnose the atmosphere. I will not summarize here any algebraic details; these can all
be found in our preceding work (Thomas, 1973; Cannon and Thomas, 1975,
1977). Here, let me simply show, schematically, how this empirical approach to modeling based on a general nonEquilibrium thermodynamic framework, proceeds.

a. The framework:

The three general descriptive equations for the matter, the energy associated with it, and its coupling to the radiation field, assuming spherical symmetry, can be writen, schematically:

matter: Ψ (U, ρ)   = 0   (1)

thermal energy: $F_1(U,\Sigma) + F_2(J_\nu, S_\nu, n_a, a_\nu, T_e)$ = 0   (2)

nonthermal energy: $G_1(U,\Sigma) + G_2(p, \rho, T_e, F_\nu, g)$ = 0   (3)

I divide the equations in this way, so that $F_2 = 0$ and $G_2 = 0$ represent the usual thermal-model solutions, corresponding, respectively, to radiative-equilibrium and hydrostatic-equilibrium. Equation (1) does not exist, in the thermal model: being simply 0 = 0. The symbol Σ simply signifies that the nonthermal terms, F_1 and G_1, depend upon the thermal parameters as well as upon the systematic velocity U.

Now, so long as U is less than 20 or 30% of the 1-dimensional thermal velocity, q, these thermal equations, ignoring the effect of U and its derivatives, suffice, under stellar atmospheric conditions (thermal viscosity and heat-conduction small) to give an adequate representation of ρ (t or τ) and T_e(r or τ). We note, in the low atmosphere and sub-atmosphere, before sphericity effects become of interest, and emphasizing this assumption of spherical symmetry, that (1) becomes:

$$U\rho = U_o\rho_o = F_M \qquad\qquad (1a)$$

where F_M is the mass-flux (per unit area) and subscript o denotes values at some arbitrary point. So, if one wants to ask what systematic velocities he can admit in the stellar atmosphere, without affecting his thermal model, all he needs do is to choose some U at some level and use the thermal model density distribution to compute U(r or τ), to determine where U ~ q and the model fails. Clearly, any choice of U_o, which gives an observable velocity effect anywhere in the atmosphere, will reach q somewhere before leaving the atmosphere and entering the ISM, noting: (i) ρ for the ISM corresponds to some 1 particle/cm^3, while for photospheres, ρ corresponds to $10^{17} - 10^{11}$ particles/cm^3; (ii) classically T_e and q reach asymptotic values near T_{eff}. Thus, the only consistent thermal models set U_o = zero. Any model, no matter how it is produced --- speculatively or empirically --- where $U_o \neq 0$, produces U ~ q somewhere, perturbs the equations (1) - (3) there, and heats, mechanically, the atmosphere, from this systematic velocity gradient alone. There may also be mechanical heating from nonsystematic velocities, such as acoustic waves, which affect (1) - (3) essentially through terms equivalent to viscosity, but macroscopic rather than thermal. All these terms do, is to produce a mechanical heating lower in the atmosphere than does the mass-flux term alone. q then rise above its "thermal-atmosphere" value; the distribution of ρ (r or τ) changes; but we can still use (1a) to determine U(r or τ), if (2) and (3), modified by these "viscosity" terms, do not depend upon U.

b. The Modeling Process:

So we have three choices of procedure to produce these nonthermal models:

(1) We solve completely the model of a subatmosphere nonthermal storage --- pulsation, rotation, convection, whatever --- leaving free the two parameters of mass-flux and mechanical energy flux. Whether the amplitudes of the subatmospheric modes depend sensitively upon these two fluxes remains to be seen: no solutions yet exist. This alternative is the most satisfactory one; as yet, no single solution to any such problem exists.

(2) We take the observed mass fluxes, and thermal-model density distributions, to ask where, in the photosphere, the value of U associated with this F_M from (1a) first reaches q, in that thermal model. We did this, some years ago, using available data (Thomas, 1973). No star, for which data existed, showed U to reach q more exterior than

$$R(q)/R(\text{photosphere base}) < 1.10 \tag{4}$$

This "outer" limit was for an Mo supergiant. For OB stars, the value of the limit was more nearly 1.01. Prediction: chromospheres would exist and start no higher than this value. The OVI observations of Tau Sco were hardly a surprise, unless you believed Tau Sco had zero mass-flux. Then, given this upper limit on the beginning height of the chromosphere [upper, because it ignores heating due to mechanical energy fluxes that do not transport mass], one surveys the list of 'abnormal' features, which we have classed as nonthermal [see footnote 1], to ask whether a chromosphere-corona beginning at that depth would remove their 'abnormality'. An example is the cited OVI observations in Tau Scorp, plus the other ions incompatible with radiative equilibrium models. The problem on which we have thus far focused is that of an explanation of emission lines in terms of chromospheres beginning not higher than these limits, as opposed to emission lines arising in extended atmospheres under RE: we consider the problem to be the same in the cool [T Taur] and hot [Be, Of] stars. Our preliminary results on hydrogen Balmer-α fluxes, and central regions of their profiles, in the cool stars, are encouraging. We note that it is considerably easier to introduce a time-variation in emission line profile coming from a low-lying chromosphere than from an extended-atmosphere 'photosphere'. [3] We adopt a procedure intermediate to [1] and [2]. Thermal, closed-system, models take T_{eff} [or F_{rad}] and g as parameters; we assert that nonthermal, open-system models need, in addition to these two parameters, two others: mass-flux, and nonthermal kinetic energy flux. The values of all these four are specified at the base of the atmosphere; and the atmospheric model is specified at the base of the atmosphere; and the atmospheric model is specified by outward integration. We note that the radiation-pressure models discussed above contend that T_{eff} and g suffice, still, to discuss these mass-fluxes, which are computed, from values of these two thermal parameters and by imposing the equivalent of the ''perfect wind-tunnel'' condition at the

''throat''. We further note that the ''coronal'' models discussed above
contend that, in addition to these two thermal parameters, it suffices
to add a value of the nonthermal kinetic energy flux at the base of the
atmosphere, and impose the equivalent of the ''perfect wind-tunnel'' condition at the ''throat''. The mass-flux is then predicted. So, the
range of models discussed in this Colloquim essentially comes down to
how many parameters, and what kind of conditions at the ''throat'', the
authors consider are needed. I repeat: our ''imperfect wind-tunnel''
model is just that: conditions at the ''throat'' are not imposed, but
investigated: essentially, the ''throat'' marks the end of the ''trans-
sonic'' flow region, within which the chromosphere-corona have been steadily heated. Work is in progress, on this model, to compare the two cases:
T Tauri [cool stars] and Be [hot stars] for which we have the best data.

[1] emission lines, superionization and excitation, BaC and IR excesses,
abnormally-broad spectral lines, "symbiotic" properties in the visual
spectrum; plus all those "symbiotic" features in the "rocket-UV", exem-
plified in the Sun.

[2] see footnote 1.

[3] eg. WR, Of, Oe, Be, Ae and Ap, T Taur, dMe, etc.

DISCUSSION FOLLOWING THOMAS

Hearn: I do find it a little surprising, Dick, that you mention diagnostics as being crucial. The three preceding speakers have made just such an effort to discuss the observations, whereas you have not mentioned them. I'm inclined to give you naught out of ten for answering the questions proposed to the panel.

Thomas: Tony, you must be careful. We were the first to predict the solar temperature distribution from the data. In the Cannon-Thomas theory we gave a velocity law; we predicted chromospheres in hot stars, and none of you people believed it was worth anything, so don't tell us we haven't predicted anything. You can take a vote if you want to, but I couldn't care less... . We were discussing hot winds linked to chromospheres, and both to emission lines when the rest of you were talking about cold winds and no chromospheres. Your public opinion polls leave me cold.

Heap: How do you account for the observations which show that two totally different kinds of O stars -- the young, massive ones, and the old, planetary nuclei -- having totally different interior conditions still have similar stellar winds?

Thomas: I'll comment if you tell me why two stars each of solar mass can show different spectral lines; one shows absorption and one shows emission lines.

Heap: I am not really looking for another question as an answer. You say it's very important to consider the interior conditions.

Thomas: The subatmosphere.

Heap: Could you clarify what you mean by the subatmosphere?

Thomas: Let's compare the Sun and T Tauri. In my eyes, to have a variation of the Hα profile from absorption to emission, there must be a hemisphere variation of the Balmer continuum. On the Sun I only observe variations in features the size of 1000 km or so, giving variability in the H and K lines. These Ca II lines depend only on quasi-local conditions. In the Sun, one thought for years that the convective zone was the whole answer to everything. On the basis of a statistically steady convection it was very hard to understand how you get the patchiness over the solar surface. Results of Deubner and calculations of others lead to a belief in non-radial pulsations, as also being important in the subatmosphere. With this, I can get all kinds of variation. In T Tauri it looks as though you have the entire hemisphere behaving in this manner. Now how do I have such different behavior from the Sun? In the same way in some other stars, how do I get similar behavior? That's what we are trying to understand from the observations. We must wait until we get a good aerodynamic theory to predict what are the motions of the subatmosphere. Look, honey, I have a choice, I have either

convection and/or maybe I have non-radial pulsation. A priori, I don't
know. The theory doesn't exist, but I can use comparative observations
between normal and peculiar stars even if the theory is not very good
to try to distinguish. I'd like enough new observations that can distinguish between sizes of mass flux and functions of the type of nonthermal kinetic energy flux (radiative flux is already present). Hopefully I can then use these observations to infer what kind of kinetic
energy (macroscopic kinetic energy) is present in the subatmosphere.
Are you with us? Geophysicists, by looking at the types of wave propa-
gation in the earth's surface, decide density distribution with height.

Morton: Are you trying to say that for your model we should stop
doing theory and obtain more data? If so, what observations do you suggest?

Thomas: I don't consider any of what has been discussed by this
panel to be theory. We need good enough resolution observations, to be
able to concentrate on the diagnostics. For example, if I have Hα emission, does it come from a hot or cold extended atmosphere or does it
come from a low lying chromosphere? The superficial response has been
that it's extended but this doesn't fit observed high dispersion profiles. One can quickly do a reasonably theory, with radiative transfer
calculations, and for a deep lying chromosphere and corresponding
velocity field, and get a reasonable fit. Another quick answer would
be to get your friends to observe whether such profiles vary in time.
If it varies in a few minutes, it's hard to believe the atmosphere is
extended. So that justifies our going a little bit more sophisticated
into the radiative transfer with the velocity field and energy dissipation, and calculations for a model of the upper photosphere of T Tauri.
So please don't stop doing theory, but do empirical theory. Take some
observations as a guide.

GENERAL DISCUSSION

Conti: I would like to address the variability question. As an observer of O type stars for 10 years and of WR type stars for a couple of years, the most important conclusion that one can make about variability is that it's very small. You have to look hard to see it. Earlier, Anne Underhill reminded us that Wolf and Rayet in 1867 described the spectra of three WR stars. These are as extreme an example of mass-losing stars as you can find and they looked more-or-less like they do 100 years later.

I would like to first talk about the observations given earlier by Dr. Vreux on variability: The star BD+40°4220 went from a very strong absorption to a very strong Hα emission in a day, an appreciable fraction of the binary period of 6.6 days. This kind of behavior is well known in other close binary systems and not very well understood, but I would suggest that this kind of extreme behavior has something to do with the double nature of the star. More observations on this system, which is quite faint, would be very, very important. As to the B supergiant, according to Jean-Marie, the scale on that spectrogram was the same as the scale on the BD star. The spectral change was only from weak emission to weak absorption. If this profile was in λ4686 (not only Hα) it would be called an O(f) type. It's a star where the emission and absorption processes in Hα are just about in balance and maybe changes can occur in a few minutes. The fact that it's central emission may be even telling us that it is more like a Be star rather than a star which has a P Cygni profile. This kind of variability is not going to be found in many stars because in most stars the Hα line is either very strongly in emission or very strongly in absorption.

Unknown Voice: Bull ...

Conti: The brickbats are finally arriving .... Anyway, I should point out that Andrillat and Vreux did observe three other stars and that they saw no changes. I would say in most stars, most of the time, even if you look very hard you see no changes above the 10% level. Substantial emission line variability is rare. Niemela and I have a recent paper on ζ Puppis where changes are pretty convincing: You pull the plate out of the hypo and it looks a little different than it did a few months previously.

I would like to give a brief summary as to what I believe are the variability timescales. Photographic methods only show changes above the 10% level: What we usually see is a "pimple" rather than a gross change of the profile. ζ Pup and λ Cep appear to have optical emission line variations in periods of <u>days</u>. That variability you can see when you pick up the plate and look at it (more than 10%). Their variability may be due to rotation but we don't understand the connection.

Delta Ori seems to have a time change in the UV lines of <u>hours</u>, mostly a disturbance in the P Cygni line of N V that propagated outward

227

P. S. Conti and C. W. H. de Loore (eds.), Mass Loss and Evolution of O-Type Stars, 227–234.

(a density enhancement according to York et al. 1977). There may be a
lot more of these kinds of data found as more Copernicus observations
are made.

And finally you heard earlier about HD 193793, the variable infrared source. This Wolf-Rayet star ejected a dust shell in the timescale
of months. These are variabilities we know of now that are relatively
large effects. None of these are periodic as far as we know.

Underhill: About observations, which Conti brought out, in connection with the points that Dick has brought up, the problem is to observe something significant. Changes are significant but perhaps very
difficult to observe. Up to now we have only had the ground-based observations. It is our chief source of supply and will continue to be.
Space observations give us some clues. You have to ask what portions
in the star are visible in the available lines or continuum and how easy
it is to observe changes. Now the infrared free-free continuum, and anything longer than 6000 Å in a hot star, is almost insensitive to the details of a model, so you can fit it with anything. (This is the reason
I could get angular diameters.) Emission lines are more sensitive, different ones in different ways, to electron densities, temperatures and
velocity distributions. Where these come from is Dick's problem and my
problem is "how do you get the heating." Dick quite reasonably pointed
out that one must start on the inside of the star.

Now to make this more evident I call your attention to the "pimples"
Peter referred to earlier; if any of you have teen-age children, you will
know that "pimples" can appear, they look very small, they cause great
social problems, but they can be due to quite deep lying disturbances.

Conti: Yes, but the teen-ager is still a teen-ager.

Underhill: The really exciting thing is that our observations
allow us to see, fairly accurately, significant photometric changes in
line shapes and continua that we could not measure before. How can you
account for them? The standard wind models are radiation-driven models.
You cannot start a flow with radiation, but it may keep one going. Another interesting point which the UV line analysis always brings up is
that the envelope temperature is 50,000 K or greater.

Zeta Pup (O4ef) is one of the very few stars whose effective temperature is near 50,000 K. Most of the stars known to have supersonic
winds have effective temperatures of less than 32,000 K. You can look
at any models: at 30,000 K, half the flux comes out longward of 1400 Å.
Very little comes out shortward of 900 Å. I do not believe that radiation pressure by itself will be very important for getting the high
velocities. Where does the wind originate?

Another really interesting problem is the difference in rate of
mass flow between a Wolf-Rayet star, an O9.5 supergiant and a late O or
a BO near main-sequence star. There is a very large difference, a factor

of 10^3 or more. Many of these stars have about the same effective
temperature, $30,000 \pm 1000$ K. Typically M_{bol} of most of these stars
differs by less than 2 magnitudes, see the following table which has
been derived from the material which Divan, Doazan, Prévot-Burnichon
and I reported here and from data in the literature on Wolf-Rayet stars.

Type of Star	Typical M_{bol}	Typical T_{eff}	$\dot{M}$ ($M_\odot$ yr^{-1})
Wolf-Rayet (not WN7 or WN8)	-7.7	31,000 K	$\geq 10^{-5}$
O9.5 supergiant (α Cam, δ Ori A, ζ Ori A, HD 188209)	-8.8	30,800 K	$\approx 10^{-6}$
B0 main-sequence (κ Aql, λ Lep, 1 Cas)	-7.4	29,300 K	$< 10^{-8}$

How do these slightly differing values of M_{bol} and T_{eff} account for mass-
flow rates differing by 10^3? There has to be another more significant
factor. The only way that I can understand it is to postulate that the
rate of mass loss is connected somehow with those subatmospheric velocities that Dick Thomas is talking about, or with magnetic fields. The
proposed X-rays, which Cassinelli so beautifully showed could account
for many of our observations, have got to come from somewhere. These
effects may be only observed as "pimples" but they're defining almost
everything we're observing.

Thomas: Peter has explained that he considers that if Hα goes from
absorption to emission he thinks it a minor perturbation. Also in T
Tauri, for sodium D to change from absorption to emission in a few
minutes, he would say that's a minor change. For me it's a major thing
in life.

Conti: Yes, Dick, but O-stars are not T-Tauri stars.

Abbott: I would like to answer the point raised by Anne Underhill
concerning whether sufficient flux is available to drive the wind by
line radiation pressure. Although I have not treated the problem dynamically, I can say that on the range of T_{eff} from 30,000 K to 50,000 K,
enough flux and the lines needed to absorb it are theoretically available
to get the observed mass loss rates.

The important parameter is not the flux but 2 times the flux. For
example, in a 30,000 K star, this product peaks at ~1200 Å. For cooler
stars, the radiation force presumably becomes progressively less important.

Morton: Did you consider just resonance lines, or do you include
also subordinate lines?

Abbott: Both. However, the major contribution comes from lines for which there are no downward transitions. These are sometimes at relatively high excitations.

Morton: Some subordinate lines are not observed to have velocity shifts (i.e., they are photospheric only). They presumably would not drive the wind.

Abbott: Not necessarily. As an example, a T_{eff} = 40,000°K star with $\dot{M} \sim 10^{-6}$ M yr^{-1} at the point $v(r) = (1/2)v_{\infty}$ has 60% of the force coming from 100–200 optically thick lines, and 40% coming from the cumulative contribution of very many optically thin lines.

Sreenivasan: I have noted that essentially all the models proposed by the panel necessitate heating of the envelope in one form or another. Could anyone suggest the source of this heating?

Castor: One possibility is that there are radiatively driven sound waves that are unstable. These are then dissipated in the wind and produce the heating. However, this is highly uncertain.

Sreenivasan: Would including this increase the mass loss rate?

Castor: It would not. Essentially the mass loss rate is determined by the high velocity part of the flow, a balance between radiation force in the lines and gravity. In this region the gas pressure then is very small. Mass loss rates are not sensitive to the temperature, at least up to values of a few times 10^6 degrees.

Hearn: This, of course, is in the Castor model. In the hot coronal model, the mass loss is determined by the energy transport in the coronal levels. Similarly, I would think, for Dick's models.

Thomas: No, by convection in the subphotosphere.

Snow: I'd like to raise a somewhat different point about velocity laws. Most of the panel's models have velocity laws which are more-or-less single valued, as evidenced by the P Cygni profiles. However, some stars are observed to have detailed structure in some ions (narrow absorption components). These can be interpreted as "plateaus" in the flow, which are, interestingly enough, relatively stable with time. Can anyone comment on how this could come about?

Hearn: That's a very interesting question. I have no suggestions.

Lamers: If these plateaus are occurring where the radiation pressure is driving the wind, then we must conclude that the radiation force is changing. How could this happen? One possibility would be a spatial variation of the ionization balance. You might reach some critical points in ionization balance. It's a very interesting observation which does need some more work.

Cassinelli: In our model, since the X-rays are the cause of the ionization, I would not expect a strong dependence of ionization with radial distance.

Van Blerkom: Dick Thomas earlier raised the point about rapid changes observed in emission line profiles in a few cases. Since the timescale is a few minutes in one instance this implies something happening very close to the star. To me, this brings about a difficult conceptual problem. We observe P Cygni lines with violet-shifted absorption line velocities extending to 2000 km s^{-1} from line center. These must occur very many stellar radii from the star. If any of these P Cygni lines we've observed change substantially in the duration of a few minutes, I don't think it could be understood.

Thomas: This worries us very much also. However, we suggest that maybe this short timescale variability must mean the entire profile is formed close to the star (a "close coronal" model). Ann Boesgaard has shown me some P Cygni profiles of T Tauri stars. Out of five plates, one showed an _inverse_ profile (redward displaced absorption).

Van Blerkom: So it is a problem.

Thomas: The entire topic is a problem.

Lamers: I think it's only a problem if the variations occur at large velocities. Within a few hundred km s^{-1}, it could well be "chromospheric" variations close to the star.

Hutchings: I'd like to clarify the observations. I also have extensive data on optical P Cygni lines and I have seen no evidence for short timescale variations in these lines. The timescale over which things are observed to change is invariably longer than the transit time for material to leave the surface and reach the terminal velocity. At small velocities close to the star, things can change quicker. I don't think one has to worry about this.

Thomas: What do you mean that you don't have to worry about it?

Hutchings: The observational evidence does not show short timescale variations occurring at large velocities, while the radius is substantial.

Noerdlinger: Some winds in QSO's have substantial variability in the absorption spectrum, which often seems short compared to sound speed travel time. I think Margaret Burbidge observed this in PHL 5200. Since the absorption feature is that material seen in projection, the projected cross section can be quite small compared to the distance to the object itself. So in stars, you could get a disturbance of the material at high velocity which changes in a small timescale. If the emission portion changes rapidly, then you might be in trouble.

Conti: The point I was trying to make earlier was that the short timescale absorption to emission variability that Jean-Marie showed occurs, in fact, in the center of the line. It's not in a P Cygni feature, and not at a high velocity.

Seggewiss: I have observed changes in P Cygni profiles in a WN star, which moved from -600 to -1100 km s^{-1} in a few days.

Hutchings: This kind of timescale is not really a problem.

Van Blerkom: Is the star a binary?

Seggewiss: No, it's not.

Thomas: I'd like to say again that any kind of variability needs some mechanism to bring it about.

Heap: I have been directed by Rudi Kippenhahn to a theoretical paper by R. Connon Smith [M.N.R.A.S. 148, 275 (1970)] on rotation, circulation currents, and shear. This could be a source of the heating and/or instabilities we have been discussing here.

Does anyone know anything about this?

Ebbets: I'd like to comment on this macroturbulence point. I spent about a year doing my Ph.D. thesis on the question of line widths in O-type stars. By doing a Fourier analysis of the profile, and also using the observed profile as is, it is easy to separate rotation from macroturbulent broadening. In O-type supergiants, with an assumed Gaussian profile for the macroturbulence, I found values of 25-30 km s^{-1}. In some main sequence O-type stars, I find only upper limits of some 13 km s^{-1} for macroturbulence.

Bidelman: τ Sco must have quite a low turbulence to judge from its spectra lines which are quite sharp.

Lamers: The photospheric turbulence is quite small, but that in the wind may be larger.

Stalio: I'd like to ask John Castor about his future work. Do you intend to put expansion and rotation in your wind?

Castor: Yes.

Sreenivasan: With respect to Sally Heap's question, with differential rotation such a physical input seems very natural.

Lamers: With respect to questions about crucial observations, I'd like to say that Cassinelli's coronal model works quite well on all the data. However, τ Sco still has a nagging problem: The ionization decreases outwards. Well, one could say that τ Sco is not a normal star and has a mass loss mechanism different from O and B supergiants. How unique is the observation that ionization decreases outwards? Do any other stars show this behavior?

Underhill: The distinction between ionization constant or dropping off outwards may depend critically on the density. The cooling depends closely on the aerodynamic solutions. The density could well drop off faster in the main sequence star, resulting in a decrease in the ionization in these objects, compared to supergiants.

Morton: Is it correct to say the models of Castor, Cassinelli, and Lamers are identical except for the ionization? All seem to have arbitrary ways to get the O VI lines. Dick Thomas is apparently waiting to see what he puts in.

Thomas: Castor's initial assumption was radiative equilibrium, whereas the initial assumption of the coronal models is non-radiative equilibrium and chromospheric-coronal mechanical heating. So there is a fundamental distinction. John has to put in an arbitrary parameter, which has nothing to do with a sub-atmosphere. We don't put in any such arbitrary parameters to get O VI once the sub-atmospheric non-thermal fluxes are fixed.

Morton: Your key point seems to be a connection of the mass flux to the sub-atmosphere.

Thomas: Exactly.

Castor: I am in there with the arbitrary folks. My model is like Henny's, except that I use a small elevation of the temperature to increase photoionization in place of a larger increase that makes collisional ionization important.

Noerdlinger: Dick Thomas pointed out that he believes turbulence must come from the subphotosphere. But, really, any kind of energy input would suffice. Is that correct?

Lamers: Yes.

Thomas: Let's be careful here. We are not talking about how you generate turbulence. A solar analogue clearly comes from the sub-atmosphere. Tony's coronal model is generated by a perturbation which is amplified by the radiation field. Castor needs some arbitrary heating, which he separates from the radiative force. I think we do need mechanical heating somewhere in any case (acoustic waves).

Underhill: Where do the acoustic waves come from?

Conti: I'd like to ask Joe Cassinelli about his coronal model. Is it true that your high temperature region is so thin in extent that it gives you no other observable features other than the required X-rays to ionize the wind?

<u>Cassinelli</u>: We checked to see if Fe XIV would be present but it is essentially not observable.

<u>Conti</u>: You wouldn't have any trouble with He I further out in the wind?

<u>Cassinelli</u>: No, the ionization balance is dominated by the photo-spheric field and by the cool temperature of the wind, so plenty of He I can be there.

<u>Hearn</u>: The crunch for that model will depend ultimately on its soft X-ray detection.

An observer who cut her big toe
Needed blood to make up for the flow
When the nurse asked "what type"
She replied without hype
"If you're working for Conti its O".

SESSION 5

a) THEORY AND MASS LOSS RATES
b) BINARY STARS
c) MISCELLANEOUS TOPICS

Chairman: J.M. MARLBOROUGH

a) 1. D.C. ABBOTT: The domain of radiatively driven mass loss in the HR diagram.
2. A.G. HEARN and I.M. VARDAVAS: Stellar corona models.
3. P.B. KUNASZ: Synthetic line profiles in early-type stellar winds. I. H and He^+.
4. D. VAN BLERKOM: The $H\alpha$ profile in Zeta Puppis.
5. P.D. NOERDLINGER: The mass loss rate of γ Velorum.
6. G. OLSON: Wind models for ζ Orionis.

b) 1. C.D. GARMANY: Binary frequency among the O-type stars.
2. K.C. LEUNG and D.P. SCHNEIDER: Contact binaries of spectral type O.
3. P.C. MASSEY and P.S. CONTI: The O-type spectroscopic binary system HD 149404.
4. N.D. MORRISON and P.S. CONTI: The O-type spectroscopic binary system HD 93206.
5. Y. KONDO, G.E. McCLUSKEY, Jr. and J. RAHE: Mass flow and evolution of UW Canis Majoris.
6. V.S. NIEMELA and J. SAHADE: The spectroscopic binary γ^2 Velorum.
7. V.S. NIEMELA: The binary orbit of HD 92740.
8. C.T. BOLTON: Results from the 1977 coordinated observing campaign on HD 226868=Cygnus X-1.

c) 1. L. CARRASCO, G.F. BISIACCHI, R. COSTERO and C. FIRMANI: The nature of the runaways: old disk population OB stars?
2. G.F. BISIACCHI, L. CARRASCO, R. COSTERO, C. FIRMANI and J. RAYO: The distribution in luminosity of OB stars and evolutionary time scales.
3. W.P. BIDELMAN: H-deficiency and mass loss.

THE DOMAIN OF RADIATIVELY DRIVEN MASS LOSS IN THE H-R DIAGRAM

David C. Abbott
Washburn Observatory
University of Wisconsin - Madison
Madison, WI   53706

INTRODUCTION

Previous work by Castor, Abbott, and Klein (1975) presented a self-consistent model of a steady-state stellar wind. They also showed qualitatively that for O stars at least a static atmosphere could not exist. This paper extends that result by calculating in detail the minimum luminosity as a function of effective temperature required for the line radiation force to exceed gravity. Within the observational and theoretical uncertainty there is a one-to-one correspondence between a star's calculated ability to self-initiate a stellar wind by radiation pressure alone and the observed presence of outflowing material in the UV resonance lines.

COMPUTATIONS

Two cases were considered: (i) static stars and (ii) stars which were assumed to have a wind initially. The line list used for both cases was a synthesis of the extensive tabulations of Kurucz and Peytremann (1975) and Abbott (1977, 1978), supplemented by more accurate experimental and theoretical oscillator strengths for the resonance lines of the important ions (e.g., Morton, 1978). The radiation force was calculated for each line and then summed over all lines in the list to give the total force.

Static Case

In the static case the force was calculated using the two level atom plus overlapping hydrogen continuum approximation. Castor (1974) has given expressions for the line radiation force in terms of the line source function for this case. The line source function was calculated following the prescription of Hummer (1968). The ionization balance was calculated using the Saha equation at density and temperature (n,T). (n = n_e, T = gas temperature) was used for points where the overlapping hydrogen or helium continua were optically thick, while (n = n_e/W = $2n_e$, T = radiation temperature) was used for optically thin points. The run of density

237

P. S. Conti and C. W. H. de Loore (eds.), Mass Loss and Evolution of O-Type Stars, 237-240.

and temperature with depth and the radiation temperature with frequency
were taken from model atmospheres of Mihalas (1972). The maximum radiation
force occurs when all lines are optically thin. However, the force continued to increase with decreasing density even beyond this point because
of ionization effects. The calculations were therefore truncated at the
density at which the drift velocity of the predominant ion equaled the
proton thermal velocity, on the assumption that at smaller densities the
ion could not transfer its absorbed radiative momentum to the rest of
the gas.

Wind Case

    For each effective temperature stellar wind models were calculated
as described in Caster, et al. (1975), except for the use of the line
list described above. The core photospheric fluxes were taken from
models of Kurucz, Peytremann, and Avrett (1974). The maximum radiation
force occurs when all lines are optically thin, however, a steady-state
solution does not exist unless at least one important line is optically
thick through the critical point. Since the density of the wind decreases with decreasing luminosity, the minimum luminosity or "wind
limit" is defined uniquely by the condition that all lines become optically thin exterior to the critical point.

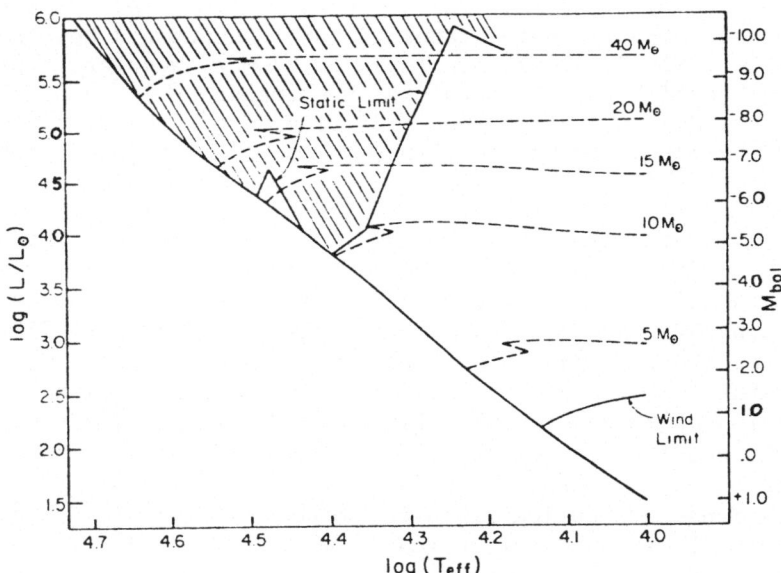

Figure 1 – Minimum luminosity required to initiate (static limit) and
           to sustain (wind limit) a radiatively driven wind.

RESULTS

The results are shown in Figure 1. Three regions are delineated.
(i) The shaded region indicates luminosities greater than the "static
limit". For these stars a static atmosphere is dynamically unstable,
i.e., a radiatively driven wind is self-initiating. (ii) For stars whose
luminosity lies between the "static" and "wind" limit both a hydrostatic
atmosphere and a stellar wind are possible equilibrium solutions. In
this region radiation pressure can sustain an existing wind but cannot
initiate a wind. (iii) For stars beneath the "wind limit" a stellar wind
can neither be initiated nor maintained by a line radiation force.

Region (i) agrees with the observed distribution of stars having
winds which Lamers and Snow (1978) found was bounded by $M_{bol} \sim -6$. This
correspondence also seems to indicate that in region (ii) stars have
chosen to remain static. This could be either because there is no other
mechanism which can initiate a wind or because in this region the wind
is an unstable equilibrium solution. Of particular interest for dis-
criminating between these possibilities are stars whose evolutionary
tracks have crossed the "static limit".

On the other hand, region (ii) has not been carefully observed in
the UV. Since the rates of mass loss are small, winds in these stars
may not be obvious. As an example, in the Copernicus U2 scans of 20 Tau
and ζ Dra the C II λ1335 and C III λ1176 lines appear to have very weak,
violet-shifted absorption extending several Angstroms beyond the photospheric profile, which may indicate outflow. Also, since even these low
rates of mass loss would overwhelm diffusion, the coincidence of the
"wind limit" with the onset of the peculiar B_p stars may indicate that
winds exist for stars in region (ii).

A paper in preparation will describe more fully these results and
their implications for peculiar Ap stars and Be stars. This work was
supported by NSF grant No. AST76-15448 to the University of Wisconsin.

REFERENCES

Abbott, D.C.: (1977), Thesis, University of Colorado.
Abbott, D.C.: (1978), J. Phys. B., in press.
Castor, J.I.: (1974), M.N.R.A.S., 169, p. 279.
Castor, J.I., Abbott, D.C. & Klein, R.I.: (1975) Ap. J., 195, p. 157.
Hummer, D.G.: (1968) M.N.R.A.S., 138, p. 73.
Kurucz, R.L. & Peytremann, E.: (1975), "SAO Special Report No. 362".
Kurucz, R.L., Peytremann, E. & Avrett, E.H.: (1974), "Blanketed Model
   Atmospheres for Early-Type Stars", Smithsonian Inst., Washington, D.C.
Lamers, J.G.L.M. & Snow, T.P.: (1978), Ap. J., 219, p. 504.
Mihalas, D.: (1972), NCAR Technical Note TN/STR-76.
Morton, D.: (1978), Ap. J., 222, p. 863.

DISCUSSION FOLLOWING ABBOTT

Underhill: Saul Adelman and I have identified ultraviolet lines in ζ Dra, B6III and τ Her, B5IV (Ap.J.Suppl.
1977). We found no need to hypothesize faint extended shortward wings in CII 1335 or CIII 1176 in these stars. Stalio
has done identifications for types near B8. The presence
of possible neighbouring lines must be carefully evaluated
before concluding a shortward extended absorption wing is
present.

Wolff : The Hg Mn stars increase both in frequency and
in the conspiciousness of their peculiarities up to a sharp
boundary somewhere between 15000 and 16000 K. This boundary
coincides fairly well with your lower limit for radiation
supported winds. However, there are certainly a few - but
a very few - peculiar stars, particularly those with enhanced
lines of phosphorus and gallium, that are hotter than your
boundary. Regrettably, 20 Tau itself shows a marginal enhancement in its manganese line strengths.

Stalio: I want to mention that the Hg Mn star α And
seems to have displaced resonance line components for SiII
(from U1 Copernicus data). The terminal velocity of the displaced components is of the order of -100 km/s.

de Loore: Concerning the Ap stars, the Si-Eu-Cr stars
are not included in your comments. Do they fit also in this
picture or not?

Abbott: These are stars observed to have magnetic fields.
Magnetic fields are effective in inhibiting a wind, which
means this picture probably does not apply to these stars.

Vreux: Could you comment on the influence of a corona
on your results?

Abbott: To the extent that a corona raises the state of
ionization it will sharply decrease the force. To my knowledge there is as of yet no observational evidence that main
sequence late B and early A stars have coronae.

A.G. Hearn and I.M. Vardavas
Sterrekundig Instituut, Utrecht, The Netherlands

A new method of calculating models for stellar coronae is being
developed with the aim of providing models of coronae around hot
stars capable of explaining the observed mass loss as a stellar wind
from a hot corona. The method gives a stationary solution to the
coupled equations of motion, continuity and energy balance. The
equation of motion contains only the gradient of the gas pressure and
the acceleration due to gravity. Radiation pressure is not yet included. The continuity equation assumes spherical symmetry. The energy
equation includes heating by weak shock waves, heating by the absorption of photospheric radiation through continuous opacity,
radiation losses, heating or cooling by conductivity and the effect on
the energy balance of the velocity of the expanding atmosphere. The
model is assumed to be optically thin. The velocity distribution is
calculated from the solution going through the critical point.

Previously these equations have always been solved as initial
value problems. But to do this one must specify at the beginning of
the calculation the mechanical flux, the pressure, the temperature and
its gradient and the velocity and its gradient, and these are
specified at the bottom of the transition region. Such a method of
solution contradicts an important conclusion of the minimum flux
corona theory (Hearn 1975) that only the mechanical flux heating the
corona is a free parameter, and that the pressure of the transition
region, the average temperature of the corona and its mass loss are
parameters specified by the heating of the corona.

In the new method the temperature structure of the corona is
calculated iteratively in a two boundary value solution. The boundary
condition at infinity is that the temperature tends to zero. The
beginning of the solution is set at a density that is so high that the
dissipation of the weak shock waves is negligible and that the
temperature of the atmosphere is determined by a simple radiative
equilibrium. If these conditions are met the only free parameter for
the solution of the equations for a given star is the initial flux of
mechanical energy. Further the position and temperature structure of

241

P. S. Conti and C. W. H. de Loore (eds.), Mass Loss and Evolution of O-Type Stars, 241–242.
Copyright © 1979 by the IAU.

the transition region is determined entirely by the calculation itself.

This method is still under development but some provisional results have been obtained for coronae with small mass loss. Firstly the solution is unique, even though the mechanical flux is the only free parameter of the solution. If the final solution is perturbed, the solution converges back again to 4 significant figures. Secondly, models have been obtained which have a second transition to an outer region which is in radiative equilibrium. In one calculation with a mass loss of only 10^{-12} $M_\odot$ yr^{-1} the corona was reduced to 4 stellar radii in thickness. The outer region in radiative equilibrium results from the large radiative losses which prevent thermal conduction from maintaining coronal temperatures in the outer layers.

REFERENCE

Hearn, A.G.: 1975, Astron. Astrophys. 40, pp. 355-364.

DISCUSSION FOLLOWING HEARN AND VARDAVAS

Snow: In the solution where you found a corona extending out to 4 R✗, what was the temperature of the corona?

Hearn: Two million degrees.

Cassinelli: Are the thermal velocity and flow velocity at the top of your coronal zone related in a simple way to the escape speed of the star?

Hearn: This is determined by the critical point of the Parker solution. In our present model calculations the critical point lies at about 3.5 R✗.

SYNTHETIC LINE PROFILES IN EARLY-TYPE STELLAR WINDS. I. H AND He[+]

P. B. Kunasz
High Altitude Observatory, National Center for Atmospheric
Research* and Joint Institute for Laboratory Astrophysics,
University of Colorado and National Bureau of Standards,
Boulder, Colorado 80309 USA

INTRODUCTION

Statistical equilibrium calculations for the first ten levels of H and He[+] were carried out for several model atmospheres of a star with $T_{eff} = 40,000$ K and $\log g = 3.5$, using the method of Mihalas and Kunasz (1978). Atomic level populations and line profiles were computed for "model C" calculated by Klein and Castor (1978) and for variants of this model in which material density and mass loss rates were scaled down by factors of 2.0, 2.2, 2.5 and 4.0. We will refer to these variant models as D, E, F, and G. The resulting line profiles are discussed. Two variants of the Klein-Castor model C are generally successful and produce He II λ4686 in emission with λ3204 in absorption. All normal Of stars observed with He II λ4686 in emission have λ3204 in absorption. In one of the models, Hα is in emission, while Hβ and the higher Balmer lines remain in absorption, as is commonly observed in Of spectra.

The basic approach to the stellar wind problem taken here is that of semi-empirical modeling, in which empirically required adjustments are made on approximate dynamical models. A sophisticated approach to the multi-level ion statistical equilibrium problem in an early-type atmosphere allows diverse models of such atmospheres to be tested. In the method used here the transfer equation is solved directly in the co-moving frame of the gas, and the approximations of the escape probability technique are not made.

The model hydrogen atom and helium ion make use of the collisional cross sections used by Klein and Castor (1978), and ten bound levels are assumed in each species.

---

*The National Center for Atmospheric Research is sponsored by the National Science Foundation.

243

P. S. Conti and C. W. H. de Loore (eds.), Mass Loss and Evolution of O-Type Stars, 243–247.

RESULTS

In models C, D and E λ4686 is a pure emission profile in the sense that although a self-reversal feature grows with decreasing mass-loss rate, it does not lower the intensity below continuum values. Profiles with this shape have been observed in several stars and fall in Beals' (1951) type IV classification. In F and G the absorption feature is stronger than the emission. Hence, finding a model in which λ4686 is in emission and λ3203 is in absorption was possible when the mass-loss rate was reduced, keeping the assumed T(r) and V(r) constant. Model E best satisfies the criterion.

The transition from emission to absorption, which occurs at λ3204 in D and E, occurs at different lines in the series in different models. In model C it occurs at λ2733, λ3204 being in P Cygni emission. In F, λ4686 is a P Cygni line, while λ3204 is in absorption. In G the emission component of λ4686 is just barely visible.

Quantitative comparisons with observations of ζ Puppis (Heap 1972, Morrison 1975) show that λ4686 and λ3203 should be somewhat stronger in emission and absorption, respectively. The effects of Stark broadening, which were not included in the calculation, will be to increase the strength of the absorption lines in the He II spectrum, while leaving the strong emission lines, which are quite broadened by the wind, relatively unchanged.

As model E is the most successful with regard to λ4686 and λ3203, a set of He II profiles for model E is presented in Fig. 1. These profiles are influenced by the effects of both wind velocity and geometric extension of the line-formation region. The higher lines in the various spectral series are relatively narrow because the line-formation zone does not lie in the fast-moving layers, and emission features are absent because this zone does not protrude into the large volume of the extended layers. For the strongest of the lines presented the line-formation zone extends several photospheric radii into the wind. The line-formation zones for the resonance lines, which (being unobservable) are not presented, extend to at least ten photospheres in the models used here.

A series of models with terminal velocities and mass-loss rates identical to those in models C,...,F were tested with a linear velocity law V(r), as applied successfully to P Cygni by Kunasz and Van Blerkom (1978). Very strong and pervasive emission in the He II spectrum proved this law too gradual.

Comparisons with observations indicate that more strength is needed in most of the absorption lines. Application of Stark broadening is likely to correct this deficiency.

Model E is the most successful in the H spectrum as well as in He II. It is the only model of those tested in which Hα is in emission

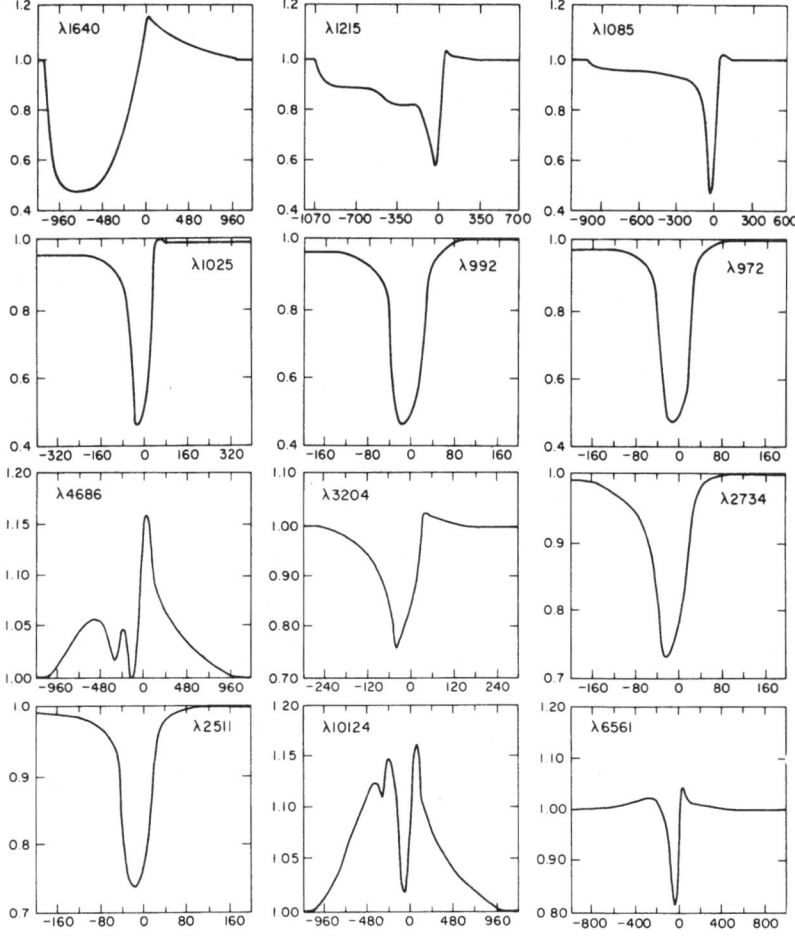

Figure 1

while the higher Balmer lines are in absorption. Further work in which
Stark broadening is included and in which other species, such as N III,
are treated will put model E to a more rigorous test.

This work was supported by National Aeronautics and Space Administration
grant NAS5-22833 through the University of Colorado.

REFERENCES

Beals, C.S.: 1951, Pub. Dom. Ap. Obs. 9, p. 1.
Heap, S.R.: 1972, Astrophys. Letters 10, p. 49.
Kunasz, P. and Van Blerkom, D.: 1978, Astrophys. J., in press.
Klein, R.I. and Castor, J.I.: 1978, Astrophys. J. 220, p. 902.
Mihalas, D. and Kunasz, P.B.: 1978, Astrophys. J. 219, p. 635.
Morrison, N.D.: 1975, Astrophys. J. 202, p. 433.

DISCUSSION FOLLOWING KUNASZ

Snow: The He II λ4686 profiles you calculated in models
C and E look very much like the observed profile of this line
in ζ Pup, on spectra taken one day apart. Hence your models
may indicate that these observations could be explained by
significant density fluctuations in the extended atmosphere
on a time scale of one day.

Heap: Could you say what it would take to get He II
λ1640 fully in emission? My memory of the few UV spectra of
Of stars I have seen is that He II λ1640 is an emission line.

Kunasz: λ1640 is not in pure emission in any of the
models I have assumed. However, the trend for λ1640 to come
.more into emission as the atmospheric density (i.e. the mass
loss rate) is increased indicates that for an atmosphere with
greater mass loss than Castor's model "C", but not a factor
two greater, λ1640 will be in emission with no absorption
component. Perhaps velocity fields unlike those of Castor,
Abbott and Klein, or warm regions in the wind, could enhance
emission in λ1640. However we won't know until such structures are modelled.

Stalio: Could you give an interpretation of those
whiggles that one sees in your Hα profiles?

Kunasz: In principle they could be a result of the complicated isovelocity surface in an expanding atmosphere. In
reality they are probably due to numerical truncation in the
"observer's frame step" of the comoving frame calculation.

Hutchings: Is the self-reversal at λ4686 real or computed
noise? Is it at zero velocity? It is probably an important
diagnostic.

Kunasz: It is due mostly to absorption occurring in low-
velocity layers near the photosphere. As the wind is weakened
in the sequence of models presented here the absorption feature in λ4686 becomes dominant, and eventually, for the weakest winds, totally dominates the small remaining emission
from the expanding envelope. λ4686 is then a pure absorption
line.

Vreux: Have you any predictions for Paschen lines? We
do observe P6 and P7 in emission in a few O stars.

Kunasz: I have treated a ten-level model of the hydrogen
atom. Hence Paschen 6 and 7 were included. They are the
transitions 3→9 and 3→10. Since level 10 is poorly determined due to the fact that no higher levels are treated,

transitions to level 10 are suspect in the results. To some
degree this statement may also apply to level 9. Hence, the
two lines were not included in the line-profile step of the
calculation.

   Noerdlinger: Perhaps the weakness of $\lambda1640$ is due to
the self-absorption you found in $\lambda4686$. This means that
trapped radiation depopulates the n=3 state. If you thin
out the flow $\lambda1640$ may get stronger, to fit the observations
in question.

   Kunasz: Yes, this is possible.

David Van Blerkom[*]
Joint Institute for Laboratory Astrophysics, University of
Colorado and National Bureau of Standards, Boulder, CO   80309

The Hα profile in the spectrum of Zeta Puppis is shown in Figure 1 (Conti and Frost 1977). It is of the P Cygni type, with a violet displaced absorption component that extends to about 300 km s^{-1} from line center. An unusual feature is the broad wing emission which can be seen out to 1500 km s^{-1} on both sides of the line. Of the roughly 2.5 Å equivalent width of Hα, 1.8 Å is due to velocity displacements $|\Delta v| > 300$ km s^{-1}. The presence of wing emission shortward of the violet absorption edge has been noted before in a Wolf-Rayet star, where it is attributed to non-coherent electron scattering (Castor, Smith and Van Blerkom 1970). In the present case, an optical depth in electrons of nearly 1.5 is required to produce the extensive wings by scattering a normal P Cygni profile. This exceeds by a factor of three the electron scattering optical thickness through the entire envelope, assuming published values of the mass loss rate, terminal velocity of the wind and photospheric radius. Thus an alternative explanation is required.

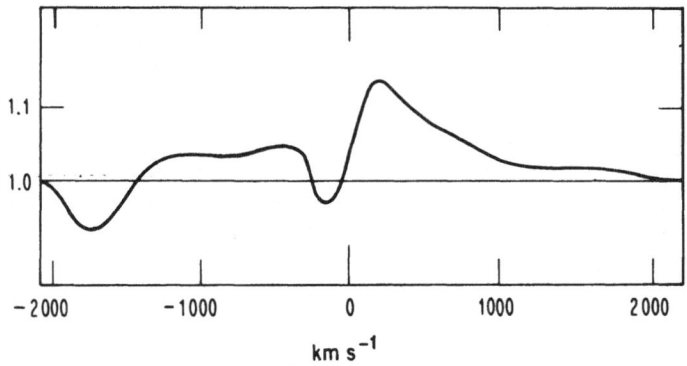

Figure 1

[*]Visiting Fellow, 1977–78, on leave from Department of Astronomy,
University of Massachusetts, Amherst, MA   01002.

249

P. S. Conti and C. W. H. de Loore (eds.), Mass Loss and Evolution of O-Type Stars, 249-252.

In a rapidly outflowing envelope, the line optical depth directly in front of the continuum emitting core $\tau(r)$ at a radius r is proportional to the line absorption coefficient $k(r)$ and inversely proportional to the velocity gradient dv/dr. If at some radius r^* the velocity gradient becomes very large, $\tau(r)$ will be considerably reduced for $r \geq r^*$. It is possible that $\tau(r) > 1$ for $r < r^*$ but $\tau(r) << 1$ for $r > r^*$. In such a situation, the envelope becomes nearly transparent, and any absorption component will terminate at a velocity displacement $\Delta v \approx -v(r^*)$. The envelope emission is also affected by the change in $\tau(r)$, but calculations show that the wing can revert to emission shortward of the absorption component. This may be the explanation of the Hα profile in ζ Pup.

Recent work by Castor and Barlow in the infrared continuum of ζ Pup finds a velocity distribution

$$v(r) = v_o(r-r_o)$$

which is difficult to reconcile with radiatively driven stellar wind theory. This latter gives a velocity law of the form

$$v(r) = v_\infty(1 - 1/r)^{1/2} \quad .$$

Since infrared measurements probe the inner part of the wind, it is possible that a transition in velocity field occurs from the first to the second type at some radius r^*. This would produce the large velocity gradient postulated above.

In order to test this hypothesis, a preliminary model of ζ Pup has been constructed with runs of the source function and optical depth given a priori. Future models will solve the statistical equilibrium equations. Figure 2 shows the profile found in one calculation. One observes the broad emission wing extending beyond the violet absorption. There is also an apparent red absorption due rather to a decrease in envelope emission at the transition radius. Possibly, too abrupt a transition from one velocity law to another has been used in these preliminary models.

Further study will show whether a region, or regions, of rapid acceleration can exist in stellar winds, and whether the Hα or other emission lines can be used as a diagnostic of this behavior. At present, we have no other explanation of the Hα profile.

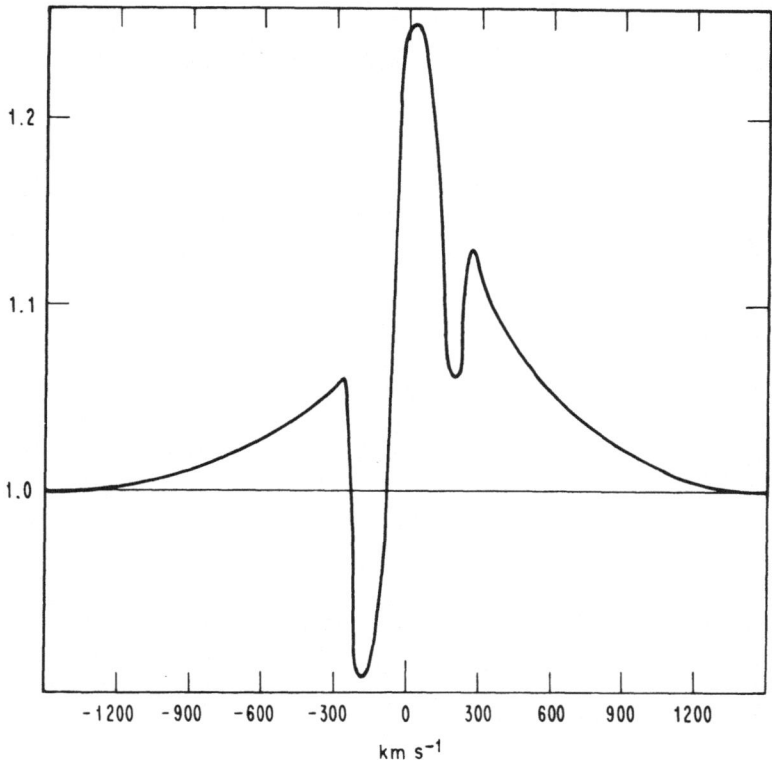

Figure 2

REFERENCES

Castor, J.I., Smith, L.F. and Van Blerkom, D.: 1970, Ap. J., 159, p. 1119.
Conti, P.S. and Frost, S.A.: 1977, Ap. J., 212, p. 728.

DISCUSSION FOLLOWING VAN BLERKOM

Underhill: You show a rapid acceleration in flow velocity rather close to the photosphere. What causes the abrupt change in the applied outward force at this point in the atmosphere? If the dominant force is radiation pressure, an abrupt change in the integral of $F_\nu K_\nu$ over the whole spectrum is implied to occur at the indicated radius. Here K_ν is the total monochromatic opacity due to all sources, lines, continua and electrons. What is the cause of this change? What happens to the state parameters of the atmosphere to cause such an abrupt change?

Van Blerkom: I have not calculated the hydrodynamic
state of the envelope including radiation pressure, but can
only refer you to calculations which have been done by others.
In particular, John Castor has shown that when overlapping
of spectral lines is properly treated, a region of slow velocity rise exists close to the photosphere and is followed by
a region of much more rapid acceleration. The infrared analysis presented earlier by Castor also indicates a region
near the photosphere in which the velocity rise is approximately linear with radius.

Conti: The Hα line profile of ζ Pup shown here is, interestingly, not very typical of most Of stars. The line is
similar to λ Cep, another Of star with relatively large rapid
rotation. I suspect that this rotation, possibly including
shear energy input, has something to do with this absorption
feature. This is the first time I have seen a theoretical
explanation for this kind of emission line.

Hearn: How certain are you of the continuum level in
this line profile?

Conti: I am quite confident that the continuum is drawn
correctly. The absorption feature on the left handside of
the figure is λ6527 of the He II (5-n) series. Leftward of
that is a very nice continuum. The Hα absorption is not ab-
solutely certain to go below the continuum but may by a few
percent as shown.

Hutchings: The profiles with emission on both sides of
the absorption are found in Be stars and you should seriously
consider rotation as an important contribution to them.

Van Blerkom: The scale of the drawing, which extends to
1500 km s^{-1} on both sides of line center may be misleading.
The absorption component is not at line center, but extends
roughly 300 km s^{-1} to the violet. While rotation is no
doubt important, I believe the violet displacement of this
component is suggestive of absorption in an expanding envelope.

Cassinelli: What did you do about the underlying photospheric line?

Van Blerkom: A photospheric absorption profile was taken
from a T_{eff} = 50,000 K, log g = 4 non LTE stellar atmosphere
model of Mihalas and used as the boundary condition for the
incident radiation field.

Cassinelli to Conti: There could not be a 4Å shift of
the underlying photospheric line could there?

THE MASS LOSS RATE OF γ VELORUM

PETER D. NOERDLINGER
Department of Astrogeophysics, University of Colorado
and High Altitude Observatory of NCAR, Boulder Colorado

1. INTRODUCTION

The CIII λ 1909 absorption feature (deepest at V=-1600 Km s^{-1}) in this star, observed by Burton et al (1975) and Johnson (1978), was predicted by Castor and Nussbaumer (1972). It implies a copious flow of CIII ions, because the oscillator strength is only f_{ik}= 1.6x 10^{-7} .

The rate of mass loss can be found two ways: (1) there must be enough CIII to produce the absorption, but not so high an n_e as to fill it in or exceed the observed emission wings by collisional excitation. This limits the H/C and He/C ratios, as well as CII and CIV. (2) One uses radio emission (Wright and Barlow 1975; Seaquist 1976; Morton and Wright 1978). The optical method (1) is mildly sensitive to the velocity law and assumed ionization structure. The radio method (2) is sensitive to the assumed mean atomic mass per ion μ, the rms ionic charge Z, and the number of electrons per ion, γ. It proved advisable to construct a large number of computer models combining the two methods. For clarity, I first estimate flow parameters analytically, then give computer results. All methods lead to these conclusions:

The mass loss rate is in the range 1.1-1.4 x 10^{-4} M$_\odot$ yr^{-1}. The hydrogen content is very small, and the ratio He/C by mass is < 30, more probably 2-4 (Nugis 1975). The CIII absorption region does *not* surround the O star, but lies within 2-3 stellar radii of the WR star. The O star is therefore deficient in UV near λ1909. I have even examined the data of Ganesh and Bappu (1967) to see if the photospheric He II λ4200 usually attributed to the O star (Conti and Smith 1972) could instead follow the WR star velocity. A statistical test showed, in spite of the great scatter, that the chance is less than 6% that λ4686 and λ4200 are in phase. Thus, the O star is peculiar, rather than actually being a B star, which would be simpler. In Section 2, I clear up some discrepancies on the radio method. Sections 3 and 4 consider CIII absorption and emission, and Section 5 combines methods.

P. S. Conti and C. W. H. de Loore (eds.), Mass Loss and Evolution of O-Type Stars, 253-256.

2. OVERVIEW OF THE RADIO METHOD

Wright and Barlow (1975) showed for a uniform temperature wind, the mass loss rate is (all such rates will be given in solar masses/yr)

$$- \dot{M} = 0.095 \ \mu V_t \ S^{3/4} \ D^{3/2} / (Z \gamma^{\frac{1}{2}} g^{\frac{1}{2}} \nu)$$

where V_t is the terminal velocity in Km s^{-1}, S the radio flux in Jy, g the Gaunt factor (=6.9), D the distance in pc (=450), and ν the frequency. Seaquist appears to have taken the mean *ionic* mass $\mu = 1/2$, which does not hold even for hydrogen ($\mu=1$). Using $V_t=1000$, he got a low $\dot{M}=-2 \times 10^{-5}$. Morton and Wright used the newer $V_t=2900$, but took $\mu= 1.3$, again low , finding $\dot{M}= -4 \times 10^{-5}$. I find μ in the range 4-9, and within this range the factors Z and γ compensate so as to yield a small range of mass loss rates $\gamma = n_e/n_i$.

3. OPTICAL CONSIDERATIONS: CIII ABSORPTION

Aside from the interesting CIII λ1909 P Cygni feature, this star has a strong CII λ1335 P Cygni profile with a flat top, showing depletion of CII at low velocity. The strong CIV λ1550 P Cygni profile peaks at V=0 and extends over the whole velocity range, which is one reason to think that not much HeII becomes neutral in the wind. The f_{ik} for these lines are so strong that no interesting lower limit is set on their abundances in the flow. He I λ10830 is in strong emission (Barnes et al 1974), but this does not require a predominance of HeI over HeII. The Pβ emission is probably from the O star wind. I shall now show that the CIII absorption ($r_\nu \sim 0.35$) although not washed out by the O star, cannot result from material overlying it. To understand this, consider the CIII number density N_3, the velocity gradient $v'= dV/dr$, and the optical depth τ where the deepest λ1909 absorption is formed. In cgs units, Castor (1970) showed $\tau=0.0265\lambda f_{ik} N_3/v'$. (More complete dependences on ray angle are included in the computer program). We can estimate N_3/v' as $N_3 \Delta r/\Delta V$, with Δr the thickness of the absorption region and ΔV the spread in velocity there. From the spectra, $\Delta V > 4 \times 10^7$ cm s^{-1}, so for $\tau > 2$ the column density $N_3 \Delta r$ must exceed 10^{21} ions cm^{-2}. Then $N_3 > 10^{21}/\Delta r$. If we take $\Delta r \sim 0.1 \ r$, the outflow of CIII ions is $4\pi r^2 V N_3 > 2 \times 10^{43} \ r_{12}$ ions s^{-1}, where $r_{12} = r/10^{12}$cm. At the photosphere $r_{12} \sim 1$, but in the absorption region it could be 1.5-4.0 (region around WR star only) or $r_{12} > 30$ (region surrounding both stars). In the latter case the mass loss rate in CIII would exceed 2×10^{-4}, to which one must add He, CII, and CIV. Besides its implausibility, such a high choice of mass loss would force the radio and CIII emission to exceed observation; thus $r_{12} < 4$ in the CIII absorption region.

4. OPTICAL CONSIDERATIONS: CIII EMISSION

The CIII collisionally excited emission must not fill in the absorption, nor produce emission wings in excess of the observed 7x10^{47}

photons s^{-1}. By using the excitation rates of Flower and Launay (1973) and some methods from Osterbrock 1970) one can show that in the T and n_e ranges of interest there are about 2-3 x $10^{-9}n_e$ excitations per ion per second. The computer model considers de-excitation and escape probability, with full angle dependence. The conclusion is that, up to a model-dependent uncertainty of a factor 2 or 3, there cannot be more than 3 electrons per CIII ion and still permit the absorption feature; about $5/r_{12}$ electrons per CIII ion are right to produce the emission wings. Thus, r_{12} is again expected to be a small number, say of order 1.5-3.0. Beyond this radius, CIII must recombine to CII , or the temperature T drop below 20,000 K.

5. MASS LOSS. FINAL RESULTS.

A grid of computer models, with velocity law $V=V_t(1-r_{core}r^{-1})^p$, 0.25<p<1.5 would fit all the radio and optical data only if the O star continuum at 1909 $\overset{o}{A}$ was rather less than that of the WR star, and the ratio of He/C by mass was <30. By contrast, if a star with cosmic abundances burned all its H to He, it would have He/C=300. This is forbidden by the data. There would be too many electrons in the CIII absorption region, and in addition, almost all the He would have to become neutral by the time it reached the radio region. It would not do to just reduce the CIII density to compensate the increased He, and so suppress extra radio and CIII emission, because then ther CIII optical depth would be too small to produce the observed absorption. All viable models, with He/C as low as 1 or as high as 30, gave mass loss rates in the range 1.1-1.4 x 10^{-4} solar masses yr^{-1}.

6. ACKNOWLEDGEMENTS

I am indebted to Drs. Peter Conti, R. Viotti, A. Hearn, R.Takens, H. Lamers, and D. Morton for useful conversations, and to Dr. E.P.J. van den Heuvel for encouragement at the University of Amsterdam, where much of the work was done. The National Science Foundation provided support under grant AST 76 05766 to Michigan State University, my permanent address.

7. REFERENCES

Barnes, T.G., Lambert,D.L. and Potter,A.E.:1974,Astrophys.J. 187,pp.73-82.
Burton,W.M.,Evans,R.G.,and Griffin,W.G.:1975,Phil.Trans.R.Soc. London
    A 279, pp. 355-369.
Castor,J.I.:1970, Mon.Not.Roy.Astron.Soc. 149, pp.111-127.
Conti,P.S. and Smith,L.F.:1972,Astrophys.J. 172,pp. 623-630.
Ganesh,K.S. and Bappu,M.K.V.:1967,Bull.Kodaikanal Obs A 183,pp.177-191.
Flower,D.R. and Launay,J.M.:1973,Astron.and Ap. 29,pp.321-326.
Johnson,H.M.:1978, Astrophys.J. Suppl. 39, pp.217-240.
Morton,D.C. and Wright,A.E.:1978,Mon.Not.Roy.Astr.Soc. 182,pp.47P-51P.
Nugis,T.:1975,in B.Sherwood & L.Plaut(ed),Variable Stars and Stellar
    Evolution, D. Reidel Co., Dordrecht, pp. 291-296.
Osterbrock,D.E.:1970, Astrophys.J. 160, pp. 25-29.
Seaquist, E.R.: 1976,Astrophys.J.(Letters) 203, pp. L35-L37.
Wright,A.E. and Barlow,M.J.:1975,Mon.Not.Roy.Astr.Soc. 170,pp.41-51.

DISCUSSION FOLLOWING NOERDLINGER

Abbott: What is the velocity of the CIII 1909 absorption fetaure?

Noerdlinger: It runs to about 1800 km s^{-1}

Castor: To reconcile the total mass loss rate with the CIII loss rate and the C/He ratio, you need 100% CIII. Is that a problem? I would expect C to be mostly CIV.

Noerdlinger: I allowed a little in the mass loss rate for CIV, which is seen in the spectrum. There cannot be much present or there will be too many electrons and the CIII 1909 absorption will fill in, also the emission part will become too strong.

Morton: γ Vel has the strongest CII P Cygni resonance line that I have seen in any star. Since single O and hot B stars do not have such CII, it must originate in the WR wind. In any model of this system, it is important to explain the CII feature, particularly the small amount of residual light at the bottom of the CII absorption.

Noerdlinger: I had missed this point, but the CII oscillator strength is so large that CII/CIII can be small and you still get a P Cygni line.

Willis: At what binary phase were your CIII 1909 observations made? The observations of γVel show the strengths and profiles of many lines undergoing strong variations as a function of binary phase. We believe that at some phases absorption of the O-star light in the WC8 stellar wind can take place, and can depress the absorption below the O-star continuum level, even though the O-star is the brighter component.

Noerdlinger: The phases were all near 0.65. At that phase there is no possibility for the WR CIII absorption region to overlie the O-star, or the mass loss rate would come out much too big, and the CIII emission too strong.

Willis: The CII 1335 Å line in γ^2Vel is known to undergo phase dependent variations in its profile, which is believed to be the result of absorption of the O-star in the WC8 wind. We thus think the CII is associated with the WC8 star.

Noerdlinger: This is a valuable observation and I would expect it is quite reasonable; the CII oscillator strength is much bigger.

WIND MODELS FOR ζ ORIONIS

Gordon L. Olson
Free University of Brussels

INTRODUCTION

Several models for the winds of O stars have been proposed to explain the unexpected presence of high ionization potential ions such as N^{+4} and O^{+5}. Lamers and Snow (1978) proposed that the winds of stars showing N V and O VI lines have elevated temperatures near $4\pm2 \times 10^5$K while cooler stars with anomalous Si IV lines have $T_e \approx 7\pm3 \times 10^4$K. Alternately, Cassinelli and Olson (1978, CO) and Olson (1978) have explained the presence of these ions by showing that a thin corona at the base of a cool wind ($T_e \leq T_{eff}$) can produce these ions by the Auger photoionization process where a single x-ray photon causes the ejection of two electrons. A third possibility is that the winds are at only slightly elevated temperatures (40 000 to 60 000K) and photoionization in an optically thick wind produces the unexpected ions. The problem of determining the local radiation field in such a wind is discussed by Klein and Castor (1978). The purpose of the present analysis is to test the ability of these three wind models to fit the observations of ζ Orionis A O9.7 Ib.

THE OBSERVATIONS AND THEIR ANALYSIS

The ultraviolet resonance lines of ζ Ori have been taken from the Copernicus catalog by Snow and Jenkins (1977) and from Snow and Morton (1976). The continuum on both sides of each line was examined to determine the continuum level which was then linearly extrapolated through the line profiles. These normalized profiles were then fitted with theoretical profiles calculated with the Sobolev approximation. The procedure used here for line profile fitting has followed the approach of Olson (1978). The result for each ion with a resonance line is a number that gives the mass loss rate times the relative ionic abundance. Since models will predict the ionic abundances, the mass loss rate of the wind can be determined.

P. S. Conti and C. W. H. de Loore (eds.), Mass Loss and Evolution of O-Type Stars, 257–260.

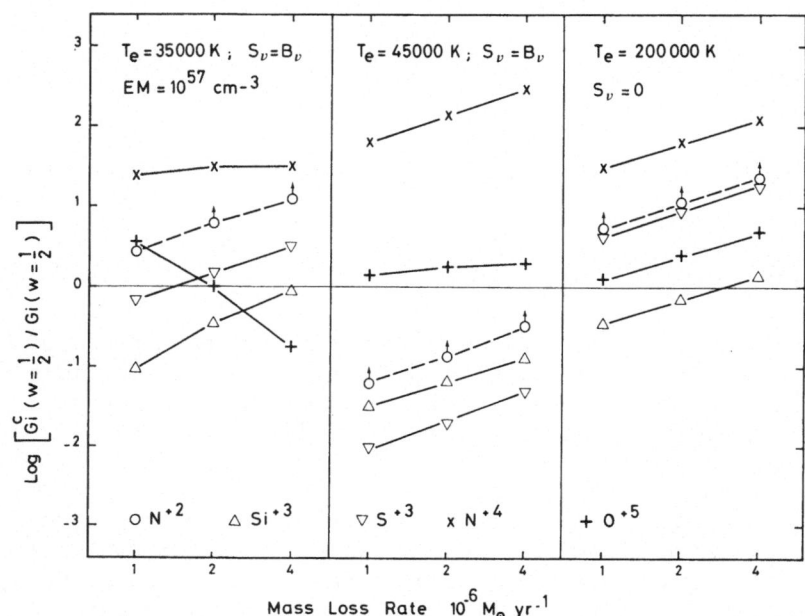

Figure 1. Comparison of computed (G_i^c) to observed (G_i) ionic abundances.

DISCUSSION OF WIND MODELS FOR ζ ORIONIS

Given the temperature and velocity structure of the wind, the ionization balance for the elements of interest can be computed by including the following processes: collisional ionization and photoionization from the ground level, Auger ionization of K-shell electrons, radiative recombination, and dielectronic recombination (see Olson 1978 and CO for details). The radiation contributing to photoionization includes the photosphere, the corona and the local wind radiation at wavelengths where the wind is optically thick. The velocity structure of the wind enters into the calculation primarily through scaling the optical depths. However, since the temperature differences among the three models affect the optical depths much more than the uncertainties in the velocity structure, there will be no discussion of velocity structure effects presented here. The same velocity law will be used for all three models.

A comparison for five ions between observations and theory for the "warm" wind model is shown in the right most panels of Figures 1 and 2 for two gas temperatures. The comparison is made for a range of mass loss rates considered appropriate from Hα analyses. The radiation from the wind has been neglected in the photoionizations because at these elevated temperatures the Non-LTE thermal source function, S_ν, is quite small. Clearly, from these figures the N^{+2} limit (shown as a limit with arrows because the observed N III doublet appears to be photospheric and can only put an upper limit on N^{+2} in the wind) requires a gas tempera-

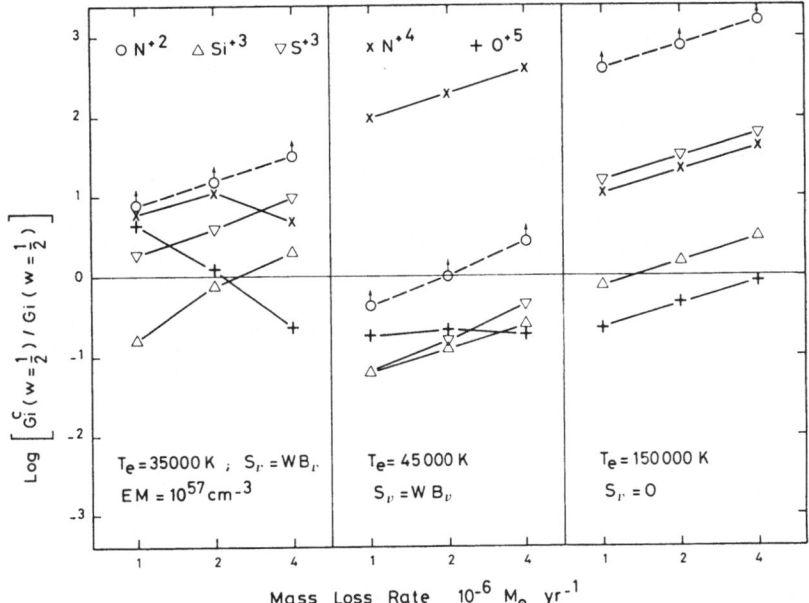

Figure 2.   Same as Figure 1.

ture greater than 200 000K. However, at such a temperature this would predict that the N V line should be saturated in the blue shifted absorption, whereas the observed line goes down to only 0.4 of the continuum level. Though Si^{+3} and O^{+5} agree nicely between 150 000K and 200 000K, none of the other ions agree well.

For computing the models with cooler winds, the radiation from the wind has been included in two ways: i) $S_\nu = B_\nu(T_e)$, a blackbody at the wind temperature, ii) $S_\nu = WB_\nu(T_e)$ a blackbody modified by geometrical dilution, W. Case (i) represents a completely optically thick wind and case (ii) a thin wind where the source function is a blackbody modified by the ground level departure coefficient which is inversely proportional to the dilution: $S_\nu = B_\nu/b_1 = WB_\nu$. The true situation is between these two approximations and quite difficult to compute (see Klein and Castor, 1978).

Figure 1 shows that without a corona the wind temperature must be raised to 45 000K to produce the observed abundance of O^{+5}. In this "tepid" wind model N^{+2} is not a problem; however, N^{+4} is overproduced by a factor of 100. Also Si^{+3} and S^{+3} are strongly depleted by this radiation field. Reducing the radiation field by using $S_\nu = WB_\nu$ (Figure 2) lessens this latter problem, but then not enough O^{+5} is produced.

With a coronal model ($T_c = 5 \times 10^6$K) the volume emission measure of the coronal zone, $EM \equiv \int N_e^2 \, dV$, is adjusted to provide enough soft x-rays to produce the observed abundance of O^{+5} by Auger photoionization.

For ζ Ori this requires an emission measure near 10^{57} cm^{-3}. Then the wind temperature can be adjusted to minimize the disagreement between the other ions and observations. The results with T_e = 35 000K are shown in the figures. Though the nitrogen ions are still difficult to fit, this model clearly matches the observation better than the other models. Part of this improvement is of course due to the fact that there are two adjustable parameters (T_e, EM) rather than one (T_e).

If nitrogen were underabundant in ζ Ori by a factor of ten, all the models would match the observations much better (especially the corona -cool wind model). However, this is unlikely. The present rough model calculations indicate that the corona-cool wind model fits the observations of ζ Ori better than either the "warm" or "tepid" wind models. However, much more detailed calculations need to be completed before one or more models can be definitely eliminated.

REFERENCES

Cassinelli, J.P., and Olson, G.L. 1978, Ap.J., submitted, CO.
Klein, R.I., and Castor, J.I. 1978, Ap.J., 220, p.902.
Lamers, H.J.G.L.M., and Snow, T.P. 1978, Ap.J., 219, p.504.
Olson, G.L. 1978, Ap.J., 226, pp.124-137.
Snow, T.P., and Jenkins, E.B. 1977, Ap.J.Suppl., 33, p.269.
Snow, T.P., and Morton, D.C. 1976, Ap.J.Suppl., 32, p.429.

DISCUSSION FOLLOWING OLSON

Morton: The behaviour of the NIII resonance line in various OB stars is very puzzling. In many cases where the CIII λ977 is present with a velocity displacement, the NIII, which requires higher ionization energy, is an undisplaced photospheric line. ζ Pup is a notable exception with a well developed P Cygni NIII profile.

BINARY FREQUENCY AMONG THE O-TYPE STARS

Catharine D. Garmany
Joint Institute for Laboratory Astrophysics, University of
Colorado and National Bureau of Standards, Boulder, CO 80309

A great deal of work has been done on the theory of mass loss and
evolution in close binaries, and numerous individual systems have been
discussed in this connection, but the general question of the binary
frequency of O-stars, and in particular, the initial binary mass ratio
frequency or distribution of secondary masses, has not been completely
answered. In general, we know that about half·of all O-type stars are
binaries; the most recent determination by Conti, Leep and Lorre (1977)
found 58% of their sample to be certain or probable binaries. However,
many of these stars were judged to be variable on the basis of only a
few spectra from different sources, and therefore require further study.
Another point to be examined concerns the binaries with available orbits: two thirds of these are double line systems. Figure 1 shows a
plot of the semi-amplitude versus orbital period for all known systems,
along with some theoretical curves for different mass ratios. Not only
is the lack of single line systems obvious, but low amplitude systems
are almost completely missing. This would appear to be only an observational selection effect, although it is to be noted that low amplitude double line Wolf-Rayet systems have been detected. If the effect
is real, it implies that O-type binaries with mass ratios (m_1/m_2)
greater than about three do not exist.

This type of effect is not observed among the B-type stars.
Studies by Abt and Levy (1978), Wolff (1978) and Blaauw and Van Albada
(1974) show no lack of single line, low amplitude systems, and even
allowing for a smearing effect in sin i, there are certainly B-type
systems with mass ratios greater than three. When a preliminary study
of some of the stars listed as variable by Conti, Leep and Lorre failed
to detect any low amplitude, short period binaries (Bohannan and Garmany
1978) we began a systematic search of the O-stars.

To eliminate the observational selection effects of Fig. 1, we
have examined all the O-stars brighter than 7th magnitude and north of
-50°, a sample of 76 stars. In the cases of known binaries with published orbits we have made no further observations, but any star judged
variable by Conti, Leep and Lorre has been reobserved by Conti and

261

P. S. Conti and C. W. H. de Loore (eds.), Mass Loss and Evolution of O-Type Stars, 261–264.
Copyright © 1979 by the IAU.

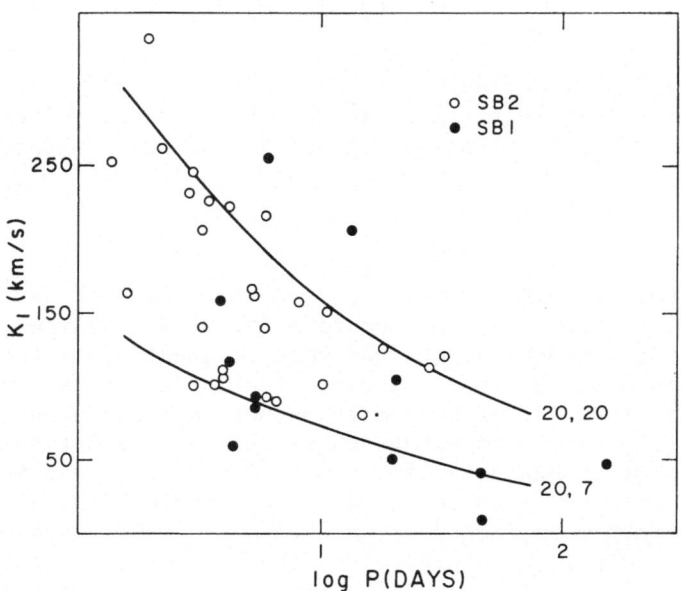

Figure 1.   Semi-amplitude versus orbital period
             for all O-type binaries.

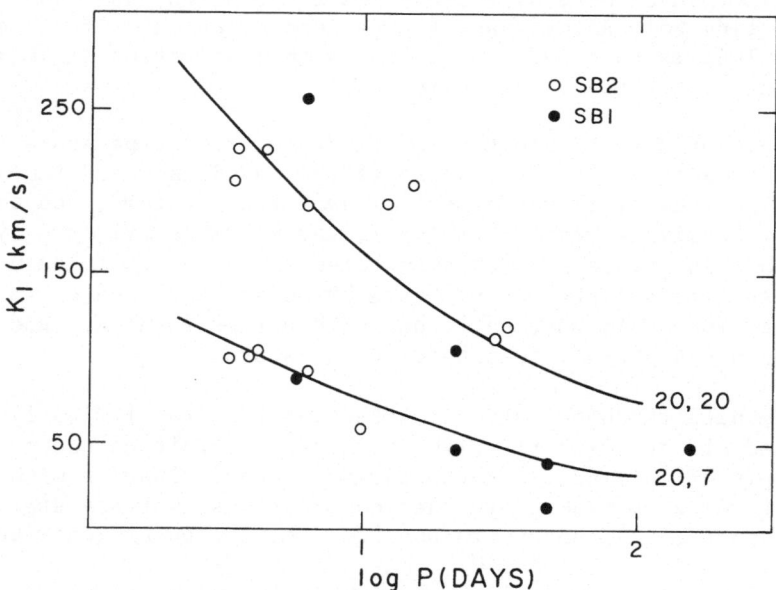

Figure 2.   Semi-amplitude versus orbital period for
             the O-type binaries corrected for obser-
             vational selection effects.

members of his group using the coude spectrograph at Kitt Peak National
Observatory.   Morrison and Massey will report on the new binaries.   Despite the use of analysis of variance tests (Tryon and Garmany, in preparation) to measure radial velocity variations, we have not found any
short period, low amplitude binaries although we expect to detect semi-
amplitudes as low as 15 km/sec. Preliminary results are shown below in
comparison with early B-stars (Blaauw and Van Albada 1974) and middle
B-stars (Abt and Levy 1978):

PRELIMINARY BINARY FREQUENCY OF O AND B STARS

	O4–O9	O9–B3	B2–B5
Single	53%	50%	51%
SB2	25%	15%	11%
SB1	11%	21%	25%
Triple	11%	14%	13%

The percentage of single stars does not vary, but the proportion of
double line to single line binaries is much higher among the O-stars.
Figure 2 shows the semi-amplitude versus orbital period for all the O-
binaries in our sample, and even with no correction for sin i, there
seem to be no systems with mass ratios greater than about three.

There are several possible physical mechanisms which might be
responsible for this effect. The O-stars may be unable to form stable
systems if the primary star reaches the main sequence much in advance
of the secondary, and thus no systems would exist with large mass ratios.   Another interesting possibility which would explain the mass
ratios of the most massive O-stars concerns the evolutionary consequences of moderate mass loss. Recent calculations by Vanbeveren, De
Greve, van Dessel and de Loore (1978) show that with the assumption of
main sequence mass loss, the mass ratios of O-type binaries will tend
towards one, and they predict an overabundance of mass ratios smaller
than 1.4 for unevolved massive close binaries.

This work was supported in part by National Science Foundation
grant AST76-20842 to the University of Colorado.

REFERENCES

Abt, H.A. and Levy, S.G.: 1978, Astrophys. J. Suppl. 36, p. 241.
Blaauw, A. and Van Albada, T.: 1974, unpublished.
Bohannan, B. and Garmany, C.D.: 1978, Astrophys. J. 223, p. 908.
Conti, P.S., Leep, E.M. and Lorre, J.J.: 1977, Astrophys. J. 214, p. 759.
Vanbeveren, D., De Greve, J.P., van Dessel, E.L. and de Loore, C.: 1978,
     (preprint).
Wolff, S.C.: 1978, Astrophys. J. 222, p. 556.

DISCUSSION FOLLOWING GARMANY

Underhill: Two selection effects may be present in the
published information on O-type binaries: (1) Single-spectrum
O-type stars with variable radial velocity were rarely observed intensively because the probable yield of information
did not justify the observing time which would be required
to find an orbit with the available equipment, and (2) the
strong absorption lines which could be measured on the low-
dispersion prismatic spectra which were available were chiefly those of H, He I and He II, all of which are broadened by
Stark effect. The net effect is that two spectra can be
separated only if K is large; most cases of moderate or small
K would be reported as a broad-lined, constant-velocity star.
The plate files at the Dominion Astrophysical Observatory
may contain much useful information for your project. The
early prismatic spectra are wide and they are obtained on a
fine-grained, blue-sensitive emulsion.

Garmany: Yes, we have considered these selection effects.
Our new data, however, are all from 18 Å/mm Coudé plates
taken at KPNO by Conti and others in his group. We have
examined both suspicious velocity variables and a certain
number of stars reported as constant in velocity, and have
performed an analysis of variance test to handle both line
to line and plate to plate variations (details of method in
Bohannan and Garmany, Ap.J. 1978, 223, 908). We believe that
we can detect variability if the semi-amplitude is greater
than 15 km/s. The case of unresolved double lines is not
completely satisfactory, but this does not change our results
concerning binaries with mass ratios greater than about three.

Bohannan: No matter what the observational selection
effects are, the mass ratio of an O-type binary system is
physically defined by the fragmentation which results from
too much angular momentum and by the radiation pressure of
the more massive star as it achieves main sequence luminosity.
The latter process should be readily evaluated by existing
radiation driven mass loss theories.

KAM-CHING LEUNG
Behlen Observatory, Department of Physics and Astronomy,
University of Nebraska; Office of Energy Research,
U. S. Department of Energy, Washington, D.C.

DONALD P. SCHNEIDER
Hale Observatories, California Institute of Technology,
Carnegie Institution of Washington

The eclipsing binaries UW CMa, AO Cas, and V729 Cyg have been systems of great interest for over fifty years. The light curves are complex and suffer significant changes on a time scale of months, but the primary attraction of these systems is that both components have O-type spectra; thus they present us with some of the few possibilities for direct measurement of absolute dimensions of very massive stars. Much effort has been expended on these systems, but no really consistent model has emerged.

Most of the difficulties are due to the closeness of the components; tidal distortions and reflections complicate the picture greatly. To handle this problem we adopted the Wilson and Devinney (1971) approach to photometric solution. Published photoelectric observations of high quality (± 0.01 mag) were employed in our analysis. Despite attempts to make detached and semi-detached models fit, all three systems were best represented by contact configurations. The computed light curves fit the observations considerably better than any previous models, but there still are minor discrepancies due primarily to asymmetries in the light curves (the models are symmetric).

Absolute dimensions were calculated by combining the newly determined photometric parameters with the published spectroscopic orbits; the results are summarized in Table 1. [A description of our procedure is given in our papers on these systems, Leung and Schnieder (1978a,b) and Schneider and Leung (1978)]. The primaries of UW CMa and V729 Cyg are among the most massive stars ever directly measured.

The evolutionary state of these systems can be seen from the mass-radius diagram (Fig. 1, which includes the known contact systems of spectral type B). Both UW CMa and AO Cas are still in the hydrogen burning phase of evolution but have evolved off the ZAMS, they are probably examples of case A mass exchange. The status of V729 Cyg is

P. S. Conti and C. W. H. de Loore (eds.), Mass Loss and Evolution of O-Type Stars, 265-269.

TABLE 1

CONTACT O SYSTEMS

Star	Period (days)		$R_1(R_\odot)R_2$		$M_1(M_\odot)M_2$		% Overcontact
UW CMa	4.39	O7+O7f	20	18	46	34	23
AO Cas	3.52	O8.5III+O8.5III	13	14	25	29	3.4
V729 Cyg	6.60	O7fIa+OfIa	33	17	59	14	31

not as clear, except it appears to be considerably more evolved and may represent a system having undergone (undergoing?) case B exchange.

For components in a contact system:

$$\frac{R_1}{R_2} \sim \left(\frac{M_1}{M_2} \right)^{0.46} .$$

Thus a line connecting them (as in Fig. 1) will be nearly parallel to the ZAMS and TAMS lines for stars of high mass. Therefore both components will appear equally evolved. This problem of evolution into contact is indeed an exciting albeit an extremely difficult one.

As one might expect, each system exhibits unusual and confusing features. UW CMa has a very complex light curve with a displaced secondary maximum, and spectroscopic studies have found mass ratios ranging from 0.75 to 1.3! AO Cas has an inclination of 51°, the light variation is due almost entirely to its ellipsoidal shape. V729 Cyg has a very small mass ratio (0.237) and a 10,000° K temperature difference between the components. The largest discrepancy, which all three systems have to some degree, lies with the relative luminosity of the components. The contact model can differ wildly from the observed spectroscopic luminosity ratio, for example, the spectroscopic luminosities of the components of V729 Cyg are roughly equal, while the contact model predicts a ratio of eight! If our models are correct the clearly line formation in massive contact systems is due to a new, poorly understood mechanism.

Future work planned by the authors includes a search for further examples of massive contact systems (DH Cep appears to be the best candidate) by K.C.L. and improved spectroscopic observations using the photon counting device built by Steve Shectman (D.P.S. with P. J. Young). Other areas of needed investigation are monitoring of period changes in these systems for evidence of mass transfer along with the aforementioned evolution and line formation problems.

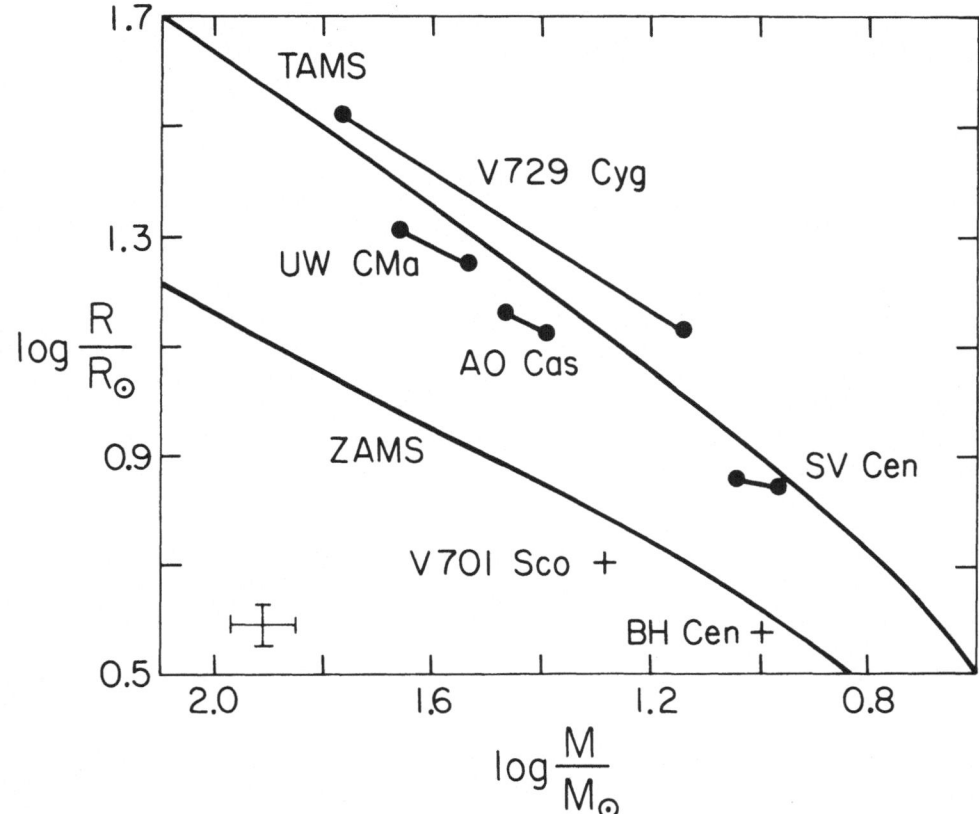

Figure 1. Mass-radius diagram for early-type contact systems. The zero-age main sequence (ZAMS) and terminal-age main sequence (TAMS) of Stothers (1972) are shown as smooth curves. The estimated errors for the mass and radius (assuming the mass ratio is reliable) are shown in the lower left-hand corner.

REFERENCES

Leung, K. C., and Schneider, D. P.: 1978a, Astrophys. J. 222,
     pp. 924-930.
Leung, K. C., and Schneider, D. P.: 1978b, Astrophys. J. 224, (in press).
Schneider, D. P., and Leung, K. C.: 1978, Astrophys. J. 223, pp. 202-206.
Stothers, R.: 1972, Astrophys. J. 175, pp. 431-452.
Wilson, R. E., and Devinney, E. J.: 1971, Astrophys. J. 166, pp 605-619.

DISCUSSION FOLLOWING LEUNG AND SCHNEIDER

Hutchings: In the case of 29 CMa (= UW Cma) the spectroscopic data give a consistent picture of a e>0 orbit. Since
both spectroscopic and photometric analyses may give spurious
results we must understand why, before we say which is correct (if either). Similar problems occur with AO Cas. Your
analyses do not allow consideration of 1) e≠0, 2) deviations
from Roche geometry by rotation, mass flow or radiation pressure.

Schneider: A contact system cannot have an orbital eccentricity: if an eccentricity is certain it would be fatal to
our model. Several of our contact systems (UW CMa, AO Cas,
V1073 Cyg) do have measured eccentricities ranging up to
0.12. In an extremely close or contact system a "pseudo-
eccentricity" can arise due to tidal distortions, eclipses,
and reflections. In all three systems we find that when these
effects are allowed for the contact models fit the observations very well. One troublesome point with UW CMa is your
measurement of enhanced mass transfer when the stars are
closest (in the eccentric orbit model). The contact model
says any enhanced rate would be due to the geometry of the
system, your results are an uncomfortable coincidence. As to
the effects of radiation pressure I think Bob Wilson has some
enlightening thoughts.

Wilson: The radiation pressure question can be put in
some perspective by noting the following points. Because of
the von Zeipel gravity darkening law, we expect the local
gradient of radiation pressure to be co-linear with the vector of gravitational acceleration. Thus in regard to a component's own self-radiation, radiation pressure force acts
only to reduce gravity, and thus has no effect on the figures
of stars of a fixed size. The problem comes in with the ir-
radiation from the other star. Even here one can find a case
in which the effects should be fairly small. This case is
that in which the system is known to be overcontact. Then
only a part of the inner-facing surface is irradiated and,
more important, it is then reduced by a considerable projection effect. However I do not want to underplay the effect
of radiation pressure on the figures of detached components,
since it is probably quite important for such cases of spectral type O.

Abbott: How certain is your mass of 46 M_O for the primary of UW CMa?

Schneider: The main problem in determining the masses
for UW CMa lies in the extreme difficulty of detecting the
secondary. Spectroscopic studies have reported values for

the mass ratio (secondary/primary) ranging from 0.75 to 1.3.
We believe the mass ratio is less than one because of the
faintness of the secondary; if the mass ratio was greater
than one our models have the secondary more luminous. As
to the specific value of q we are not certain, but we think
that the primary's mass probably lies between 35 and 50 M_o.
A more reliable estimate will require better spectroscopic
observations.

THE O-TYPE SPECTROSCOPIC BINARY SYSTEM HD 149404

Philip Massey and Peter S. Conti[*]
Joint Institute for Laboratory Astrophysics, University of
Colorado and National Bureau of Standards, and Department of
Astro-Geophysics, University of Colorado, Boulder, Colorado
80309

HD 149404 (HR 6164) was noted to have double lines by Conti, Leep
and Lorre (1977). The spectrum has been previously classified as O9Ia,
and shows strong Si IV λ4089, 4116 absorption, He I stronger than He II,
N III λ4634,41 weakly in emission, He II λ4686 "filled in," and all the
Balmer lines through H13 in absorption except Hα, which is strongly in
emission.

PSC and Dr. Nancy Morrison obtained 40 coude spectrograms (18 Å/mm)
of this star from Kitt Peak and Cerro-Tololo. Of these, 13 showed
double lines. The orbit solution shown in Fig. 1 is based only on these
double-lined plates, and only on the absorption lines. The formal elements are given below and were derived from a differential correction
technique after the period was found using a program written by R. J.
Wolff and N. Morrison based on the Lafler-Kinman (1965) search routine.

Elements of HD 149404	$P = 9.^\mathrm{d}813$ (adopted)	
	$e = 0$ (assumed)	
	$T = $ JD 2442498.7	
	Primary	Secondary
Spectral type	O8.5I	O7III(f)
γ (km/s)	-37 ± 1	-28 ± 2
K (km/s)	101 ± 1	60 ± 2
a sin i ($\times 10^7$ km)	1.37 ± 0.02	0.81 ± 0.01
$m \sin^3 \underline{i}$ (solar masses)	1.6	2.7
mass ratio	1.68	

[*]Visiting Astronomer, Kitt Peak and Cerro Tololo Inter-American Observatories, which are supported by the National Science Foundation under
Contract No. AST74-04128.

271

P. S. Conti and C. W. H. de Loore (eds.), Mass Loss and Evolution of O-Type Stars, 271-275.

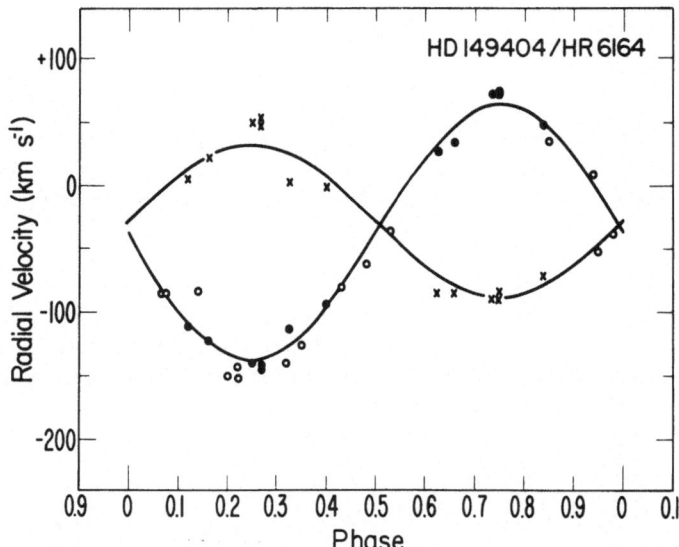

Fig. 1.   Velocity curve HD 149404. The filled circle
          (primary) and the crosses (secondary) are
          from the double-lined plates; the open circles
          are from the single-lined plates and were not
          used in computing the orbit.

     The brighter star is a late-O supergiant, which we call the pri-
mary; the secondary is a somewhat earlier giant. The N III λ4634,41
emission clearly shifts with the secondary. The primary is the less
massive of the two. It should also be noted that the primary's γ
velocity is blue-shifted relative to that of the secondary. This
effect was seen in HDE 228766 and BD+40°4220 (Massey and Conti 1977)
and is readily interpreted as evidence that mass is moving outwards
even down deep in the line-forming photospheric regions.

     Since the absorption lines are quite blended even at quadrature,
we must consider how large an effect this has on the minimum masses.
Wilson (1941) developed a simple method for obtaining the mass ratio
of a double-lined binary without an orbit solution. When this approach
is applied to the two sets of absorption lines, we find a mass ratio of
1.64, in excellent agreement with the formal solution as expected from
the fact we used the same velocity data. However, when we use the
secondary's N III λ4634,41 emission we find a mass ratio of 1.0. This
would drive the minimum masses up to 4 for each star. It should be
emphasized that this emission can occur even in a plane parallel geometry (Mihalas, Hummer and Conti 1972) and can be a good velocity indicator as shown by some WN stars (Conti, Niemela and Walborn 1979).
Unfortunately, only 5 plates have measurable N III emission so an
orbit solution is not possible.

We see evidence of mass loss from the supergiant in that (1) its γ velocity is more negative than its companion's, (2) it may be under-massive for its luminosity, and (3) there is very strong, broad Hα emission, indicating that mass is being lost somewhere in the system. We must now ask how the mass is being lost: does the supergiant fill its Roche lobe? We can argue that it does as follows:

We know the absolute luminosity of the system from its cluster membership in ARA OB1a (Humphreys 1978); after correcting for the con-tribution of the secondary we find $M_v \sim -6.8$. Using the effective temperature scale and bolometric corrections of Conti (1973), we compute that the supergiant should have radius of 27 R_o. We may estimate the size of the Roche radius r_1 following Paczyński (1970):

$$ r_1 \sin i = (a_1 + a_2) \sin i \left(0.38 + 0.2 \log \frac{m_1}{m_2} \right) \quad , $$

or $r_1 \sin i = 13 \ R_o$ from our orbital elements. Therefore, if the star did not fill its Roche lobe, $i < 30°$. If the actual masses are about 20 M_o, then $i \geq 30°$. Furthermore, if the star is rotating synchronously in its orbit its equatorial velocity should be 140 km/s; we estimate $v \sin i$ from the line widths to be 120 ± 30 km/s, implying that $i \gtrsim 40°$. These arguments are slightly strengthened if the mass ratio indicated by the N III emission is more nearly correct. We conclude that the supergiant probably fills its Roche lobe.

It should be mentioned that the Hα emission is usually double peaked, but is variable in the sense that it becomes single during phases near to quadrature; e.g., when the absorption lines become double. We do not know where the emission is produced; however, if we interpret it as coming from the supergiant we compute a mass loss rate from the models of Klein and Castor (1978) of $6 \times 10^{-6} \ M_o$/yr.

This research has been supported by the National Science Foundation under grant AST76-20842 to the University of Colorado.

REFERENCES

Conti, P.S.: 1973, in H II Regions and Related Topics, ed. T. L. Wilson and D. Downes (Berlin: Springer-Verlag), p. 207.
Conti, P.S., Leep, E.M. and Lorre, J.J.: 1977, Ap. J. 214, p. 759.
Conti, P.S., Niemela, V.S. and Walborn, N.R.: 1979, Ap. J. (in press).
Humphreys, R.M.: 1978, Ap. J. Suppl. (submitted).
Klein, R.I. and Castor, J.I.: 1978, Ap. J. 220, p. 902.
Lafler, J. and Kinman, T.D.: 1965, Ap. J. Suppl. 98, p. 216.
Massey, P. and Conti, P.S.: 1977, Ap. J. 218, p. 431.
Mihalas, D., Hummer, D.G. and Conti, P.S.: 1972, Ap. J. 175, p. L99.
Paczyński, B.: 1970, in Mass Loss and Evolution of Close Binaries, ed. K. Gyldenkerne and R. M. West (Copenhagen Univ. Publ.), p. 139.
Wilson, O.C.: 1941, Ap. J. 93, p. 29.

DISCUSSION FOLLOWING MASSEY AND CONTI

Hutchings : Is there a light curve for this system?
It should be an ellipsoidal variable, and the light curve
should be analysed.

Massey : Yes, Nancy Morrison has done photometry for
this system.

Morrison : I have obtained photometry of this star
on the uvby photometric system, and the results are consistent with the brighter star filling its Roche lobe. The
light curve is roughly sinusoidal between phase 0.0 (where
the brighter star is in front) and phase 0.5, with maximum
light at phase 0.25, where the velocity separation is a
maximum, as would be expected on the basis of an ellipsoidal
model. Phases are computed on the basis of the ephemeris
from the velocity curve. At phases greater than 0.5, the
light curve may show intrinsic variability, but the data are
too fragmentary for definite conclusions to be drawn. The
amplitude of the ellipsoidal variation is 0.03 mag.

Bidelman : How do you know whether so are the NIII
emission lines on the absorption lines to determine the
mass ratio in this system?

Massey : Well, we shouldn't be surprised if the absorption lines from the fainter stars are shifted towards the
lines of the brighter star - this is the well known effect
of pair blending. On the other hand, NIII $\lambda4640$ emission
has been shown to be a surprising good velocity indicator.
In HDE 228766, Conti and I found that the NIII emission from
the Of star gave the same center-of-mass velocity as that
determined from the absorption lines of the other star -
a result we misinterpreted as the time. Niemela's work on
the Wolf-Rayet star HD 97420, and Conti's work on other Of
stars also support this contention. I believe that it is
well known that NIII $\lambda4640$ can be formed in emission in a
plane parallel atmosphere - it is not formed in the wind.

Bolton : In order to do pair-blending corrections
properly it may be necessary to take a variety of photographic effects into account. If this not done, the correction will probably be underestimated. In other words,
I doubt that it is possible to do the pair-blending correction properly for photographic data.

Massey : Let me just say that I do not believe blending
corrections can be done correctly in any event. I personal-
ly don't think that photographic effects on IIa-O's taken at
Coudé dispersions can be very significant, but I have not
looked into it.

Cowley : In the X-ray binaries it is fairly common to
find the emission lines indicate a smaller mass ratio than
the true value. This is because the emission is partially
formed between the stars. In this system perhaps the NIII
emission gives a similarly too small velocity amplitude.

Massey : Our NIII emission gives a larger velocity
amplitude than the absorption, as was also the case in
HDE 228766. It is easy to see why the absorption line
amplitude might be too small, due to the effects of blending. It is hard to see why the emission amplitude should
be too big.

THE O-TYPE SPECTROSCOPIC BINARY SYSTEM HD 93206

Nancy D. Morrison[1] and Peter S. Conti[1]
Joint Institute for Laboratory Astrophysics, University of
Colorado and National Bureau of Standards, Boulder, CO 80309

The star HD 93206 (=QZ Carinae) is a double-lined (Conti et al. 1977), eclipsing (Moffat and Seggewiss 1972) binary with a period of 6 d. Walborn (1973) classified it O9.7Ib:(n). Since the star is probably a member of the cluster Collander 228 (which is near η Carinae), its distance can be assumed to be 2600 pc. In principle, one can determine the masses of the components of HD 93206 from observations of the radial velocities and the light curve, and a spectroscopic orbit is the object of this investigation. A mass determination for an evolved star such as this one is especially important for checking recently computed evolutionary tracks with mass loss for massive stars (de Loore et al. 1977, Chiosi et al. 1978, Dearborn et al. 1978).

Between 1974 March and 1977 April, we obtained 29 blue spectrograms of HD 93206 with the No. 1 coudé camera of the 1.5-m telescope at the Cerro Tololo Inter-American Observatory. All have dispersion 17 Å mm^{-1} and are widened to 0.6 or 0.8 mm. We measured them for radial velocity in both forward and reverse directions with a Grant oscilloscope comparator. For the orbital analysis, we used lines of He I. We traced six of the spectrograms and, using the method of Petrie (1940), we obtained the light ratio and the individual spectral types.

We used the technique of Lafler and Kinman (1965) to find the period of the velocity variation of the fainter star, which we henceforth call Star B. This period turns out to be essentially that of the eclipsing binary (Moffat and Seggewiss 1972). We then used a version of the program by Wolfe et al. (1967) to perform a differential-correction orbital solution. Figure 1 shows the radial velocities of both stars, plotted in the period we found for Star B, and, as a full curve, the theoretical velocities predicted from the orbital elements for Star B, which are listed in Table 1. The fit to the observations of Star B is reasonable in view of the internal errors (about 10 km s^{-1}), but the velocities of Star A show no indication of orbital motion with this period. In an attempt to fit the velocities of Star A, we searched for periods in the range 0.9 to 1.1 d and 15 to 40 d; the best-fitting period is 20.72 d. Figure 2 shows the velocities of Star A plotted in

277

P. S. Conti and C. W. H. de Loore (eds.), Mass Loss and Evolution of O-Type Stars, 277-280.
Copyright © 1979 by the IAU.

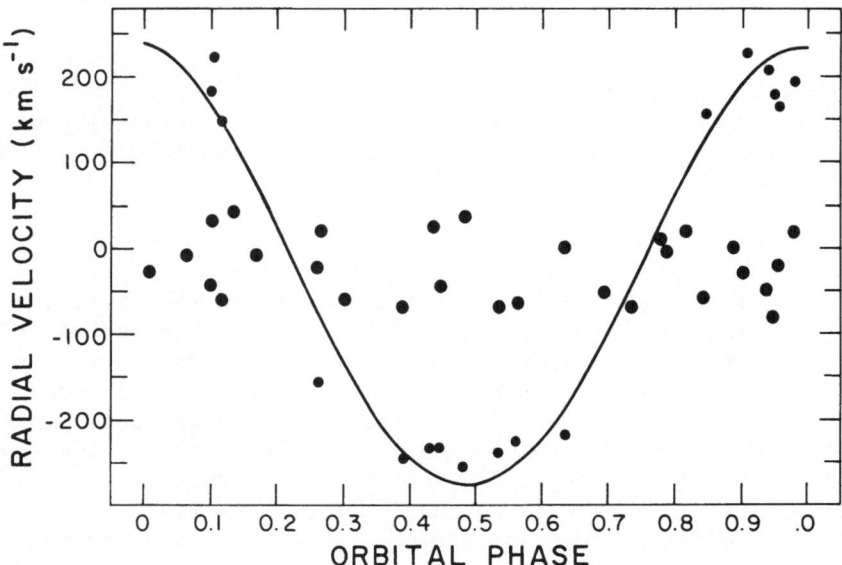

Figure 1.   Radial velocities for HD 93206 AB as a function of orbital
phase, which is defined by the orbital elements for Star B in Table 1.
Large dots: Star A (the brighter star); small dots: Star B; full curve:
theoretical velocities for Star B.

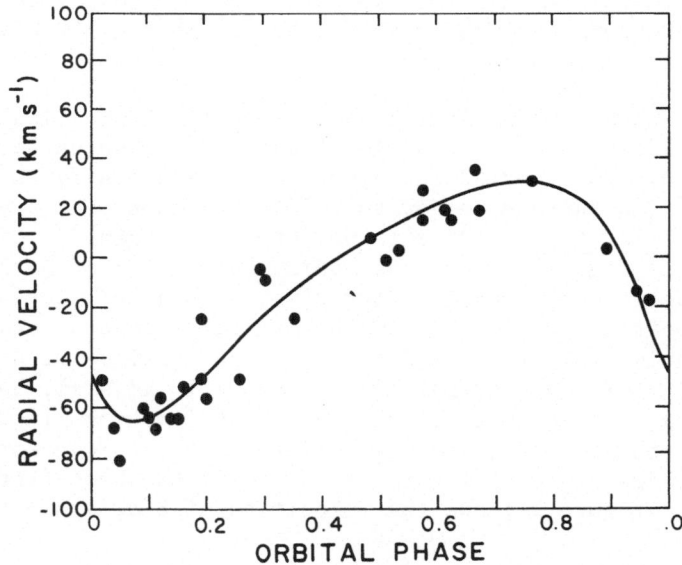

Figure 2.   Radial velocities for HD 93206 A as a function of orbital
phase, which is now defined by the elements given for Star A in Table 1.
Full curve: theoretical velocities for Star A.

Table 1. Orbital Elements for HD 93206

	A	B
Period (d)	20.72 ± 0.02	5.9965 ± 0.0015
K (km s^{-1})	48 ± 2	256 ± 10
V_0 (km s^{-1})	-8 ± 2	-26 ± 7
e	0.34 ± 0.04	0.04 ± 0.04
ω (rad)	2.2 ± 0.2	0.09 ± 0.83
a sin i (10^6 km)	13.0 ± 0.6	21.1 ± 0.8
f(m) (m$_\odot$)	0.20	10.5
Spectral type	O9.5I	O9III
J.D. of zero phase	2442529.8	2443235.9
	(periastron)	(maximum positive velocity)

this period, along with the theoretical velocity curve computed from
the orbital elements in Table 1.

We conclude that the two stars we observe do not belong to the same
binary system, but probably to two widely separated binary systems, with
Star A the primary of one and Star B of the other (which eclipses). The
data do not allow us to state whether the two systems are gravitationally
bound to each other. Given that multiple systems are common among later-
type spectroscopic binaries (Batten 1973), it is not surprising that an
O-type binary should turn out to be multiple. When orbital analyses for
this spectral type are complete, the theory of formation of massive stars
will benefit from a comparison of the incidence of multiple systems with
that at later types.

This research was supported in part by National Science Foundation
Grant No. AST76-20842 to the University of Colorado.

[1]Visiting Astronomer, Cerro Tololo Inter-American Observatory, which is
supported by the National Science Foundation under Contract No. AST74-
04128.

REFERENCES

Batten, A.H.: 1973, Binary and Multiple Systems of Stars (New York:
    Pergamon), p. 59.
Chiosi, C., Nasi, E. and Sreenivasan, S.R.: 1978, Astron. Astrophys. 63,
    pp. 103-124.
Conti, P.S., Leep, E.M. and Lorre, J.J.: 1977, Astrophys. J. 214, pp.
    759-772.
Dearborn, D.S.P., Blake, J.B., Hainebach, K.L. and Schramm, D.N.: 1978,
    Astrophys. J. 223, pp. 552-556.
Lafler, J. and Kinman, T.D.: 1965, Astrophys. J. Suppl. 11, pp. 216-222.
de Loore, C., De Grève, J.P. and Lamers, H.J.G.L.M.: 1977, Astron.
    Astrophys. 61, pp. 251-259.
Moffat, A.F.J. and Seggewiss, W.: 1972, IAU Inform. Bull. No. 681.
Petrie, R.M.: 1940, Publ. Dominion Astrophys. Obs. 7, pp. 205-238.
Walborn, N.R.: 1973, Astrophys. J. 179, pp. 517-525.
Wolfe, R.H., Horak, H.G. and Storer, N.W.: 1967, in Modern Astrophysics,
    ed. M. Hack (New York: Gordon and Breach), pp. 251-273.

DISCUSSION FOLLOWING MORRISON AND CONTI

Underhill: What spectral lines did you measure? The possible interpretations may be affected by your choice, for some lines may be formed in a surrounding disk or gas stream. I think of famous systems such as HD 47219, Plaskett's star, and β Lyrae and the many interpretations that have been given for them.

Morrison: The velocities I showed are determined from lines due to He I. Velocities from an average of Si IV and N III are well correlated with these velocities; hence, the scatter that is shown when the velocities of star A are plotted in a 6-day period is not observational scatter.

Leung: There seems to be a very large eccentricity in your radial velocity curve but the photometric light curve of Tony Moffat showed that the secondary minimum occurred at phase 0.5. Thus, yours must have a very special orientation.

Morrison: The light curve refers to the 6-day star, which according to our analysis has zero orbital eccentricity. Hence, the two sets of data are consistent.

Cowley: In the case of your star "A", you show both a large velocity amplitude and a rather long period which implies substantial masses. Could you comment on these values?

Morrison : The masses implied are not particularly large, since the mass function is only 0.20 M_0. A rough estimate for the mass of the companion can be derived from the assumptions than sin i = 1 and the mass of the primary of the 20-day component is 30 M_0. Then the mass of the secondary is 6 M_0.

Moffat: I should point out that Seggewiss' and my slightly higher dispersion, 12 Å/mm Coudé spectrograms show fairly sharp Si IV 4089 absorption for each component (A and B) implying that star B is also a supergiant (BOIb). The mass function and the 6-day light curve (deeper minimum when the BOIb star is in front) leads to a mass of the unseen 6-day companion of star B which is larger than star B.

Morrison: The difference between our luminosity classifications for star B might be due to the difference in dispersion or to a different drawing of the continuum. The larger mass for the companion of star B is also consistent with the spectroscopic data and is very interesting.

MASS FLOW AND EVOLUTION OF UW CANIS MAJORIS

Yoji Kondo
Goddard Space Flight Center, Greenbelt, Md./U.S.A.

George E. McCluskey, Jr.
Division of Astronomy, Lehigh University, Bethlehem
Pa./U.S.A.

Jürgen Rahe
Remeis-Sternwarte Bamberg - Astronomical Institute
University of Erlangen-Nürnberg/F.R.G.

ABSTRACT

The far-UV spectrum of the eclipsing binary UW CMa
(O7f + O-B) had earlier been utilized to derive a mass-loss
rate of about 10^{-6} to 10^{-5} solar mass per year. The mass
flow seems to be basically in the form of a stellar wind
emanating from the O7f primary component, with radiation
pressure as the controlling factor. The main characteristics that make UW CMa a possible progenitor of a Wolf-
Rayet system are discussed.

1. INTRODUCTION

The close eclipsing binary UW CMa consists of an O7f star
and a late O or possibly B component. It has repeatedly
been studied by a number of workers, the latest very
detailed investigations are the photometric study by Leung
and Schneider (1978), using the Wilson and Devinney (1971)
approach, and the spectroscopic analysis by Hutchings
(1977). But there are still many unresolved problems associated with this system.

The far-ultraviolet spectrum of the O7f star was first observed in 1973 by McCluskey et al. (1975) near phase 0.75,
and then again in 1975 by McCluskey and Kondo (1976) around
phase 0.25. Both times the U2 spectrometer of the Copernicus
satellite, in the wavelength region from about 950 to 1560 Å
at a resolution corresponding to 0.2 Å was used. Recently,

P. S. Conti and C. W. H. de Loore (eds.), Mass Loss and Evolution of O-Type Stars, 281-286.

new IUE observations were made by Kondo and McCluskey in
the longer wavelength region between 1200 and 3000 Å. They
complement the far-UV Copernicus and the optical observations and thus may help clarify some of the still unresolved problems connected with this system.

2. MASS FLOW

The dominant features in the far-UV region are P Cygni
lines of SIV, PV, FeIII, CIII, CIV, NV, and SiIV. When
comparing the measurements obtained about phase 0.75 with
those obtained at phase 0.25, it appears that the absorption line velocities do not change; they yield radial
velocities of about - 500 km/sec. The peaks of the emission components, on the other hand, are shifted by several
hundred Km/sec and thus significantly larger than the projected orbital velocity of about 200 km/sec. The radial
velocities of the UV absorption lines are stationary and
not phase-related; they probably originate in a gas cloud
surrounding the entire binary system. Ground-based data
obtained and analyzed by Hutchings (1977), on the other
hand are related to the orbital phase and show a variable
flow from the primary in a non-circular orbit.

From the strength of the absorption component of these
lines, a mass loss rate of about 3×10^{-6} M_O per year was
deduced (McCluskey et al., 1975). This result is in good
agreement with the value found by Hutchings (1976) in a
ground-based mass-loss survey of hot supergiants.

The effects of radiation pressure on the Roche equipotential
surfaces in UW CMa have been discussed by McCluskey and
Kondo (1976). It is emphasized that radiation pressure is
the controlling factor, not the Roche lobe around the primary. Vanbeveren (1977) has pointed out that the opening
of the Roche lobe behind the secondary star (the star with
no radiation pressure) could be fallacious as the model
neglects the shadowing effect of the secondary star. However, the effect of radiation pressure on the critical lobe
surrounding the primary star is probably not far from the
original predictions, in which the critical surface, the
characteristic figure eight, is not present when the
radiation pressure force is nonzero. Castor, Abbott, and
Klein (1975) have shown that radiation pressure in the
spectral lines of abundant elements can easily give a
value of the ratio of the radiation pressure force to the
gravitational force for the O7f star, that is equal to or
greater than unity, indicating that the mass flow is determined by the nature of the stellar wind emanating from the
O7f primary star. The mass does not necessarily flow from
component one to component two through the inner Lagrangian

point, L_1, to form a gaseous ring surrounding the secondary, but has access to a much larger volume than it has in the case of negligible radiation pressure. The forces giving rise to the stellar wind are apparently at least as import- ant as gravitational, Coriolis or Centrifugal forces, and a well defined model must take these forces and their interactions into account. The use of conventional Roche lobe theory to derive orbital parameters for systems, in which a stellar wind is present, may lead to serious errors.

Applying accretion theory to UW CMa, one finds that the secondary component may accrete from 0.25 to 1% of the mass lost by the primary in its wind. This means that during the stellar wind phase, essentially all of the mass lost by the primary, leaves the system.

3. EVOLUTION

It is well known that most Wolf-Rayet stars appear to be in binary systems, and it has been argued that actually all these stars are binaries.

According to Kuhi (1973) and Bohannan and Conti (1976), a typical Wolf-Rayet system shows redshifted emission lines of +90 to +190 km/sec, a W-R star with an average mass of about 10 M_O, which is overluminous for its mass, and a companion with a spectral type O5 to BO; the mass ratio, M_{WR}/M_{OB} tends to be between 0.2 and 0.4.

If most - or perhaps even all - W-R stars are in binaries, then there must also be immediate progenitors to that state. In fact, a few close binary systems have already been suggested as being in the process of evolving into Wolf-Rayet binaries.

Massey and Conti (1977) have suggested that the O-type binary HDE 228766 is becoming a Wolf-Rayet binary. Bohannan and Conti (1976) made a similar suggestion con- cerning BD + 40° 4220.

HDE 228766 consists of an O5.5f and an O7.5 I star, each star having about 16 solar masses. The Of star is losing mass at a rate of about 10^{-5} solar mass per year. As the hydrogen envelope is lost from the Of star, the mass ratio will decrease and eventually this star will become a Wolf-Rayet star with a more massive O-type companion.

Bohannan and Conti (1976) note that in the system BD + 40° 4220, which consists of two Of supergiants with minimum masses of 47 solar masses for the primary and 16 for the secondary, the secondary is overluminous for its

mass and has all of the characteristics of a Wolf-Rayet
star except for the spectrum. The emission lines are redshifted by about +200 km/sec, and the mass loss is estimated to be about 10^{-5} M_O/yr. The mass ratio of 0.2 is
presently the same as that of typical Wolf-Rayet binaries.
The authors conclude that BD + 40° 4220 is an immediate
progenitor of a Wolf-Rayet binary system.

The characteristics of UW CMa match those of HDE 228766
to some extent. Mass is being lost by the Of star at a
rate of 10^{-6} to 10^{-5} M_O/year. The initial mass of the
Of star was probably at least 30 solar masses and may have
been significantly more. The mass ratio is not certain due
to the difficulty of detecting the spectrum of the secondary component but is probably close to unity. The Of star
is overluminous for its present mass and also in this
respect similar to the Wolf-Rayet star in a Wolf-Rayet
binary system. The spectral type of the secondary is late
O-type or possibly BO. The emission lines are redshifted
by several hundred km/sec.

These characteristics make UW CMa a possible progenitor of
a Wolf-Rayet system. As suggested by several investigators,
the Wolf-Rayet star may then become a supernova leaving a
neutron star or black hole. The companion O-star will then
evolve and an X-ray binary might develop.

REFERENCES

Bohannan, B., and Conti, P.S.: 1976, Ap. J. 204, 797.

Castor, J.I., Abbott, W.N., and Klein, M.J.: 1975, Ap.J.
        195, 157.

Hutchings, J.B.: 1976, M.N.R.A.S. 203, 438.

Hutchings, J.B.: 1977, PASP 89, 668.

Kondo, Y. and McCluskey, G.E., Jr.: 1976, in Structure and
        Evolution of Close Binary Systems, eds. P. Eggleton
        et al. (Reidel), p. 277.

Kuhi, L.V.: 1973, Wolf-Rayet and High Temperature Stars,
        ed. M.K.V. Bappu and J. Sahade (Boston: Reidel),
        p. 205.

Leung, K-Ch. and Schneider, D.P.: 1978, Ap.J. 222, 924.

Massey, P. and Conti, P.S.: 1977, Ap.J. 218, 431.

McCluskey, G.E.,Jr., Kondo, Y., and Morton, D.C.: 1975,
        Ap.J. 201, 607.

McCluskey, G.E., Jr., and Kondo, Y.: 1976, Ap.J. <u>208</u>, 760.

Vanbeveren, D.: 1977, Preprint.

Wilson, R.E., and Devinney, E.J., Jr.: 1971, Ap.J. <u>166</u>, 605.

DISCUSSION FOLLOWING KONDO, McCLUSKEY and RAHE

<u>Hutchings</u>: It is important to be clear that your non phase-related UV data refer to a circumstellar wind from the system. The ground based data I showed are phase related and show that there is a variable flow from the primary in a non-circular orbit.

<u>Rahe</u>: We hope that the new IUE data obtained at the wavelength region of 1200 to 3000 Å will complement the far-UV Copernicus (950 to 1560 Å) and the optical measurements, and help to clarify this problem.

<u>Wilson</u>: It is important to distinguish between the role of radiation pressure in establishing the figures of binary components and its role in circumstellar gas flows. The important characteristics of the interesting surfaces are quite different for these two problems. For example, the analog of the inner Lagrangian point will be shifted due to radiation pressure at the "L1-point". However, since the von Zeipel law predicts that the radiation flux drops to zero at the L1-point, there will be no radiation pressure to include in considering the maximum size a binary component can have, unless there is also radiation from the other component. Thus, the"L1-point" will be unshifted and the "Roche lobe" will be the classical Roche lobe.

<u>Abbott</u>: In the envelope of UW CMa, the line radiation force greatly exceeds the continuous radiation force and this should be accounted for in equipotential calculations.

<u>Rahe</u>: You are quite right. The radiation pressure effect increases substantially if the influence of the line radiation force is taken into account. The main point is that as mentioned in Mc Cluskey and Kondo (1976, Ap.J. <u>208</u>, 760), the mass flow parameters are not determined by the Roche equipotentials, but by the nature of the stellar wind emanating from the O7f star.

<u>Underhill</u>: In talking about the radiation pressure driving the wind, what happens to the photon's momentum?

Castor: The photon's momentum is not lost, only the direction is changed. At each scattering an impulse is given to the gas equal to the change in the vector momentum. After one scattering the photon can fly across the envelope and be going outward again before the next scattering, thus giving in each case a net outward impulse. The magnitude of momentum goes down only on account of the Doppler shift, but this effect is negligible.

Vanbeveren: I have tried to show that the conclusion of Y. Kondo and McCluskey about the shape of their critical surfaces is strictly dependent on their assumptions for the radiation force (they consider a spherical symmetric approximation). However, taking into account gravitational darkening and using the von Zeipel law, it was shown that we get total different results for the critical surface. Concerning the absorption coefficient, I really don't want to say something definite about it. In any case, I don't think that one has to use the line opacities appearing in the theory of Castor because there already a velocity gradient is present and equipotential surfaces apply to hydrostatic material.
I just want to remark that one has to be very careful with the word critical surface. A Roche lobe is totally different from a surface where the gravitational forces are balanced by the radiation force: the mode of mass loss for both processes is totally different.

Editor's Note : Dr. Yoji Kondo has requested that the following statement of his be added to this discussion: "It is very unlikely that von Zeipel's theorem is applicable to this problem. The conditions under which this theorem would be useful are almost certainly inapplicable here".
(in answer to the remark of Wilson).

THE SPECTROSCOPIC BINARY γ 2 VELORUM

Virpi S.Niemelä [*]
51 y 11,Villa Elisa,Buenos Aires,Argentina

and

Jorge Sahade [**]
Instituto de Astronomía y Física del Espacio,
Buenos Aires,Argentina

γ2 Velorum is the brightest binary system with a Wolf-Rayet component;it has been classified as a O9I+WC8 pair (Conti and Smith 1972). The orbital elements of this system were determined earlier by Ganesh and Bappu (1967) from spectrograms of 125 Å/mm dispersion,on the basis of the radial velocities from CIII-IV λ4652 Å emission for the WC component,and on the radial velocities from H absorption only,for the O star.

In this paper we report on the preliminary results from a new spectrographic investigation of the γ2 Velorum system,based primarily on a large number of 42 Å/mm spectrograms secured by one of us (J.S.)at the Córdoba Observatory,Argentina,in the interval 1948-1962.A few additional coudé spectra were obtained by V.S.N. at the Cerro Tololo Inter-American Observatory,Chile,between 1971-1977.In addition we have used 76 old objective prism spectrograms obtained by C.D.Perrine between January and July,1919,at the Córdoba Observatory,Argentina,in order to improve the value of the orbital period.

All suitable absorption and emission lines on our spectrograms were measured for radial velocity by V.S.N.with the Grant engine at the Cerro Tololo Inter-American Observatory.

---

[*] Visiting Astronomer,Cerro Tololo Inter-American Observatory,supported by the National Science Foundation under contract No. NSF-C866.

[**] Member of the Carrera del Investigador,Consejo Nacional de Investigaciones Científicas y Técnicas,Argentina.

P. S. Conti and C. W. H. de Loore (eds.), Mass Loss and Evolution of O-Type Stars, 287–289.

We have then derived a new orbital solution for the γ2 Vel system. The radial velocities of the O star were derived from the measurement of a number of absorption lines,about 13 per plate.The radial velocities from the emission features show a large scatter,and those from CIV and CIII were preliminarily used only to derive the semiamplitude of the velocity curve of the WC8 component.Thus we obtained the following preliminary orbital parameters,namely,

$$P = 78.5002 \pm 0.0001 \text{ days}$$

$$\gamma = +12 \pm 1 \text{ km/s}$$

$$K_{O9} = 68 \pm 2 \text{ km/s}$$

$$K_{WC} = 130 \pm 10 \text{ km/s}$$

$$e = 0.40 \pm 0.01$$

$$w = 256° \pm 3°$$

$$M_{O9}\sin^3 i = 32 \pm 6 \quad M_{\odot}$$

$$M_{WC}\sin^3 i = 17 \pm 2 \quad M_{\odot}$$

The radial velocities from the absorption lines and of the C emissions are plotted in Figure 1,where the dashed line represents the above listed orbital solution.It is clear that an orbital solution based on the velocities from the C emissions only,would suggest a smaller eccentricity and further analysis will be carried out so as to understand such a behaviour.

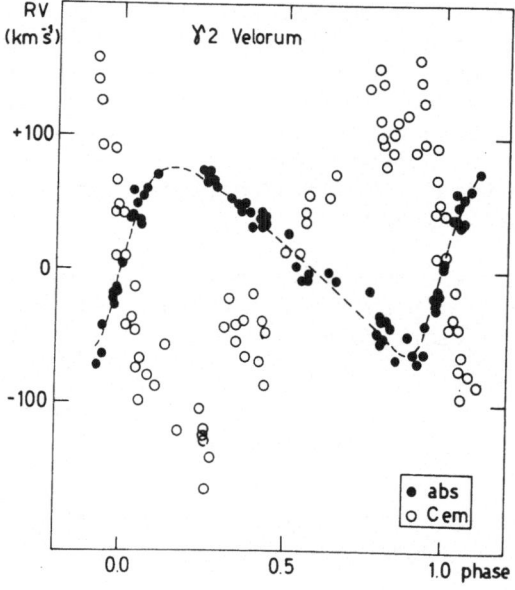

Figure 1. Velocity curve of γ2 Velorum.

The minimum masses that correspond to our orbital parameters have
a very different ratio than the one derived earlier by Ganesh and Bappu
(1967).If we adopt the value of i suggested by Moffat's (1977) photometric investigation,namely 70°,the masses of the components turn out
to be M_{O9} = 38 $M_\odot$ and M_{WC} = 20 $M_\odot$,respectively.It is worth pointing out
the large value that we have derived for the mass of the WC star,as
compared with the usually accepted value for Wolf-Rayet objects,namely
10 $M_\odot$,clearly an unappropriate extrapolation from the results in the
case of the WN+O system V 444 Cygni.

In regard to the violet-shifted absorptions that flank the emission
lines,they follow the orbital motion of the WC component.The mean radial velocities of these absorptions are correlated with the excitation
potential of the upper level of the corresponding line,as shown in Figure 2.A correlation with increasingly negative radial velocity with
decreasing excitation potential is generally interpreted as an outwards
accelerated velocity field in the stellar atmosphere.That this should
be true in the WC8 component of γ2 Vel agrees with what we know about
the velocity gradients in other Wolf-Rayet atmospheres (e.g.Kuhi 1972;
Seggewiss 1975:Niemelä 1975).

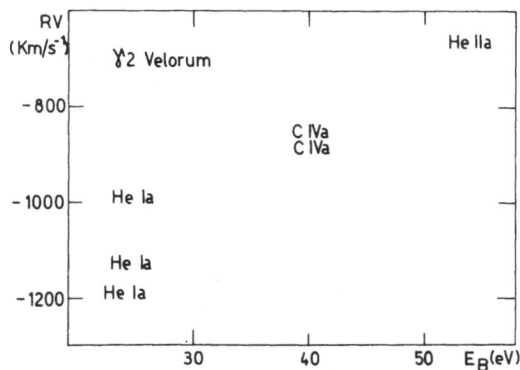

Figure 2. Radial velocities of the violet shifted absorptions
in function of the excitation potential of the upper level of
the corresponding line.

References:

Conti,P.S. and Smith,L.F. 1972,Ap.J.,172,623.

Ganesh,K.S. and Bappu,M.K.V. 1967,Kodaikanal Obs.Bull.,No.183.

Kuhi,L.V. 1972,Ap.J.,180,783.

Moffat,A.F.J. 1977,Astr.Ap.,57,151.

Niemelä,V.S. 1975,in Physics of Movements in Stellar Atmospheres,Coll.
        No.250 of C.N.R.S.of France,eds.R.Cayrel and M.Steinberg,p.467.

Seggewiss,W. 1975,in Variable Stars and Stellar Evolution,eds.V.Sher-
        wood and L.Plaut,(Dordrecht:Reidel),p.285.

THE BINARY ORBIT OF HD 92740

Virpi S.Niemelä [*]
51 y 11,Villa Elisa,Buenos Aires,Argentina

HD 92740 is a star located in the Carina nebula showing a Wolf-Rayet spectrum of type WN7.Faint absorption lines of the upper Balmer series of hydrogen,and also of the Pickering series of HeII are present in the spectrum,in addition to the WN emissions.Although absorption lines present in a Wolf-Rayet spectrum are generally assumed to arise in a companion OB star,a previous study (Niemelä 1973) of the radial velocities of HD 92740 showed that the absorption and emission lines followed the same orbital motion.Subsequent spectral observations of this star have been carried out at the Cerro Tololo Inter-American Observatory,Chile,and at the Córdoba Observatory,Argentina,during four years;the observational data are listed in Conti,Niemelä and Walborn (1978).These observations showed that the true period is 8 times longer than the initially derived period of 10 days,and that doubtlessly the absorption lines belong to the WN star.

A new orbital solution has been derived for HD 92740 with the program based on the method of Lehmann-Filhés,published by Bertiau and Grobben (1968).The mean radial velocities of the narrow emission lines of NIII,NIV and SiIV define the best orbit,i.e.that with smaller probable error.The mean radial velocities of the hydrogen upper Balmer series and HeII Pickering series absorption lines yield a similar orbital solution with slightly higher probable error,which may be partly due to that not all the mean values included the same number of lines,or that the steep velocity gradient present on these lines is somewhat variable. Both orbital solutions are given in Table 1.

The observed mean radial velocities of the narrow emissions and the absorption lines are plotted in Figure 1,which shows clearly the same orbital motion of the absorption and emission lines.The dashed

---

[*]Visiting Astronomer,Cerro Tololo Inter-American Observatory,supported by the National Science Foundation under contract No. NSF-C866.

291

P. S. Conti and C. W. H. de Loore (eds.), Mass Loss and Evolution of O-Type Stars, 291–295.

curve represents the orbital solution of Table 1 corresponding to the
narrow emissions.

Table 1

Orbital elements of HD 92740		
	narrow emission	absorption
P (days)	80.34 $\pm$ 0.01	
e	0.61 $\pm$ 0.02	0.64 $\pm$ 0.02
K (km/s)	74 $\pm$ 2	79 $\pm$ 3
γ (km/s)	-28 $\pm$ 1	-152 $\pm$ 2
w	276° $\pm$ 3°	248° $\pm$ 4°
T_0(2440000+)	728.9 $\pm$ 0.2	726.6 $\pm$ 0.4

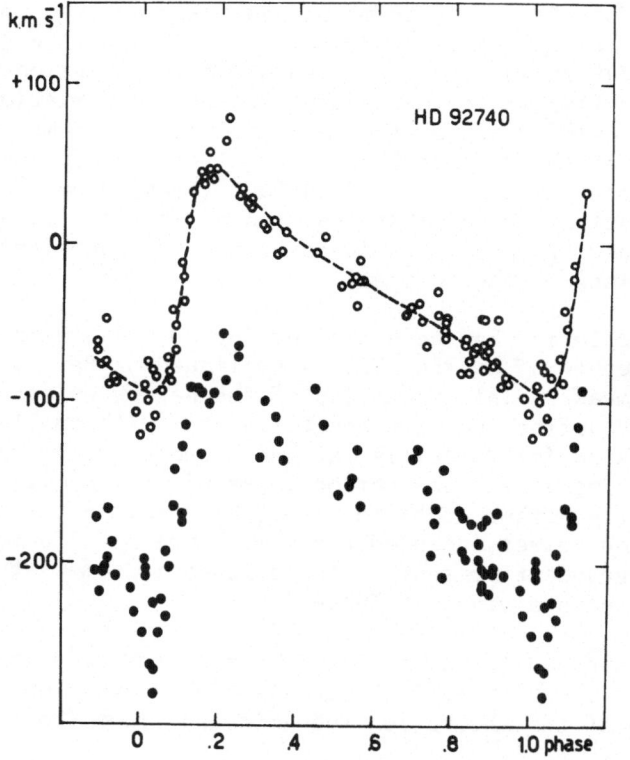

Figure 1. Velocity curves of HD 92740 from the narrow emissions (open circles) and upper Balmer and HeII Pickering absorption lines (filled circles).

On spectrograms obtained in 1977 using the III aJ emulsion,when the WN star has the highest positive radial velocity most absorption lines appear double.This may happen due to a contribution of the secondary component of the binary system with the highest negative velocity to the blueshifted WN absorptions.Some very faint features on other spectrograms also suggest they may be absorption lines from an O type companion.Figure 2 shows a plot of the radial velocities of these faint absorptions with respect to the orbit derived from the narrow emission lines.

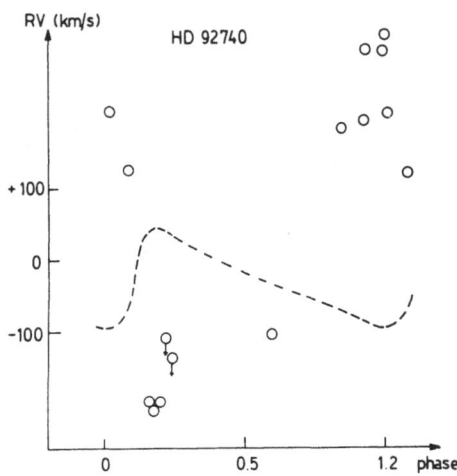

Figure 2. Distribution of the radial velocities of the fainter absorption lines.The dashed curve represents the orbital solution from the narrow emissions of the WN star.

If these absorption features actually exist and belong to the companion star,then from Figure 2 it is clear that the WN component would be the more massive of the system.The corresponding minimum masses would result aproximately

$$M_{WN7}\sin^3 i = 64\ M_\odot$$
$$M_O\ \sin^3 i = 24\ M_\odot$$

This would be the first case when in a binary system the Wolf-Rayet component is the more massive,since in all other cases the reverse is true (e.g.see tabulation in Kuhi 1973).

References:

Bertiau,F.C. and Grobben,J. 1968,Ricerche Astr.,8,No.1,Specola Vaticana

Conti,P.S.,Niemelä,V.S. and Walborn,N.R. 1978,Ap.J.,in press.

Kuhi,L.V. 1973,in Wolf-Rayet and High Temperature Stars,eds.M.K.V.Bappu
        and J.Sahade,(Dordrecht:Reidel),p.208.

Niemelä,V.S. 1973,Publ.A.S.P.,85,220.

DISCUSSION FOLLOWING NIEMELA

Morton: What period did you find for γ Vel? How does
it compare with the value derived with Ganesh and Bappu?

Niemela: 78.5002 days, Ganesh and Bappu used 78.5 d.

Bolton: Dorothy Fraquelli of David Dunlap Observatory
and Jiri Horn of Ondrejov Observatory have redone the orbit
of the WN 4.5+O9I star HD 190918 using high dispersion
spectra. They find a much larger period and velocity amplitude than previous investigators. They find an eccentricity
of 0.4 and ω near 220°. There is evidence for large variations in the emission line intensities and for changes in
absorption line profiles. These effects may be distorting
the velocity curve, but their work on this problem is just
beginning.

van den Heuvel: Both orbits seem to have very large
eccentricities, and have ω near 270°. This seems rather
strange to me, and seems to suggest that there is a strong
distortion of the radial velocity curve of some sort. Do
you have an explanation for this?

Niemela: No, it may be casual (but see the remark of
Dr. Bolton).

Noerdlinger: At a previous Wolf-Rayet Conference you
said that the He I 3888 emission was peaked in the violet
when the O star is in front and to the red when it is behind. I had interpreted this to mean He I comes from the
O star: when it is behind, the WR star cannot interfere
with the outflow behind it, but when it is in front the WR
star may ionize some of the He I behind the O star and also
divert its flow to the side. Have the new observations
given different results, or have you another interpretation?

Niemela: This is partly ldue to the orbital motion of
the H8 absorption belonging to the O9I star. My interpretation of the violet emission peak was that it may arise
in a gas stream from the WC towards the O star. It may
also be due to the asymmetry of the WR envelope. We have
no quantitative measures of this intensity change yet, but
the radial velocity variations of the He I emission follow
the WC orbit, therefore it probably comes from the envelope
of this star.

Massey: For γ Vel, you said you got the W-R velocity
curve from the carbon emission lines. Do you get the same
amplitude curve when you look at He II λ4686?

Niemela: Yes, but with different γ-velocity.

Conti: I would like to emphasize the very important result presented here that the absorption and the emission lines in this system are in phase and in the WN star. The detection of the secondary is not yet certain but of course would give a very important result for the WR mass.

Moffat: Moffat and Seggewiss (1978, A & A in press) have studied HD 92740 using 12 Å/mm Coudé spectrograms in the blue. The results here confirm ours except that we find the velocity amplitude of the Balmer absorption lines to be half that of the emission lines of the WR star, while He II and N V absorption lines are the same. This may be due to blending of a just unseen OB companion which may have quite wide, unresolvable photospheric absorption lines, the strongest being those of the Balmer series. Also we find no indication of the anti-phase, large amplitude absorption lines from a possible OB companion.

Niemela: If the lower amplitude you find were due to a companion, then it should be a quite peculiar object with all absorption lines shifted to a -150 km/s systemic velocity.

Moffat: Our mean (systemic) γ-velocity of the observed Balmer lines (H8,9,10) is -132 ± 10 km s^{-1} (not -150 km s^{-1}) while their width is over 400 km s^{-1}. Thus, there is sufficient margin to allow a normal companion OB star with Balmer lines varying in antiphase around a low γ-velocity.

RESULTS FROM THE 1977 COORDINATED OBSERVING CAMPAIGN ON HDE 226868 = CYGNUS X-1

C.T. Bolton
David Dunlap Observatory, University of Toronto
P.O. Box 360, Richmond Hill, Ontario, Canada

A coordinated campaign of radio, optical, and x-ray observations of the bright x-ray source Cygnus X-1 took place August 7 - 21, 1977 under sponsorship of Commission 42 of the International Astronomical Union. Radio flux measurements, optical spectra, photometry, and polarimetry were obtained during this period by ten groups from Canada, Great Britain, the Soviet Union, and the United States. The x-ray flux was monitored continuously by the SAS-3 satellite between August 11.0 and 17.0.

A review of the campaign data and previously obtained data by the campaign participants in March, 1978 led to the following conclusions. 1) Except a tendency for x-ray absorption events to occur near superior conjunction of the x-ray source, no unambiguous correlation between x-ray and optical behaviour has been seen. 2) There is a correlation between x-ray and radio variations if two kinds of x-ray transitions are recognized. 3) At least 2 percent of the total light of the system in the B band is modulated on time scales different than the 5^d6 period. It is not clear at this time whether this "extra" modulated light is a continuum or in one or more very broad emission lines. 4) The emission line velocity curves are not well-behaved, and estimates of the mass ratio based on these are not reliable. 5) The linear polarization behaviour of the system is complicated and its interpretation is presently unclear. 6) There is some evidence for a $78^d/39^d$ modulation of the x-ray flux, U-band flux, and U- and V-band polarization. The existence and possible interpretation of this phenomenon is controversial. 7) A good model for the complicated phenomena seen in this system has not yet emerged, but a consensus is developing that explanations of the behaviour will be easier if the inclination of the orbit is large with the x-ray source nearly grazing the limb of the visible star at superior conjunction. 8) The estimated mass of the x-ray source is still well in excess of upper limits on the mass of a conventional neutron star.

297

P. S. Conti and C. W. H. de Loore (eds.), Mass Loss and Evolution of O-Type Stars, 297–298.
Copyright © 1979 by the IAU.

DISCUSSION FOLLOWING BOLTON

Morton: What is the current estimate for the upper limit on the luminosity of the secondary of Cyg X-1?

Bolton: The secondary contributes less than 2% of the light of the primary at λ4026 He I (2σ limit). There are distortions in the He I lines λλ4713, 5876, 6678 which could be interpreted as due to the secondary, although I suspect they are due to some other effect (mass loss?). These distortions are < 5% of the total light of the system.

THE NATURE OF THE RUNAWAYS: OLD DISK POPULATION OB STARS?

L. Carrasco, G.F. Bisiacchi, R. Costero and C. Firmani
Instituto de Astronomía, Universidad Nacional Autónoma de México

We propose that most of the OB runaway stars are Old Disk Population objects in the same evolutionary phase as the hot UV-bright stars in globular clusters. Bimodal Gaussian fits to the peculiar radial velocity distribution are computed for 386 0-type and 1093 B-type stars. Both samples independently yield one Gaussian with $\sigma \simeq 13$ km s^{-1}, a value typical of extreme Population I objects, and a second one with $\sigma \simeq 28$ km s^{-1} which is characteristic of the Old Disk Population. The fraction of stars under the high velocity-dispersion distribution (HVD stars) is 47% of the 0 and 23% of the B stars. We analyze the kinematics of the sample of OB stars divided into low peculiar radial velocity, $|Vrp| < 20$ km s^{-1}, and high-velocity stars, $|Vrp| > 45$ km s^{-1}. The results for the solar motion and mean peculiar velocity of the groups are the expected ones for the extreme Population I objects in the case of the low-velocity group ($U_\odot = 9.1 \pm 0.1; V_\odot = 14.8 \pm 0.1$ km s^{-1}) and an asymmetric drift of approximately 20 km s^{-1} for the high velocity stars ($U_\odot = 2.7 \pm 1.4; V_\odot = 32.7 \pm 1.4$ km s^{-1}). This lag behind circular motion also corresponds to Old Disk Population objects.

We conclude then that there is good evidence supporting our interpretation for the runaway stars, and suggest the existence of an important number of old, low mass OB stars in the galactic disk without marked spectroscopic differences between them and the young OB stars. The estimated number density of HVD stars in the galactic plane indicates that these objects probably outnumber the planetary nebulae by about two orders of magnitude. A rough estimate for the lifetime of the HVD stars yields 10^5 yr, in agreement with the timescales resulting from evolutionary tracks calculated for the UV-bright stars.

An extended paper on this subject has been submitted to The Astrophysical Journal.

P. S. Conti and C. W. H. de Loore (eds.), Mass Loss and Evolution of O-Type Stars, 299.

THE DISTRIBUTION IN LUMINOSITY OF OB STARS AND EVOLUTIONARY TIMESCALES

F. Bisiacchi, L. Carrasco, R. Costero, C. Firmani and J.F. Rayo
Instituto de Astronomía, Universidad Nacional Autónoma de México

We have obtained the observed fraction of supergiant (luminosity classes I and II), giant (III) and dwarf (IV-V) stars of spectral types B2 and earlier. The stellar sample used was formed with all the stars with bi-dimensional spectral classification listed in the Catalogue of Galactic O stars by Cruz-González et al. (1974), the unpublished compilation of B0 and B0.5 stars by J.F. Rayo, and the B1-B2 stars listed by Morgan et al. (1955). The latter sample is by far the least complete one. The results are listed in Table I, together with the total number of stars (in parenthesis) considered in each spectral interval. A prominent conclusion is drawn from the table: The fractions remain approximately constant all over the spectral range considered.

TABLE I

Observed Fraction of Stars by Luminosity Class for Different Spectral Types

Sp.T. Class	O3–O5.5 (22)	O6–O7.5 (90)	O8–O8.5 (60)	O9–O9.5 (276)	B0 (544)	B0.5 (752)	B1–B2.5 (462)	O3–B2.5 (1744)
I-II	0.23	0.20	0.28	0.32	0.26	0.28	0.31	0.28
III	0.18	0.26	0.17	0.22	0.26	0.23	0.19	0.24
IV-V	0.59	0.54	0.55	0.45	0.48	0.49	0.50	0.49

The observed relative fractions cannot be explained by the classical evolutionary models of massive stars without mass loss. Adopting the mass function by Prentice and ter Haar (1969) and the luminosity calibration by Conti and Alschuler (1971), the predicted fraction of supergiant to dwarf, late O- and early B-type stars is smaller by at least an order of magnitude than the observed one. The disagreement is mainly due to the failure of the classical models to produce low enough surface gravities during the core hydrogen burning phase. However, models where mass loss

301

P. S. Conti and C. W. H. de Loore (eds.), Mass Loss and Evolution of O-Type Stars, 301-304.

is considered (e.g. the ones presented by Chiosi and coworkers, and by de Loore and his group in this symposium) do extend the core hydrogen burning phase into the region of the log g, log T_{eff} plane where supergiant stars lie, particularly those with the highest mass loss rates.

Adopting the mass function mentioned above, the surface gravity and effective temperature calibrations for the O-type stars by Conti (1973) and those for the B-type stars by Morton and Adams (1968), we have computed the expected fractions of supergiant, giant and dwarf stars from the models by Chiosi et al. (1978) with $\alpha = 0.96$ (the highest mass loss used in the paper). The resulting fractions are, respectively, 0.24, 0.28 and 0.48 for the O9-B0 spectral range and 0.36, 0.27 and 0.36 for the B0-B1 one. In deriving these numbers we have assumed equal timescales for segments of equal length of the track (in the log g, log T_{eff} plane) for a star of a given initial mass.

The above estimated fractions are in reasonable agreement with the observed ones. However, these models fail in predicting a roughly constant value for the relative fractions as observed for the whole O3 to B2 spectral range. Models with even higher values of α might be able to reach better agreement with the observed fractions for the hottest stars. An alternative explanation for the large fraction of supergiants observed could be the possible presence of hot UB-bright stars in the galacatic disk (Carrasco et al., 1976). In the log g, log T_{eff} plane the evolutionary tracks for UV-bright stars by Pacyński (1971) and Gingold (1974) are almost parallel to those by Chiosi et al. (1978) for massive star with high mass loss rates, and the evolution takes place at almost constant pace with log T_{eff}. Hence, the difference between the fractions of supergiants predicted for the UV-bright stars and those for massive stars will depend largely on the masses of the progenitors of the former stars, their mass function and the mass lost during both red giant branches previous to the UV-bright stage. Adopting the total mass lost for solar-type stars as a function of initial mass given by Fusi-Pecci and Renzini (1975,1976), we found that the fraction of supergiants among the UV-bright stars is comparable to the one observed for OB stars in general.

We conclude that the observed fractions of supergiants, giants and dwarfs for OB stars can be explained by the evolutionary models of massive stars with mass loss and/or the presence of hot UV-bright stars in the galactic disk.

REFERENCES

Carrasco, L., Bisiacchi, F., Cruz-González, C. and Firmani, C.: 1976, Bull. Am. Astron. Soc., 8, p. 536.
Chiosi, C., Nasi, E. and Sreenivasan, S.N.: 1978, Astron. Astrophys., 63, pp. 103-124.
Conti, P.S.: 1973, Astrophys. J., 179, pp. 181-188.

Conti, P.S. and Alschuler, W.R.: 1971, Astrophys. J., 170, pp. 325-344.

Cruz-González, C., Recillas-Cruz, E., Costero, R., Peimbert, M. and Torres-Peimbert, S.: 1974, Rev. Mexicana Astron. Astrof., 1, pp. 211-259.

Fusi-Pecci, F. and Renzini, A.: 1975, Astron. Astrophys., 39, pp. 413-419.

Fusi-Pecci, F. and Renzini, A.: 1976, Astron. Astrophys., 46, pp. 447-454.

Gingold, R.A.: 1974, Astrophys. J. 193, pp. 177-185.

Morgan, W.W., Code, A.D. and Whitford, A.E.: 1955, Astrophys. J. Suppl., 2, pp. 41-74.

Morton, D.C. and Adams, T.F.: 1968, Astrophys. J., 151, pp. 611-621.

Pacyński, B.: 1971, Acta Astron., 21, pp. 417-435.

Prentice, A.J.R. and ter Haar, D.: 1969, Monthly Notices Roy. Astron. Soc., 146, pp. 423-444.

DISCUSSION FOLLOWING BISIACCHI, CARRASCO, COSTERO, FIRMANI and RAYO

Ovenden: This paper is clearly very important, and its conclusions must be taken seriously. Taking them seriously consists, in part, of asking if other explanations are possible. Regarding the velocity residual histograms, it must be remembered that the residuals are found relative to an assumed global galactic rotational velocity field. Is there a possibility that the adopted model field is wrong? One possibility is that dynamical effects of spiral arms must be included. We cannot do this well at present.
Also, the stars involved have life-times less than the phase-mixing time for galactic orbits (~ 10^8 years), so that the velocity distribution might show relics of the dynamical processes involved in star formation. Finally, I would like to emphasize that since early-type stars play an important role in the investigation of the kinematics of the galaxy, this investigation is very important for the study of stellar kinematics.

Carrasco: Yes, I agree this point requires further investigation.

Garmany: I would like to mention that in our study of the brighter O-stars, we have studied the radial velocities of a number of run-away O-stars, and so far have found indications for variations only in α Cam, unlike the predictions made by Beckenstein and Bowers. In addition, I would like to ask how our discovery of a Balmer gradient in several run-away stars (Bohannan and Garmany, Ap.J., 1978, 223, 908) affects your conclusions.

Carrasco: There may be some run-away stars which are
actually not Pop II, low mass stars, but this would only
add to our sample some stars and then our estimates of the
number of run-aways of Pop II are only upper limits. No
theory about the origin of run-aways via ejection and/or
velocity gradient can explain the asymmetric drift derived
by us.

van den Heuvel: From the UV spectrum there is a simple
way to distinguish between low-mass, low-radius halo O-type
stars and population I O-stars, as the terminal wind velocity is always a few times the escape velocity V_{esc}. V_{esc} is
given by $\sqrt{2.g.R}$ where g is the surface gravity and R is the
stellar radius. A spectral type gives us g and T_{eff}, so g
is the same in both cases. The halo O-stars are expected to
have a 50 to 100 times smaller radius than the population I
O-stars, so one expects their wind outflow velocities to be
some 7 to 10 times lower. With IUE such a difference must
be easy to see.

Carrasco: Yes, in fact we have written a proposal to
observe this effect with the IUE. However one should not
expect effects on the terminal velocities as high as factors
of 7 to 10, since the radii of the UV bright stars may be
comparable to 1 R_{0}, if so then the factors to observe may
be only of about 1.5 to 3.

Heap: Can you say what the typical luminosity and mass
of these high-velocities stars are?

Carrasco : Typical luminosities should fall in the
$L = 10^{3} L_{0}$ to $10^{4} L_{0}$ range, while the masses should be in
the 0.5 to 1.4 M_{0} range.

Heap: How do you account for so many Pop II OB stars
needed by your interpretation of run-aways?

Carrasco: They are not many, in fact our estimates of
the number density of run-aways are in good agreement with
both the theoretical evolutionary time scales of UV-bright
stars by Gingold and the number of this kind of objects
observed in globular clusters.

HYDROGEN DEFICIENCY AND MASS LOSS

William P. Bidelman
Warner and Swasey Observatory-Case Western Reserve University

Abstract. Some conclusions concerning the luminous hydrogen-deficient stars are presented, as is also a list of 45 such objects.

In a symposium such as this I am rather surprised that no one has yet discussed perhaps the most obvious examples of the effects of stellar mass loss: namely, the stars that exhibit substantial hydrogen deficiency in their atmospheres (Hack 1967, Hunger 1975). Though the phenomenon undoubtedly also exists among stars of lower luminosity, I am thinking here of stars in the range of visual absolute magnitude from -3 to -5 or so. Table 1 contains those objects that I now consider to belong to this class. A few Wolf-Rayet stars probably could be added, but the evidence is not too clear on this point. Also there are several additional objects in the Large Magellanic Cloud that could have been included.

While the objects listed in Table 1 display a remarkably wide variety of observable features, nonetheless some regularities can be discerned:

1. All of the cooler stars exhibit carbon bands, implying C/O ratios of larger than 1. That is, there is no such thing as a hydrogen-poor M-or S-type star. Further, no hydrogen-deficient carbon star shows C^{13}.

2. Most of the cooler stars are variables of the R CrB type (Feast 1975), surrounded by infrared-emitting envelopes (Feast and Glass 1973).

3. In only two cases do we actually see the presumably fairly recently ejected envelope surrounding the stars: V348 Sgr, where the nebulosity is seen only when the central star is faint, and V605 Aql, the "unique variable," where a bubble-like envelope remains clearly visible even though the central star has long since disappeared from view. In both cases the envelopes appear to have plenty of hydrogen.

4.Most of the hotter stars are not variable; however, V348 Sgr and MV Sgr are dramatic exceptions to the general rule.

5. In general, the objects seem to be single stars. However, at least two are unquestionably binaries, KS Per and υ Sgr. These stars do not share the usual excessive abundance of carbon of most of the rest of the group, but appear to be nitrogen-abundant instead. The

305

P. S. Conti and C. W. H. de Loore (eds.), Mass Loss and Evolution of O-Type Stars, 305–308.
Copyright © 1979 by the IAU.

TABLE 1

LUMINOUS HYDROGEN-DEFICIENT STARS

Name	α (1900) δ		Mag.	Type	References and Notes
+37°442	1 52	+38	10.0	O	CR 262, 1105 (H.P. 8, No. 19)
XX Cam	4 0	+53	8.2-10.3	F(C)	A+A 31, 203
KS Per	4 41	+43	7.6-7.8	A-F	PASJ 24,495; SB: P=363^d; HD 30353; br Hα
W Men	5 27	-71	13.8-<18.3	F(C)	MNRAS 158, 11P; LMC member
SU Tau	5 43	+19	9.1-16.0	F(C)	MKK Atlas
+37°1977	9 18	+37	10.1:	O	CR 278, 227
+10°2179	10 33	+10	10.0	B	A+A 37,87
-58°2721	10 44	-58	10.5	B	JSD; br Hα
UW Cen	12 37	-53	9.6-<13	K(C?)	MNRAS 161, 293
Y Mus	12 59	-64	10.5-12.1	F(C)	IBVS 1453
-37°9248	14 7	-37	8.8:	A-F	PASP 84, 388; br Hα
HD 124448	14 8	-45	10.0	B	A+A 37, 87
S Aps	14 59	-71	9.6-15.2	C	MNRAS 158, 11P
HD 137613	15 21	-24	7.6	C	MNRAS 137, 119
-48°10153	15 31	-48	11.5	B	ApJ Lett. 179, L31
R CrB	15 44	+28	5.8-14.8	F(C)	MNRAS 137, 119
RT Nor	16 15	-59	11.3-16.3	C	HB 920, 32
-9°4395	16 23	-9	10.6	B:	PASP 84, 388
RZ Nor	16 24	-53	11.1-<12.7	F(C)	IAU Symp. 67, 135
LR Sco	17 20	-43	10.9-12.3	F(C)	IBVS 1453
-35°11760	17 31	-35	9.8	B	JSD: HDE 320156; br Hα
V2076 Oph	17 36	-17	9.9	O	Liege No. 357, 337; HD 160641
LS IV -1°2	17 46	-1	11.0	B	JSD

TABLE 1 (Cont.)

Name	α (1900) δ	Mag.	Type	References and Notes
$-1°3438$	17 58 -1	10.4	B:	PASP 84, 388
WX CrA	18 2 -37	11.0-<16.5	C	HB 902, 6
VZ Sgr	18 8 -29	11.8-<14.0	F(C)	Vistas 2, 1428; WPB
RS Tel	18 11 -46	9.3-<13.0	C	HC 224
HD 168476	18 14 -56	9.3v	B	A+A 37, 87
GU Sgr	18 18 -24	11.3-15.0	C	W.&S. 1, No.4
V348 Sgr	18 34 -23	10.6-17	O-B	BAC 19, 265; br C II
MV Sgr	18 38 -21	12.0-15.6	B	A+A 53, 23; br Hα, FeII, etc.
HD 173409	18 40 -31	9.5	C	MNRAS 137, 119
V CrA	18 40 -38	8.3-<16.5	C	HC 224
HD 175893	18 52 -29	9.3	C	MNRAS 137, 119
LS IV$-14°$109	18 54 -14	11.2	A	JSD
SV Sge	19 3 +17	11.8-16.2	C	ApJ 117, 25
RY Sgr	19 10 -33	6.2-13.7	F(C)	MNRAS 158, 305
V605 Aql	19 13 +1	11-20:	C	BAAS 5, 442
υ Sgr	19 16 -16	4.3-4.4	B-A	PASJ 19, 564; SB: P=138^d; br Hα
HD 182040	19 17 -10	7.0v	C	MNRAS 137, 119
LS II$+33°$5	19 41 +33	10.4	B	ApJ Lett. 223, L29
V482 Cyg	19 55 +33	11.8-<15.5	F(C)	Vistas 2, 1428; WPB
$+1°4381$	20 46 +1	9.6	B-A	JSD
U Aqr	21 57 -17	10.5-<14.4	F(C)	Vistas 2, 1428; HEB
UV Cas	22 58 +59	11.8-16.5	F(C)	Vistas 2, 1428; WPB

LS refers to the Case-Hamburg Northern Luminous Stars Survey. Other designations are BD or CoD except for $-58°2721$ which is CPD. Under-lined magnitudes are photographic. WPB = W.P. Bidelman, HEB = H.E. Bond, JSD = J.S. Drilling.

strong Hα emission seen in these stars may perhaps originate from hydrogen supplied by the other component of the binary.

It is clear that any theories of mass loss and stellar evolution
will have to explain the existence of these fascinating objects.

And, in conclusion, though it has nothing to do with this subject,
I would like to comment that we should be prepared for some surprises
in the interpretation of our early mass-loss observations. It is not
clear to what extent some of our favorite stars are typical of the
general stellar population. P Cygni and η Carinae are certainly unusual
objects, and one might also be somewhat suspicious of such high-mass-
loss stars as ρ Leo and τ Sco. Are we certain that these objects are
what we think they are?

REFERENCES

Feast, M.W.: 1975, in V.E. Sherwood and L. Plaut (eds.), "Variable Stars
      and Stellar Evolution," IAU Symp. 67, 129.

Feast, M.W. and Glass, I.S.: 1973, Monthly Notices Roy. Astron. Soc.
      161, 293.

Hack, M.: 1967, in M. Hack (ed.), "Modern Astrophysics," Gordon and
      Breach, New York, p. 163.

Hunger, K.: 1975, in B. Baschek, W.H. Kegel, and G. Traving (eds.),
      "Problems in Stellar Atmospheres and Envelopes," Springer,
      New York, p. 57.

DISCUSSION FOLLOWING BIDELMAN

van den Heuvel: I was somewhat surprised to hear you
mentioning τ Sco was a peculiar star.

Bidelman: I did not mean to imply that τ Sco is actually
peculiar; it is certainly generally considered perfectly
normal. Nonetheless, there are several evolved (α Sco) and
spectroscopically peculiar (3 Cen and others) objects in the
same cluster and, with its very sharp lines, it might perhaps
not be a typical B0 V star in all respects.

Bolton: Some of the OBN stars are also H-deficient.
HDE 235679 is definitely so, and there are comments in the
literature that suggest mild H deficiencies for HD 72754 and
V453 Sco. All show evidence that mass transfer has influenced
their evolution.

GENERAL DISCUSSION

Hutchings: Do you have any comment on the existence or origin of the λ5300 line reported in HD 153919 by Dupree and tentatively attributed to Fe XIV?

Olson: A broad emission feature at 5294Å was reported by Baliunas, Dupree and Lester (1977, Bull. A.A.S., 9, 298) and tentatively identified as an Fe XIV fine structure transition. However, this emission was seen in an X-ray binary (HD 153919, 4U 1700-37) and not in any of the single O stars they examined. Therefore, at the present time, this observation is not relevant to single star wind models.

Lester: Coudé spectra taken at Cerro Tololo in May 1977 at the same phases at which the (Fe XIV) was previously reported show no emission which can be attributed to this ion to a limit of < 1% of the continuum. There is also no variability in this spectral region with phase.

Seggewiss: Let me come back to Wolf-Rayet stars with intrinsic absorption lines. Beside HD 92740 there are some other WN7 stars in the η Carinae region showing O-type absorption lines which very likely belong to the WR star : HD 93162 (Moffat, 1978, Astron.Astrophys.) and HD 93131 (Moffat and Seggewiss, 1978, Astron.Astrophys, in press). According to our Coudé spectroscopy these stars don't show velocity variations. They may be binaries only if the orbital inclination is fortuitously small and/or the companion is a very low mass star. This appears contrived and more likely they are single WR stars with intrinsic absorption lines like the photospheric absorption lines in Of stars.

Underhill: I would like to emphasize that the spectra of WN7 and WN8 stars are very similar to Of spectra, the chief differences being the greater intensity of the emission lines in the WN stars. The line widths are comparable. Wolf-Rayet stars of all other spectral subtypes have significantly broader emission lines. They form a different morphological group. At the time the type WN7 was first recognized, the type Of was barely known owing to the weakness of the defining emission lines. Then all hot stars with strong NIII and He II emission lines were put automatically into the Wolf-Rayet class.

Abbott: How much geometric extension is needed to produce emission in lines?

Kunasz: For the kinds of models I have worked with, if the line formation region extends as far as 1.3 stellar radii some emission is expected.

309

P. S. Conti and C. W. H. de Loore (eds.), Mass Loss and Evolution of O-Type Stars, 309–310.
Copyright © 1979 by the IAU.

Castor: I want to point out a numerical coincidence
that may be significant for the coronal-plus-cool wind
model. That is that the 1-2 KeV X-ray flux required in
Cassinelli's and Olson's model of ζ Pup is just what is
produced if all the mass of the wind passes through a shock
wave with a strength of about 600 km/s. The emission
measure of the cooling region behind the shock is the same,
to a factor 2, as that required in the coronal wind model.
The shock strength 600 km/s will produce a hot region with
a temperature of 5×10^6 K. We can only guess what might
produce such a shock; we might get an oblique shock if the
radial flow in a rotating star developed a non-axisymmetric
instability. I have not found a really convincing picture
of this, but it could work.
I have talked with Nelson about what his and Tony's instability would do. It is important to notice that the instability exists only for length scales less than a mean free
path of the lines that drive the flow. That is 10^8 or 10^9
cm - very small. With a mixing-length-type description of
the finite amplitude motion, we find velocity amplitudes
like a few kilometers per second.

There was an astronomer whose spectra
Were of hot supernova ejecta
But her lack of detection
Of semi-convection
Made de Loore and Chiosi reject her.

SESSION 6

EVOLUTION WITH MASS LOSS: SINGLE STARS

Chairman: A. COWLEY
Introductory Speaker: C. DE LOORE

1. C. CHIOSI, E. NASI and G. BERTELLI: Theoretical evolution of massive stars with mass loss by stellar wind.

2. D.S.P. DEARBORN and J.B. BLAKE: Critical rates of mass loss.

3. S.A. LAMB: Supergiant mass loss and the Cassiopeia A progenitor.

4. S.R. SREENIVASAN and W.J.F. WILSON: The role of rotation in the evolution of massive stars losing mass.

5. H.J. FALK and R. MITALAS: Evolution of a 30 M_o star with mass loss.

6. T.J. MAZUREK: Mass conservation and rapid mass loss on the main sequence.

EVOLUTION OF SINGLE STARS WITH MASS LOSS

C. de Loore
Astrophysical Institute, Free University of Brussels

1. EVIDENCE FOR MASS LOSS

A. Mass loss has been observed by means of UV, optical, IR and radio observations.
B. There exists also indirect evidence that stars lose matter during their evolution.

1.1. Mass determination of binaries

For stars on or near the ZAMS the masses agree well with the mass luminosity relation (see for instance Stothers, 1972). The effective temperatures and luminosities correspond with models of these stars for luminosity class V. For some evolved systems the masses and luminosities are such that mass exchange and mass loss might have occurred, since one of the members does not fit the mass luminosity law (Examples : Ly Aur (Andersen et al., 1974), HD 47129 (Hutchings and Cowley, 1976), HDE 228766 (Massey and Conti, 1977), BD+40°4220 (Bohannan and Conti, 1976)). For this last star the mass loss could have been as large as 60 M_0 in 3.10^6 years, corresponding with a mass loss rate of $\sim 2.10^{-5}$ $M_0 yr^{-1}$ (Conti, 1978).

1.2. Overluminosity of supergiants and Of-stars

The luminosities of early type supergiants and of Of-stars seem to be considerably higher than expected from the observed masses, derived from evolutionary tracks or the mass luminosity relation (Hutchings, 1976, Conti, 1976). Also the optical companions of massive X-ray binaries are overluminous; this was ascribed to mass loss by a strong stellar wind during the hydrogen burning stage by de Loore, De Grève, Lamers (1977), de Loore, De Grève, Vanbeveren (1978), Ziolkowski (1977). This has important repercussions on the evolution and the evolutionary status of massive X-ray binaries and WR stars.
More evidence is furnished by the mechanism for the production of the X-rays; these may either be produced by the accretion of matter expelled by a stellar wind acting on the hot companion (Lamers, van den Heuvel,

P. S. Conti and C. W. H. de Loore (eds.), Mass Loss and Evolution of O-Type Stars, 313–336.
Copyright © 1979 by the IAU.

Patterson, 1976) or by a stellar wind enhanced by Roche overflow from
the hot companion of the neutron star.
The surface composition of supergiants can show products of nuclear
burning; it might be that these products were brought from the interior
to the surface by convection or that the outer layers were expelled, so
that the inner regions which have undergone nuclear burning are revealed.
Early type supergiants do not have convective envelopes hence in the
stars of this group showing evolved surface abundances probably considerable mass loss occurred (Walborn, 1976; Dearborn and Eggleton, 1977).
Another possibility to bring processed material to the surface, meridional mixing may be ruled out, since at no time during the evolution the
envelope convection penetrates through the hydrogen burning shell (Lamb,
Iben, Howard, 1976; Lamb, 1978).

1.3. Features of the HRD which cannot be explained by usual evolutionary
     computations

a) an uninterrupted distribution of stars with spectral types O through
   AO for $\log L/L_0 < 5.3$
b) a relative lack of stars with spectral types later than B3 for log
   $L/L_0 > 5.3$
c) the absence of stars of spectral type M for $\log L/L_0 > 5.3$.

If instead of using a conservative assumption upon the mass, mass loss
is taken into account, most of these characteristics may be explained :
the strange features as overluminosity (or undermassiveness) disappear
and a new mass luminosity relationship may be derived, the X-ray production in binaries can be explained at least qualitatively, the position
in the HRD becomes clear, certainly for the points b and c, i.e. the
thinning out of more luminous stars with spectral type later than B3 and
the absence of stars of spectral type M for $\log L/L_0 > 5.3$. Different
assumptions for the mass loss may be adopted. The effect of these as-

assumptions	opacities	
	Cox Stewart	Carson
no mass loss	no	no
mass loss only important during late type supergiant phase	no	no
sudden mass loss at some criti-cal effective temperature		no
mass loss occurs continuously in all parts of the HRD	can account for the presence of OBN and WN stars and the absence of very luminous M supergiants	
	no satisfactory explanation of WN stars with large H-defi-ciencies	effective temperature for the ZAMS too low to account for the ob-servation of blue su-pergiants of the lowest luminosity

Table 1. Survey of various assumptions concerning mass loss and their
influences on the features of the HRD. "No" means that the observed features of the HRD are not in agreement with the assumption.

sumptions is shown in Table 1. It turns out that only mass loss occurring continuously in all parts of the HRD explains these characteristics,
and neither mass loss occurring only during the late type supergiant
stage, nor sudden mass loss starting at some critical effective temperature.

2. EVOLUTION OF SINGLE STARS WITH MASS LOSS

Although evidence for mass loss in stars of different spectral types was
discovered many years ago (Morton, 1967; Hutchings, 1976), with mass loss
rates between 10^{-7} and 10^{-4} $M_o yr^{-1}$ for luminous stars, mechanisms for the
explanation of this phenomenon have been developed recently, or are still
being developed. It is clear that mass loss rates of this order will
affect the evolution of the stars. Evolution of stars decreasing in
mass was computed by Massevich (1958) in order to explain statistics for
stars of the galaxy and the assumption that the mean molecular weight
changes along the main sequence. She assumed that the star loses mass
continuously owing to corpuscular radiation and calculated sequences of
equilibrium models with decreasing mass. A mass loss rate, $\dot{M} = -kLR/M$
was proposed by McCrea in 1962. Tanaka (1966) calculated the evolution
of two massive stars ($\sim$ 15 M_o and $\sim$ 47 M_o) from ZAMS to core H exhaustion.
Hartwick (1967) used the McCrea relation with k such that for main sequence stars $\dot{M} = 10^{-6}$ $M_o yr^{-1}$. Chiosi and Nasi used the same relation
with various k for the calculation of the evolution of stars with initial
masses of 20 and 40 M_o. The tracks are shown in Figure 1. The most
striking features are the following :

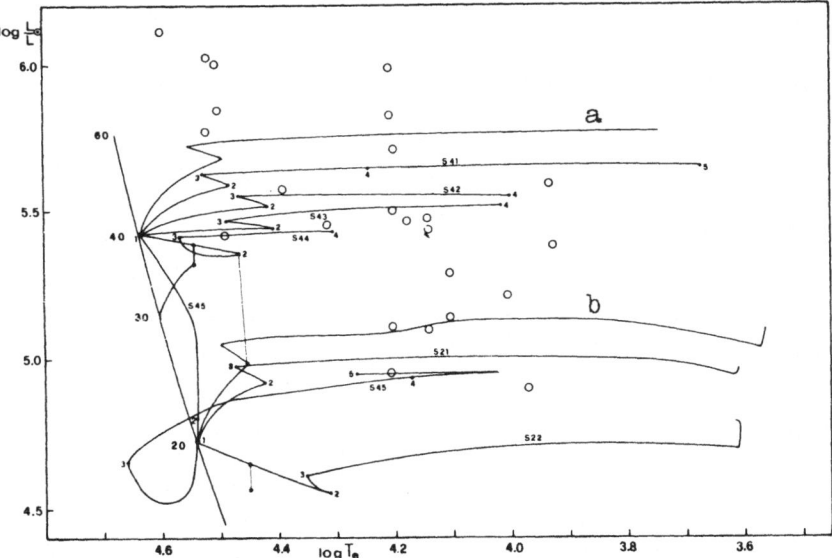

Figure 1.   Evolutionary sequences with mass loss for initial masses of
20 and 40 M_o. The tracks a and b are tracks without mass loss (Chiosi
and Nasi (1974).

1. for moderate $\dot{M}$ the evolutionary tracks are similar to those at constant mass but with decreasing luminosity for higher $\dot{M}$.

2. for very high $\dot{M}$ the evolutionary tracks show similarities with those of the mass losing primaries in massive close binaries with Roche lobe overflow. Dearborn and Eggleton (1977) paid especially attention to the problem of CNO abundances in stars that lost matter. According to their computations a 32 M_O star, losing mass at a rate of ~ 1.5 10^{-6} $M_O yr^{-1}$ can be converted into an OBN star (Walborn, 1976) : the burning products of the CNO cycle show up at the surface. This is particularly interesting in connection with the problem of the WR abundances. The status of evolutionary computations with mass loss is given in Table 2.

Table 2. Survey of evolutionary computations including mass loss.

Massevich 1958	mass loss by corpuscular radiation
McCrea 1962	mass loss rate $- \dfrac{dM}{dt} = kLR/M$ (L,R,M in solar units, k in $M_O yr^{-1}$)
Tanaka	evolution of 46.8 M_O 15.6 M_O
Hartwick	evolution of 15 M_O
Chiosi & Nasi 1974	evolution of 20 and 40 M_O
de Loore, De Grève, Lamers 1977	20-50 M_O
de Loore, De Grève, Vanbeveren 1978	50-100 M_O
Dearborn & Eggleton 1977	
Chiosi & Nasi, Sreenivasan 1978	20-100 M_O
Sreenivasan, Wilson 1978	15 M_O
de Loore, De Grève, Vanbeveren (Dec.1978)	20-100 M_O
Stothers & Chao wen Chiu	
Czerny, Michal (Poland)	40-180 M_O
Falk, Mitalas	30 M_O

3. COMPUTATIONS OF EVOLUTIONARY TRACKS WITH MASS LOSS

3.1. Mass loss rates

A considerable number of observations on mass loss exists but the number of stars with accurately determined rates is less extended (Snow, T., these proceedings). For the determination of the theoretical mass loss rates we calibrated the values of Hutchings (1976) and of Barlow and Cohen (1977) by means of 7 standard stars. Rather large errors (up to 1 magnitude) may be expected since discrepancies up to a factor of 10

might exist for the standard stars, and since extrapolation is necessary
(standard stars of spectral type O9f to B1a). Barlow and Cohen (1977)
derive an expression for the mass loss rate (from the IR excess of 34
OBA supergiants and 10 Of and Oe stars, together with a velocity law
derived from the observations of P Cygni). For our computations we
started with the equations proposed by Barlow and Cohen :

$$\dot{M} = \alpha (L/L_O)^\beta$$

$\alpha = 6.8 \ 10^{-13} \ M_O yr \qquad \beta = 1.1 \pm 0.06$ for the O-stars
$\alpha = 5 \quad \ 10^{-13} \ M_O yr \qquad \beta = 1.2 \pm 0.08$ for the B and A

These values are about a factor of 2 smaller than those quoted for supergiants in the literature. Also the values of Sterken (1977), derived
from the Hα profile of B1a and B2a supergiants agree within a factor of
2. β is approximately 1, hence we assumed the mass loss nearly proportional to the luminosity, and expressed the mass loss rates in terms of
a parameter N defined as

$$N = -\dot{M} \ c^2/L$$

The N values derived from the mass loss rates of Barlow and Cohen are
shown in Figure 2. In the figure are also given the values of ζ Pup and
τ Sco and the 6 supergiants of Sterken (underlined). The lower limit
for mass loss for O and B stars (Snow and Morton, 1976) is $M_{bol} = -6$.
In the late B and A stars Rosendhal (1973) found a lower limit at $M_{bol} = -7.8$. From the figure it may be seen that M is of the order of 100±50
throughout the whole spectral range. The figure does not show variations

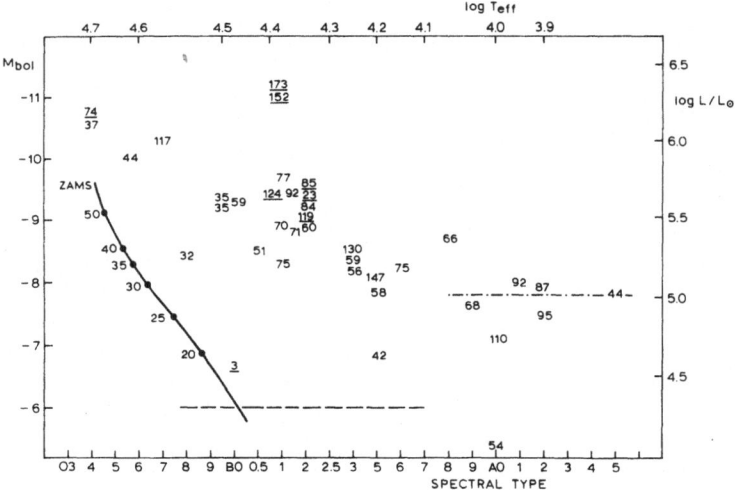

Figure 2. Observed mass loss rates converted into N-values. The lower
limits for significant mass loss observed in the far UV (dashed line)
and in the visual (dash-dotted line) are indicated also.

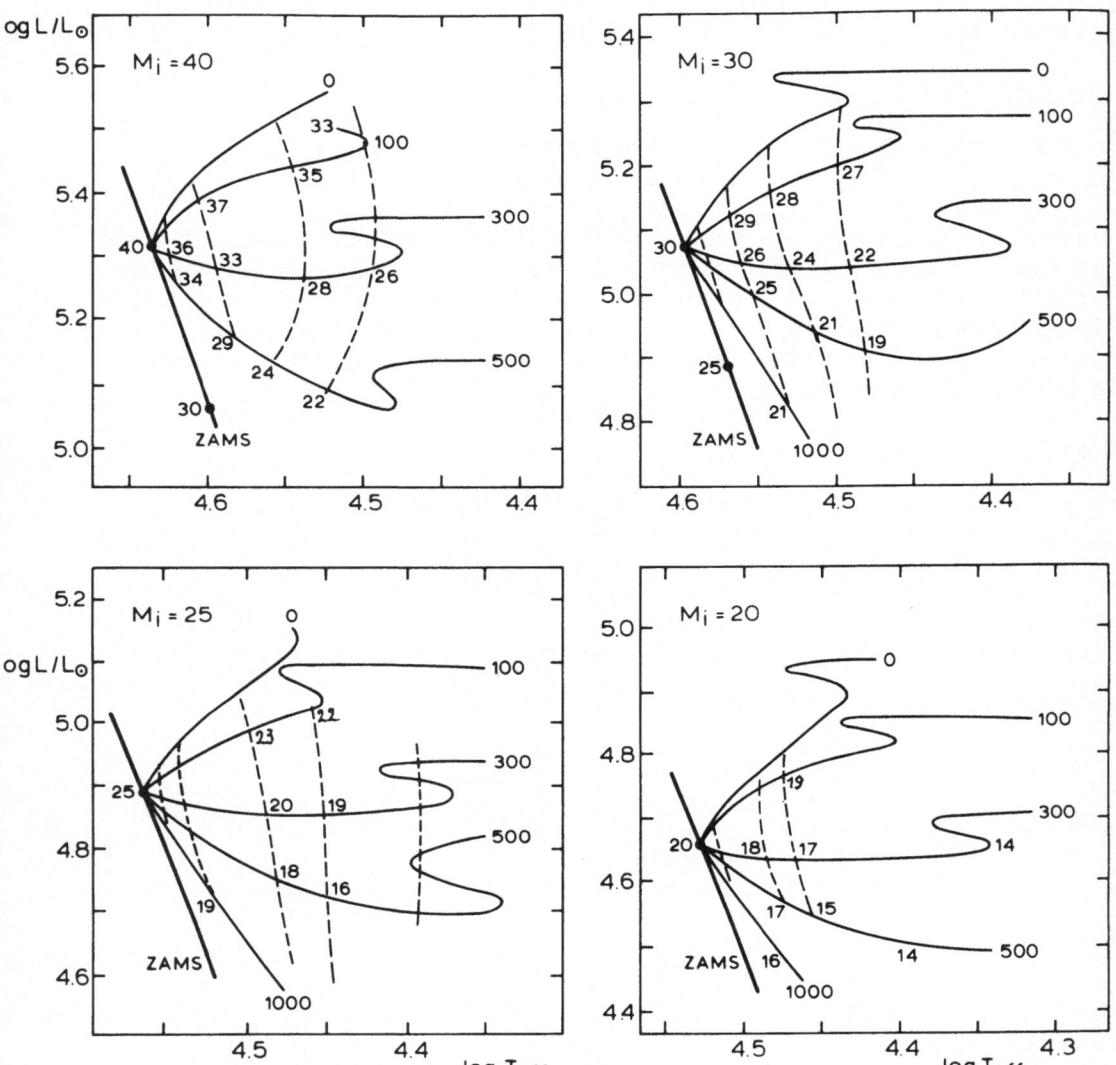

Figure 3. Evolutionary tracks in the HR-diagram of mass losing stars
with initial masses of 40, 30, 25 and 20 M_o. The evolutionary tracks,
in full line, are labeled with the N-values. The dashed curves are
isochrones, given for intervals of one million years. Near the isochrones the stellar masses are indicated.

of N with luminosity or spectral type and this was the reason why we computed our tracks with constant N. We used N-values ranging between 100
and 300 (cfr. Fig.2), keeping in mind that the Barlow and Cohen values
are about a factor 2 smaller than those of other observers.

3.2. Evolutionary tracks

For the computation of evolutionary tracks for massive single stars we
adapted our code for the evolution of primaries of close binaries. Sets
of atmospheres were calculated for various luminosities, effective temperatures, masses and mass loss rates in domains containing the pre-
estimated point on the presumed track in the HRD. Entropy losses were
taken into account. An initial chemical composition of X=0.7, Z=0.03
was used. Evolutionary tracks for masses between 20 and 50 M_0, for
N=0, 100, 300, 500 and 1000 are shown in Figure 3. The tracks for different values of N (100-1000) for 20 M_0 and 40 M_0 may be compared with
those of Chiosi and Nasi (1974); the agreement is rather good. Since
they used a changing N instead of a constant value this indicates that
the general conclusions concerning the supergiant stage are independent
of the precise behaviour of N. Striking features are :

    1. the luminosity drops are more pronounced for larger N-values;
for N=1000 the evolutionary tracks move downwards along the ZAMS (this
was also noted by Chiosi and Nasi)

    2. for increasing N-values the luminosity at the first turn point
steadily decreases together with the effective temperature; hence the
hydrogen burning phase covers a wider strip in the HRD.

After hydrogen exhaustion in the core the star follows a track similar
to the conservative case (i.e. at nearly constant luminosity). If we
compare the values of the luminosities in these regions with the luminosities of single stars at the end of the main sequence without stellar
wind we find that the stellar wind remnants are overluminous by a factor
1.3 (N=100) to 5 (N=300)(Figure 3). Another consequence of the mass
decrease is an increase of the hydrogen burning phase of about 30%
(N=300) to 10 à 15% (N=100) compared with evolution at constant mass.
Beyond the turn point the influence of the mass outflow on the evolution
is only marginal due to the small time scale involved with core contraction prior to helium ignition (10^4 years). Stars with initial
masses between 30 and 50 M_0 lose about 50% of their mass during the core
hydrogen burning phase (N=300) or ~ 20% (N=100). For stars in the range
60-100 M_0 the luminosity is higher (of the order of 10^{-5} $M_0 yr^{-1}$), so
that more hydrogen rich layers are expelled and helium rich layers
appear at the surface already during hydrogen burning. The evolutionary
tracks for this mass range (de Loore, De Grève, Vanbeveren, 1978) are
shown in Figure 4. The figure shows the decreasing masses and the hydrogen abundance at the surface. These massive stars attain their turning point during core hydrogen burning and turn to the left. The most
massive ones (80 and 100 M_0) even cross the ZAMS before evolving towards
the red giant region. The time between red and blue main sequence points

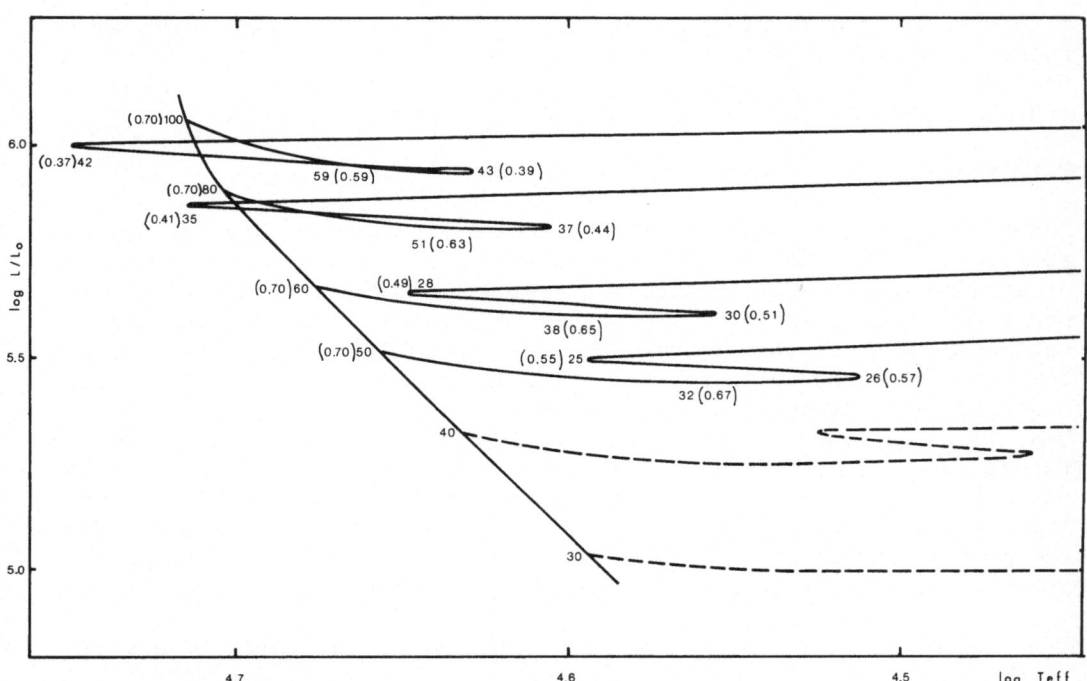

Figure 4. Evolutionary tracks for mass losing stars of 100, 80, 60 and
50 M_0 (N=300). The numbers at the tracks are the remaining masses in
M_0. Between brackets is indicated the hydrogen content of the atmosphere.

is some 10^4 years. Figure 5 shows the masses at the end of the core
hydrogen burning as a function of the initial masses. For the three N-
values used for the computations a linear relation between final mass
and initial mass may be obtained. The limiting value for the mass corresponding with vanishing mass loss is ~ 13 M_0 (where UV observations
point to ~ 15 M_0 (Snow and Morton, 1976)).
Chiosi, Nasi and Sreenivasan (1978) have calculated evolutionary tracks
with mass loss using the mass loss rates predicted by Castor, Abbott and
Klein (1975).

$$\dot{M} = \frac{L}{cv_{th}} \cdot \frac{\alpha}{T} \left|\frac{1-\alpha}{1-\Gamma}\right|^{\frac{1-\alpha}{\alpha}} (K\Gamma)^{1/\alpha}$$

with L, M stellar luminosity and mass, c the light velocity, v_{th} the
thermal velocity of random motion, Γ the ratio of the luminosity to the
Eddington luminosity.
($\Gamma = \dfrac{\sigma_e L}{4\pi GM_c}$ with σ_e the mass scattering coefficient for free electrons,

K and α were used as adjustable parameters). The α-value of 0.76 used
by Chiosi et al. seems to correspond with our case N=100, and the value
of α=0.9 is slightly smaller than our N=300 case. The agreement between

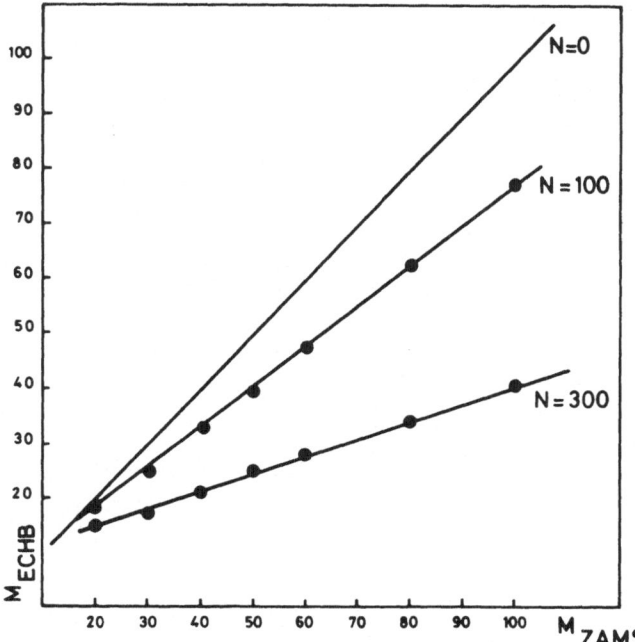

Figure 5. Mass at the end of core hydrogen burning as a function of
the initial mass for three N-values.

the two sets of computations (de Loore et al. and Chiosi et al. (1978))
for the hydrogen burning stage is very good. Chiosi et al. have treated
semi-convection in a very extended way, while in the computations of de
Loore et al. only convection, but not semi-convection, is treated properly. However this has little effect on the evolution as the layers involved are removed. Chiosi et al. (1978) have also incorporated mass
loss during the red giant phases when the evolutionary tracks cross the
acoustic flux dominated region. As already calculated before (de Loore,
1970; Castellani et al. 1971) the acoustic flux is a sharply peaked
function of T_{eff}, so that the switch from predominantly radiation
pressure driven mass loss to acoustic flux driven mass loss occurs
quite suddenly. The mass loss rate is given by

$$\dot{M} = \varepsilon \, \frac{L_{ac}}{v_e v_c}$$

with v_e the escape velocity of the star, v_s the sound speed, L_{ac} the
acoustic flux and ε is the efficiency parameter (Fusi-Pecci and Renzini,
1975a,b, 1976). Chiosi et al. adopted a value of 10^{-4} for ε.
Evolutionary tracks were computed for initial masses of 20 - 100
M_o, lowered by stellar wind losses ($\alpha=0.9$) with further mass loss due
to acoustic fluxes when log T_{eff} equals 3.84, 3.83 and 3.795 respectively. The mass loss rates involved have values of 5 to 6 10^{-3} $M_o yr^{-1}$,
5 10^{-3} $M_o yr^{-1}$ and 2.5 10^{-4} $M_o yr^{-1}$ respectively. This is shown in Fig.
6 and 8.

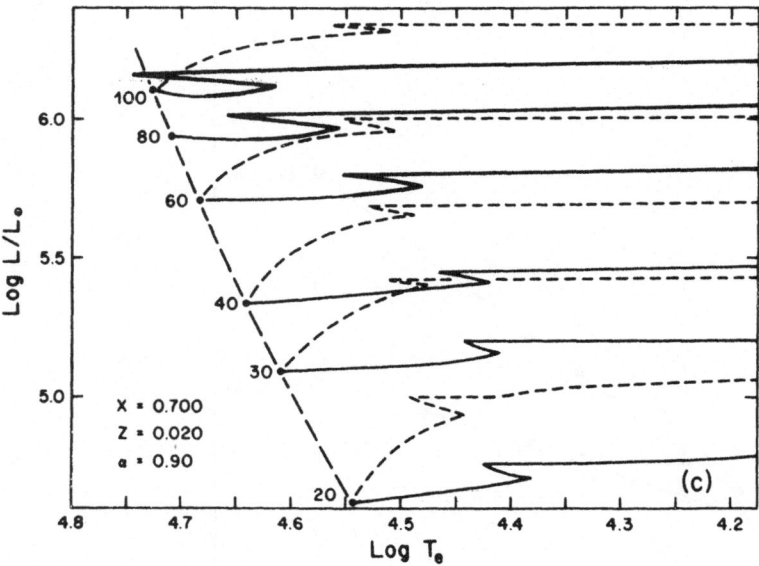

Figure 6. Theoretical HR diagram of the 20, 30, 40, 60, 80, 100 M_O
stars with mass loss by radiation pressure (α=0.76) during the main
sequence and shell H-burning phases (Chiosi, Nasi, Sreenivasan, 1978).

The same set of calculations was repeated for α=0.76 (shown in Figure 7).
The tracks show the following characteristics :

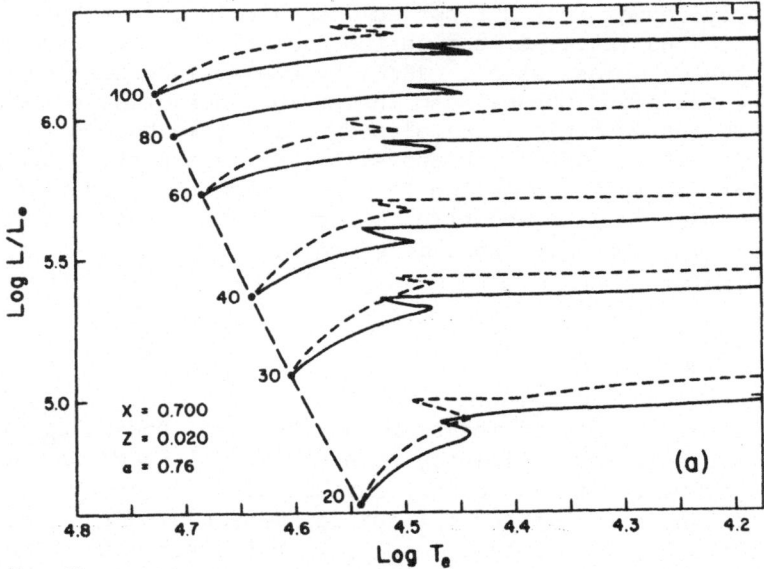

Figure 7. Theoretical HR diagram of the 20, 30, 40, 60, 80, 100 M_O
stars with mass loss by radiation pressure (α=0.90) during the main
sequence and shell H-burning phases (Chiosi, Nasi, Sreenivasan, 1978).

1) as long as layers with the initial chemical composition are expelled the stars move towards lower effective temperatures. The acoustic
flux increases hence the mass loss rate becomes higher. This stage ends
when layers that have undergone internal mixing or CNO processes appear
at the surface. This stage does not occur in higher masses since CNO-
processed layers reached already the surface before

2) two competing effects occur now, leading to loops in the HRD :
on the one hand gravitational core contraction and consequent He-burning
causes an expansion of the envelope and a decrease of the effective temperature; the star tends to move to the right. On the other hand He-
enriched layers will tend to shift the star to move to the left. The
result is that the star describes loops (Figure 8).

3) as a consequence of the mass loss and the H-burning shell moving
outwards, the outer layers are eaten from two sides. Consequently the
relative He-core mass increases. When this value exceeds values of 0.6-
0.7 the star moves towards the left, and again the high temperature mass
loss mechanism takes over.

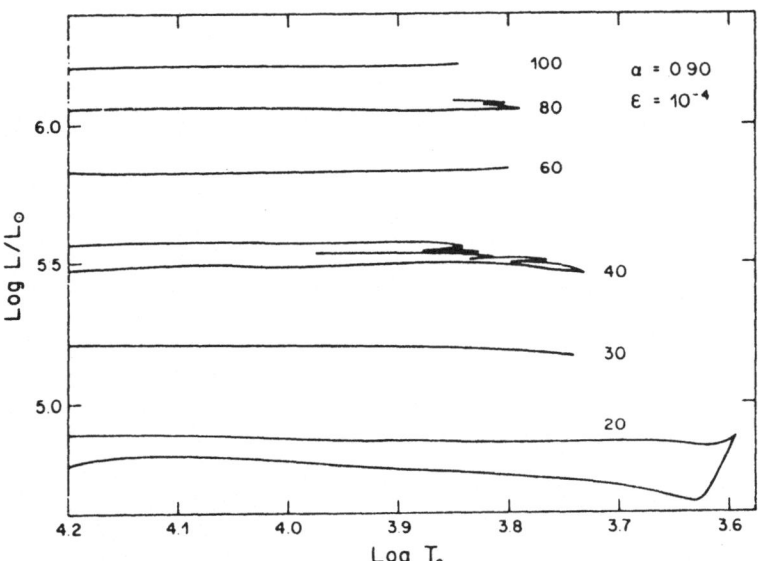

Figure 8. Theoretical HR diagram for the set α=0.90 during the phase of
acoustic flux driven wind (ε=10^{-4}) and early stages of central He-burning.
Each sequence is labelled by the initial mass in solar units (Chiosi,
Nasi, Sreenivasan, 1978).

Recently Stothers and Chao-wen Chiu (1978) carried out evolutionary computations including mass loss for different assumptions

1) mass loss occurs in all regions of the HRD, according to (McCrea,
1962)

$$\dot{M} = -kLR/M$$

(for k a value of 10^{-11} $M_\odot yr^{-1}$ is used).

2) mass loss only occurs during the late type supergiant stage
(log T_{eff} < 3.85)

3) sudden mass loss occurs at a critical effective temperature
(T_{eff} ~ 5000K) (Bisnovatyi-Kogan and Nadezhin, 1972) for M > 20 M_0.

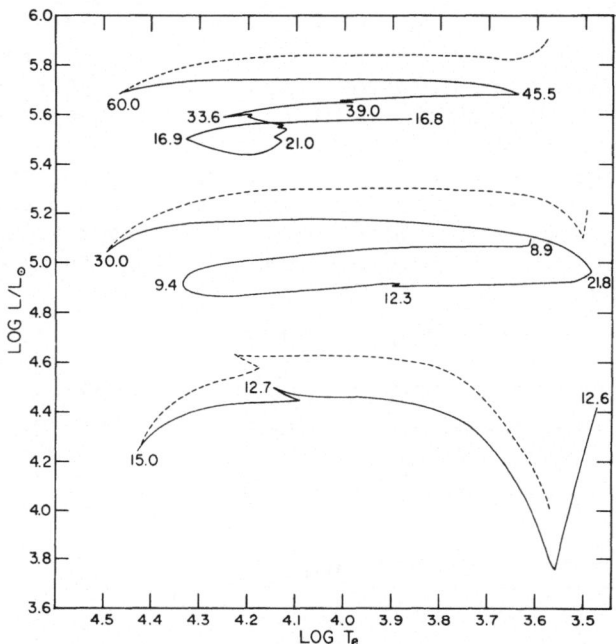

Figure 9. HR diagram showing the evolutionary tracks for initial masses
of 60, 30 and 15 M_0. The dashed lines represent conservative tracks,
the solid lines represent the case with mass loss occurring continuously
in all parts of the diagram with $\dot{M}$ given by the McCrea expression
(Stothers and Chao-wen Chiu, 1978).

Evolutionary tracks are shown in Figure 9. The results are similar to
those of Tanaka (1966), Chiosi and Nasi (1974), Dearborn and Eggleton
(1977), de Loore et al. (1977), Sreenivasan and Wilson (1978), Chiosi
et al. (1978), de Loore et al. (1978).
The evolution of a 15 M_0 star is shown in Figure 10, with the assumption
that mass loss is only important among late type supergiants. The star
executes a long blue loop and ends as a star of 3.6 M_0. The hydrogen
envelope has X=0.36. For very high masses the evolution is similar to
the case of sudden mass loss : the star describes loops in the HRD (see
Figure 10, 2 upper tracks) with rapid leftward motion and motions towards the right on a nuclear time scale. Much of the time is spent in
an unstable area, bordered by log T_{eff} = 3.63 and log T_{eff} = 3.73, forming a yellow supergiant region. In contrast to this, a star losing
mass continuously spends most of the time as a blue supergiant.

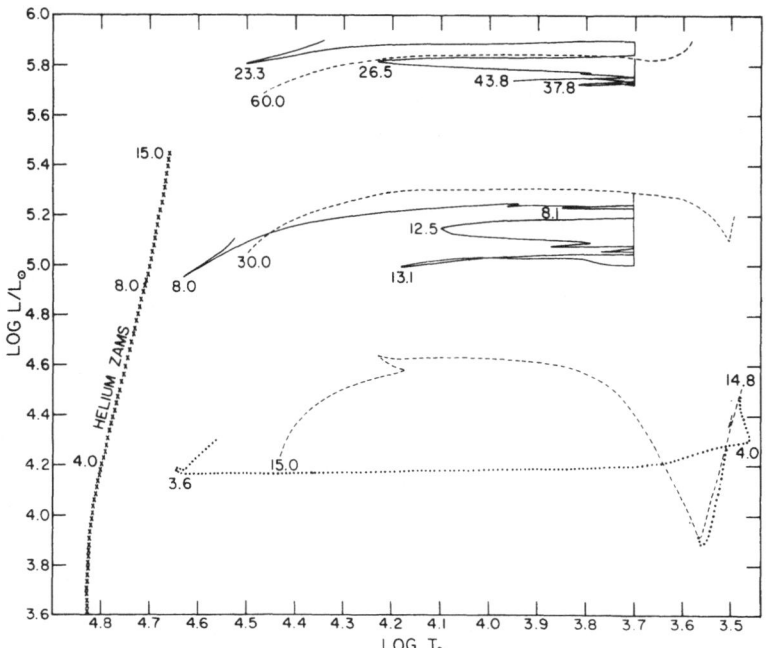

Figure 10. HR diagram showing the evolutionary tracks up to the stage
of helium exhaustion for the conservative case (dashed curves), for the
case of mass loss during the late supergiant stage (dotted curves) and
for the case of sudden mass loss in the region of yellow supergiants of
very high luminosity (Stothers and Chao-wen Chiu, 1978).

5. COMPARISON WITH OBSERVATIONS

5.1. Mass-luminosity relation

a) Masses of stars in the core hydrogen burning phase
A careful compilation of masses, luminosities and temperatures of a
large number of OB-stars was carried out by Snow and Morton (1976);
these data can be used for a comparison with the results of our calculations, and since the computations of Chiosi et al. (1978) and Stothers
and Chao-wen Chiu (1978) are comparable, also with those. In Figure 11
are shown equal mass curves (EMC). These EMC are not identical with
evolutionary tracks as is the case for conservative evolution. The EMC
are in the case N=300 much steeper than the EMC for N=0. During shell
burning the evolution is so fast that although the mass loss rates are
considerable the stars lose only marginal fractions of their mass; for
this stage the EMC may be considered as evolutionary tracks. In Figure
11 are shown also the stars of Snow and Morton's list, of classes V, IV
and III with masses, derived by comparison with Stothers' conservative
tracks. These values are compared with the EMC derived from our calculations for N=300. The agreement is very good. At the end of core H-
burning the EMC are again very close to the conservative evolution tracks
for the same mass. Consequently here again the mass estimates for both

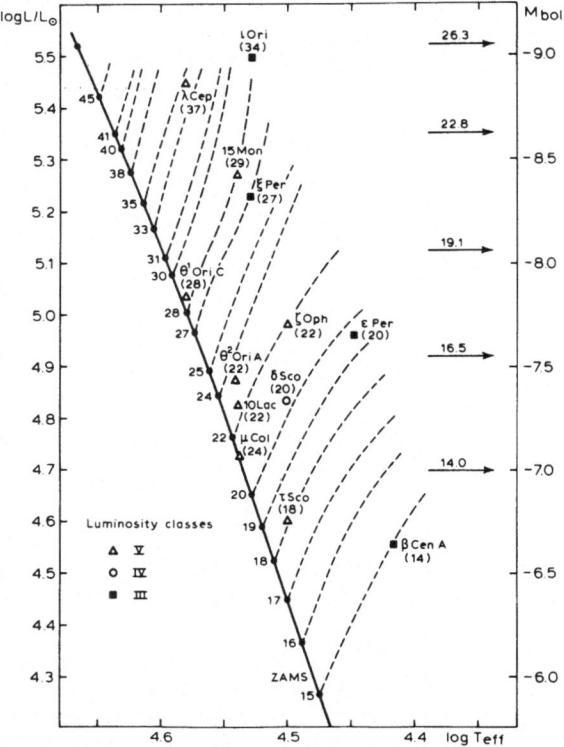

Figure 11. The equal mass curves for hydrogen core burning (dotted lines)
for the case N=300 are compared with masses of stars of luminosity classes
III, IV and V derived from the observations by using conservative evolutionary tracks (Stothers, 1972). The arrows indicate the luminosity of
the equal mass curves during hydrogen shell burning.

cases are similar. The largest differences are found near the start of
the contraction phase. Hence mass estimates of stars in the core hydrogen burning phase based on conservative tracks will differ only slightly
from estimates based on tracks with mass loss, even if the mass loss is
large (N=300).

b) OBA-Supergiants
During hydrogen shell burning the luminosity of the star remains nearly
unchanged as well for mass losing tracks as for the conservative case.
This enables us to establish a mass-luminosity relation for supergiants
not dependent on the spectral type. No masses of supergiants are known
accurately enough; only 6 supergiants could be found for which a reaso-
nable accurate mass can be derived (ζ Ori, ε Ori, η CMa, o^2CMa, HD 7583,
HD 33579). The masses and luminosities of these stars, with 1σ error
bars are plotted in Figure 12. The uncertainty in the masses is too
large to draw any firm conclusion about the exact value of N.

c) The lack of very luminous stars with M_{bol} < -9 (log L/L_0 > 5.5) and
log T < 4.25 cannot be explained by a luminosity decrease during H shell

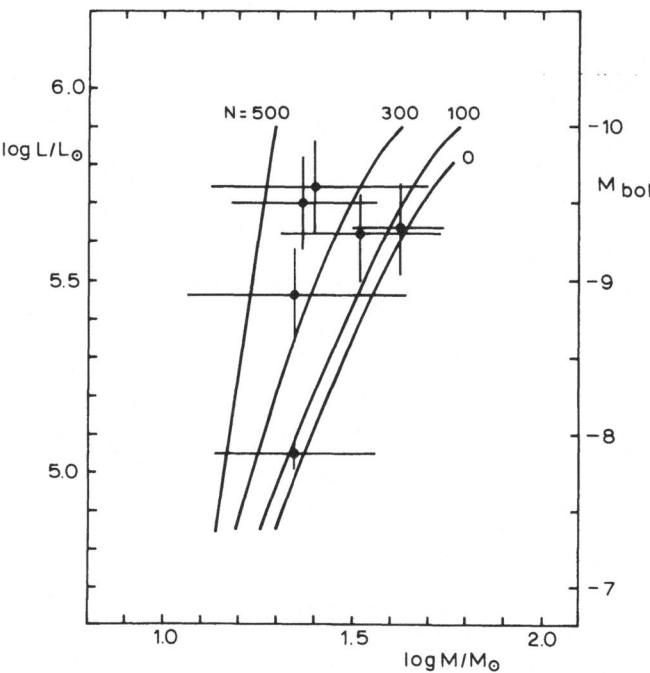

Figure 12. The observed masses and luminosities of six supergiants (with 1σ error bars) compared with the predicted mass-luminosity relation for different mass loss rates (N=0, 100, 300, 500).

burning since this phase is too short to allow a significant mass decrease, nor can this be explained by a speeding up of the evolution on the horizontal track (de Loore et al. 1977).

d) The assumption of a strong stellar wind has the consequence that in the main sequence band a variety of stellar remnants are produced that should be observed as "overluminous"(= undermassive), helium enriched or nitrogen rich. Among the population I stars of the galaxy, helium stars, OBN stars and Wolf Rayet stars show these characteristics. This is valid for Cox-Stewart opacities as well as for Carson opacities (Stothers and Chao-wen Chiu, 1978). Table 3 shows the changing atmospheric abundance and the N(H)/N(He) ratio for various evolutionary stages and different initial masses (N=300). Figure 13 shows again the evolutionary tracks for massive stars and the regions of the HRD where the WN 7 and 8 stars are located, the WR (WC) stars as well as the position of the optical companions of 5 massive X-ray binaries (Conti, 1978). As can be seen from the figure, single WN 7 and 8 stars might have evolved in a natural way from O stars with masses between 80 and say 120 M_O with strong stellar winds. The formation of other WR stars, with more helium in the atmosphere cannot be explained by this mechanism. The evolutionary scenario of Conti (1976) implying that Of-stars would evolve into transition WR stars and further into WR stars is only partially confirmed. It is not easy to explain how such hydrogen deficient

X_{at}	$N(H)/$ $N(He)$	100 M_0			80 M_0		
		$t/10^6$	log L	log T_{eff}	$t/10^6$	log L	log T_{eff}
0.7	10.37	0	6.06	4.72	0	5.90	4.71
0.6	6.5	1.785	5.95	4.67	2.253	5.80	4.63
0.5	4.25	2.430	5.94	4.64	2.858	5.80	4.61
0.4	2.8	2.882	5.94	4.63	3.230	5.92	4.27
0.36	2.5	2.992	6.06	4.34			
		60 M_0			50 M_0		
		$t/10^6$	log L	log T_{eff}	$t/10^6$	log L	log T_{eff}
0.7	10.37	0	5.67	4.68	0	5.52	4.66
0.6	6.5	2.954	5.59	4.59	3.726	5.45	4.53
0.5	4.25	3.657	5.61	4.56			
0.4	2.8						
0.36	2.5						

Table 3. Atmospheric hydrogen abundance (X_{at}) and the ratio N(H)/N(He) for various evolutionary stages and different initial masses (N=300).

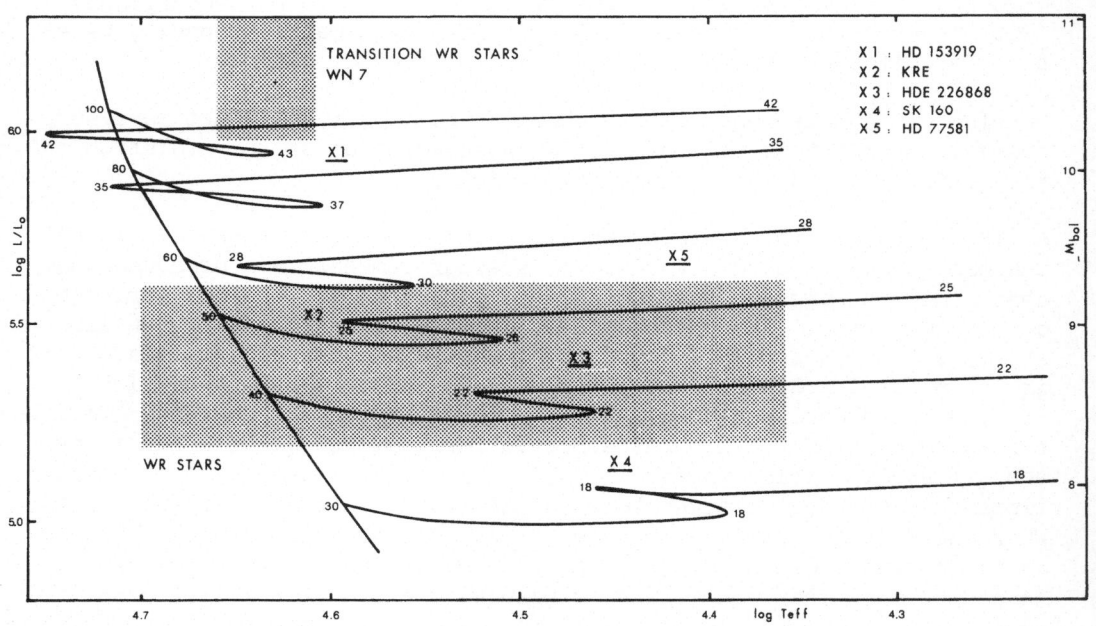

Figure 13. Evolutionary tracks for stars with initial masses between 30 and 100 M_0 losing mass, and the position of WN7 stars, WR stars and companions of massive X-ray binaries.

stars could be formed by single star evolution : if the mass estimates
for these stars are correct, they cannot have had as ancestors stars
with masses larger than 60 $M_\odot$; for lower masses however the mass loss
rates are not sufficiently high to allow the rich layers to show up at
the surface. Two alternatives remain for these WR stars :

a) they are binaries (Kuhi, 1973) : for massive binaries it is perfectly possible to remove 70% of the stellar mass and to form pure He
stars (Paczynski, 1967). We would therefore argue in favor of the
picture that even the presumed "single" WR stars are in fact binaries
composed of a pure He-star and a neutron star companion (van den Heuvel
and de Loore, 1973; Massevich, Tutukov, Yungel'son, 1976; de Loore et al.
1975, 1977; van den Heuvel, 1976). Although it would be very difficult
to detect the binary character of such systems, a clue could perhaps be
found in the distance determinations : indeed as a consequence of the
supernova explosion involved in the formation of the neutron star, such
systems would have undergone an extra kick, and should on the average
be farther away from the galactic plane than the other WR binaries composed of a luminous star and a helium star. Tentative observations
carried out by Stenholm and Lundstrom (1977) seem to give some evidence
to this idea. In this context, HD 197406 should be mentioned : this
star is situated at z ~ 1 kpc from the galactic plane; it is probably
a runaway star, a single line spectroscopic binary with a low mass companion, accelerated out of the galactic plane ~ 5.10^6 years ago (Moffat
and Seggewiss, 1978).

b) Another solution of the problem is suggested by Chiosi et al.
(1978). After the acoustic flux regime, when the relative mass of the
core is 0.65-0.68 the star moves towards the left in the HRD. The lifetime spent in the region of large mass loss decreases with the initial
mass loss. Average mass loss rates in the acoustic flux phase increase
with luminosity. A consequence of the combination of these two effects
is that the largest mass loss occurs during the lifetime spent in the
yellow-red supergiant stage. Optimum mass loss occurs for stars with
initial masses between 18 and 35 $M_\odot$. These stars could account for the
box of WR stars. In this picture the occurrence of single WC stars
would correspond with the fraction of the core He-burning phase after
the removal of the H-rich envelope and before He-exhaustion.
However the large mass loss rates required are not in agreement with
the values of Reimers (1977) for late supergiants. Moreover, the sense
of the tracks, when rather large mass loss rates are applied, seem to
go into the wrong sense : it is hard to understand that the luminosities
will increase when extremely large mass loss rates are used. We have
calculated a sudden mass loss, for a stellar wind remnant of 17.8 $M_\odot$,
remnant of an initial 30 $M_\odot$ star. A mass loss rate of 0.004 $M_\odot yr^{-1}$ was
adopted when log T_{eff} attained a value of 3.85. Entropy losses were
taken into account. Figure 14 shows the track. The luminosity drops
and the star evolves towards the blue part of the HRD. As pointed out
by Willis and Wilson (1978) the evolution is critically dependent on
the mass loss rate. They argue that large mass loss rates for O type
supergiants and Of stars could have been higher in previous stages. In
the evolutionary computations with mass loss carried out with Carson
opacities (Stothers and Chao-wen Chiu, 1978) the mass loss rates needed

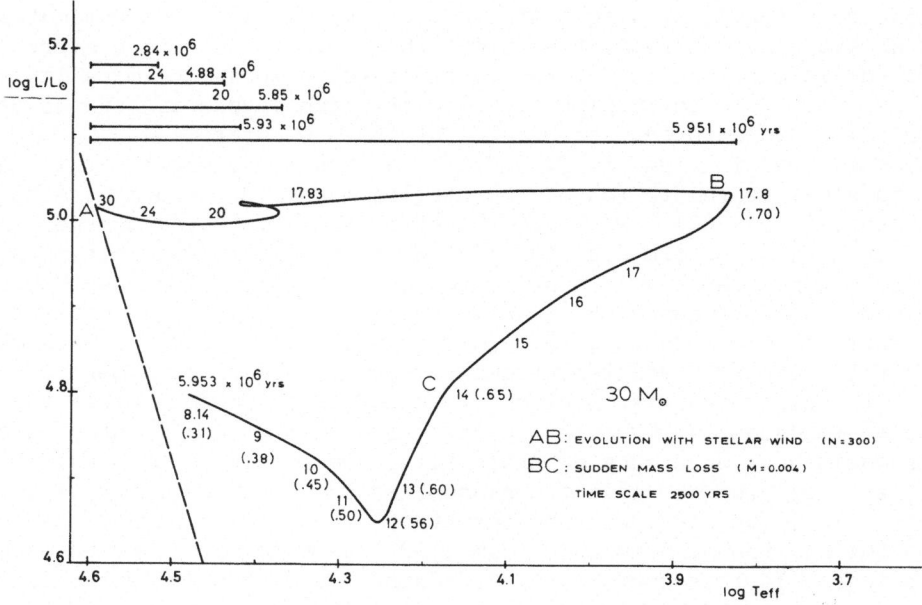

Figure 14. The evolutionary track for an initial mass of 30 M_0, first losing mass by stellar wind, leaving a remnant of 17.8 M_0. During the stage of red supergiant when log T_{eff} reached a value of 3.85 a sudden mass loss $(4.10^{-3} M_0 yr^{-1})$ was adopted, causing the star to move towards the left, with decreasing luminosity.

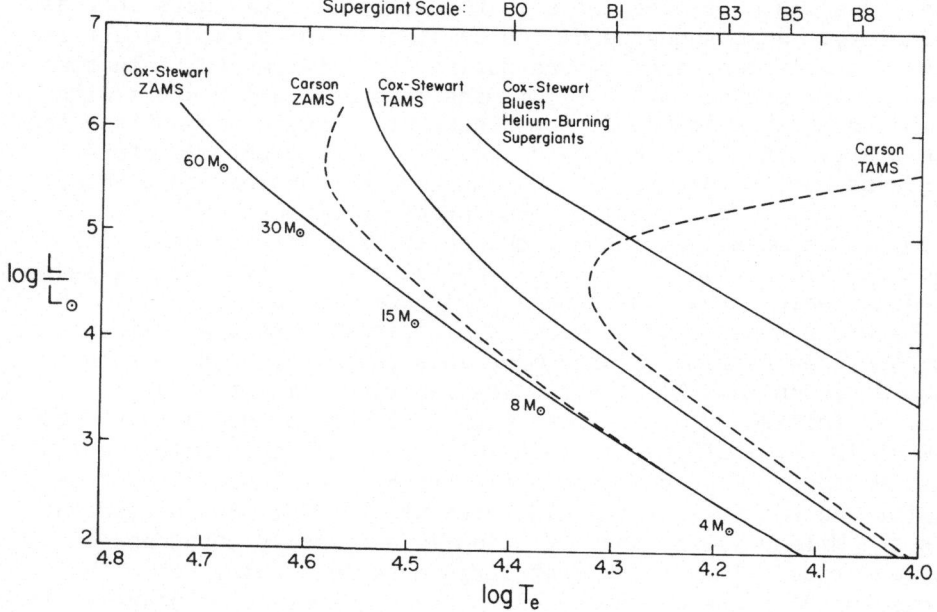

Figure 15. The ZAMS, the end of the hydrogen burning stage and the place of the bluest He-burning supergiants (Stothers, 1977).

for the explanation of the low number of very luminous blue supergiants
of spectral type later than B3 are smaller. However they lead to effective temperatures for the ZAMS that seem to be rather cool (Figure 15)
(log T_{eff} ~ 4.58). The observed spectroscopic and Zanstra effective
temperatures of O stars seem to favor smaller opacities; the Carson opacities fail to account satisfactorily for the observations of the blue
supergiants of lower luminosities.

5. EVOLUTIONARY COMPUTATIONS FOR LATE STAGES

Evolutionary computations for a 15 M_O and 25 M_O star were carried out by
Lamb, Iben and Howard (1976) from main sequence through core carbon
burning. The results of the processes of nuclear burning are dependent
on the physical conditions in the star (temperature, density as function
of radius, energy transport, and neutrino emission (pair, plasma, photo)).
A very detailed treatment of convection is required; this was done by
Lamb et al. (1976), not only for the later phases but also for previous
burning stages. It turns out that a star of 15 M_O evolves across the
Hertzsprung gap burning carbon as a red supergiant, while a 25 M_O star
finishes its C-burning at higher effective temperatures, and does not
reach the red-supergiant region. This explains not completely the minimum of M_{bol} ~ -9 for red supergiants; it would be very interesting to
repeat the calculations, including mass loss.

6. INFLUENCE OF ROTATION

The influence of rotation on the evolution of stars losingmass was computed by Sreenivasan and Wilson (1978) for a 15 M_O star. Rotation was
considered as a factor leading to enhanced mass loss by assuming that
gravity is weakened by the centrifugal force. Loss of angular momentum
is carried through in the computations and conservation of energy and
angular momentum are used to derive the spin down of such stars. Normally the effect of mass loss for an initial 15 M_O star is marginal.
Inclusion of rotation increases the mass loss rates to 3-4 10^{-7} $M_O yr^{-1}$
(ZAMS) and 4 10^{-9} $M_O yr^{-1}$ (red giant phase). The mass decreases to ~10 M_O.

7. INFLUENCE OF CONVECTION

One of the main features caused by mass loss is an increase of the main
sequence lifetime. In that respect the treatment of convection, causing
mixing and hence more efficient hydrogen burning is important. Very
recently Roxburgh (1978) developed a new convection theory, retaining
the kinetic energy flux and assuming the viscous dissipation to be small.
The calculations show that with this new treatment the convective cores
contain now 50 to 70% more mass than other models and the stars have
slightly lower luminosities. The net effect is that the extension of
the main sequence of stellar evolution is about 70% larger than in previous determination, hence the main sequence lifetime is larger by ~ 70%.

ACKNOWLEDGEMENT

The author is indebted to E.P.J. van den Heuvel, C. Chiosi, P. Conti,
J. Ziolkowski and J.P. De Grève for discussions and constructive remarks.
Drs. Chiosi, Nasi, Sreenivasan, Stothers and Chao-wen Chiu are acknow-
ledged for the permission to reproduce figures and for sending preprints.

LITERATURE CITED

ANDERSEN, J., BATTEN, A.H., HILDITCH, R.W. : 1974, Astron.Astrophys.31,1.
BARLOW, M.J., COHEN, M. : 1977, Astrophys.J. 213, 737.
BISNOVATYI-KOGAN, G.S., NADEZHIN, D.K. : 1972, Astrophys.Space Sci.15,353.
BOHANNAN, B., CONTI, P.S. : 1976, Astrophys.J. 204,797.
CASTELLANI, V., PUPPI, L., RENZINI, A. : 1971, Astrophys. Space Sci.10,136.
CASTOR, J.I., ABBOTT, D.C., KLEIN, R.I. : 1975, Astrophys.J. 195,157.
CHIOSI, C., NASI, E. : 1974, Astron.Astrophys. 34,355.
CHIOSI, C., NASI,E., SREENIVASAN,S.R. : 1978, Astron.Astrophys.63,103.
CONTI, P.S. : 1976, Mém.Soc.Roy.des Sci.de Liège, 6e Série,Tome IX,193.
CONTI, P.S. : 1978, Ann.Rev.Astron.Astrophys.
CONTI, P.S. : 1978, Astron.Astrophys.63,225.
CZERNY, M. : 1978, preprint.
DEARBORN, D.S.P., EGGLETON, P.P. : 1977, Astrophys.J. 213,448.
FUSI-PECCI, F., RENZINI, A. : 1975a, Mém.Soc.Roy.Sci.Liège, p.383.
FUSI-PECCI, F., RENZINI, A. : 1975b, Astron.Astrophys. 39,413.
FUSI-PECCI, F., RENZINI, A. : 1976, Astron.Astrophys. 46,447.
HARTWICK, F.D.A. : 1967, Astrophys.J. 150,953.
VAN DEN HEUVEL, E.P.J., DE LOORE, C. : 1973, Astron.Astrophys. 25,387.
VAN DEN HEUVEL, E.P.J. : 1976, in "Structure and Evolution of Close
    Binary Systems",p.35 ,eds.P.Eggleton et al.,Reidel,Dordrecht.
HUTCHINGS, J.B. : 1976, Astrophys.J. 203,438.
HUTCHINGS, J.B., COWLEY, A.P. : 1976, Astrophys.J. 206,469.
KUHI, L.V. : 1973, in "Wolf-Rayet and High-Temperature Stars", IAU Symp.
    No.49, eds. M.K.V. Bappu, J. Sahade, 205, Dordrecht, Reidel.
LAMB, S.A., IBEN, I., HOWARD, W.M. : 1976, Astrophys.J. 207, 209.
LAMB, S.A.: 1978, Proc. IAU Symp. No.80.
LAMERS, H.J.G.L.M., VAN DEN HEUVEL, E.P.J., PETTERSON, J.A. : 1976,
    Astron.Astrophys. 49,327.
DE LOORE, C., DE GREVE, J.P., DE CUYPER, J.P. : 1975, Astrophys.Space
    Sci. 35,241.
DE LOORE, C., DE GREVE, J.P., LAMERS, H.J.G.L.M. : 1977, Astron.Astro-
    phys. 61,251.
DE LOORE, C., DE GREVE, J.P., VANBEVEREN, D. : 1978, Astron.Astrophys.
    67,373.
DE LOORE, C., DE GREVE, J.P., VANBEVEREN, D. : 1978b,Astron.Astrophys.
    Suppl.Series, 34, 363.
DE LOORE, C. : 1970, Astrophys.Space Sci. 6,60.
MASSEVICH, A.G. : 1958, Proceedings IAU Symp. No. 10, Moscow.
MASSEVICH, A.G., TUTUKOV, A.V., YUNGEL'SON, L.R. : 1976, Astrophys.Space
    Sci. 40,115.
MASSEY, P., CONTI, P.S. : 1977, Astrophys.J. 218,431.
McCREA, W.H.:1962, Quart.J.Roy.Astron.Soc. 3,63.

MOFFAT, A.F.J., SEGGEWISS, W. : 1979, these proceedings.
MORTON, D.C. : 1967, Astrophys.J. 150,535.
PACZYNSKI, B. : 1967, Acta Astron. 17,355.
REIMERS, D. : 1977, Astron.Astrophys. 61,217.
ROSENDHAL, J.D. : 1973, Astrophys.J. 186,909.
ROXBURGH, I. : 1978, Astron. Astrophys. 65,2.
SNOW, T.P., MORTON, D.C. : 1976, Astrophys.J.Suppl. 32,429.
SREENIVASAN, S.R., WILSON, W.J.F. : 1978, in press.
STENHOLM, B., LUNDSTRÖM, I. : 1977, personal communication.
STERKEN, C. : 1977, Ph.D.Thesis, Brussels.
STOTHERS, R. : 1972, Astrophys.J. 175,431.
STOTHERS, R., CHAO-WEN CHIU : 1978, preprint.
STOTHERS, R. : 1976, Astrophys.J. 109,800.
TANAKA, Y. : 1966, Publ.Astron.Soc.Japan 18,47.
WALBORN, N.R. : 1976, Astrophys.J. 205,419.
WILLIS, A., WILSON, R. : 1979, these proceedings.
ZIOLKOWSKI, J. : 1977, in "Highlights of Astronomy", IAU General Assem-
   bly, Grenoble.

DISCUSSION FOLLOWING DE LOORE

Marlborough: Do the evolutionary calculations including
mass loss, calculated by Stothers using Carson's opacities
agree with evolutionary calculations including mass loss by
other workers such as Chiosi et al.?

de Loore: They show the same overall characteristics
as our tracks and those of Chiosi et al. in the sense that
models with both sets of opacities can explain some special
features of the HRD. However, there are differences in the
interpretation : the Carson opacities used by Stothers and
Chao-wen-Chiu are so dominant that other parameters as
thermodynamic functions and even the mass losses are of
minor importance. The influence of the Carson opacities
is so overwhelming that all other factors are not significant, and the stars expand to their largest possible configuration.

Garmany: I want to ask a question about the physics
behind the increased length of time a star spends on the
main sequence in the case of mass loss compared to the conservative case. Is the reason for this simply the decreased
mass of the star, or is it more complicated?

de Loore: It is just the decreased mass of the star.

Garmany: I ask this because I have seen the times for
the cases of different rates of mass loss, and there seems
to be a very big difference between the conservative case

and the cases of different mass loss rates.

Massey: In the slide you showed of real stars on your HR diagram, you said that the numbers listed were the "observed" masses - since at least some (if not all) of those stars are single, could you clarify what you meant by "observed"?

de Loore: The stars are selected from the list of Snow and Morton, and from spectral type and luminosity the masses are derived.

van den Heuvel: The masses in your diagram must be semi-empirical ones (i.e. not purely theoretical ones, derived from evolutionary tracks, because then you would be using a circular argument) obtained from a fine-analysis of the spectrum, for stars for which the distance is known. For such stars we have L, T_{eff} and $g = \frac{GM}{R^2}$ which allows one to determine M.

Hutchings: Please clarify what supergiant masses you used to determine N. (The X-ray primaries are all under-massive by ~ 50%, and the mass of Cen X-3 is certainly not well known at present).

de Loore: The supergiant masses were derived from effective temperatures and g-values (or radii) from ζ Ori, ε Ori, η CMa, HD 7583 and HD 33579. The optical companions of X-ray binaries were not used for this purpose.

Conti: I would like to show another numerical determination of the parameter N that Bert discussed in his paper. Stars are placed on a L,M diagram thus determining a value for N. In my paper on the companions to five X-ray binaries (Astron.Astrophys. 63, 225, 1978) I was able to determine L in four cases by appeal to the X-ray eclipse duration to give an estimate of R, and the spectrum to give an estimate of T_{eff}. The combination of these parameters gives the luminosity. The masses are found very accurately in those three cases where a pulsating neutron star is present, and somewhat less accurately in the other two by appeals to mass ratios. Placing these five stars on the L,M diagram is illustrated in the accompanying figure. The arrow for HD 153919 indicates that I may have overestimated R, hence L, if the stellar wind does indeed contribute to the X-ray eclipse duration (e.g., de Loore et al. 1978). Taken at face value, this figure indicates that for these binary objects, the value of N is near to 400 and larger on the average than shown by the single stars discussed by Bert in his paper. If this effect is real, it indicates that binaries

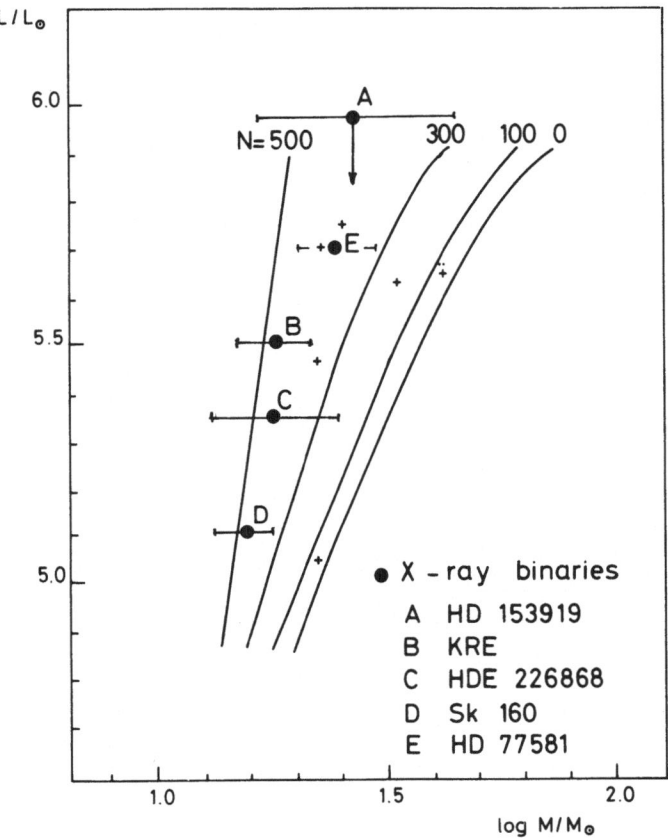

may well loose mass by a stellar wind at a more rapid rate
than single stars. This has been also suggested by Hutchings
using other kinds of data. Note that the mass loss suffered
by those four stars with neutron star companions cannot have
transferred to the collapsed object, it must have been removed from the system.

    Ziolkowski: Would you like to comment on the point we
discussed privately?

    de Loore: The idea of Dr. Ziolkowski is that N-values
of the order 100 are better in agreement with the observations than higher values. However the observed overluminosity can only be explained by rather large mass loss rates
during the hydrogen burning stage.

     <u>Bisiacchi</u>: From the observations of the number of O-type supergiants to dwarfs, we infer that only evolutionary models with large mass loss rates are able to reach, while burning hydrogen, the low gravity region of the log g, log T_{eff} plane where the supergiants lie. Our data agree with the tracks by Chiosi with $\alpha = 0.90$, which apparently corresponds to your tracks for $N = 300$.

THEORETICAL EVOLUTION OF MASSIVE STARS WITH MASS LOSS BY STELLAR WIND.

C. Chiosi, E. Nasi[†], G. Bertelli[*]
Istituto di Astronomia di Padova
† Osservatorio Astronomico di Padova, and Istituto di Astronomia di Bologna
* Unità di Ricerca G.N.A./C.N.R. di Asiago-Padova

INTRODUCTION

Although it was known since a long time that very luminous blue and red stars may show evidence of mass outflow,
it was the advent of the Copernicus satellite that clearly
ascertained all OB supergiants (and also the most luminous
main sequence stars) are losing mass at rates that are significant for their evolutionary history. The observational information for blue luminous stars (O, Of and WR) has been recently reviewed by Conti (1978), who discussed in some detail
the data available for different spectral regions. The rates
of mass loss inferred from these observations are estimated
to be in the range 10^{-7} to 10^{-5} $M_\odot$/yr for OB stars, and from
10^{-5} to 10^{-4} $M_\odot$/yr for WR stars. Several theoretical models
have been proposed to explain those high mass-loss rates, and
the high terminal velocities ranging from 1000 to 3000 km/sec.
The two basic models are: the cool radiation pressure model,
originally proposed by Lucy and Solomon (1970) and elaborated
by Castor et al. (1975), in which the envelope is accelerated
by momentum transfer from radiation to ions due to the ultraviolet line absorption; and the coronal model of Thomas
(1973), and Hearn (1975), where the wind is sustained by gas
pressure in a hot corona around the star. As both models do
not completely account for the observations, several complementary modifications have been suggested. The nowaday situation is reviewed by Cassinelli and Lamers (1978), and Conti (1978).

On the side of red supergiants, both the observational
and theoretical understanding of mass loss phenomenon unfortunately have not yet received the detailed systematic scru-
tiny afforded for luminous OB stars. The presence of circumstellar absorption lines in the spectra of cool stars is cu-
stomarily thought of as indicator of mass outflow.

For lack of knowledge about the specific mechanism responsible of the mass loss, empirical mass-loss rates have

337

P. S. Conti and C. W. H. de Loore (eds.), Mass Loss and Evolution of O-Type Stars, 337–348.
Copyright © 1979 by the IAU.

been estimated, Reimers (1975). However, the many uncertainties involved in the transformation of a circumstellar line
profile into a mass-loss rate are such that the current esti
mates for particular stars may differ by order of magnitude,
Bernat (1977).

   Amongst the others, one possible mechanism was suggested by Fusi-Pecci and Renzini (1975), who supposed the wind
be originated by a hot corona produced by a non thermal process, such as the dissipation of the acoustic energy generated in the outer convective envelope.

   The effect of mass loss, whatever might be the specific
mechanism, on the evolutionary history of massive stars was
argument of several independent theoretical investigations,
Chiosi and Nasi (1974), Chiosi et al. (1978a,b), Czerny
(1978), Dearborn et al. (1978), de Loore et al. (1977a,b).
Most of these computations are limited to the central and
shell H-burning, whereas only one attempt was made to follow
the model evolution up and through the core He-burning. For
lack of a completely satisfactory theory of mass loss in all
evolutionary phases, semiempirical and parametrized approaches were adopted, and sequences of models whose mass simply
decreases with time were computed. As long as the occurrence
of mass loss is taken into account by such a hydrostatic
treatment, the model computations are roughly equivalent even
though different specific analytical relations for the mass-
loss rate are adopted. Nevertheless, the sensitivity of the
evolutionary results to the total amount of mass removed from
the models during a given phase rises the problem of the con
sistency of the mass-loss rate with the evolutionary phase
under consideration. This is particularly true when semiempi
rical mass-loss rates, inferred from the observational data,
are used in computing the models. As an example of this, the
adoption of mass loss rates for supergiant stars, presumably
in a post main sequence stage, to compute losing mass models
in central H-burning would certainly lead to overestimate
the amount of mass lost during this phase.

   In this note we will discuss the results of Chiosi et
al. (1978a,b), and perform a comparison with those of other
authors. Moreover, by comparing the general features of the
HR diagram of luminous OB stars we try to put lower and upper limits to the mass-loss rate that has to be used in the
model computations. The estimated mass-loss rate is confined
within a range much narrower than the observational uncertainty. Finally we outline the current theoretical understan
ding about the evolutionary status of the transition WR, Con
ti (1976), and cast some light on the possibility that single stars might become single WR stars.

1. DESCRIPTION OF THE EVOLUTIONARY RESULTS

Since quite exhaustive descriptions of the results about evolutionary models incorporating mass loss already exist in the literature, in the discussion below we will report only on the most salient points.

In the model calculations of Chiosi et al. (1978a,b) the mass-loss rate was calculated from one of the three relations. At high effective temperature the evolutionary sequences were calculated using rates based on the theoretical study of Castor et al. (1975)

$$\frac{dM}{dt} = \frac{L}{cv_{th}} \quad \frac{\alpha}{\Gamma} \left[\frac{1-\alpha}{1-\Gamma}\right]^{\frac{1-\alpha}{\alpha}} (K\Gamma)^{1/\alpha} \qquad (1)$$

where the symbols have their usual meaning, α was taken as an adjustable parameter varying from 0.76 to 0.90, Chiosi et al. (1978a). In the same range of effective temperature, Chiosi et al. (1978b) adopted a mass-loss rate based on the empirical formulation of Barlow and Cohen (1977)

$$\frac{dM}{dt} = a(L/L_{\odot})^b \qquad (2)$$

Two choices for a have been explored, namely a=6.8 10^{-13} $M_{\odot}$/yr and two times this value, b was kept fixed and equal to 1.10. It is evident that the mass-loss rate of O stars of Barlow and Cohen (1977) was adopted. At low effective temperature, roughly below 3.8 in the logarithm, the mass-loss rate adopted by Chiosi et al. (1978a) is based on the theoretical suggestion of Fusi-Pecci and Renzini (1975) of an acoustically driven wind. The rate is given by

$$\frac{dM}{dt} = \epsilon \frac{L_{ac}}{v_{esc} v_s} \qquad (3)$$

which is based on the momentum conservation. In the above relation L_{ac} is the acoustic luminosity in solar units, v_{esc} and v_s are the escape and sound velocities respectively, ϵ is a free parameter (10^{-4}), Chiosi et al. (1978a). It must be said that the rates actually used in the computations were larger than those estimated by Reimers (1975) for less luminous (massive) stars, although close to the recent evaluation of Bernat (1977). Semiconvection during core H-burning, shell H-burning, and He-burning was taken into account adopting the Schwarschild and Härm (1958) condition ($\nabla_R = \nabla_a$). More details about the input physics and the computational procedure are in Chiosi et al. (1978a) and references quoted the-

rein. Several sets of evolutionary sequences were computed
for models of initial mass in the range 20 $M_\odot$ to 100 $M_\odot$ and
with Pop I initial chemical composition (X=0.700, Z=0.02).
All of them cover the core and shell H-burning phases, but
with different laws for the mass-loss rate, either (1) or
(2), while only three evolutionary sequences (20, 40, 80 $M_\odot$
of initial mass, and mass-loss rate given by eq. (1) with
α=0.9) were carried through the acoustic flux driven wind re
gime for explorative purposes.

1.1. The HR Diagram of the Core H-Burning Models

Several novel features of the HR diagram of massive
stars undergoing mass loss are evident by comparison with the
evolutionary sequences at constant mass. With moderate mass-
loss rates (10^{-7} $M_\odot$/yr for a 20 $M_\odot$ to 5 10^{-6} $M_\odot$/yr for a 100
$M_\odot$/yr on the main sequence), the core H-burning models are
less luminous than they would have been without mass loss,
and cover a wider range of effective temperature. With slight
ly larger mass-loss rates the most massive stars, approximately from 60 $M_\odot$ to 100 $M_\odot$, invert their path in the HR diagram, and long before the central H-exhaustion stage shrink
towards the main sequence. This fact is mostly due to the
lowering of the opacity in the outer layers when He-rich CNO
processed material is brought to the surface by mass loss.
This behaviour of the evolutionary sequences leads to quite
a natural explanation of the observed upper boundary of the
luminosity of OB supergiants, decreasing with decreasing effective temperature , Hutchings (1976), Snow and Morton (1976).
It goes without saying that owing to the very short time sca
le of the shell H-burning phase, the area of highest observa
bility in the HR diagram is enclosed between the zero age
main sequence and the locus of lowest effective temperature
during the core H-burning phase. As those stars are very near
to the main sequence, the location of the standard occupation
area for core He-burning models in this range of mass (60 to
100 $M_\odot$) is too red to account them, Stothers and Chin (1976).
The discussion below will clarify that, even though in
contrast with the conservative evolution, the core He-burning
of losing mass models in this range of mass is found to take
place near the main sequence, the rather short lifetime cannot account for the number of these stars which is comparable
to what it would have had on the basis of a standard initial
mass function. Moreover, the speculation is that core He-burn
ing models of this initial mass should not appear as O, Of
but rather as WN7 stars.

1.2. Internal Structure and Semiconvection

Several features of the internal structure of the core
H-burning models are worthy of mention because of their im-

plications on subsequent evolution. They are: the mass exten
sion of the semiconvective regions, the mass of the He core
existing at the time of the core He-ignition, the profile of
chemical composition throughout the models.

As it was firstly noticed by Chiosi and Nasi (1974), and
later discussed to some extent by Chiosi et al. (1978a,b), no
ne of the evolutionary sequences computed taking into account
mass loss shows the presence of semiconvection during the en
tire core H-burning phase. Semiconvection and full intermedia
te convection are restored later, although to a much less
extension than it would have occurred in constant mass models.
The analysis of Chiosi et al. (1978b) also clarifies that
mass-loss rates as low as those given by Barlow and Cohen
(1977) for O stars, likely constitute an upper limit for the
existence of semiconvection in massive stars. With the occur
rence of mass loss, the mass of the He core is significantly
smaller than for constant mass models, and decreases with in
creasing mass-loss rate. This result is relevant for the eva
luation of the yield of heavy elements per stellar generation and correlated galactic chemical evolution, Arnett (1978).
Adopting the new $M_\alpha(M_i)$ relation for losing mass models, Chio
si and Caimmi (1978) and Chiosi (1978a) estimated a significantly lower yield of heavy elements. It must be also emphasized that this result depends only on the occurrence of mass
loss during the main sequence phase, and is not fraught with
the many uncertainties inherent to mass loss during the red
phases.

1.3. The Problem of the Mass Determination

In presence of mass loss the evolutionary tracks do not
coincide with lines of constant mass in the HR diagram, and
the current procedure of assigning the mass to a single star
by means of either a mass-luminosity relation or the use of
evolutionary tracks is no longer valid. This problem was discussed to some extent by Chiosi et al. (1978a). No attempt
however was made to re-estimate the mass of individual stars
in the list of Snow and Morton (1976). The possibility of a
substantial difference between the actual mass of a star in
presence of mass loss and that it would have been assigned
by a standard procedure was taken into account by Abbott
(1978) discussing the relationship between the terminal and
the escape velocity. Amongst the others, the adoption of more realistic masses for stars undergoing mass loss revealed
that terminal velocities can be fitted by the empirical rela
tion $v_\infty \simeq 3 v_{esc}$, in agreement with the predictions from the
theory of radiatively driven stellar winds of Castor et al.
(1975). Finally, Chiosi et al. (1978a) showed how the uncertainty in the mass determination is not very severe for stars
close to the main sequence and of rather low mass, approxima
tely below initial 45 $M_\odot$, but can be as large as 40-50 % in

the case of the most luminous stars.

1.4. The Isochrones

The determination of the age of star clusters by means
of the isochrones in the HR diagram involves the bolometric
magnitude M_b and/or the spectral type Sp of the turn off
point of the core H-burning sequences, Barbaro and Chiosi
(1973). These two parameters are known to correlate primarily with the age, and secondly to the initial chemical composition X and Z. The effect of mass loss is to lower the ages
for the same M_b and Sp and fixed chemical composition below
those allowed by constant mass evolution. This fact could
partially remove the well known discrepancy between the kine
matic expansion ages and nuclear ages (estimated from evolutionary tracks).

1.5. The Generator Function of the Mass-Loss Rate

The comparison of the present results with those of other
authors, Czerny (1978), and de Loore (1977a,b) was performed
by Chiosi et al. (1978b). To this purpose several basic rela
tionships among fundamental variables $[\text{Log} <L/L_\odot> \sim \text{Log} <\dot{M}>$,
$\text{Log}(M_f/M_\odot) \sim \text{Log} <\dot{M}>$, and $\text{Log } t_H \sim \text{Log} <\dot{M}>]$ were constructed,
aimed to provide a theoretical network to which compare the
observations. Owing to this, the area covered in the HR
diagram by losing mass models in core H-burning, the amount
of mass lost by the models during the core H-burning phase,
the variation of the core H-burning lifetime, and the surface chemical abundance have been compared as functions of the
initial mass, mass-loss rate, and mass-loss rate relation.
This analysis allowed us to put plausible constraints on the
mass-loss rate and its dependence upon fundamental stellar
parameters to reproduce the observed distribution of losing
mass OB stars in the HR diagram. A good agreement is achieved
by models having mass-loss rates intermediate to those of the
set $\alpha=0.83$ and $\alpha=0.90$ of Chiosi et al. (1978a). The comparison of the theoretical rates to those inferred from the observations, Barlow and Cohen (1977),is made introducing the
concept of the generator function of the mass-loss rate dependence. In fact, it goes without saying that rate and luminosity of losing mass stars are conditioned by the occurrence of mass loss in their previous history. Hence the lumi
nosity to mass-loss rate dependence, such as suggested by the
observations, might be somewhat different from the one actual
ly involved in the mass loss process, and responsible of the
observed features. In virtue of this, and taking into account
that the maximum probability of observing a losing mass star
is expected at some intermediate stage of central H-burning,
Chiosi et al. (1978b) correlated the experimental rates of
Barlow and Cohen (1977) to theoretical values intermediate

between those of the sets α=0.83 and α=0.90 of Chiosi et al.
(1978a) when the mass-loss rate of ζ Pup is taken as a basic
constraint.

1.6. The Core He-Burning Phase

Three evolutionary sequences of the set α=0.90, namely
with initial masses 20, 40, 80 M$_\odot$ were continued through the
red stages in the HR diagram, taking into account mass loss
by acoustic flux mechanism of Fusi-Pecci and Renzini (1975),
and carried up to advanced stages of central He-burning,
which takes place at high effective temperature, Chiosi et
al. (1978a). Although the mass-loss rates of the acoustic
flux driven wind phase were higher than those extrapolated
from the empirical relation of Reimers (1975) for lower lumi
nosities, still the results below can be used as exploratory
indication of the effect of mass loss on core He-burning models.

Current ideas about the problems of the core He-burning
phase of conservative evolution have been recently reviewed
by Chiosi (1978b). The location in the HR diagram of conservative models in stationary central He-burning is known to
depend primarily upon the intermediate convective instability through its effect on the chemical profile of the models,
and secondly on details of the input physics. In the case of
the Schwarzschild and Härm (1958) condition, the results can
be schematically described as follows. In the range of mass
20 to 40 M$_\odot$ the models ignite the central He-burning as blue
supergiants, slowly moving redwards in the HR diagram. Two
areas of stationary burning with nuclear time scale, can be
identified, blue and red separated by a yellow region of se-
cular instability, thermal time scale. The relative lifetime
spent in the blue and red stages is not unique with the stel
lar mass, but is known to depend also on several factors of
the input physics, Chiosi (1978b).

Models whose mass is greater than roughly 40 M$_\odot$ do not
show the blue phase, but after a rapid run toward low effective temperature regions spend the whole core He-burning as
red supergiants. This picture was understood mostly in terms
of the effect of the H-profile on the model location in the
HR diagram, Chiosi (1978b) and references quoted therein. The
occurrence of mass loss during core H and He-burning phases
somewhat affects the above schematization, and the models of
Chiosi et al. (1978a,b) allow to draw several preliminary re
marks.

Models with significant loss of mass during the core H-
burning phase, no matter of their initial mass, move redward
on a very short time scale. Further mass loss at high rate
($\sim 10^{-4}$ M$_\odot$/yr), via the specific mechanism of acoustic flux
driven wind in our computation, in the region of low effecti
ve temperature forces the models back into the blue supergiant

area. The bluewards movement seems to take place once the
fractionary mass of the He core, M_{He}/M, gets larger than 0.6,
being the models still far from the thermal equilibrium condition. Hence the stationary He-burning in the core can initiate only after the models have reappeared as blue supergiants. Later on they slowly move to higher and higher effec
tive temperature as the central He-content is burnt out. The
fate of models with smaller mass-loss rate at low effective
temperatures, being not yet computed is still unknown, although we might expect a qualitatively similar behaviour.

Moderate mass loss during the phase of central H-burning
gives quite a different picture. Models with initial mass
greater than ~40 $M_\odot$ do not start the stationary central He-
burning at high effective temperature, but on a short time
scale move redward. Their subsequent evolution in presence of
mass loss resembles to that of the previous case. On the con
trary models with initial mass ≤40 $M_\odot$ have the stationary co
re He-burning as blue supergiants,while slowly moving redward.
The phase of low effective temperature mass loss should be
reached for these models in a very advanced stage of central
He-burning. Their later evolution not yet computed is still
uncertain.

We would like to outline as the most salient result of
these computation the discriminatory effect of mass loss on
the movement direction in the HR diagram for core He-burning
models. Moderate mass loss (blue and red stages) causes the
models in stationary central He-burning to move from blue to
red. On the contrary high mass loss (blue and red stages) ma
kes the models in stationary He-burning to move bluewards.
The preliminary nature of these computations makes the core
He-burning phase of losing mass models still fraught with ma
ny uncertainties, and demands a deeper investigation.

1.7. The Blue, Yellow and Red Supergiants

The comparison of the predicted zone of occupation for
stars evolving at constant mass and in core He-burning phase
with the experimental distribution of blue, yellow and red su
pergiants in the HR diagram was widely discussed by Stothers
and Chin (1976), and references quoted therein. It was clear
from that analysis that in the case of the Schwarzschild and
Härm criterium the models were too cool compared with the
bulk of blue stars, and also they predicted too many red supergiants of mass greater than 30 $M_\odot$. On the contrary, one of
the most remarkable features of the HR diagram of supergiant
stars in the galaxy is the lack of the very luminous red supergiants (M_b ≤ -8). Stothers and Chin (1976) suggested that
either different opacities or inclusion of mass loss in model
computations could remove the above difficulties. Models of
Chiosi et al. (1978a) showed that the core He-burning phase
of losing mass models firstly covers a broader area of the HR

diagram, greatly improving the agreement with the observation, secondly that stars more massive than 50 $M_\odot$ on the main sequence, which suffer substantial mass loss during the core H-burning and later phases, do not reach the Hayashi line, thus never becoming red supergiants. It is worth noticing that this result is due to the effect of mass loss mostly during core H-burning rather than later phases. Alhtough preliminary, these results lend some support to the idea that mass loss from massive stars could actually be a clue to achieve a satisfactory understanding of the distribution of blue, yellow and red supergiant stars in the HR diagram.

2. THE WR STARS

The majority of WR stars are known members of binary systems, but whether or not they all are binaries is not yet firmly established. The current theoretical understanding of confirmed binary WR stars recognizes in the mass exchange process the mechanism responsible for producing a WR object although many questions are still left unsolved. However, as a number of well studied objects (e.g. HD 192163, HD 50896) has no indication of binary nature, we must consider the possibility that some of the WR stars are single objects and seek other processes than mass exchange. Conti (1976) made the suggestion that O stars would become Of's if substantial wind were to exist. With increasing strength of the mass outflow, Of's would transform into "transition WR stars". Later, if further mass loss were still to occur, as suggested by the observations, Conti (1978) and references quoted therein, the transition WR's would evolve into classical WR stars. It is evident that a temporal evolutionary sequence is suggested to exist among O, Of and all WR types. The discussion below will propose a slightly different picture which is perhaps able to remove part of the difficulties present in Conti's (1976) suggestion. Reviews of the general properties of WR stars are by Conti (1976) and Paczynski (1973). However a few basic points are worth mentioning here, as they are relevant to our discussion. The position of WR stars in the HR diagram was discussed by Conti (1976). The majority of WR's are located in the box $-8 > M_b > -9$, $4.4 < \mathrm{Log}\ T_e < 4.7$, (the bolometric corrections and the color index-effective temperature scale may however introduce a considerable uncertainty), on the contrary, the transition WR's otherwise classified as WN7/WN8 are clearly apart from other subclasses as they are situated at higher luminosities and have average hotter effective temperatures. As far as the chemical composition of the surface is concerned, there is little evidence of hydrogen in most WR's, the $N(H)/N(He)$ ratios being in the range 1.-2. for WN7/WN8 types, roughly 0.4 for WN4, WN6 and very close to zero for WN5. It appears also that WN stars have more N, whereas WC stars have more C

and O at the surface. Elementary considerations suggest that
layers coming from regions of internal nucleosynthesis (CNO
for WN stars, and 3α for WC stars) are to be exposed at the
surface to explain those chemical compositions. Paczynski
(1973) suggested the peeled onion skin model that relies on
the existence of mass loss. Very presumably WC's should be
more evolved than WN's. The statistics about the frequency of
WR's among OB stars was elaborated by Smith (1973), and by
Moffat and Seggewiss (1978) for the transition WR's in parti
cular. Chiosi et al. (1978a) in the light of their results on
the evolution of massive stars with mass-loss in blue and red
phases rediscussed the problem of the theoretical interpreta
tion of WR stars.

Before proceeding further it must be emphasized that
Chiosi's et al. (1978a) scenario was aimed to suggest a plau
sible explanation of single WR's, and is not in conflict with
the mass exchange mechanism in binary systems. In their scena
rio, all single WR stars are in the core He-burning phase,
whereas the existence of different classes is interpreted in
terms of the progenitor mass and overall effect of mass loss.
As far as the progenitor star is concerned, single stars
with initial mass smaller than 20-25 $M_\odot$ would never undergo
a WR phase as a consequence of mass loss by stellar wind in
the course of their evolution. The computations in fact indi
cate that very little mass is lost on the whole, preventing
the stars from showing very low H content at the surface.
They should behave as standard main sequence and supergiant
stars but with minor changes caused by the existence of mass
loss (e.g. anomalies in the surface chemical composition,
Luck 1977a,b; 1978).

Single stars with initial mass in the range 20-25 $M_\odot$ to
40-50 $M_\odot$ would first appear as WN and later as WC during the
most advanced stages of central He-burning. This is due to
the combined effect of mass loss during the core H-burning
(radiation pressure) and He-burning (acoustic flux driven
wind, and radiation pressure). When the stationary He-burning
in the core is initiated, the models run blueward in the HR
diagram, still having a rather thin H-rich envelope at the
top of the He core. This envelope can be easily lost by the
star on a time scale shorter than the He-burning lifetime if
mass loss, presumably by radiation pressure, were to occur
at a rate compatible with the current estimate for WR stars
(10^{-5} to 10^{-4} $M_\odot$/yr). It was also suggested that single WC
stars would correspond to the fraction of the core He-burning
lifetime after the removal of the whole H-rich envelope and
before the central exhaustion stage. WN stars would occur of
course before the H-rich envelope is lost and after the models
have already moved to high effective temperature (Log $T_e \gtrsim 4.3$).
In fact, although the models were not carried until the cen
tral He-exhaustion stage, still they went very near to the
area of classical WN and WC stars. Finally, stars of initial

mass greater than 40-50 $M_\odot$ would also undergo a WR phase during their core H-burning phase but appearing only as transition WR's, as they would die before the H-rich envelope is
completely lost. The ultimate motivation of this picture
rests on the high efficiency of mass loss during core H-burning phase in massive stars. The numerical computations of
Chiosi et al. (1978a) seem in fact to indicate that when the
mass of the H-rich envelope reduces below 40% of the total
mass no red models are possible; mass loss at low effective
temperature loses part of its importance. A relative significant H-rich envelope is left and the time scale necessary to
remove it is longer than the core He lifetime. Later phases
are unrelevant to this purpose owing to their extremely short
lifetime in presence of the neutrino cooling. Our predictions
for the transition WR's is somewhat supported by the observational analysis of Moffat and Seggewiss (1978). In conclusion,
the transition WR's have evolved from progenitors of higher
mass than all other WR stars and moreover in view of their
distinct luminosity gap in the HR diagram very presumably
they would not evolve into other subclasses.

A further speculation suggested by the above picture is
the ultimate fate of stars in the range of mass 20-25 $M_\odot$ to
40-50 $M_\odot$. If we ask the question of the optical appearance
of a SN produced by the explosion of stars deprived of their
H-rich envelope, their spectrum could resemble to SNI spectrum. At the time being there is no strong observatipnal evidence of this category of type I supernovae, and further investigation is necessary to elucidate the point.

REFERENCES

Abbott, D.C.: 1978, preprint
Arnett, D.W.: 1978, Astrophys. J. <u>219</u>, 1008
Barbaro, G., Chiosi, C.: 1973, Astron. Astrophys. <u>28</u>, 7
Barlow, M.J., Cohen, M.: 1977, Astrophys. J. <u>213</u>, 737
Bernat, A.P.: 1977, Astrophys. J. <u>213</u>, 756
Cassinelli, J.P., Lamers, N.J.G.L.M.: 1978, preprint
Castor, J.I., Abbott, D.C., Klein, R.I.: 1975, Astrophys. J.
        <u>195</u>, 157
Chiosi, C.: 1978a, preprint
Chiosi, C.: 1978b, in Proc. I.A.U. Symposium N° 80, Washing-
        ton D.C., in press
Chiosi, C., Caimmi, R.: 1978, preprint
Chiosi, C., Nasi, E.: 1974, Astron. Astrophys. <u>34</u>, 355
Chiosi, C., Nasi, E., Sreenivasan, S.R.: 1978a, Astron.
        Astrophys. <u>63</u>, 103
Chiosi, C., Nasi, E., Bertelli, G.: 1978b, submitted to
        Astron. Astrophys.
Czerny, M.: 1978, preprint
Conti, P.S.: 1976, Mem. Soc. Roy. Sci. Liège 6° Serie 9, 193

Conti, P.S.: 1978, Ann. Rev. Astron. Astrophys., in press
Dearborn, D.S.P., Blake, J.B., Haineback, K.L., Schramm, D.N.:
        1978, Astrophys. J. in press
de Loore, C., De Grève, J.P., Lamers, H.J.G.L.M.: 1977a,
        Astron. Astrophys. 61, 251
de Loore, C., De Grève, J.P., Vanbeveren, D.: 1977b, preprint
Fusi-Pecci, F., Renzini, A.: 1975, Mem. Soc. Roy. Sci. Liège
        6° Serie 8, 383
Hearn, A.G.: 1975, Astron. Astrophys. 40, 355
Hutchings, J.B.: 1976, Astrophys. J. 203, 438
Luck, R.E.: 1977a, Astrophys. J. 212, 743
Luck, R.E.: 1977b, Astrophys. J. 218, 752
Luck, R.E.: 1978, Astrophys. J. 219, 148
Lucy, L.B., Solomon, P.M.: 1970, Astrophys. J. 159, 879
Moffat, A.F.J., Seggewiss, W.: 1978, preprint
Paczynski, B.: 1973, in Wolf Rayet and High Temperature Stars,
        Ed. M.K.V. Bappu and J. Sahade (Dordrecht, Reidel)
        p. 143
Reimers, D.: 1975, Mem. Soc. Roy. Sci. Liège 6° Serie 8, 366
Schwarzschild, M., Härm, R.: 1958, Astrophys. J. 128, 348
Smith, L.F.: 1973, in Wolf Rayet and High Temperature Stars,
        Ed. M.K.V. Bappu and J. Sahade (Dordrecht, Reidel)
        p. 15
Snow, T.P., Morton, D.C.: 1976, Astrophys. J. Suppl. 32, 429
Stothers, R., Chin, C.W.: 1976, Astrophys. J. 204, 472
Thomas, R.N.: 1973, Astron. Astrophys. 29, 297

DISCUSSION FOLLOWING CHIOSI, NASI and BERTELLI

de Loore: How can you explain that during the acoustic
flux regime mass loss the luminosity increases? I should
presume that due to entropy losses the luminosity drops. I
would refer to our computations for a 30 $M_\odot$ star, lowered to
17.8 $M_\odot$ owing to stellar wind, where at the red giant stage
a mass loss rate of 4.10^{-3} $M_\odot yr^{-1}$ was applied. There the
track goes down as it should. I think that energy losses
should be taken into account.

Chiosi: I agree with you and I believe that a fully
hydro-dynamical approach would be much more appropriate.

Mazurek: Type I supernovae have large expansion velocities, implying a relatively low mass (~ 1 $M_\odot$). You are making
Type I supernovae out of massive (~ 50 $M_\odot$) stars. How can
you justify this?

Chiosi: In my picture I was trying to suggest that a new
kind of Type I supernovae might be originated by rather massive stars say between 30 to 60 $M_\odot$. This suggestion is however
very speculative as we do not yet have any observational support for this. The statistics should be in any case rather poor

CRITICAL RATES OF STELLAR MASS LOSS

D. S. P. Dearborn and J. B. Blake
University of Arizona and The Aerospace Corporation

INTRODUCTION

Many of the effects of mass loss on OB stars have now been explored. Mass loss will cause a star to be overluminous for its mass (though less luminous than a star of its original mass) and, for moderate mass-loss rates, the luminosity decreases at the same rate as the mass contained in the convective core decreases causing the main sequence lifetime to remain unchanged (Chiosi and Nasi 1974, 1978, Deloore, DeGreve and Lamers 1977, Dearborn, Blake, Hainebach and Schramm 1978). Mass loss can also expose layers where ^{14}N has been enhanced via the CNO tricyle (Dearborn and Eggleton 1977) and, in extreme cases, can produce a stripped helium core resembling a Wolf-Rayet Star (Hartwick 1967). While many of these phenomena (in particular the composition change) are more sensitive to the total mass removed than the formalism used to represent the mass loss, significant differences will result for the same average mass-loss rate depending on whether the mass was removed early (near the ZAMS), or late (near core hydrogen depletion). In addition, there appears to be a critical mass loss rate which depends on initial mass and separates those models which continue to evolve in a relatively normal (though subluminous) manner, and those models which evolve to a Wolf-Rayet configuration.

CALCULATION

In earlier work (Dearborn et at. 1978) over thirty stellar tracks were evolved with different rates of mass loss for stars of 15, 30, and 60 $M_{\odot}$. A mass loss formula similar to that of McCrea (1962) was used:

$$M = -\mu \, \frac{\phi(T_e) \, L \, R}{M} \qquad \phi(T_e) = \frac{\pi}{\sigma T_e^4} \int_o^{\lambda_c} B(\lambda, T_e) d\lambda$$

where L R and M are the luminosity, radius, and mass in solar units. The constant μ is used to scale the mass loss rate and B is the Planck function. This formula differs from McCrae's in that it assumes only the luminous flux shortward of λ_c (2000Å) is effective at driving mass loss.

349

P. S. Conti and C. W. H. de Loore (eds.), Mass Loss and Evolution of O-Type Stars, 349–356.
Copyright © 1979 by the IAU.

The evolution of a number of the more interesting models has now been followed into carbon burning. At this point, the time remaining prior to core collapse (Arnett 1973) and resulting supernova is too short for mass loss to play any further role.

RESULTS

In order to better understand our calculations we begin by constructing a framework on which to compare the models. For this it is necessary to consider three timescales, the main sequence timescale without mass loss (T_{ms}), the actual time spent on the main sequence (T), and the average mass-loss timescale ($T_{\dot{m}} = T M_i/(M_i-M_f)$, where M_i and M_f are the masses at the beginning and end of the main sequence). Figure 1 is a diagram on which the vertical axis (T/T_{ms}) indicates how much mass-loss extends the life a model, and the horizontal axis ($T_{ms}/T_{\dot{m}}$) is an indicator of the mass-loss rate. There is to the right of this diagram an excluded region requiring total depletion of the mass of the star. The curved line on the diagram then shows the effects of different initial mass-loss rates on a 30 $M_\odot$ model.

As the mass-loss rate increase from zero the main sequence lifetime remains approximately the same until a critical mass-loss rate of

$$M_{cr} = 7.65 \times 10^{-10} \left(\frac{M}{M_\odot}\right)^{2.35} M_\odot /yr$$

is reached. At this point the lifetime of the model begin to rapidly increase. Attempts to increase the mass-loss rate beyond this point cause the luminosity

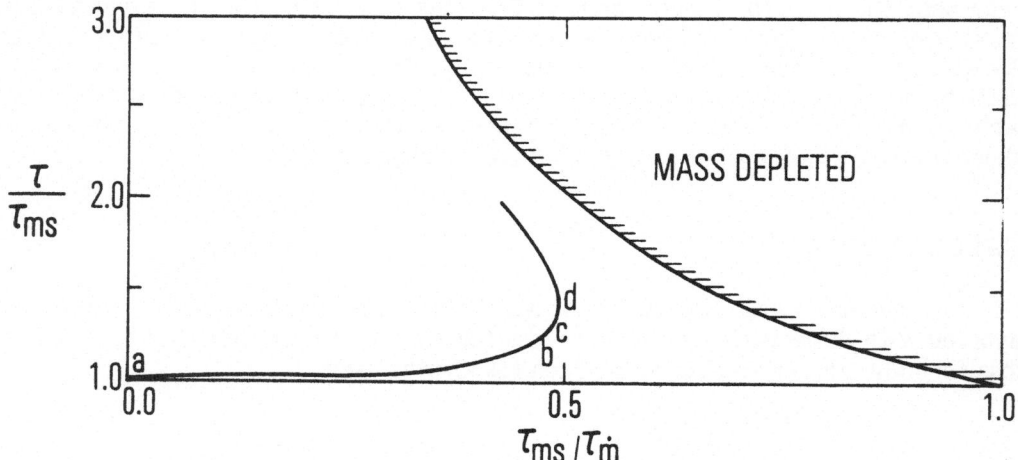

Fig. 1.   This figure compares the mass-loss rate ($T_{ms}/T_{\dot{m}}$) to the lifetime of the mass losing star (T/T_{ms}) for a 30 $M_\odot$ star. For low to moderate rates of mass loss, the main sequence lifetime is unaffected. Eventually critical mass-loss rate is reached at which the lifetime begins to change. The positions of models a, b, c and d are marked.

of the star to decrease resulting in a lower mass-loss rate. If the mass-loss rate were not allowed to decrease, mass-loss rates higher than critical mass-loss rate would deplete the star of all mass.

In our case where mass-loss rate is allowed to decrease with luminosity, mass-loss rates initially higher than the critical mass-loss rate lead to the upper branch of the curve in Figure 1 which parallels the mass depleted region. The models on this upper branch are essentially stripped helium cores. This critical mass-loss rate then divides the models which evolve to Wolf-Rayet like configurations from those which evolve more normally.

Table 1 shows the physical parameters of some 30 $M_\odot$ models which were evolved to a presupernova configuration. Given are the scaling constant μ, the initial mass-loss rate $\dot{M}_i$, the final mass-loss rate (and the maxmum in case b) $\dot{M}_f$, the main sequence lifetime (T), the resulting core mass M_c, and the final stage. According to Conti (1978), the mass-loss rates used here to produce a Wolf-Rayet configuration are too low to produce a signature in the visible spectrum of a star, and therefore while these mass-loss rates are high, they are not unreasonable.

TABLE I

Model	μ	$\dot{M}_{initial}$	$\dot{M}_{final}$	$\overline{M}$	T	M_c	Final State
a	$0(\times 10^{-1})$	$0(\times 10^{-6})$	$0(\times 10^{-6})$	$0(\times 10^{-6})$	$6.0(\times 10^{6})$	11.5	Red Giant
b	2	1.7	4.3	2.5	7.2	8.2	Red Giant
c	3	2.5	2.8	2.6	8.4	4.5	Wolf-Rayet
d	4	3.4	1.7	2.7	9.2	3.8	Wolf-Rayet

Mass loss in the red region was not considered, and may alter the final evolution of models a and b. To have a significant effect however, the mass-loss rate must be very high ($\sim 10^{-4}$ $M_\odot$ /yr) and the star must spend sufficient time at this mass-loss rate to remove the envelope. The amount of time a star spends in the red prior to core collapse depends sensitively on the treatment of semi-convection, and is not a settled question. It seems however entirely possible that mass loss in the red could strip the remaining envelope off of model b resulting in a Wolf Rayet configuration.

Figure 2 shows the evolutionary tracks of models a, b, and c. Model b is subluminous compared to model a and the surface abundance of ^{14}N is enhanced 3 to 4 times its original value. Model c is substantially lower in luminosity than model a. In addition it undergoes two stages of ^{14}N enhancement in layers which are exposed where ^{12}C and then ^{16}O were converted to ^{14}N. This model begins as an O star, and evolves through an OB-N stage, an OB-N stage with ^{16}O deficient, a WN stage, a WC stage and finally a supernova. The temperature predicted by the models during the Wolf Rayet stages are unrealistically high because of the assumed boundary conditions. This does not affect the evolution of the models, but the surface temperatures are uncertain. Models c and d did supernova in the blue, and, not having a 10^{14} cm envelope to filter the energy, may have produce an x-ray or hard UV supernova such as has been proposed for the progenitor of the Cas A SNR (Chevalier 1976, Lamb 1978).

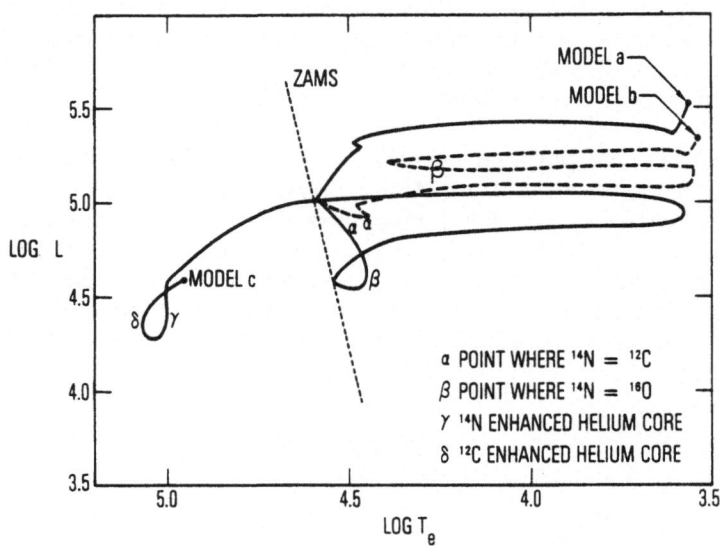

Fig. 2.   This figure shows tracks in an H-R diagram for models a, b and c. The points are marked at which composition changes occur.

In addition a 60 $M_\odot$ star was evolved with a mass-loss rate sufficient to place it on the upper branch in Figure 1. It evolves directly to a Wolf-Rayet configuration without ever evolving to the red.

CONCLUSIONS

A critical mass-loss rate exists above which models result in a Wolf-Rayet configuration. Figure 3 shows the relation between initial mass and resulting helium core mass for stars losing mass at this critical rate. This shows that without mass-loss in the red, a 40 $M_\odot$ star is required to produce a 10 $M_\odot$ Wolf-Rayet star. Of course if sufficient mass is removed in the red, a 10 $M_\odot$ stripped core can result from a lower initial mass star which has lost mass at a rate below the critical rate during its main sequence lifetime.

In whatever manner a Wolf-Rayet star is produced, mass-loss rates of $\sim10^{-5}$ $M_\odot$/yr such as have been reported here will have significant effect on the evolution of the star. Over the helium burning lifetime such a rate will remove the ^{14}N-rich helium envelope and expose ^{12}C-rich material once contained in the helium-burning core convection zone.

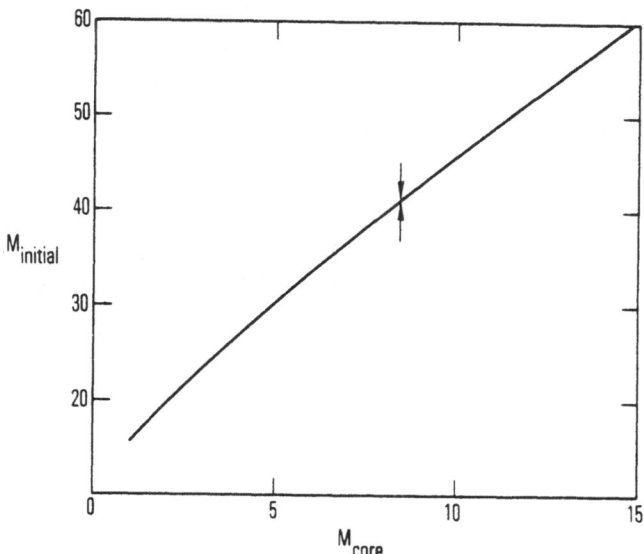

Fig. 3.   This figure shows the resulting core mass for stars losing mass at the
critical rate. A mass-loss rate of 10^{-5} $M_\odot$ /yr would expose ^{12}C
rich layers in all of these cores over their helium burning lifetime.
In fact such a mass-loss rate would cause stars less massive than
those indicated by the arrow to be completely depleted of mass
over their helium burning lifetime.

These ^{12}C-rich zones also can contain significant enhancements of ^{16}O
from ^{12}C (α,γ) ^{16}O. In the WC models we evolved, the ^{12}C achieved a mass
fraction of ~0.08, while in the WN models the ^{14}N was enhanced only about 10
times normal.

Mass loss has a number of interesting effects on the nucleosynthesis in
massive models. Because mass-loss reduces the core mass, less ^{14}N is trapped in
the core for additional processing. Models which lose sufficient mass to become
stripped cores yield about two times as much ^{14}N as a conservative ($\dot{M}=0$) model.
The smaller core results in a smaller yield of metals to helium,because
relatively more metals are trapped in the collapsed core during supernova. Model c yields $\Delta Y/\Delta Z$ of 3.0 in the ejected material instead of 0.5
obtained in case a. Due to both the smaller core size and the fact that
mass loss causes the convective helium burning core to retreat, the C/O
ratio is enhanced. Also when mass loss produced a WC star, it ejects a
region which is ^{18}O enhanced (via ^{14}N (α,γ) ^{18}O). Such an ^{18}O rich region
exists in all massive stars, but there is some question as to whether it
can survive the supernova shock without additional processing (Truran
1976, Dearborn, Tinsley and Scramm 1978). In WC stars we are probably
seeing this material ejected.

Quantitative observations of the composition in OB stars can yield useful
information on the amount of mass that has been lost. A ^{14}N enhancement of 3

to 5 times indicates a region has been exposed in which ^{12}C has been converted to ^{14}N. Higher ^{14}N enhancements requires much more extensive mass-loss in order to expose material originally in the convective core. It would be particularly interesting to look for ^{14}N in x-ray binaries where it is believed significant amounts of mass are lost. It is however important to make quantitative determinations in order to know at what level we are seeing (or not seeing) an anomaly.

One of the authors (DD) wishes to acknowledge the Smithsonian Foreign Currency Program for support during part of this work. In addition we wish to thank Kitt Peak National Observatories for providing part of the computing time. This work was supported in part by The Aerospace Corporation on company-financed funds.

REFERENCES

Arnett, W. D.: 1973, Explosive Nucleosynthesis, ed. Schramm and Arnett, Univ. of Texas Press, Austin.

Chevalier, R.: 1976, Ap. J. 208, 826.

Chiosi C. and Nassi, E.: 1974, Astr. and Ap. 34, 355; 1978; Astr. and Ap., 63, 103.

Conti, P.: 1978, preprint.

Dearborn, D. and Eggleton, P.: 1977, Ap. J. 213, 448.

Dearborn, D., Blake, B., Hainebach, K. and Schramm, D.: 1978, Ap. J. July 15th, in press.

Dearborn, D., Tinsley B., and Schramm, D.: 1978, in press.

DeLoore, C., DeGreve, J., and Lamers, H.: 1977, Astr. and Ap., 61, 251.

Hartwick, F.: 1967, Ap. J., 150, 953.

Lamb, S.: 1978, Ap. J., 220, 186.

McCrae, 1962, Quart. J. Roy. Astr. Soc., 3, 63.

Truran, J.: 1976, CNO Isotopes in Astrophysics, ed. J. Audouze, D. Reidel Publishing Co. Dordrecht.

DISCUSSION FOLLOWING DEARBORN

Vanbeveren: Did I understand you correctly: did you
follow really the CNO cycle step by step, layer by layer?

Dearborn: Yes.

Vanbeveren: In this case, how long does it take to reach
the equilibrium abundances in the core during hydrogen core
burning?

Dearborn: ^{12}C-^{14}N equilibrium is achieved throughout the
core within 5% of the main sequence lifetime. The ^{16}O approaches equilibrium over a longer period. The envelope of
course contains CNO processed material, but not in equilibrium.

Sreenivasan: Could you tell us the rate of mass loss you
used in the track which makes two somersaults in the HR diagram?

Dearborn: The mass loss rate varied from $2.5 \ 10^{-6}$ to
$2.8 \ 10^{-6} \ M_o yr^{-1}$ with a time averaged mass loss rate of
$2.6 \ 10^{-6} \ M_o yr^{-1}$.

Lamb: Given the rates for the triple-α and $^{12}C(\alpha,\gamma)^{16}O$
reactions wouldn't you expect the abundances of ^{16}O to be
higher than that of ^{12}C in the later phases of mass loss from
your model stars?

Dearborn: What you say would be correct for complete
helium burning, but the region exposed here has only partial
converted ^{4}He into ^{12}C and ^{16}O. In helium burning, ^{12}C is
initially enhanced strongly over ^{16}O. Only after the ^{12}C is
enhanced and ^{4}He begins to deplete does ^{16}O begin to overtake
^{12}C.

Chiosi: Would you please specify what θ^2_α was used in
the computations of the 3α reaction network.
Secondly, I would like to know the fraction of lifetime spent
by your models between the stage of complete removal of the
H-rich envelope and the stage when 3α processed material is
brought to the surface.

Dearborn: I used $\theta^2_\alpha \approx 0.06$. The fraction of the lifetime spent after the removal of the hydrogen-rich envelope
depends on the mass loss rate. A higher mass loss rate produces a smaller core which has a longer helium-burning lifetime.

Van Dessel: I noticed you situated the WR stars in the region log T_{eff} = 5.0. This kind of high effective temperature is not confirmed by observational data. Could you comment?

Dearborn: The high effective temperatures are due to the static atmospheric model used as a surface boundary contition. A more accurate hydrodynamic approximation would result in lower effective temperatures. As this does not effect the nucleosynthesis and evolution of the interior I have not worried about it.

Moffat: You state that WN stars can evolve into WC stars. Does this also apply to WN 7/8 stars too, which appear to lie in a region of the HR diagram quite distinct from the remaining WN stars?

Dearborn: In any core helium burning star the carbon enhanced region is not that far below the hydrogen-helium discontinuity. If WN 7/8 stars are in this stage of evolution, the observed rates of >10^{-5} $M_\odot$/yr should expose ^{12}C rich material in less than the core helium burning lifetime.

Massey: What are the lifetimes you get for the WN's compared to the WC's?

Dearborn: The relative lifetime of these two stages depends on the mass loss rate adopted. In this case the star spent six times as long as a WN star than as a WC star. The mass loss rate was however much lower than those talked of here. A higher mass loss rate could easily cause the WC lifetime to equal the WN lifetime.

SUPERGIANT MASS LOSS AND THE CASSIOPEIA A PROGENITOR

Susan A. Lamb
University of California, Los Angeles

Observations of the chemical abundances in young supernova remnants may be used, in some circumstances, to place constraints on the evolution of the progenitor stars. For example, if a progenitor was massive ($M > 10$ $M_\odot$), the presence of high $^{14}N/^1H$ ratios (that is, more than five time the solar value) in the supernova remnant can imply that substantial mass loss took place during the star's early evolution, that is, while it was an early-type supergiant.

An example is the supernova remnant Cassiopeia A, which has a non-solar composition and a $^{14}N/^1H$ ratio which is very large in some regions. A comparison of the observed abundances in Cas A with the abundances predicted from theoretical calculations of the evolution of massive stars indicates, firstly, that the progenitor star was massive and, secondly, that substantial mass loss took place while it was an early-type supergiant. The rate of mass loss required in this case is consistent with the rates observed in early-type supergiants.

I. INTRODUCTION

The observed chemical abundances in young supernova remnants can be used to place constraints on the masses and evolution of the progenitor stars. Detailed stellar evolutionary calculations have now been performed for a wide range of initial masses, and in some cases, the evolution has been followed through much of the stars' life. Although the composition in a supernova remnant is likely altered to some extent by the supernova explosion itself, there are circumstances in which a comparison between the compositions predicted by stellar evolution theory alone and those observed can be meaningful.

P. S. Conti and C. W. H. de Loore (eds.), Mass Loss and Evolution of O-Type Stars, 357–365.

Only the abundances in those supernova remnants that
have swept up little interstellar material can be compared
directly with stellar evolution theory, and very few such
objects are known in our galaxy, Cas A and Kepler being
the best examples. Perhaps the best studied remnant
abundances are those in the Crab and the Cas A supernova
remnants (see Davidson, 1978, and Chevalier and Kirshner,
1978). Arnett (1975) has explored some of the implications
of these two supernova remnants for the nature of the
progenitor stars. He compared his models for 4 $M_\odot$ and
8 $M_\odot$ 'helium cores', which correspond to stars with masses
of approximately 15 $M_\odot$ and 25 $M_\odot$, respectively, with earlier
abundance determinations for these objects by Davidson
(1973) and by Peimbert (1971). He found, for example, that
the abundances in some regions of the Cas A supernova
remnant (namely, the "fast-moving knots") are consistent
with the abundances obtained in his 8 $M_\odot$ 'helium core'
model as it nears the end of its quasi-static evolution.
The Cas A supernova remnant is particularly interesting
in that it has at least two distinct types of optically
emitting regions. The 'fast moving knots' are observed to
be traveling outwards from the center of the region at
about 10^4 km/sec, whereas the 'quasi-stationary flocculi'
have only a small outward velocity $\sim$ 150 km/sec superimposed
on large random velocities (see Kamper and van den Bergh,
1976). The chemical compositions of these two types of
emitting region are very distinct. The 'fast moving knots'
have no observed hydrogen but are enhanced in oxygen and
the burning products of oxygen, while the 'quasi-stationary
flocculi' are overabundant in nitrogen and helium (see
Chevalier and Kirshner, 1978).

It has previously been suggested that the material in
the 'fast moving knots' was ejected from the stellar interior
during the supernova explosion, and that the 'quasi-
stationary flocculi' are formed from material that was shed
from the surface of the star prior to the supernova
explosion (see Peimbert and van den Bergh, 1971, and
Chevalier, 1976). The dynamics of the 'quasti-stationary
flocculi', as determined by Kamper and van den Bergh (1976),
imply that the material was lost from the stellar surface
at least 10^4 years ago.

The abundances in the 'fast moving knots' imply that
the progenitor star had an initial mass of at least 9 $M_\odot$
and possibly was much more massive (see Arnett, 1975,
Chevalier and Kirshner, 1978, and Lamb, 1978a).

Using these suggestions concerning the origin of the
material in the Cas A remnant and the abundances as
determined by Chevalier and Kirshner (1978), Lamb (1978a)

has constructed a consistent evolutionary picture for the
Cas A progenitor star which involves two periods of mass
loss, one while the star was an early-type supergiant
(during its early evolution) and the other when the star
was considerably more evolved.

The argument for mass loss while the progenitor was
an early-type supergiant hinges on the high N/H ratios
observed in the 'quasi-stationary flocculi', as is explained
in Section II, where a comparison between this observed
ratio and those predicted by evolutionary calculations of
massive stars is presented. In Section III we estimate
the required mass loss rates and compare them with those
observed for early-type supergiants. Finally, in Section
IV we give a brief summary of our conclusions.

II. THE EARLY EPISODE OF MASS LOSS

If the 'quasi-stationary flocculi' consist of material
shed from the progenitor star prior to the supernova
explosion, then one would expect their chemical composition
to be very similar to that at the surface of the appropriate
star at some phase in its evolution. Concentrating on the
firmest abundance determinations for the 'quasi-stationary
flocculi', which indicate that the N/H ratio is approximately
10 - 20 times the 'solar' value and that the He/H ratio
is also approximately an order of magnitude larger than
the 'solar' value (Chevalier and Kirshner, 1978), we find
that at no time during the evolution of massive stars do
such high relative abundances of nitrogen and helium appear
at the stellar surface (see Lamb, Iben, and Howard, 1976),
if there is no mass loss.

In massive stars of constant mass, 'non-solar'
abundance ratios only appear at the surface once a convective
envelope has formed. In the 15 $M_\odot$ stars of Endal (1975)
and Lamb, Iben, and Howard (1976), this occurs near the
end of core helium burning, as the star evolves to the red
across the Hertzsprung gap. However, in the 25 $M_\odot$ model
of Lamb et al., no convective envelope develops during
the evolution through core-carbon burning. Even in the
15 $M_\odot$ model, the N/H ratio does not climb to more than five
times the solar value in the convective envelope as the
convection eats down into layers partially processed through
core-hydrogen burning.

To find a region within a massive star where the
nitrogen and helium abundances with respect to that of
hydrogen are consistent with those found in the 'quasi-
stationary flocculi' one must look further back in the

evolution, to the period of core hydrogen burning, that is
to the main sequence evolution. At this time deposits of
hydrogen depleted material enriched in nitrogen and
helium are laid down outside the contracting convective
core (see Lamb, Iben, and Howard, 1976, and Lamb, 1978a).

An example of the composition profiles through a
massive star at a time near the end of core-hydrogen
burning is given in Figure 1, which shows a 50 $M_\odot$ star
(Lamb, 1978b) with an initial composition of $Y = 0.28$ and
$Z = 0.02$. In the region between $M_r \sim 32\ M_\odot$ and $M_r \sim 41\ M_\odot$,
nitrogen and helium have been built up at the expense of
carbon, oxygen, and hydrogen. As the evolution progresses,
the abundances of nitrogen and helium in this region climb
with respect to that of hydrogen, which continues to be
depleted. Eventually the N/H ratio rises to approximately
ten times the solar value in a region which expands inwards
from $M_r \sim 41\ M_\odot$. (Analogous composition profiles for
15 $M_\odot$ and 25 $M_\odot$ stars are presented in Figures 1 and 2
of Lamb, 1978a, for a time shortly after the end of core-
hydrogen burning; by this stage the N/H ratio has risen
to approximately ten times the 'solar' value.) If this
nitrogen and helium enriched material is to appear at the
stellar surface undiluted by the overlying hydrogen
envelope, this outer hydrogen layer must be shed from the
star prior to the development of a convective envelope.
This means that the mass loss epidsode must have taken
place to the blue of the Hertzsprung gap, that is, at the
time when the star was successively an O, B, and A type
supergiant. In the next section we estimate the mass loss
rates which are required to remove the outer hydrogen
envelope in the available time and compare these with
mass loss rates determined from observations of early-type
supergiants.

III. MASS LOSS RATES

A comparison of the abundance profile for a 50 $M_\odot$ star
with those for 15 $M_\odot$ and 25 $M_\odot$ stars (see Figures 1 and 2
of Lamb, 1978a) indicates that in all three stars the
amount of material which must be shed from the stellar
envelope to expose the nitrogen enriched material is
approximately 9 $M_\odot$. Mean mass loss rates can be computed
using the known times available for the mass loss. The
relevant evolutionary time scales for the 15 $M_\odot$ and 25 $M_\odot$
stars are given in Lamb, Iben, and Howard (1976) and yield
mean mass loss rates of $\sim 6 \times 10^{-7}\ M_\odot\ yr^{-1}$ and $\sim 1 \times 10^{-6}$
$M_\odot\ yr^{-1}$, respectively. The time available to the 50 $M_\odot$ star
to lose the 9 $M_\odot$ of material is approximately 4.5×10^6
years, which yields a mean loss rate of $\sim 2 \times 10^{-6}\ M_\odot\ yr^{-1}$.

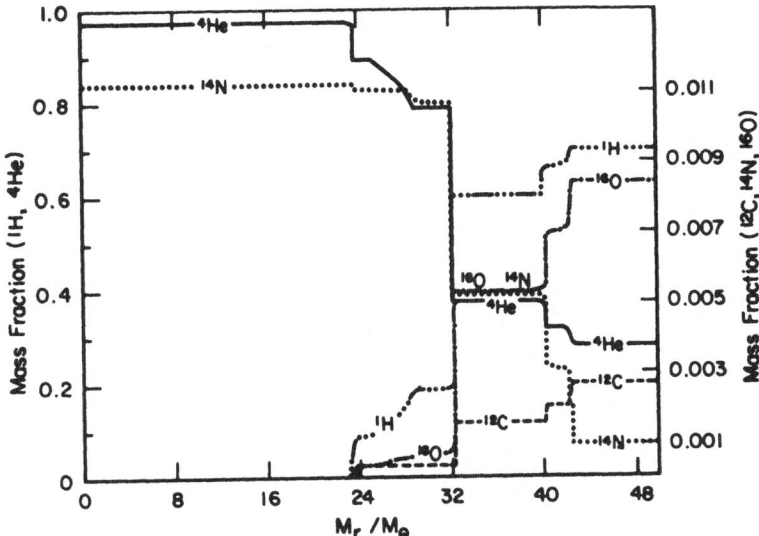

FIG. 1. - Compositional profiles within a 50 M$_\odot$ star near the end of core hydrogen burning. Abscissa, Lagrangian mass co-ordinate; ordinate, mass fraction.

These mass loss rates can be compared with those found observationally for early-type supergiants.

Estimates of mass loss rates for a large sample of O and B supergiants have been obtained optically by Hutchings (1976), who found a large spread in mass loss rates within his sample. For the region of the H-R diagram through which stars in the mass range 15 M$_\odot$ to 50 M$_\odot$ are likely to evolve, namely between M$_{bol}$ ∿ - 6.0 and M$_{bol}$ ∿-9.5, he found mass loss rates ranging from < 10^{-7} M$_\odot$ yr^{-1} to ∿ 7 x 10^{-6} M$_\odot$ yr^{-1}. In another study, Barlow and Cohen (1977) have obtained mass loss rates for a sample of O, B, and A supergiants using infrared observations of the stars together with a velocity law for the mass outflow from P Cygni. The latter was derived from previously published radio and infrared data. A least squares fit to their mass loss rates for O stars yielded the expression

$$\dot{M} = 6.8 \times 10^{-13} \, L^{1.10 \pm 0.06} \, M_\odot \, yr^{-1}.$$

When applied to the maximum luminosity attained by the 50 M$_\odot$ model of Lamb (1978b), this expression yields a mass loss rate of 2.2(+2.8,-1.2) x 10^{-6} M$_\odot$ yr^{-1}. Using Barlow and Cohen's expression for the rate of mass loss for B and A supergiants and the 15 M$_\odot$ and 25 M$_\odot$ models of Lamb, Iben, and Howard (1976) Lamb (1978a) obtained mass loss rates of

1.9 (+2.6, -1.1) x 10^{-7} M yr^{-1} and 1.0 (+ 1.6, - 0.6) x
10^{-6} M yr^{-1}, respectively. Thus, the mass loss rates
required for any of the three model stars (15 M, 25 M,
and 50 M) are consistent with those observed for the
relevant types of supergiants.

The loss of $\sim$ 9 M or more of material from the outer
envelope of any massive star during core-hydrogen and core-
helium burning will significantly affect the star's
evolution. Evolutionary calculations for stars including
various amounts of mass loss have been calculated by
Hartwick (1967), de Loore, De Grève, and Lamers (1977),
Chiosi, Nasi, and Sreenivasan (1978), among others. These
studies indicate that mass loss extends the stellar lifetime,
as well as altering the internal structure of the star. The
semiconvective region which occurs above the convective core
during core-hydrogen burning is reduced in extent, and the
convective core itself is smaller than would otherwise be
the case. This implies that the total amounts of ^{14}N and
^{4}He left outside the contracting convective core during core-
hydrogen burning are smaller when there is mass loss, and
that the distance to which the deposits extend outward is
also smaller. However, the modest mass loss rates required
to explain the Cas A abundances are not expected to
drastically alter either the internal structure or composition.

IV. SUMMARY

The Cas A progenitor was probably a massive star that
lost at least 9 M of its outer envelope during the core-
hydrogen and core-helium burning phases of its evolution,
that is, while it was an early-type supergiant. This mass
loss uncovered nitrogen and helium rich material at the
stellar surface. Later mass loss from the star (which
nevertheless took place at least 10^{4} years prior to the
supernova explosion) supplied the material which is now in
the 'quasi-stationary flocculi'. Thus two periods of mass
loss are required to form a consistent evolutionary picture
of the Cas A progenitor. The material lost in the first
mass loss episode had a composition very close to that of
'solar' material and hence did not produce compositional
anomalies in the region surrounding Cas A. However, a higher
than average density in the region surrounding the remnant
is consistent with X-ray observations of Cas A (see Charles,
Culhane, and Fabian, 1977).

We conclude that a signature of early mass loss in the
massive progenitor star of a supernova may be a high ^{14}N/^{1}H
ratio in the remnant. Thus it would be of considerable
interest to investigate this abundance ratio in as many young

supernova remnants as possible.

REFERENCES

Arnett, W.D.: 1975, Astrophys. J., 195, 727.
Barlow, M.J., and Cohen, M.: 1977, Astrophys. J., 213, 737.
Charles, P.A., Culhane, J.L., Fabian, A.C.: 1977, Mon. Not.
    R. astr. Soc., 178, 307.
Chevalier, R.A.: 1976, Astrophys. J., 208, 826.
Chevalier, R.A. and Kirshner, R.P.: 1978, Astrophys. J.,
    219, 926.
Chiosi, C., Nasi, E., and Sreenivasan, S.R.: 1978, Astron.
    Astrophys., 63, 103.
Davidson, K.: 1973, Astrophys. J., 186, 223.
Davidson, K.: 1978, 'Emission-line Spectra of Condensations
    in the Crab Nebula (preprint)
de Loore, C., De Grève, J.P., and Lamers, H.J.G.L.M.: 1977,
    Astron. Astrophys., 61, 251.
Endal, A.S.: 1975, Astrophys. J., 197, 405.
Hartwick, F.D.A.: 1967, Astrophys. J., 150, 953.
Hutchings, J.B.: 1976, Astrophys. J., 203, 438.
Kamper, K. and van den Bergh, S.: 1976, Astrophys. J.
    Suppl., 32, 351.
Lamb, S.A.: 1978a, Astrophys. J., 220, 186.
Lamb, S.A.: 1978b, 'Evolution of a 50 $M_\odot$ star' (in
    preparation).
Lamb, S.A., Iben, I., and Howard, W.M.:1976, Astrophys. J.,
    207, 209.
Peimbert, M.: 1971, Astrophys. J., 170, 261.
Peimbert, M. and van den Bergh, S.: 1971, Astrophys. J.,
    167, 223.

DISCUSSION FOLLOWING LAMB

Kwok: You said that the N and He were ejected during
the F or G phase of the star, if so why is the ejection
velocity 150 km/s as we know the ejection velocity in the
late stage is ~ 10 km/s? Do you then mean that the progenitor of Cas A did not go from an OB supergiant to a supernova in 10^4 yrs? If N and He were ejected during the OB
supergiant phase, could you explain why the gas velocity is
not $\geq$ 1000 km/s?

Lamb: I said that the material enriched in nitrogen and
helium was possibly ejected during the F and G supergiant
phase of the star. It could also have been ejected while
the star was a red supergiant. The observed velocity of the
nitrogen and helium enriched material is not likely to reflect the ejection velocity from the star, as the supernova
explosion itself could have imparted momentum to the circumstellar material.
It seems most likely that the Cas A progenitor became a
supernova after at least an oxygen core had formed. This
occurs considerably more than 10^4 years after the end of
core helium burning, the epoch when the transition from OB
supergiant to red supergiant is thought to take place for
stars in the approximate mass range $10 \leq M_0 \leq 25$. For stars
more massive than 25 M_0 the motion in the HR diagram is less
certain at present.
It would seem very unlikely that the nitrogen and helium
were ejected during the OB supergiant phase because:
a) the observed velocities are not $\geq$ 1000 km/s, and
b) the material would now be outside the present supernova
remnant.

de Loore: Can you comment on the fact that your 25 M_0
star goes back to the blue part of the HRD at a rather blue
point. Is this due to your treatment of convection and has
this not as consequence a large or too extensive mixing?

Lamb: We find that the 25 M_0 star does not cross the
Hertzsprung gap prior to core Ne-burning. It is not obvious
that this is due to our treatment of convection, and the
matter requires further investigation. Consistent with the
star not becoming a red supergiant no convective envelope
forms, rather the convective shell present above the H-
burning shell during early core He-burning in all massive
stars persists throughout the evolution followed, for our
25 M_0 model. The subsequent convective mixing in this model
is thus not as extensive as that in the 15 M_0 model, which
acquires a convective envelope.

Chiosi: Concerning the location in the HR diagram of core He-burning models of massive stars, I would like to ask if you expect that the presence of mass loss during the previous phases (core and shell H-burning) will affect this location. More specifically, your computations show that these models get bluer as the mass of the star increases (constant mass evolution). I have the feeling that the inclusion of mass loss during the core H-burning would on contrary produce redder models in core He-burning. The reason of it might be perhaps attributed to the chemical profile of these models which is such to prevent them from reaching thermal equilibrium at high effective temperature. In any case the final answer to this question is only possible taking into account the occurrence of mass loss also at low effective temperature.

Lamb: In our calculations of the evolution of massive stars of 15 M_0 and 25 M_0 (Lamb, Iben and Howard, Ap.J. 207, 209, 1976), we found that the main core He-burning phase of the evolution takes place at slightly higher surface temperatures for the more massive star (~ 0.5 in the log of the surface temperature). As shown in the above mentioned paper this agrees with Humphrey's (Ap.Letters 6, 1, 1970) observations of the distribution of blue supergiants in the HR diagram. Significant mass loss would be expected to shift the location of the core He-burning region in the HR diagram.

Sreenivasan: I believe, in your paper with Howard and Iben (Ap.J. 1977), that there were alternate convective and radiative regions outside the shrinking core of your massive star models. You said that such structure can mimic semi-convection regions. If so, I wonder why there exists a difference regarding the question of blue versus red supergiant stages. I agree, however, that one should understand the origin of these differences more clearly.

Lamb: The region of alternating convective and radiative shells outside the shrinking core is a so called semi-convective region. The criterion for convective stability used in our calculations was the Schwarzschild criterion. The ratio of times spent as a blue versus red supergiant would be different if we had used the Ledoux criterion.

THE ROLE OF ROTATION IN THE EVOLUTION OF MASSIVE STARS LOSING MASS

S.R. Sreenivasan and W.J.F. Wilson
Department of Physics, The University of Calgary
Calgary, Alberta, Canada T2N 1N4

ABSTRACT

The role of differential and solid body rotation in the evolution of massive stars undergoing mass loss is discussed. The implications for Of, WR, β Cephei stars and shell stars are brought out.

1. INTRODUCTION

In recent years a number of investigations have been reported in which the effect of mass loss on the evolution of massive stars has been assessed (see de Loore in this volume for references) following the discovery that was made possible through ultraviolet observations with rockets and satellites (see Snow in this volume for detailed references). While these studies have clarified to a certain extent the reasons for the presence of overluminous stars, the relative rarity of red supergiants above a certain mass limit and a possible explanation of the origin of Wolf-Rayet stars, there still exist some controversies, e.g. whether WR stars are hydrogen burning objects or helium burning objects, whether they are single or double stars and some discrepancies, e.g. relative number of red versus blue supergiants. There are also many things about these stars that we do not quite understand, e.g. whether there are abundance anomalies amongst yellow supergiants, the nature of the β Cephei phenomenon, and the nature of maa loss in red supergiants.

In this progress report we shall outline briefly the role of rotation (solid body as well as differential rotation) in the evolution of massive stars undergoing mass loss. One of the reasons for this study is an attept to understand phenomenologically the observational results concerning the rotation of OB stars.

P. S. Conti and C. W. H. de Loore (eds.), Mass Loss and Evolution of O-Type Stars, 367–370.

2. EFFECTS OF ROTATION

In a recent paper (Sreenivasan and Wilson, 1978a) we have attempted to assess the effect of a centrifugal force on the amount of mass lost by a massive star due to stellar winds driven by radiation pressure acting on the outer layers due to resonance lines absorption. We concluded that the mass loss rate can be enhanced by about 20-30% over the rate for non-rotating models. We also showed that the surface rotation in these models can be reduced to zero well before hydrogen is exhausted in the core. If mass loss due to an acoustic energy flux driven wind is included it can be shown that a 15 M_O star (X = 0.70; Z = 0.03) will have about 70% of its mass left when it enters the stage of being a cepheid. It thus appears possible to understand the mass discrepancy as well as the fact that cepheids are either non-rotating or slowly rotating objects.

3. DIFFERENTIAL ROTATION

The time required for spin-down of the surface layers is different according to whether one discusses only conservation of angular momentum or conservation of angular momentum and energy (including rotational energy of a mass-losing star). Clearly, there are stars which exhibit some rotation (albeit small) in spectral stages later than B. Many supergiants show a small residual rotation of the order of 50 km/sec according to Conti and Ebbets (1977). This is clearly due to differential rotation between the surface and the interior regions as was shown by Sreenivasan and Wilson (1978a).

We have made explicit allowance for the effect of differential rotation in the form of a macro-turbulent pressure in a subsequent study (Sreenivasan and Wilson, 1978b) by including a term:

$$\frac{1}{3} \, \bar{\rho} \, \frac{(V_b - V_s)^2}{1}$$

to represent the force due to the turbulent pressure gradient in the outer layers. Here V_b is the rotational speed of the boundary of the interior model, V_s is the rotational speed at the surface, and 1 is the distance between the points at which $V = V_b$ and $V = V_s$ and $\bar{\rho}$ is the mean density of the shell of width 1. Estimating the contribution to mass loss in the same fashion as in Sreenivasan and Wilson (1978a), we find that the enhancement in mass loss rate is about a factor of 2-3 over the rate when only a centrifugal force is included. We have shown elsewhere (Sreenivasan and Wilson, 1978c) that one can explain the β Cephei phenomena as a manifestation of the differential rotation in massive stars in the mass range 15-20 M_O due to Kelvin-Helmholtz instability resulting in observed variations of the order of 0.1 magnitude.

4. SHELL STARS

It is known that β Cephei stars are confined to a narrow range in spectral type B0-B2 (Lesh and Aizenman, 1978) whereas Be stars occur with lower effective te-peratures. The core of an evolving star contracts as it exhausts hydrogen while the surface rotation drops in speed due to mass loss. This results in a faster spinning core coupled to a surface which is braking due to the stellar wind. A dynamical consequence of this could well be the ejection of a shell revealing a rapidly spinning interior. Thus one might understand why β Cephei stars are slow rotators whereas the shell stars are rapid ones. Before any firm conclusions can be drawn on these aspects, more careful quantitative work is required. However, it is clear that differential rotation in a mass losing star holds the key to a number of interesting properties of massive stars.

5. WR STARS

Finally, we should like to remark that the origin of Wolf-Rayet stars is connected to the high mass loss rates that are necessary to understand not only their spectra but the che-ical state of their interior. De Loore et al (1977) believe that WR stars are predominantly hydrogen burning stars, which are members of a binary system. Chiosi et al (1978) on the other hand argue that the WC stars could possibly be helium stars whose outer layers have been depleted by an acoustic energy flux driven wind and that only those stars for which the helium burning lifetime is larger than the time scale needed to remove the hydrogen rich outer layers are possible candidates. The effect of rotation and differential rotation tend to favour their inequality $\tau_{He} > \tau_{\dot{M}}$ and also strengthen it because a higher mass loss rate produces a larger q_{He} value. As also emphasized by them and Giannone earlier, the q_{He} value holds the key to the subsequent evolutionary pattern of stars.

Fuller details of this investigation will be published elsewhere.

6. ACKNOWLEDGEMENT

We wish to acknowledge partial support of our investigation by a grant from the Canadian National Research Council to S.R. Sreenivasan and by the University of Calgary to W.J.F. Wilson.

REFERENCES

Chiosi, C., Nasi, E. and Sreenivasan, S.R.: 1978, Astron. Astrophys. 63, 103.

Conti, P.S. and Ebbets, D.: 1977 Ap. J. 213, 438.

De Loore, C.: (1978) Review paper on Mass loss from Single Stars (this volume).

Lesh, J.R. and Aizenman, M.: 1978, Ann. Rev. Astr. Ap. 16 (in press).

Snow, T.P.: (1978) Review paper on Mass loss observations (this volume).

Sreenivasan, S.R. and Wilson, W.J.F.: 1978a, Astron. Astrophys. (in press).

Sreenivasan, S.R. and Wilson, W.J.F.: 1978b, Bull. AAS 10, 438.

Sreenivasan, S.R. and Wilson, W.J.F.: 1978c, Proc. Conf. Stellar Pulsational Instabilities, Greenbelt, MD (in press).

DISCUSSION FOLLOWING SREENIVASAN and WILSON

Vanbeveren: To derive the formula for $\dot{\omega}$, did you assume that after a mass layer has left the star, its angular momentum remains constant?

Sreenivasan: Yes, if no subsequent mass loss from star magnetic fields, other mechanisms for transport of angular momentum are not considered.

EVOLUTION OF A 30 M☉ STAR WITH MASS LOSS

H. J. Falk and R. Mitalas
University of Western Ontario

ABSTRACT. Evolutionary tracks for a 30 M☉ star with mass loss rates (0.0, 1.0, 2.5, 5.0, 10.0)x10^{-7} M☉/yr have been calculated. The effect of the different rates on the main sequence lifetime and on the effective temperature of the core He burning is discussed.

Evolutionary calculations employing the Schwarzschild convective stability criterion indicate that constant mass stars more massive than 15 M☉ begin core He burning as blue supergiants (BSG). The stars then evolve redward on a nuclear timescale, becoming red supergiants (RSG) only in the last phases of core helium burning (Simpson, 1971; Stothers and Chin, 1976). Helium ignition as a BSG is attributed to the H burning shell encountering a convective shell which appears just beyond the hydrogen exhausted region (Iben, 1966; Barbaro et al. 1971). Since mass loss affects the convective stability of the regions outside the burning core (Chiosi and Nasi, 1974; Chiosi et al. 1978), it is pertinent to investigate the effects of mass loss on this convective shell and on the stars' position in the HR diagram during core helium burning.

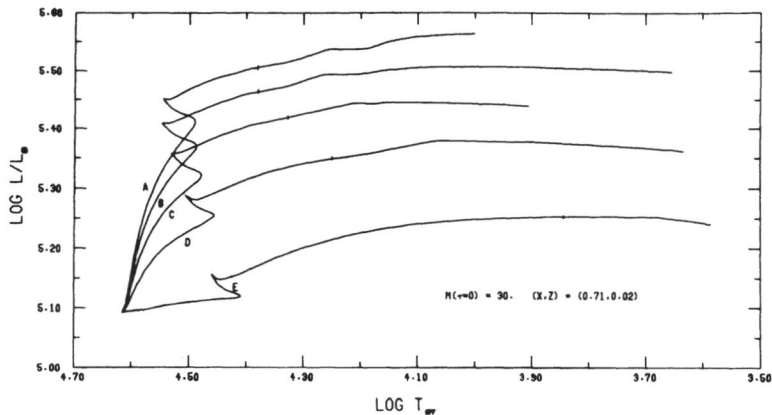

Fig. 1 Theoretical HR diagram for 30 M☉ star with initial mass loss rates of (0.0, 1.0, 2.5, 5.0, 10.0)x10^{-7} M☉/yr for tracks A,B,C,D,E,

371

P. S. Conti and C. W. H. de Loore (eds.), Mass Loss and Evolution of O-Type Stars, 371–374.

respectively. Core He ignition is indicated by a tick on each track.

Evolutionary tracks with different initial mass loss rates were
constructed for a star with initial mass of 30 $M_\odot$ and $(X,Z)=(0.71,0.02)$.
The Schwarzschild criterion for convective stability was used; semi-
convection was treated in a manner similar to that of Robertson (1972).
McCrea's (1962) mass loss algorithm, $\dot{M}=kLR/M$, was adopted, where k was
chosen to give the desired mass loss rate on the ZAMS.

Fig. 1 shows the evolutionary tracks for the 5 different initial
mass loss rates listed in table 1. As shown by Chiosi and Nasi (1974),
de Loore et al. (1977), and Chiosi et al. (1978) the tracks become

TABLE 1

Track	Main Sequence				X_{sc}	M_{sh}	Helium Ignition			Final Model			
	$\dot{M}$ $(10^{-7}$ $M_\odot/yr)$	τ_{ms} $(10^{+6}$ yr)	$\frac{M}{M_\odot}$	Semi-convection		$\frac{}{M_\odot}$	τ $(10^{+6}$ yr)	$\frac{M}{M_\odot}$	Log T_{eff}	$\frac{M}{M_\odot}$	Y_c	Log T_{eff}	$\dot{M}$ $(10^{-6}$ $M_\odot/yr)$
A	0.0	5.804	30.00	Important	0.44	9.609	5.808	30.00	4.379	30.00	.096	4.002	0.0
B	1.0	5.707	28.66	Important	0.43	8.798	5.710	28.66	4.380	22.98	0.024	3.654	4.8
C	2.5	5.677	26.87	Still significant	0.41	8.194	5.680	26.87	4.328	20.77	0.322	3.897	3.6
D	5.0	5.734	23.89	Not important	0.34	7.455	5.738	23.87	4.251	19.15	0.899	3.634	20.
E	10.0	6.093	17.51	Totally unimportant	0.22	6.089	6.099	17.38	3.845	17.17	.980	3.585	37.

less luminous as the mass loss rates increase, due to the decreasing
mass of the star. The final masses after the hydrogen burning phase
for each track appear in column 4 to illustrate the amount of the main
sequence mass loss associated with each initial rate. Because the
luminosity is reduced in mass losing stars, the rate of hydrogen con-
sumption is also decreased. Hence, one would expect the main sequence
lifetime (τ_{ms}) to increase. However, as shown in table 1, this is not
exactly true. With small mass loss rates, τ_{ms} actually decreases while
for large rates it increases. The semiconvective zone, which is
attached to the convective core during most of the core H burning phase,
adds hydrogen to the burning region and thus extends the τ_{ms}. The decreasing importance of semiconvection as higher mass loss rates are
imposed is manifested in a decrease in τ_{ms}. For the 30 $M_\odot$ case, it was

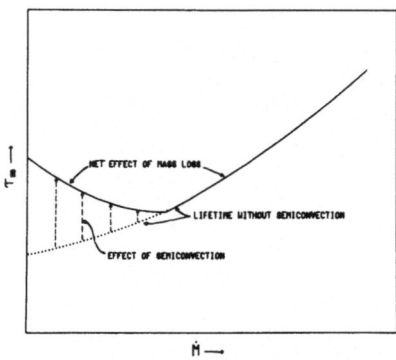

Fig. 2 Schematic illustration of the effects of semiconvection and
mass loss on the main sequence lifetime.

found that for mass loss rates greater than about 5.0×10^{-7} M$_\odot$/yr, semiconvection was completely suppressed. The competing effects of semiconvection and mass loss are illustrated schematically in fig. 2.

The hydrogen abundance of the convective shell which breaks out at the end of core hydrogen burning is listed in column 6. The effect of mass loss on the convective shell is shown in the reduced hydrogen abundance. The mass position of the hydrogen shell source at the time of He ignition is also listed.

Following core H exhaustion, the low mass loss rate models (tracks A,B,C) evolve redward in non-thermal equilibrium with the hydrogen shell source locked onto the H discontinuity formed by the convective shell. These stars ignite He as BSGs. Then they evolve to the red on a nuclear timescale, becoming RSGs only at the end of core He burning. Evolution for a 30 M$_\odot$ star with low initial mass loss rates ($\leq 2.5 \times 10^{-7}$ M$_\odot$/yr) behaves much the same as for the constant mass star, except for the occurence of He ignition at a slightly lower effective temperature. The relation between thermal balance and the location of the H burning shell in a massive star as it crosses the HR diagram has been discussed by Barbaro et al. (1971). For the highest mass loss rates (track E) the models never achieve thermal equilibrium as blue or yellow supergiants. These stars evolve across the HR diagram on a thermal timescale and ignite He as RSGs. The hydrogen shell source approaches but never reaches the compostion discontinuity of the convective shell before the stars become RSGs. They will remain RSGs during core He burning unless blue loops occur. Track D is an intermediate case. With this mass loss rate, a star ignites He as a BSG, like the lower mass loss rate stars. However, subsequently it evolves across the HR diagram on a timescale between the thermal and nuclear timescales.

REFERENCES
Barbaro G., Chiosi C., and Nobili L. 1971, Proceedings of the Trieste
          Colloquium on Supergiant Stars. Ed. M. Hack, p.334.
Chiosi C. and Nasi E. 1974, Astron.Astrophys. 34, p.355.
Chiosi C., Nasi E. and Sreenivasan R. 1978, Astron.Astrophys. 63, p.103.
de Loore C., de Greve J.P., Lamers H.J.G.L.M. 1977,
          Astron.Astrophys. 61, p.251.
Iben I. 1966, Astrophys. J. 143, p.516.
McCrea W. H. 1962, Astron. J. 71, p.172.
Robertson J. 1972, Astrophy. J. 177, p.473.
Simpson E. 1971, Astrophys. J. 165, p.295.
Stothers R. and Chin C.-w. 1976, Astrophys. J. 204, p.470.

DISCUSSION FOLLOWING FALK and MITALAS

Underhill: How do your results compare with those of
Hartwick who used a similar method for a 15 M_o star?

Chiosi: Perhaps I could reply to this remark by Dr.
Underhill. The reason why the present results differ from
those of Hartwick (1967) although they both assume the same
mass loss rate dependence, can be explained by comparing the
time scale involved in mass loss and the evolutionary time
scale. The results in fact do not depend on the mass loss
dependence but rather on both the amount of mass removed from
the star and the evolutionary stage in which most of the mass
loss occurs. A too huge mass loss causes in models of not
high central concentration a strong decrease of the luminosity, as shown by Hartwick. An empirical reasoning suggests
that for any given initial mass only mass loss rates below
some critical value are allowed if we wish to fit the models
to a number of observational constraints (a strong decrease
of the luminosity seems in fact to be unnecessary). In this
sense the present results are consistent with the old ones.

Dearborn: Is the slight decrease in main-sequence time
scale in your models with a low rate of mass loss due to the
size of the convective core decreasing more rapidly than the
luminosity?

Falk: No. While the decreasing convective core mass
with increasing mass loss rate will cause a decrease in τ_{ms},
the accompanying luminosity reduction is more important, so
that the τ_{ms} increases. Calculations show that for models
at the same time, but with different mass loss rates (hence
different masses), the ratio of convective core mass to
total mass is the same; that is, the core mass changes at
the same rate as the stellar mass. Since τ_{ms} is roughly
proportional to the convective core mass and inversely proportional to the luminosity which varies as the cube of the
total mass, the net effect of mass loss alone is to increase
τ_{ms}. Therefore the decrease of τ_{ms} at the low mass loss
rates is due to the lessening importance of semi-convection
as the mass loss rate increases.

MASS CONSERVATION AND RAPID MASS LOSS
ON THE MAIN SEQUENCE

T.J. Mazurek
Department of Astronomy
University of Texas at Austin

The rapid mass loss observed in O stars can affect their evolution dramatically if it takes place during core hydrogen burning on the main sequence. Conti has suggested[1] that this is the case for the brightest Of stars, and that these stars evolve into Wolf-Rayet (WR) stars. Support for this scenario comes from the similarity[1] in the spectra of the Of and WR stars, and from the observed helium enrichment of Wolf-Rayet WN stars along with the correlation[2] of the latter with the most luminous O stars. If the most massive Of stars evolve to WN stars of relatively low mass, one needs to determine the range of zero-age stellar masses where such an evolution occurs. This communication abstracts some observational evidence that bears on the question of the minimum zero-age mass for rapid mass loss on the main sequence. It then summarizes the author's investigation of a model for mass loss where photospheric acoustic waves control the flow rate.

The HR-diagram of stellar models[3] can be used to estimate[1] the masses of O stars. Figure 1 shows an HR-diagram for observed massive stars[4] and the theoretical main sequence strip. Only luminosity classes IV and V and f-type stars are included. It is evident from the figure that the f-characteristics appear in O stars with $M \gtrsim 30\ M_\odot$. Of stars are present along the zero-age line, and the dense envelopes associated with these stars can appear in unevolved stars.

Empirical determinations of rates of mass loss are available for ζ Pup ($\sim 7 \times 10^{-6}\ M_\odot\ yr^{-1}$, ref.5) and τ Sco ($\sim 7 \times 10^{-9}\ M_\odot\ yr^{-1}$, ref. 6). In attempts to correlate empirically determined rates with theoretical stellar models only stars near the zero-age line should be used. For rapid mass loss, the internal structures of evolved stars is not known since the behavior of mass loss rates as evolution progresses is poorly determined. Fortunately ζ Pup and τ Sco fall near the zero-age line in the HR-diagram, and presumably have structures similar to the theoretical[3] ones. Their positions in the HR-diagram indicate masses of $\sim 100\ M_\odot$ and $18\ M_\odot$ for ζ Pup and τ Sco, respectively.

P. S. Conti and C. W. H. de Loore (eds.), Mass Loss and Evolution of O-Type Stars, 375–377.
Copyright © 1979 by the IAU.

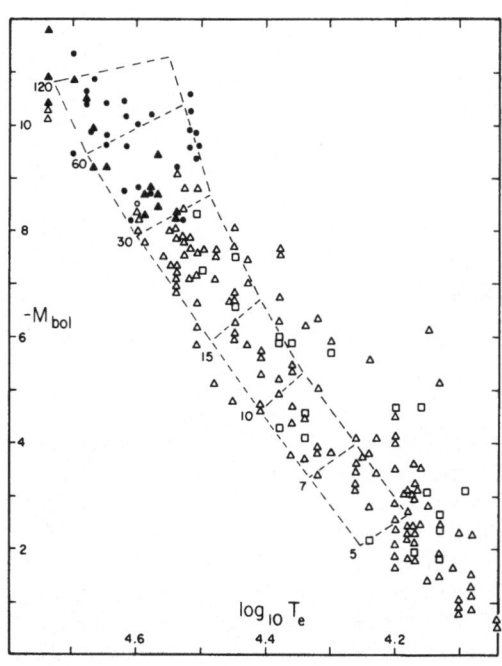

Fig. 1.--The HR-diagram (masses along ZAMS in $M_\odot$). Luminosity classes IV, V, and other are denoted by □, Δ, and o, respectively; f-type stars have filled symbols.

The proposed mechanisms for powering stellar winds have been discussed[7] extensively at this symposium. The presence of a radiation driven mechanism apparently is required to achieve the high terminal velocities that are observed. But the spectroscopically inferred presence of ions having a large range of ionization potentials indicates temperatures much greater than those expected if radiative acceleration is the sole mechanism. This implies that another wind mechanism is present. The apparent lack[8] of correlation between observed terminal velocities and stellar luminosities may indicate that radiation forces can act efficiently to expel all matter that is ejected from the stellar surface. Hence a non-radiative mechanism may actually control the rate of mass loss, while radiation pressure completes the acceleration of the wind to its terminal velocity.

A coronal model for the wind flow near the stellar surface has been presented by Hearn[9]. The sources of the energy required to maintain the corona need to be examined in detail. If surface turbulence is present in O stars, it will generate sound waves which can deliver energy to the corona. The presence of macroturbulence would explain in a straight forward manner the observed dearth of O stars with sharp lines[10]. The process producing the stellar wind may thus begin with sound waves generated in the photosphere, which then steepen into shocks giving a hot corona, and radiation forces supply the final acceleration to the flow.

The connection between acoustic wave generation and coronal rates of mass loss has been examined in detail for main sequence stars by the author. In this investigation the velocities of the surface turbulence are determined by photospheric conditions which are directly related to stellar luminosities, temperatures, and masses. Theories[11] of acoustic wave generation in turbulence are applied to determine energy fluxes from the photospheres. The energy balance between acoustic input and coronal losses is then used to determine the coronal structure, and hence the rate of mass flow. Published theoretical models[3] are then used to estimate rates of coronal flow on the main sequence. Some salient results of this investigation are the following.

The predicted turbulent velocities at the photosphere range between 7 and 15 km-s^{-1}. The turbulent velocity predicted for τ Sco ($\sim$ 7 to 9 km-s^{-1}) is in good agreement with empirical determinations[12]. The variation in velocity over the different spectral classes is consistent with the lower envelope[10] to the Doppler broadening of observed O stars. A considerable increase in the rate of acoustic energy generation above that given by Stein[11] is necessary to produce mass flow rates as high as those of ζ Pup. Turbulent magnetic fields in the photospheres can give the enhancement required. With enhanced acoustic generation, the coronal mass flow rates predicted for ζ Pup and τ Sco ($\sim 6 \times 10^{-6}$ and 12×10^{-9} $M_{\odot}$ yr^{-1}) are in good agreement with the empirical values quoted above.

For stars with M $\gtrsim$ 30 $M_{\odot}$, the radiative losses outside the corona's critical point are so large that the coronal flow in that region becomes energetically inconsistent. Radiative acceleration becomes essential for mass loss. The high cooling may give a density increase in the wind flow, and hence be the cause of the observed high densities in Of stars. For stars with M $\gtrsim$ 60 $M_{\odot}$, the rate of mass flow increases rapidly with evolution across the main sequence strip. A runaway in the rates of mass loss is indicated. These stars should then evolve into WN stars. The lower mass limit for this behavior is consistent with an empirical estimate[2]. The stars with 30 $M_{\odot} \leq$ M $\leq$ 60 $M_{\odot}$ are predicted to lose only a small fraction ($\leq$ 10%) of their mass during core hydrogen burning.

This work was supported in part by NSF (Grant No. AST-76-07629).

---

1. Conti, P.S. 1976, Mem. Soc. Roy. Sci. Liége, 9, 193.
2. Moffat, A.F.J., and Seggewiss, W. 1979, preprint.
3. Stothers, R. 1972, Astrophys. J., 175, 431.
4. For most stars, M_{bol} and T_e were determined from the (B-V)$_o$ given by G.L.H. Harris (1976, Astrophys. J. Suppl., 30, 451) with the conversions of A.D. Code, J. Davis, R.C. Bless and R.H. Brown (1976, Astrophys. J., 203, 417) and D.C. Morton (1969, Astrophys. J., 158, 629). For some stars the results of ref. 8 or those of P.S. Conti and M.L. Burnichon (1975, Astron. Astrophys., 38, 467) were used.
5. Lamers, H.J.G.L.M., and Morton, D.C. 1976, Astrophys. J., 32, 715.
6. Lamers, H.J.G.L.M., and Rogerson, J.B. 1978, Astron. Astrophys., 66, 417.
7. See in particular the contributions of J. Cassinelli, J. Castor, A.G. Hearn, H. Lamers, and R. Thomas.
8. Snow, T.P., and Morton, D.C. 1976, Astrophys. J. Suppl., 32, 429.
9. Hearn, A.G. 1975, Astron. Astrophys., 40, 277, and 40, 355.
10. Conti, P.S. and Ebbets, D. 1977, Astrophys. J., 213, 438.
11. Stein, R.G. 1968, Astrophys. J., 154, 297; and Kulsrud, R.M. 1955, Astrophys. J., 121, 461.
12. Smith, M.A., and Karp, A.H. 1978, Astrophys. J., 219, 522.

GENERAL DISCUSSION

<u>Kwok</u>: If the chairman would permit me I would like to make a comment on the effects of mass loss in the red phase of stellar evolution. We have recently been investigating metallicity as a parameter of stellar evolution. It has been shown that radiation pressure on grains is not a feasible mechanism if the metallicity of the star is below 10% of that of solar value (Kwok, Ap.J. <u>198</u>, 583). If radiation pressure on grains is in fact the mass loss mechanism for the red phase, this implies that Pop II stars will not be able to lose mass. This is in fact consistent with the absence of gas in globular clusters. Recently it has been suggested that (Kwok et al., Ap.J. <u>211</u>, L125) planetary nebulae do not represent a separate mass loss phase but instead are the result of sweeping up matter ejected during the red phase. If this is true then the underabundance of planetary nebulae in globular clusters can also be easily explained. From a study of sulphur abundances in planetary nebulae, Barker (Ap.J. <u>221</u>, 145) also concludes that planetary nebulae are produced by a predominantly metal-rich population. In conclusion metallicity may be an important parameter for it may imply that only low mass Pop II stars can become white dwarfs and any star more massive than $\sim 2~M_{\odot}$ will become a supernova.

<u>Vanbeveren</u>: I want to ask a question to the people who are doing evolutionary calculations in the red giant phase: as there exists a very extended atmosphere at the red giant phase and if we can expact excessive mass loss rates as has been assumed by Chiosi, I wonder how one can possibly treat atmospheres with a hydrostatic approximation?

<u>Sreenivasan</u>: At present there exist four "prescriptions" for removing mass in an evolutionary scheme : (1) due to McCrea; (2) Lucy-Solomon; (3) Lamers, van den Heuvel and Petterson and (4) Castor, Abbott and Klein. The first two depend upon luminosity and increase with it, the third goes through a maximum and comes down while the fourth remains at a fairly constant rate up to log $T_{eff} \sim 4.0$. In addition to this variation, one has two parameters k and α in the fourth recipe, which allows for different rates and although one can say that a good correspondence exists for k = 0.076 and α = 0.90 or 0.83 in some parts of the HR diagram, such is not the case in other parts. It is therefore necessary to use different combinations of k and α at different stages. It is important to have observational guidance to produce acceptable models of evolutionary sequences, e.g. those of Niemela concerning firm indications of mass and spectral classifications for the WR stars: WN or WC. This enables one to discriminate between the different "scenarios" of

379

P. S. Conti and C. W. H. de Loore (eds.), Mass Loss and Evolution of O-Type Stars, 379–382.
Copyright © 1979 by the IAU.

formation of WR stars - single versus binaries. I am happy
to know that WC stars of 17-20 M_O exist in binary systems
and similarly WN stars of higher masses, and that not all
WR stars are 10 M_O objects!

Hyland: I should like to address a question to Susan
Lamb. In supernova events a considerable mass of interstellar gas is swept up by the expanding shock front, which
would suggest that the (N/H) and (He/H) abundances which you
observe in the Cas A supernova remnant may be lower limits.
1) Have you estimated the mass of gas swept up in the Cas A
SNR and 2) what are the upper limits of (N/H) and (He/H)
which you can derive from your computations?

Lamb: The Cas A supernova occurred about 300 years ago
and it is thus a very young S.N. remnant. This small age
together with high densities in the "fast-moving knots"
suggest that very little interstellar material has been swept
up so far. Thus one can make a meaningful comparison between
the abundances in the S.N. remnant and stellar models. The
maximum $^{14}N/H$ ratios that are predicted by the model calculations are ~ 10 if mass loss is allowed. If mass is
conserved the maximum value that this ratio can attain is
~ 5.

Rahe: Question to Dr. Lamb: Do your calculations give
any information on isotope abundances, such as the $^{12}C/^{13}C$
ratio, in the matter lost?

Lamb: I have not calculated isotope abundances, such as
^{12}C and ^{13}C, for these models. However, these can be calculated from the models at a later time.

Leung: I would like to comment on an earlier point about
mass lost during the red-star phase. Pretty well all late-
type supergiants are variable (semi-regular variables). They
have a typical periodicity of several hundreds of days, and
with other much longer modulation periods, typically of 10
times longer. There must be a driving force to feed this
instability. Thus, the mass loss rate will be affected
somehow. The red giants are also unstable, as long period
variables.

Abbott: Since we know an accurate rate of mass loss for
at least one star, ζ Pup, I would favor those evolutionary
calculations whose rate of mass loss formula agreed with that
of ζ Pup. For de Loore et al. this implies N ~ 150 to 200
while for Chiosi et al. this implies α ~ .85.

Morton: Certainly it is desirable to extend the types
of stars for which the mass loss rate and ionisation level

have been determined. Several O type main sequence and giant stars are available with Copernicus to obtain the line profiles of CIII, NV, OIV, SiIV etc, and IUE could provide the CIV profile. It would be good to analyse some stars using the Copernicus high resolution scans, which are not subject to large background corrections and are less con- fused by interstellar lines.

Conti: I think we may have a good estimate of mass loss rates in Of stars (e.g. ζ Pup) where optical emission lines are seen, but we have only upper limits for O stars. These stars may have substantial mass loss rates, as evidented by their UV spectra. A detailed analysis of 9 Sgr, an Of star, is badly needed and will help this problem considerably.

van den Heuvel: Can anyone explain to me why some authors get massive stars into the right hand part of the HR diagram and other authors don't? I have not yet heard a satisfactory explanation for this here.

Chiosi: As for constant mass evolution we know that amongst the others one of the leading parameters determining the position in the HR diagram of models in the core burning phase is the detailed shape of the H profile and its corre- lation with the H burning shell. Any modification of the H profile due either to intermediate or external mixing or both may therefore significantly change the location in the HR diagram. Similar arguments seem to be still valid also for losing mass sequences and allow us to qualitatively un- derstand why core He-burning models get redder as the mass loss rate increases, and might also explain why slightly different rates give substantially different locations of the models in the HR diagram.

de Loore: This is an interesting point as it could ex- plain the fact that in the right red upper part of the HRD stars are lacking. I feel that it is due to the treatment of convection by Susan Lamb, which means excessive mixing, in the large number of shells used in the computations. It is not exactly the same as semi-convection; in her treatment more mixing is used.

Lamb: The physical nature of so called semi-convection is not understood, and therefore it is impossible to know if one is modelling this phenomena correctly. If important details of the evolution of massive stars are found to de- pend heavily on the precise gradient in the mean molecular weight left behind in the star after the main sequence phase of semi-convection, then we will have a large uncertainty in the evolutionary calculations.

Dearborn: In the regions that Susan wants to expose in order to obtain the large ^{14}N overabundance, the $^{12}C/^{13}C$ ratio would be near its equilibrium abundance of 3. There would however be very little carbon of any form as it was used to create the ^{14}N overabundance.

There was a young man whose veracity
Was questioned because his opacity
While given to Stothers
Was held back from others
With a singular show of tenacity.

SESSION 7

EVOLUTION WITH MASS LOSS: DOUBLE STARS

Chairman: D.C. MORTON
Introductory Speaker: I. ZIOLKOWSKI

1. A. TUTUKOV and L. YUNGELSON: Evolution of massive common
      envelope binaries and mass loss.

2. D. VANBEVEREN, J.P. DE GREVE, C. DE LOORE and E.L. VAN
      DESSEL: The influence of stellar wind mass loss on
      the evolution of massive close binaries.

3. A. DELGADO: Common envelope binaries and mass loss.

4. C. FIRMANI, G. KOENIGSBERGER, G.F. BISIACCHI, E. RUIZ
      and A. SOLAR: HD 50896: an other WR binary star.

EVOLUTION OF BINARY STARS WITH MASS LOSS

Janusz Ziółkowski
N. Copernicus Astronomical Center
Polish Academy of Sciences
ul. Bartycka 18, 00-716 Warsaw, Poland

ABSTRACT

Three situations involving mass loss from binary systems are discussed. (1) Non-conservative mass exchange in semi-detached binaries. No quantitative estimate of this mechanism is possible at present. (2) Common envelope binaries. There are both theoretical and observational indications that this phase of evolution happens to many systems, even to some that are not very close initially (orbital periods ~ years). (3) Stellar winds in binaries. Observational evidence suggests that stellar winds from components of close binaries (especially semi-detached) are significantly stronger than from single stars at the same location in the H-R diagram. Theoretical arguments indicate that in some cases stellar wind may stabilize the component of a binary against the Roche lobe overflow. In some cases there is weak evidence of an anisotropy in the stellar wind.

There are two major mechanisms of mass loss from O-type binaries. One of them is non-conservative mass exchange between the components after one of the components overflew its critical Roche lobe. The second mechanism is strong stellar wind from one or both components. The first mechanism can probably operate in any close binary, independent of its mass. The second mechanism is efficient only for massive systems, which practically limits its importance to only O-type binaries (at least one of the components has to be initially an O-type star). A special and very important case of the first mechanism occurs if a common envelope binary is formed during the process of mass exchange. In this paper we shall briefly discuss all three situations: (1) non-conservative mass exchange in semi-detached systems, (2) mass loss from common envelope binaries, (3) stellar winds in binaries.

385

P. S. Conti and C. W. H. de Loore (eds.), Mass Loss and Evolution of O-Type Stars, 385-399.
Copyright © 1979 by the IAU.

1. NON-CONSERVATIVE MASS EXCHANGE

The conservative theory of the evolution of close binaries is summarized in excellent review articles by Plavec (1968), Paczyński (1971) and Thomas (1977). The most controversial among the assumptions on which this theory is based is just the assumption about the conservation of the total mass and the total orbital angular momentum. Certainly this is not a very good assumption. We have observational evidence for mass loss from some binaries undergoing the process of mass exchange at the present moment (Kruszewski 1966, Huang 1966, Batten 1970). Unfortunately no reliable quantitative estimate of this phenomenon is available. The motivation for performing almost all theoretical evolutionary calculations with the conservative assumptions is neither our belief that mass loss is insignificant nor the need for computational conveniency. It is just our ignorance (both observational and theoretical) about the quantitative aspects of the mass loss and orbital angular momentum loss from the system that makes the conservative assumptions appear the most natural. From technical point of view, it is quite easy to incorporate an arbitrary amount of mass and angular momentum loss from the system. We can do so by introducing two parameters defined as follows:

$$f_1 = \frac{\Delta M}{\Delta M_1} \tag{1}$$

and

$$f_2 = \frac{J_{\Delta M}/\Delta M}{J/M} \tag{2}$$

Here $M = M_1 + M_2$ is the total mass of the system at a given moment, J is the total orbital angular momentum, ΔM is the amount of mass lost by the system during one evolutionary time step and $J_{\Delta M}$ is the angular momentum taken away by this matter. We assume that the mass is leaving component M_1 and that ΔM_1 represents the total amount of mass lost by this component during our time step. Part of this mass, equal to $(1-f_1)\Delta M_1$, is accreted by component M_2 and part, $f_1\Delta M_1 = \Delta M$, leaves the system. In such an approach, the parameter f_1 tells us how large a fraction of the mass lost by the M_1 component, actually leaves the system. Of course, we have $0 \le f_1 \le 1$. The parameter f_2 is describing the ratio of the average angular momentum per mass unit for the matter leaving the system to the similar average for the binary system. If we could decide on numerical values for the parameters f_1 and f_2, we could proceed with the evolutionary calculations.

Unfortunately, such a decision has to be quite arbitrary, although some observational and theoretical attempts to estimate the values of f_1 and f_2 have been made. Svetchnikov (1969) was comparing the observed pre-mass exchange and post-mass exchange binaries and tried to deduce the value of parameter f_1 averaged over the entire phase of mass transfer. He found that $f_1 \sim 0.3 \div 0.9$ but his analysis is not very conclusive, since the observational data are quite uncertain and their interpretation is by no means unique. From similar considerations, Ziółkowski (1971) found no substantial evidence of significant mass loss during the past evolution of majority of semi-detached binaries. Hall (1976) suggested that careful investigations of period changes in semi-detached binaries could give us in some cases information about present mass loss from these systems. On theoretical grounds, Drobyshevski and Reznikov (1974) estimated $f_2 \sim 3$ from analysis of the re-distribution of angular momentum in the system during the mass transfer. Flannery and Ulrich (1977) used the restricted three-body approximation for particle trajectories and found that for the matter leaving the system from the vicinity of Lagrangiant point L_2, the value of f_2 is $\gtrsim 7$.

Given the situation as described above, the best one can do is probably to make theoretical calculations for different trial values of f_1 and f_2 and in this way to estimate the uncertainties of our conservative evolutionary theory. Such calculations were first done by Paczyński and Ziółkowski (1967). A similar approach was also used by Plavec et al. (1973) and Massevitch and Yungelson (1975). All these calculations confirmed that the evolution of the mass-losing component is determined primarily by its internal structure and is not very sensitive to various assumptions about the mass loss and angular momentum loss from the system. On the other hand, the final orbital parameters are quite sensitive to the values of f_1 and f_2, in particular, the final orbital period can be much shorter in the case of the mass loss from the system.

To summarize our considerations: the mass transfer in semi-detached binaries is not conservative, but we do not know how significant the deviations from the fully conser-vative process are. The good agreement between the theore-tical evolutionary model for β Lyrae (Ziółkowski, 1976a) and the observational data indicates that in some cases the conservative approach does not lead to very bad results.

2. COMMON ENVELOPE BINARIES

Binaries with a deep common envelope are a relatively

new idea and I found that they still need some advertisement.
Therefore I shall start with a brief summary of theoretical
and observational arguments indicating that such objects are
really being formed in our Galaxy.

2.1. Theoretical arguments

Benson (1970) was the first who investigated the evolution of the mass-receiving component of a close binary and
found that already at an early stage of mass transfer the
secondary (mass-receiving) component expanded so much that
it filled its own Roche lobe and a contact system was formed.
Similar calculations were later done by many other authors
(Yungelson 1973, Webbink 1975, Kippenhahn and Meyer-Hofmeister 1976, Ulrich and Burgher 1976, Flannery and Ulrich
1977, Neo et al. 1977). The common result of all these
investigations was that even in the binaries that were
initially relatively wide (with initial periods $\gtrsim$ 10 days)
a contact system was formed after only a few percent of the
initial mass of the primary were transferred. The evolutionary tendency indicated the built-up and expansion of the
common envelope during the subsequent evolutionary phase.
Flannery and Ulrich (1977) followed the evolution of their
binary system up to the moment when the common envelope
reached the equipotential surface passing through the outer
Lagrangian point L_2. One may expect that as the surface of
the common envelope expands further beyond the L_2 point,
mass loss from the system might occur. Flannery and Ulrich
investigated this problem using the restricted three-body
approach for particle trajectories. They found that particles
ejected from a co-rotating atmosphere through the L_2 point
with low velocities would indeed leave the system, carrying
away considerable angular momentum. The parameter f_2 as
defined above is in this case given by:

$$f_2 = C(1+q)^2/q \qquad (3)$$

where $q = M_1/M_2$ and $C = 1.7 \div 2$ depending on the degree of
asynchronism and the mass ratio. A very similar formula for
f_2 was found independently by Tutukov and Yungelson (1978).
Let us note that the minimum value of f_2 is ~ 7 and that
for large mass ratios (X-ray binaries) we have $f_2 \approx 2q$.
This means that mass outflow from L_2 is a very efficient
mechanism for angular momentum loss from the system. Due
to this loss the orbit of the binary will shrink very rapidly. It is easy to show that the separation between the
mass centers A will change with the total mass of the system
according to the approximate formula:

$$d \log A/d \log M \approx 2f_2 \qquad (4)$$

It appears that under typical conditions $(f_2 \sim 10)$ the loss of only 12 percent of the mass by the system will decrease the size of the orbit by one order of magnitude! Due to this rapid shrinking, the point L_2 will sink deep into the common envelope and soon the surface of the envelope will be far outside this point and at a large distance from two rotating mass centers. At such a large distance the rotation is certainly very asynchronous and there is no longer any compelling reason for the matter to leave the system. As a result a binary with a deep, roughly spherical and still expanding common envelope forms.

2.2. Observational arguments

Observational arguments can be divided into three categories.

A) The statistics of observed Algol-type systems evolving in Case A and B and comparison with theoretical evolutionary time scales indicates a strong deficit of systems evolving in Case A. The most likely explanation of this deficit is that binaries evolving in Case A quickly form deep common envelope configurations and as such are lost from the statistics of Algol-type systems (Ziółkowski, 1976b)

B) Some systems like SV Cen and probably V 367 Cyg are now observed to be in the very process of building up a common envelope (Kreiner and Ziółkowski, 1978). These common envelopes are not very thick yet (otherwise we would not observe them as binary systems) but the observed evolutionary trend (mass transfer) indicates that they might become quite thick after only a few thousand years.

C) We observe many close binaries the origin of which can be understood only by assuming dramatic angular momentum loss (and some mass loss) from the systems during their past evolution. One can mention here a broad class of cataclysmic binaries (novae and dwarf novae) and some interesting individual systems like: (1) the binary pulsar PSR 1913+16 (e.g. Smarr and Blandford, 1976), (2) UU Sge (Bond et al., 1978), (3) V 471 Tau (e.g. Paczyński, 1976), (4) PG 1413+01 (Green et al., 1978), (5) UX CVn = HZ 22 (Schönberner, 1978), (6) LB 3459 (Dearborn and Paczyński, 1978) and (7) BD $-3°5357$ (Dworetsky et al., 1977). The common feature of all these systems is that they have very short orbital periods but contain components which are very advanced from the evolutionary point of view (white dwarfs, hot subdwarfs or neutron stars). From the theory of evolution of single stars we know that such objects could be incubated only inside red giants or supergiants of large radius and luminosity. This implies that the initial separations between the components of the

systems discussed above had to be much larger than at present and that their initial orbital periods had to be of the
order of years. For example, in the case of V 471 Tau, Paczyński (1976) found the initial orbital period of the order
of 10 years, while the present period is only 12 hours. The
only known way to decrease so dramatically the size of the
orbit is to evolve the system through the phase of a deep
common envelope which is subsequently lost by the system
(see the next section).

2.3. Mass loss from common envelope binaries.

Common envelope binaries undergo mass loss probably
during two phases of their evolution. The first of these
- discussed earlier - occurs soon after formation of the
common envelope, at the moment when its surface reaches the
L_2 point. As described earlier, some matter will escape
from the vicinity of L_2, but due to the considerable angular
momentum loss accompagnying this process, the binary orbit
will shrink rapidly and mass loss will probably have to
cease soon. The fraction of the mass lost at this phase is
unknown but might be quite small - perhaps of the order of
few percent.

The subsequent evolution of a common envelope binary
was discussed qualitatively by Paczyński (1976). Numerical
calculations of this phase of evolution were carried out
by Taam et al. (1978), Tutukov and Yungelson (1979) and
Delgado (1978) for somewhat different systems in which one
of the components is a neutron star or a white dwarf. However
uncertain the details of such evolution, the basic results
can be summarized as follows. The main mechanism responsible
for the evolution is the drag force experienced by the components of the binary system (or their dense cores) moving
through the matter of the common envelope. The action of
this drag force will transfer energy and angular momentum
from the orbital motion to the common envelope. This will
have two effects: (1) rapid shrinking of the binary orbit
and (2) increase in the luminosity of the envelope (during
most of the evolution the drag luminosity is much larger
than the intrinsic luminosity of both stars) and its further
expansion. When the luminosity and the dimensions of the
envelope are large enough, different instabilities similar
to those responsible for planetary nebulae ejections from
single stars (e.g. Iben, 1974) will appear and as a result
the common envelope will be lost. The development of these
instabilities is easier than in single stars due to large
energy generation (drag luminosity) at the base of the envelope. An additional mechanism facilitating the loss of
mass from the envelope is the generation of acoustic waves
by two dense cores orbiting each other inside the envelope.

If the two cores are not merged prior to the loss of the
envelope, then the final product will be a low-mass binary
with very short orbital period. In some cases one or both
components of such a system could be in an advanced phase
of evolution (white dwarf, neutron star). The existence of
the objects discussed in section 2.2.(C) indicates that the
evolutionary scenario as described above is indeed realized
by stars. In at least one case (UU Sge) we even observe the
remnant of the ejected common envelope, in the form of a
planetary nebula around the binary system.

3. STELLAR WINDS IN BINARIES

The evolution of single stars with mass loss due to
stellar wind was discussed in detail in previous talks by
de Loore and Chiosi. Their analysis covered wide ranges of
stellar masses and of stellar wind strengths and was based
on extensive numerical calculations. Similar calculations
for stars which are members of binary systems are rather
scarce. I know about only two papers which might be relevant:
Ziółkowski (1977) and Vanbeveren et al. (1978). The first
of these papers is essentially dealing with problems of
X-ray binaries and only the latter one is fully devoted to
evolution of binaries with stellar winds. Fortunately, the
evolution of single stars with mass loss due to stellar
wind is very similar to the evolution of the same star in
a close binary, assuming that the strength of the stellar
wind is the same. Usually this is not the case, and we shall
discuss this problem later; but if we know (or assume) the
proper rate for binaries, then we can use calculations for
single stars to describe quite adequately the evolution of
the components of binary systems. For this reason I shall
not discuss the evolution of the components (for this purpose, see papers by de Loore and Chiosi after selecting
the proper rate of mass loss) but I shall rather comment
on some problems that are specific for stellar winds in
binaries. Among these problems are: (1) the effects of stellar wind on orbital parameters, (2) the difference in stellar wind strength between single and binary stars and
(3) possible anisotropy of stellar winds in binaries.

3.1. The effects of stellar wind on the orbital parameters

As far as the changes of the orbital parameters are
concerned, the mass loss due to stellar wind can be well
approximated by the Jeans' mode of mass loss. Jeans' mode
assumes that the specific (per mass unit) angular momentum
of the matter lost from the system is equal to the specific
angular momentum of the mass-losing component. Since the
stellar wind is roughly spherically symmetric and has high

velocity, this assumption is well satisfied. It is well
known that for the Jeans' mode of mass loss, the orbital
period P and the separation of the components A are changing
with the total mass of the system according to the relations:

$$d \log P = - 2 d \log M \tag{5}$$

and

$$d \log A = - d \log M \tag{6}$$

This means that both the orbital period and the separation
of the components are increasing due to mass loss via stellar wind. It means also that the radius of the critical
Roche lobe around the mass-losing component:

$$R_1^{cr} = A(0.38 + 0.2 \log q) \tag{7}$$

will increase as well although slightly slower than A
(because q is decreasing). In the case of the binary system
discussed by Ziółkowski (1977) the strong stellar wind
decreased the mass of the primary from 35 $M_\odot$ to 13.5 $M_\odot$
by the end of the core hydrogen burning. During that time
the orbital period increased by a factor of $\sim$ 6, the separation of the components increased by a factor $\sim$ 2.5, and
the radius of the Roche lobe around the primary increased
by a factor $\sim$ 2. As we see, the expansion of the Roche lobe
may be quite substantial. Let us recall, in addition, that
strong stellar wind tends to decrease the stellar radius
starting from a certain point of the main sequence evolution.
These two effects together mean that in some systems the
presence of the stellar wind may prevent the Roche lobe
overflow which would happen otherwise. This fact has great
importance for the lifetimes of massive X-ray binaries
(Ziółkowski, 1977). An additional mechanism that might help
to stabilize massive components of X-ray binaries against
Roche lobe overflow is the "evaporative" stellar wind proposed by Basko et al. (1977). They suggested that strong
X-ray heating by the compact companion will stimulate the
stellar wind and increase the rate of the mass loss enough
to prevent the overflow of the Roche lobe.

3.2. Are stellar winds in binaries stronger than for single stars?

Asking such a question we have in mind stars that have
the same location in the H-R diagram and are either single
or members of binaries. The answer is definitely affirmative
and is based on the following arguments.

A) <u>Direct observational evidence</u>. Hutchings (1976)
found in his survey of mass loss due to stellar wind that
mass loss is systematically higher for stars that are proven

or suspected members of binaries. To give a clear example, the mass loss rate from Krzemiński's star was recently estimated (Hutchings et al., 1978) to be of the order of $\sim 5 \times 10^{-6}$ $M_\odot$/year, while the typical rate for single stars at the same location in the H-R diagram is $\sim 10^{-7}$ $M_\odot$/year (Hutchings, 1976).

B) <u>Analysis of the O-B components of massive X-ray binaries</u>. Masses and luminosities of some of these stars are known well enough to permit quantitative analysis. In every well investigated case, the O-B component appears to be significantly overluminous for its mass or significantly undermassive for its luminosity (Ziółkowski, 1977). We are reasonably sure that this mass deficit could not result from the Roche lobe overflow, because in a massive X-ray binary the mass ratio is usually so high ($q \sim 15$) that any outflow from the L_1 (or L_2) point will cause such dramatic shrinking of the orbit that a common envelope configuration will form almost immediately. It follows then that the mass deficit in the O-B components of X-ray binaries is due to mass loss via stellar wind during their past evolution. From the known evolutionary time-scales, we can estimate the stellar wind strengths necessary to produce the observed mass deficits. For two of these stars (Krzemiński's star and Sk 160) evolutionary models were constructed by Ziółkowski (1976c, 1978). He found that the initial masses of these stars had to be respectively ~ 36 $M_\odot$ and ~ 31 $M_\odot$ (present masses are about 18 $M_\odot$ for both stars) and that both stars are still in the main sequence (core hydrogen burning) phase of evolution. The average rate of mass loss during the past evolution had to be for both stars of the order of 3×10^{-6} $M_\odot$/year. This value is consistent with the estimate of the present rate of mass loss from Krzemiński's star ($\sim 5 \times 10^{-6}$ $M_\odot$/year). Now, let us compare this picture with the mass loss calculated for single stars with similar initial masses. Using observational rates determined by Hutchings (1976), Czerny (1978) found that stars with initial masses < 40 $M_\odot$ will lose less than ~ 0.3 $M_\odot$ during their entire main sequence evolution. Using another set of empirical data, given by Barlow and Cohen (1977), Chiosi et al. (1978) found that a star with initial mass of 36 $M_\odot$ will lose only 3 to 5 $M_\odot$ during its main sequence evolution. This comparison suggests that stellar winds in binaries might be stronger by an order of magnitude than those in similar single stars. A similar conclusion was reached by de Loore's group (de Loore et al. 1977, Vanbeveren et al. 1978). They determined the value of a parameter N, which is proportional to the rate of mass loss, for different observed O-B stars. For O-B components of X-ray binaries they found the value N $\sim 400 \div 500$. At the same time, inspection of Fig. 1 of de Loore et al. (1977) implies that for single stars in the relevant part of the H-R diagram the value of N is of the order of $30 \div 50$.

C) <u>Analysis of undermassive O-type components of other close binaries</u>. Altogether we know about 40 O-type binaries. Only few of them have mass determinations and in some of these (UW CMa, V 729 Cyg, LY Aur and perhaps also the others) one of the conponents seems to be undermassive (Conti 1978, 1979). In all these cases we observe evidence of strong stellar winds at present, but we cannot also exclude some large scale mass exchange in the past. For this reason, the analysis of these systems is not very conclusive.

D) <u>Theoretical arguments</u>. A semi-detached component of binary system has lower average surface gravity than the similar single star. This lower gravity (especially near the L_1 point) might increase the efficiency of whatever mechanism is responsible for the stellar wind. No quantitative analysis of this problem has been done. In X-ray binaries, the additional mechanism stimulating the stellar wind might be due to X-ray heating (the "evaporative" stellar winds of Basko et al., 1977).

3.3. Are stellar winds in binaries anisotropic?

The information concerning this problem is very scarce so far. We know one system (HD 47129) that is observed to change its rate of mass loss over the range $2 \div 8 \times 10^{-6}$ $M_\odot$/ year as a function of the orbital phase (Hutchings and Cowley, 1976). From the theoretical side, one could argue that the area on the surface of the star near the L_1 point has much lower gravity than the rest of the surface and this could produce an asymmetry in the stellar wind. Also the explanation of the large X-ray luminosity of SMC X-1 might be easier if Sk 160 (its optical companion) had an anisotropic stellar wind (Ziółkowski, 1978), but this is only one of the possible solutions.

On the other hand, it might appear that also some single stars have anisotropic stellar winds. Pismis (1979) investigated three emission nebulae excited by central stars of Of or WN-type with strong stellar winds. She found that the mass loss responsible for formation of the nebulae was anisotropic and suggested that also the present stellar winds from the central stars might be anisotropic.

REFERENCES

Barlow, M.J. and Cohen, M.: 1977, Astrophys. J. 213, p. 737.
Basko, M.M., Hatchett, S., Mc Cray, R. and Sunyaev, R.A.:
     1977, Astrophys. J. 215, p. 276.
Benson, R.S.: 1970, Bull. Am. Astron. Soc. 2, p. 295.
Bond, H.E., Liller, W. and Mannery, E.J.: 1978, Astrophys.
     J. 223, p. 252.

Chiosi, C., Nasi, E. and Bertelli, G.: 1978, preprint.
Conti, P.S.: 1978, Ann. Rev. Astron. Astrophys. 16, p. 371.
Conti, P.S.: 1979, this volume.
Czerny, M.: 1978, Acta Astron. (in press).
Dearborn, D.S. and Paczyński, B.: 1978, Monthly Notices
    Roy. Astron. Soc. (in press.)
Delgado, A.: 1979, this volume.
de Loore, C., De Grève, J.P. and Lammers, H.J.G.L.M.: 1977,
    Astron. Astrophys. 61, p. 251.
Drobyshevski, E.M. and Reznikov, B.I.: 1974, Acta Astron.
    24, p. 29.
Dworetsky, M.M., Lanning, H.H., Etzel, P.B. and Patenaude
    D.J.: 1977, Monthly Notices Roy. Astron. Soc. 181, p. 131
Flannery, B. and Ulrich, R.: 1977, Astrophys. J. 212, p. 533.
Green, R.F., Richstone, D.O. and Schmidt, M.: 1978,
    Astrophys. J. 224, p. 892.
Hall, D.S.: 1976, in "Structure and Evolution of Close
    Binary Systems", I.A.U. Symp. No. 73, eds. P. Eggleton
    et al. (Dordrecht: D. Reidel), pp. 283-288.
Hutchings, J.B.: 1976, Astrophys. J. 203, p. 438.
Hutchings, J.B. and Cowley, A.P.: 1976, Astrophys. J. 206,
    p. 490.
Hutchings, J.B., Cowley, A.P., Crampton, D. and van Paradijs,
    J.: 1978, preprint.
Iben, I.Jr.: 1974, Ann. Rev. Astron. Astrophys. 12, p. 215.
Kippenhahn, R. and Meyer-Hofmeister, E.: 1976, Astron.
    Astrophys. 54, p. 539.
Kreiner, J. and Ziółkowski, J.: 1978, Acta Astron. (in press)
Massevitch, A.G. and Yungelson, L.R.: 1975, Mem. della Soc.
    Astron. It. 46, p. 217.
Neo, S., Miyaji, S., Nomoto, K. and Sugimoto, D.: 1977,
    Publ. Astron. Soc. Japan 29, p. 249.
Paczyński, B.: 1971, Ann. Rev. Astron. Astrophys. 9, p. 183.
Paczyński, B.: 1976, in "Structure and Evolution of Close
    Binary Systems", I.A.U. Symp. No. 73, eds. P. Eggleton
    et al. (Dordrecht: D. Reidel), pp. 75-80.
Paczyński, B. and Ziółkowski, J.: 1967, Acta Astron. 17, p. 7
Pişmiş, P.: 1979, this volume.
Plavec, M.: 1968, Adv. Astron. Astrophys. 6, p. 201.
Plavec, M., Ulrich, R.K. and Polidan, R.S.: 1973, Publ.
    Astron. Soc. Pac. 85, p. 769.
Schönberner, D.: 1978, Astron. Astrophys. 70, p. 451.
Smarr, R. and Blandford, R.: 1976, Astrophys. J. 207, p. 574.
Svetchnikov, M.A.: 1969, "The Catalogue of Orbital Elements,
    Masses and Luminosities of Close Binary Stars", Ural
    University Press, Sverdlovsk.
Taam, R.E., Bodenheimer, P. and Ostriker, J.P.: 1978,
    Astrophys. J. 222, p. 269.
Thomas, H.C.: 1977, Ann. Rev. Astron. Astrophys. 15, p. 127.
Tutukov, A.V. and Yungelson, L.R.: 1978, preprint.
Tutukov, A.V. and Yungelson, L.R.: 1979, this volume.

Ulrich, R. and Burgher, H.: 1976, Astrophys. J. 206, p. 509.
Vanbeveren, D., De Grève, J.P., van Dessel, E.L. and
      de Loore, C.: 1978, preprint.
Webbink, R.: 1975, Ph.D. Thesis, University of Cambridge,
      Cambridge, England.
Yungelson, L.R.: 1973, Nauch. Inf. Astron. Sov. AN S.S.S.R.
      27, p. 93.
Ziółkowski, J.: 1971, unpublished.
Ziółkowski, J.: 1976a, Astrophys. J. 204, p. 512.
Ziółkowski, J.: 1976b, in "Structure and Evolution of Close
      Binary Systems", I.A.U. Symp. No. 73, eds. P. Eggleton
      et al. (Dordrecht: D. Reidel), pp. 321-322.
Ziółkowski, J.: 1976c, paper presented at the meeting of
      I.A.U. Commision 42, XVI I.A.U. General Assembly,
      Grenoble, France.
Ziółkowski, J.: 1977, in "Eight Texas Symposium on Relati-
      vistic Astrophysics", Annals New York Academy Sciences,
      New York, USA, 302, pp. 47-54.
Ziółkowski, J.: 1978, in "Nonstationary Evolution of Close
      Binaries", ed. A.N. Żytkow (Warsaw: Polish Scientific
      Publishers), pp. 29-54.

DISCUSSION FOLLOWING ZIOLKOWSKI

Cowley: Although the observations of the X-ray binaries
show they have undergone a very large percentage of mass loss
(more than 50%), in only one system (3U1700-37=HD 153919) do
we now see evidence of an exceptionally strong stellar wind
in the optical spectrum. This is not to say there is no wind,
but that it is not nearly large enough at present to account
for the earlier mass loss. For example SMC X-1 is the most
luminous X-ray binary, but we see none of the characteristic
signatures (as discussed by Hutchings) in the spectrum to infer a strong stellar wind. The present rate of mass loss is
typical for a single star of the same T_{eff} and L.

Conti: Mass loss deduced from the optical spectrum probably only detects the highest rates. We need the UV line
analysis to get better numbers for rates. 9 Sgr, an O4 main-
sequence star, might have a substantial rate.

Cowley: I would suppose your mass loss rates for the
companions of X-ray sources are lower limits, because some
material may have been transfered from the X-ray progenitor
to the present OB companion. This affects how much mass
needs to have been lost.

Conti: I should probably modify Ann Cowley's statement
HD 153919 (=3U1700-37) might have an anomalous carbon abundance. The λ4650 CIII emission, and λ5701, 5812 CIV lines

are stronger than in any other Of star (save one). Unfortunately there is not yet a detailed analysis as the relevant
line atomic physics is not available, nor are Of envelopes
wholly satisfactory. There exists only a possibility that
carbon is enhanced.

Morton: I do not think we can be sure that the absence
of strong visual mass loss features in a star requires that
the mass loss rate be less than from a star having such features. For example 9 Sgr (O4V) has only relatively mild visible emission features and $\lambda 4686$ is in absorption, whereas
the UV spectrum shows a highly saturated NV profile and one
of the largest terminal velocities. Clearly, some direct
estimates of $\dot{M}$ are needed from the UV spectra.

van den Heuvel: I think that for SMC X-1 and Cen X-3 you
do not need a strong wind, as the high mass transfer rate can
also be explained by Roche lobe overflow, as can in fact be
inferred partly from your own work. You have shown that due
to the mass loss the star can look like a giant while it is
still burning hydrogen. It has been shown by several authors
(McCray et al.1977, preprint; Savonije 1978, Astron.Astrophys.
          and: in press) that such a star can remain near
its Roche lobe and transfer mass at a moderate rate for as
much as 10^5 years, and in this way power a strong X-ray
source. In fact the agreement between this prediction and
the observations is very good in the cases of Cen X-3 and
SMC X-1.

Wilson: In SV Cen it appears that the envelope will
reach the outer contact surface when the masses are approximately equal. This is a coincidence which might have some
interesting consequences for the further evolution of the
system. In particular it should affect the efficiency of
ejection of mass from the outer Lagrangian point and thus the
formation or non-formation of a non-synchronous common outer
envelope. Have you looked into this point?

Ziolkowski: No, I have not.

Hutchings: V453 Sco (HD 163181) is very relevant to your
discussions. It is a B0Ia+O8V system in which the B star at
13 M_o is undermassive by a factor ~3. It shows mass transfer
via Roche lobe and stellar wind characteristics. Also, at
this level of mass loss we see CNO abundance anomalies. We
do not see such anomalies in the X-ray systems which have
lost only ~ half of their mass.

Dearborn: It should be noted that the very low mass of
Krzemenski's star, and most X-ray binaries require ^{14}N enhancements by a factor of 3 to 5. In some extreme cases where
the star is reduced to 1/3 of its original mass, a region is

reached where ^{16}O is partially converted to ^{14}N, and ^{14}N is enhanced by 10 times. Also in mass transfer systems, if mass is currently deposited on a star by one showing chemical peculiarities, both should show the same peculiarity.

Cowley: As nearly as we can tell from spectroscopic data which we have for all of the X-ray supergiant binaries, the only one which shows peculiar abundances is SMC X-1. However, although in that system the heavy elements are greatly weakened, this seems to be a general property of the Small Cloud as a whole and probably has nothing to do with the fact the system is an X-ray binary in which there has been much mass lost.

Vanbeveren: One has to be very careful with the argument about the X-ray luminosity of SMC X-1. The theoretical X-ray luminosity is very critically dependent on the value of the wind velocity in the vicinity of the neutron star.

Lamb: What is the maximum mass that can be lost from the binaries you have been discussing that will allow them to remain gravitationally bound?

Ziolkowski: I believe that if we have gradual,not instantaneous, mass loss, we can lose any amount of matter without disrupting the binary.

Plavec: Disruption of a binary system can occur in the case of a supernova explosion, where the sudden mass loss occurs on a time scale short even compared to the orbital period of the system. Dr. Ziolkowski has been talking here about an uncomparably much slower mass loss.

de Loore: Just a remark concerning your discussion about the calculation of evolutionary tracks with stellar wind: according to you one should stick as much as possible to the observations rather than using an equation for the mass loss. Nevertheless, the final products of our mass loss evolution with the Barlow and Cohen equation (for N=100) are comparable with those that were carried out with M values related immediately to the observations. So I do not see why one should not use a mass loss equation.

Sreenivasan: A few comments regarding the specific mass loss rate used for a given calculation are perhaps in order in response to what you said. I agree that it is desirable to use specific observed empirical rates when they are available. Unfortunately they are not always available. In addition, it is not appropriate to use a property of the flow to predict characteristics of that flow. It is better to employ reasonable physical arguments to estimate any para-

meters involved such as: ε or (α,k) or N in $\dot{M}=\dfrac{\varepsilon\ L}{V_{escape}\cdot V_{sound}}$;
or in Castor et al.'s rate for mass loss, which is a function
of L, Γ, k and α; or in $\dot{M}=\dfrac{NL}{c^2}$ used by de Loore et al. and then
compare the results with observations. "Appropriate" values
of these parameters are often employed by model makers to
claim that mass loss rates so used, agree with observations!
Perhaps, we should not be so afraid of the observers. It's
just probable that observed estimates are subject to uncertainties in or limitations of the measurements. They are
also often incomplete.

EVOLUTION OF MASSIVE COMMON ENVELOPE BINARIES AND MASS LOSS

Tutukov, A., Yungelson, L.
Astronomical Council of the USSR Academy of Sciences

ABSTRACT

A way of treatment of evolution of common envelope binaries based only on the laws of conservation of energy and angular momentum is suggested. It is shown that the final configuration depends on masses of components and initial period of the system, and on parameters describing friction in the envelope, and mass loss by the system. Possible final stages for massive binaries are either a Thorne-Zytkow type object for initially close binaries or a Wolf-Rayet star in pair with a relativistic compact remnant for wider ones. In the course of disruption of the latter system with orbital periods up to several hours very high space velocity (up to 500 km/s) pulsars can arise.

1. INTRODUCTION

The common envelope stage in the course of evolution of close binaries seems inevitable if we try to explain the origin of such objects as double cores of planetary nebulae, binary radio-pulsar, cataclysmic binaries, X-ray sources, because all of them probably lose the excess of mass and angular momentum of progenitor systems by mass loss from common envelopes. Also, from purely theoretical point, as was shown by Benson (1970) and Yungelson (1973), a common envelope binary inevitably forms if in a binary system the time-scale of mass exchange is shorter than the thermal time-scale of the accreting component.

The quantitative investigation of common envelope binary evolution is extremely complicated, as a lot of processes lack not only appropriate mathematical description, but even a precise physical formulation. Let us name some of them: friction between the envelope and the double

401

P. S. Conti and C. W. H. de Loore (eds.), Mass Loss and Evolution of O-Type Stars, 401–407.

core; generation of acoustic waves in the common envelope and their transformation into the shock waves in the upper part of the atmosphere; mixing of the matter in the envelope; mass loss by the system; tidal effects.

Below we suggest a simplified treatment of the common envelope binaries problem, based only on conservation laws for energy and angular momentum and on assumption that the momentum exchange between layers of a star does not occur, on the short thermal time-scale at least. The latter assumption is not valid when the secondary is in the convective part of the common envelope.

2. FORMULATION OF THE PROBLEM

We shall discuss a system consisting of a (super)giant and a compact object -- unevolved main sequence star, white dwarf or neutron star with initial mass ratio exceeding, say 3. We assume that the preceeding evolution of such systems follows the simple scenario outlined, e.g. by Paczyński (1971) or Massevitch et al. (1976). The zero-point of our discussion is the instant when the giant has overflowed the Roche lobe and just engulfed the compact object (CO). CO is surrounded by an accretion disc supported by mass transfer from the primary. We shall ignore the increase of the CO mass, which is insignificant during the period of existence of the common envelope that is close to the thermal time-scale. We assume in this paper that the radius of interaction of CO and the giant envelope is equal to the radius of CO lobe of Roche surface. We do not take into account the momentum exchange between orbital motion and axial rotation.

The orbital angular momentum of the system is

$$\mathcal{L} = G^{\frac{1}{2}} M_r \, m \, R^{\frac{1}{2}} / (M_r + m)^{\frac{1}{2}} \tag{1}$$

where M_r is the mass of the giant inside the orbit of CO with mass m, R -- distance between the centers of stars. If CO accelerates the resting element of the giant envelope to the angular velocity $\psi\omega$, where ω is the Keplerian angular velocity of CO, $\psi \leqslant 1$, then the moment transferred from CO to the giant per unit time is

$$\dot{j} = \alpha \dot{M}_r \, \psi \omega \, R^2 M_r^2 / (M_r + m)^2 \tag{2}$$

where $\dot{M}_r$ is amount of matter passing through the orbit of CO per unit time: $\dot{M}_r = 4\pi\rho R^2 (\dot{R} - \upsilon)$, where υ is the velocity of evolutionary expansion of the giant. If the system rotates rigidly, $\alpha \approx 2/3$, but varying α one may vary the efficiency of momentum transfer. Within our assumptions the only source of $\dot{j}$ is the orbital angular momentum. Then, differentiating (1) and equating the derivative to $\dot{j}$, we get

$$\dot{R}/R = \dot{M}_r/M_r \left[\frac{2\alpha\varphi M_r^2}{m(M_r+m)} + \frac{M_r}{M_r+m} - 2 \right] . \qquad (3)$$

Placing (2) into (3) and substituting variables: $y = M_r/4\pi\rho R^3$, $x = M_r/m$, where ρ is density, we get

$$\dot{R} = - \frac{\upsilon}{\frac{y}{\left[\frac{2\alpha\varphi x^2}{x(1+x)} + \frac{x}{1+x} - 2 \right] - 1}} . \qquad (4)$$

Assuming that the force of friction is proportional to the square of relative velocity we get that the transfer of momentum from CO to the spherical layer of the giant envelope is described by the equation

$$\gamma\rho(1-\varphi)^2 \omega^2 R^5 = \alpha \dot{M}_r \varphi\omega R^2 \quad , \qquad (5)$$

where the parameter γ depends mainly on the coefficient of friction and on geometrical cross-section of interaction. We can transform equation (5) into

$$\varphi = (1+\delta)(1 - \sqrt{1 - 1/(1+\delta)^2}) , \qquad (6)$$

where $2\delta = \alpha y\dot{M}_r/(\gamma\omega M_r)$ is by order of magnitude equal to the quotient of division of orbital period by the time-scale of evolutionary expansion of the giant. In the beginning of the common envelope stage CO moves in the low density layers and is able to effectively transfer angular momentum — φ is close to 1.

As the momentum is lost, CO moves into denser layers of the giant. The former interacts now with layers with large momentum of inertia. The momentum of CO is not high enough to accelerate these layers up to the velocity of CO. Because of this the loss of momentum and matter from the envelope selfaccelerates and further proceeds in the dynamical time-scale (CO "falls through the giant"). The same is true if CO moves in the convective layers of the common envelope. In this case evolutionary expansion of the giant does not play any role and from the condition of conservation of angular momentum it follows that

$$\varphi = (y+2)/(2\alpha x^2) + (y+1)/(2\alpha x) . \qquad (7)$$

Formally (7) is equivalent to the turning into 0 of the denumerator of (4). The boundary corresponding to the transition into "fall"-regime is marked by letter A in Figure 1.

If the geometrical cross-section of interaction is not fixed as now, but is determined by ram pressure, then its variations due to relative motion of CO and gas of the envelope may control the process of angular momentum transfer and the dynamical "fall"-stage may proceed rather slowly. It is possible that extremely luminous η Car and P Cyg type objects with very high mass loss rates are examples of binaries passing through such "slow fall" stage.

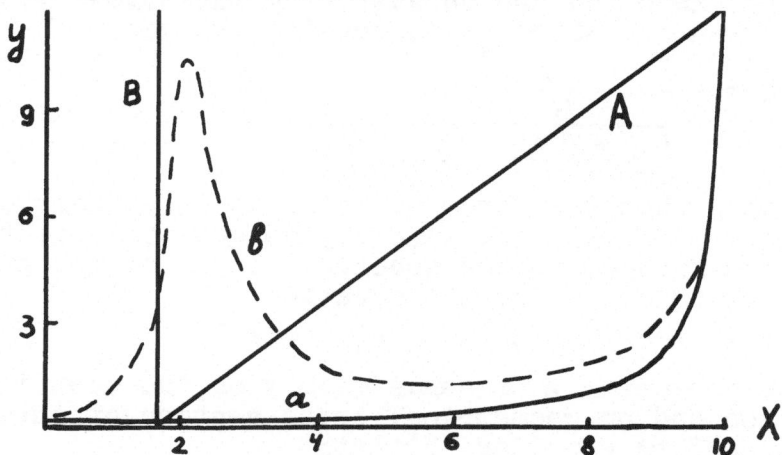

Figure 1. x-y plane. a -- curve corresponding to ZAMS star $10M$
b -- evolved $\log T_e = 3.6$, $10 M_\odot$ star, A,B -- lines
dividing regions of different regimes of spiral-in.

In the densest layers of the giant that experience expansion due to nuclear burning it is possible that CO moves away from the center. The right-side border of this region in the x-y plane is defined by the equation $2\alpha\varphi x^2/(1+x) + x/(1+x) - 2 =$ that formally corresponds to the turning into 0 of the right-side part of (3). In Fig. 1 the curve determined by the above equation for the case $\varphi = 1$ is marked as B.

The observational appearance and final configuration of systems with common envelopes depend on mass loss. Movement of a double core in a common envelope generates acoustic waves, which, propagating outward, transform into shock waves, heat the envelope and cause coronal-type mass loss. We assume that part β of gravitational orbital energy is spent on mass loss, but part $(1-\beta)$ is given off in layers immediately surrounding CO. Thus we introduce into the discussion an additional "shell-source" of energy with luminosity $L_{sh} = \ell_* + (1-\beta) G M_r \dot{R}/2/R^2$, where ℓ_* is the sum of the own luminosity of CO and the accretion-generated luminosity (equal to the critical luminosity in the upper limit). Thus the problem of evolution of common envelope binaries is completely defined by 3 parameters α, β, γ.

Let us get estimates concerning the fate of common envelope systems. Gravitational energy liberated when CO moves from the surface of the red giant to the surface of its helium core is $0.5GM_{He}m/R_{He}$. Energy necessary for ejecting of the envelope is $\psi GM^2/R$, where $\psi \approx 0.5$, because our computations show that dynamical"fall"-stage begins when the radius of the star is approximately two times greater than at the moment of contact of CO and the giant. If $0.5\beta GM_{He}/R_{He} > \psi GM^2/R$, the envelope is dispersed before the compact object reaches the boundary of the helium core. The final system is then a double core surrounded by loose remnants of the giant envelope (a double nucleus of a planetary nebula for small mass stars). For hydrogen-shell burning stars with $M \gtrsim 10 M_\odot$, $L_*/L_\odot = 10^2 (M_*/M_\odot)^{2.15}$ (Popova et al., 1978), and $R_{He}/R_\odot = 10^{-0.75}(M_{He}/M_\odot)^{0.6}$ for helium cores (Tutukov et al., 1973). This allows us to transform the above condition for energy to

$$T_e^* \lesssim (\beta/\psi)^{0.5}10^{4.5}(L_\odot/L_*)^{0.1}(m/M_\odot)^{0.5}. \qquad (8)$$

Curves determined by (8) for $m = M_\odot$ and $m = 2M_\odot$ and $\beta = 1$, $\psi = 0.5$ are drawn in Fig. 2 along with evolutionary tracks for massive stars (Popova et al., 1978). If $m/M_\odot = 10^{-1.2} M_0/M_\odot$, where M_0 is initial mass of a star (Tutukov and Yungelson, 1973), and $\beta = 1$, $\psi = 0.5$, we obtain the curve drawn in Fig. 2 by dashed line. If at the instant of

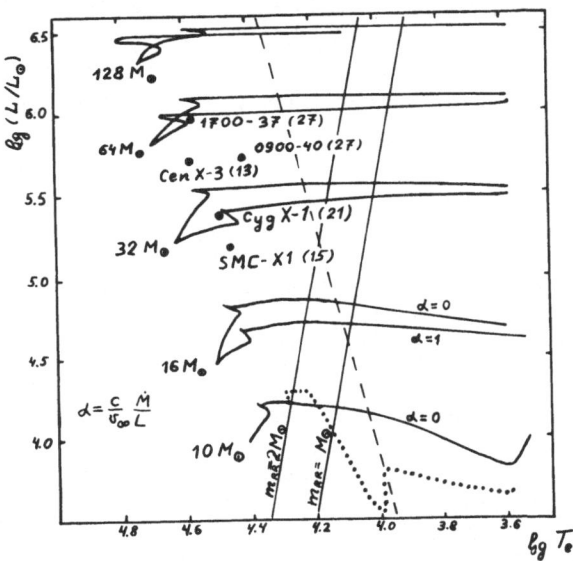

Figure 2. Evolutionary tracks of massive stars in plane. Positions of optical components of some X-ray binaries are marked. The numbers are their masses in solar units.

formation of a common envelope $T_e > T_e^*$, nuclei of the system will merge giving birth to a Thorne-Zytkow-type (1977) object. Lifetime of such objects is probably limited by mass overflow from the envelope and is unlikely to be higher than

10^7 years. Otherwise we should see $\sim 10^2$ times more red or infrared supergiants. Systems with $T_e < T_e^*$ will evolve into systems containing a hot helium star (like Wolf-Rayet stars) with a compact companion. Orbital periods of such binaries are of the order of several hours or days. Supernova explosion and disruption of the system gives birth to two single pulsars with spatial velocities up to ~ 500 km/s as the highest velocity observed for pulsars. Order of magnitude estimates are confirmed by the results of evolutionary computations of the common envelope binary $10 M_\odot + 1 M_\odot$ (Fig. 2). CO was a white dwarf with the own luminosity $\ell_* = 3{,}10^3 L_\odot$, $\beta = 0.5$. The computations were performed according to the procedure outlined above. In agreement with the above considerations three evolutionary phases were distinguished: the first -- slow, the second -- fast, the third -- again slow. In the last model computed CO was in the layer $M_r \approx 2 M_G$ $R \approx R_\odot$, mass of the giant was $7 M_\odot$. The amount of released gravitational energy was not great enough to disperse the whole envelope. The further evolution of the system will be determined by diffusion of angular momentum from the double nucleus (that was not considered in this paper).

In Fig. 2 positions of optical counterparts of some X-ray sources, according to Conti (1977), are marked. As indicated by position of stars relative to the dashed curve in Fig. 1, stars constituting the X-ray systems can merge.

REFERENCES

Benson, R.S.: 1970, Bull. Amer. Astron. Soc., 2, p.295.
Conti, P.S.: 1977, Preprint.
Massevitch, A., Tutukov, A., Yungelson, L.: 1976, Astrophys. Space Sci., 40, pp.115-134.
Massevitch, A., Yungelson, L.: 1975, Mem. Soc. Astron. Ital., 46, pp.217-230.
Paczyński, B.: 1971, Ann. Rev. Astron. Astrophys., 9, pp.183-208.
Popova, E., Tutukov, A., Yungelson, L.:1978, in preparation.
Thorne, K., Zytkow, A.N.: 1977, Astrophys. J., 212, pp.832-858.
Tutukov, A., Yungelson, L.: 1973, Naut. Inf. of the Astron. Council of the USSR Acad. of Sci., 27, pp.70-85.
Yungelson, L.: 1973, Naut. Inf. of the Astron. Council of the USSR Acad. of Sci., 27, pp.93-98.

DISCUSSION FOLLOWING TUTUKOV and YUNGELSON

Underhill: About 4 years ago Stecher and Sparks suggested that the infall of a white dwarf companion of a red giant
might cause a supernova explosion. Do your more detailed results support their conclusions?

Tutukov: In the frame of our local approach the process
of "spiral in" occurs on a thermal time scale mainly, and the
orbits of the core's components remain close to ring. Only
part of evolution can proceed on a time scale comparable with
a dynamical one. But we hope that the selfadjustment of effectiveness of the momentum loss process by ram pressure can
slow down the "spiral in" during that stage. The existence
of the binary pulsar supports our hope that "peaceful coexis-
tence" of the components of the double core inside a supergiant is possible in some cases at least.

Bidelman: For some time I have been wondering whether
the old hypothesis that cepheid variables are actually close
double stars should be resurrected.Recent investigations of
common-envelope binaries make this a far more attractive idea
than in the past. I think it is now well worthy of theoretical investigation.

Dearborn: When the compact object spirals in, and trans-
fers angular momentum to the envelope, it may produce a rapidly rotating star. Do your calculations indicate whether
this effect will be large enough to be observable.

Tutukov: The compact object spirals down in the expanding envelope of a supergiant with the increasing moment of
inertia of its envelope. So during this stage this conver-
sion of the orbital angular momentum into rotational angular
momentum almost does not influence the rotational velocity.
If in the red supergiant stage at least a part of the rigidly
rotating convective envelope is lost, the excess of angular
momentum will be lost too. But if the angular momentum loss
is not so effective, then it would be possible to find blue
supergiants with an excessively high velocity of rotation.

THE INFLUENCE OF STELLAR WIND MASS LOSS ON THE EVOLUTION OF MASSIVE
CLOSE BINARIES.

D. Vanbeveren, J.P. De Grève, C. de Loore
Astrophysical Institute, Vrije Universiteit Brussel and
E.L. van Dessel
Royal Belgian Observatory.

1. INTRODUCTION

It is generally accepted that massive (and thus luminous)
stars lose mass by stellar wind, driven by radiation force (Lucy and
Solomon, 1970; Castor et al. 1975). For the components of massive
binary systems, rotational and gravitational effects may act together
with the radiation force so as to increase the mass loss rate. Our
intention here is to discuss the influence of a stellar wind mass loss
on the evolution of massive close binaries. During the Roche lobe overflow phase, mass and angular momentum can leave the system. Possible
reasons for mass loss from the system are for example the expansion of
the companion due to accretion of the material lost by the mass losing
star (Kippenhahn and Meyer-Hofmeister, 1977) or the fact that due to
the influence of the radiation force in luminous stars, mass will be
lost over the whole surface of the star and not any longer through a
possible Lagrangian point as in the case of classical Roche lobe overflow (Vanbeveren, 1978). We have therefore investigated the influence
of both processes on binary evolution. Our results are applied to 5
massive X-ray binaries with a possible implication for the existence of
massive Wolf Rayet stars with a very close invisible compact companion.
A more extended version of this talk is published in Astronomy and
Astrophysics (Vanbeveren et al. 1978; Vanbeveren and De Grève, 1978).
Their results will be briefly reviewed.

2. THE PHASE OF STELLAR WIND MASS LOSS

Taking into account the similarity in evolution between a
single star and a binary component before the Roche lobe overflow phase,
it is clear from an observational point of view that components of
massive binary systems will lose mass at a rate of 10^{-7} $M_\odot$/yr to
10^{-5} $M_\odot$/yr. In our computations, we have used the equation for the
stellar wind mass loss introduced by de Loore (these proceedings). As
the parameter N in this formula is assumed to be constant, our results
may be considered as average results; our conclusions are independent

409

P. S. Conti and C. W. H. de Loore (eds.), Mass Loss and Evolution of O-Type Stars, 409–414.
Copyright © 1979 by the IAU.

of that assumption. We will return to this in section 4.
a) For a more massive (hence more luminous) primary we may reasonably
expect that its mass loss rate is higher than the $\dot{M}$ value of the companion. The mass ratio increases therefore with time and approaches
unity.
b) Assuming a spherically symmetric stellar wind, the angular momentum
loss can be described by a Jeans like mode (Huang, S.S., 1963). In that
case the period of the system increases. For this reason, the probability for the occurrence of a case B (Roche lobe overflow appears during
the hydrogen shell burning phase) increases if stellar wind mass loss
is included.

3. THE ROCHE LOBE OVERFLOW PHASE

        In view of the foregoing remark we considered only case B
systems. We computed different cases for the mass and angular momentum
loss (Δm and ΔH) from the system during the Roche lobe overflow. It
turns out that the final mass and luminosity of the primary are largely
independent from the choice of Δm and ΔH. The reason is tha' the moment
of He ignition is almost independent from the behaviour of the expanding envelope. However, as the formation of the He core depends on the
total mass loss by stellar wind, we get different results for different
stellar wind losses. The relation between initial and final masses can
be seen in Figure 1.

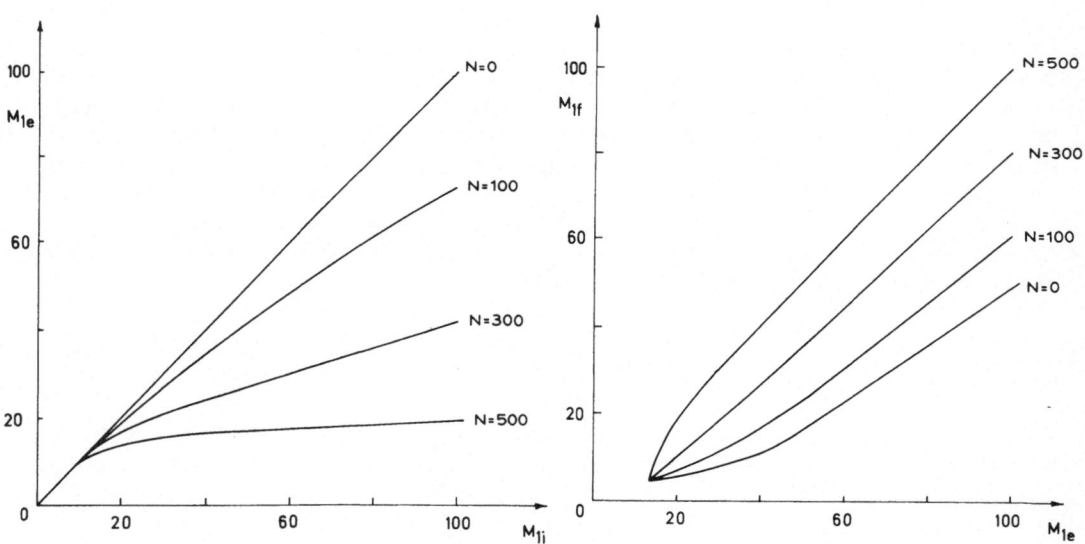

Figure 1a. The mass of the primary at the end of the core H burning
phase (M_{1e}) as a function of ZAMS mass (M_{1i}) for different stellar wind
cases.
Figure 1b. Remnant primary mass after lobe overflow (M_{1f}) as a function
of M_{1e} for various preceding stellar wind cases.

4. APPLICATION TO 5 MASSIVE X-RAY BINARIES

In this section we shall focus our attention to systems
consisting of a massive OB primary at the end of the core H burning
phase and a compact companion (neutron star or black hole).

More detailed results concerning this application to the
massive X-ray binaries Vela X-1, Cyg X-1, 3U 1700-37, SMC X-1 and
Cen X-3 will be published in Astronomy and Astrophysics (Vanbeveren and
De Grève, 1978). We will recall in this section some of the overall
conclusions of this paper.

Assuming the optical components of the massive X-ray binaries
to be normal stars at the end of their core hydrogen burning phase, it
was shown in the paper of Vanbeveren et al. (1978)(and reviewed in the
talk of C. de Loore) that the 5 systems are situated between the N=300
and N=500 tracks in the luminosity-mass diagram. Using the system parameters given by Conti (1978) we have computed the final mass after the
Roche lobe overflow phase for both values of N (Table 1). The critical
Eddington luminosity for X-rays is reached for accretion rates of the
order of 10^{-6} $M_\odot$/yr; the supergiant, when overflowing its Roche lobe
loses some 10^{-4} $M_\odot$/yr, hence most of the material leaves the system.
On the other hand, a large angular momentum loss may be expected during
the Roche lobe overflow if mass is leaving the system (Vanbeveren et
al. 1978). Therefore, considering the periods of the X-ray binaries,
the final systems will consist of a massive He star (a Wolf Rayet) and
a very close compact companion. As the observed mass loss rates during
the Wolf Rayet stage are very high, the X-rays are extinguished and the
system appears like a single Wolf Rayet star. Comparing the lifetimes
of the different evolutionary stages (t_{WR+OB}, $t_{WR+compact\ star}$, t_{X-ray})
and taking into account that about 50 % of all well studied WR stars
appear in binaries with a massive OB companion (Underhill, 1966) it
seems plausible from a statistical point of view, that a large number
of the remaining 50 % (seen as single WR stars) are in fact very close
binaries with a compact companion.
If the tracks for N=300 and N=500 in the mass-luminosity diagram may be
considered as boundary curves for the domain of the X-ray binaries, we
can have an indication about the expected N(H)/N(He) ratios for the
systems (Figure 2). It seems that for at least 4 of the 5 systems, the
N(H)/N(He) ratio is considerably lower than the normal value of 10.7
(assuming an initial composition X=0.70, Z=0.03). It would be interesting to have more observations in order to compare them with the above
model.
If N is variable, instead of constant, one finds that :
a) the mass ratio variation, the probability for the occurrence of a
case B and the final mass and luminosity of the Roche lobe overflow
phase increase compared with the N = constant computations;
b) the H abundance in the surface layers at the end of the hydrogen
core burning phase is lower.
Hence a variable N strengthens the conclusions concerning the influence
of the stellar wind on binary evolution.

System	M_{1e} ($M_\odot$)	M_{2e} ($M_\odot$)	P_e (days)	M_{1f} ($M_\odot$) N=300	M_{1f} ($M_\odot$) N=500	M_{1f} ($M_\odot$) N=0
Vela X-1	25	2	8.959	13.3	24	7.8
Cyg X-1	25-30	9-14	5.607	13.3-17.5	24-30	7.8-9.9
3U 1700-37	27-35	1.3-2.5	3.412	15-21.7	27-35	8.8-11.2
SMC X-1	~ 30	~ 1.5-4	3.893	~ 17.5	~ 30	~ 9.9
Cen X-3	~ 18	~ 0.7-1.2	2.087	~ 7.4	~ 16	~ 4.5

Table 1. Masses of primaries of massive X-ray binaries after Roche lobe overflow (M_{1f}) for different stellar wind rates. The observed data M_{1e}, M_{2e} and P_e are taken from Conti (1978).

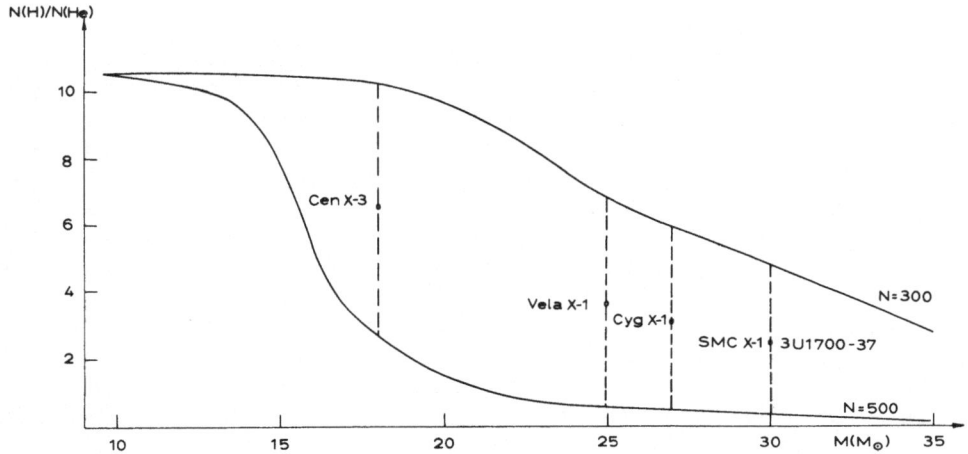

Figure 2.
The N(H)/N(He) ratio as a function of the mass at the end of core
Hydrogen burning; the locations of 5 massive X-ray binaries are indi-
cated.

REFERENCES

Castor, J.I., Abbott, D.C., Klein, R.I. : 1975, Astrophys. J. 195, 157.
Conti, P.S. : 1978, Astron. Astrophys. 63, 225.
De Loore, C. : 1979, IAU Symposium N° 83.
Huang, S.S. : 1963, Astrophys. J. 138, 473.
Kippenhahn, R., Meyer-Hofmeister, E. : 1977, Astron. Astrophys. 54, 539.
Lucy, P.B., Solomon, P.M. : 1970, Astrophys. J. 159, 879.
Underhill, A.B. : 1966, The Early Type Stars, Reidel, Dordrecht.
Vanbeveren, D. : 1978, Astrophys. Space Sci. (in press).
Vanbeveren, D., De Grève, J.P., Van Dessel, E.L., De Loore, C. . 1978,
        Astron. Astrophys. (in press).
Vanbeveren, D., De Grève, J.P. : 1978, preprint.

DISCUSSION FOLLOWING VANBEVEREN, DE GREVE, DE LOORE
and VAN DESSEL

Chiosi: I would like to comment on the adoption of
large mass loss rate (N=400) during the core H-burning phase
of the most massive component before it starts exchanging
mass. As a significant fraction of the massive stars appears
as single stars, I wonder if by adopting those high rate we
will still be able to reproduce the observed properties of
the HR diagram nearby the main sequence. It might however
be that mass loss by stellar wind is enhanced if the star is
in a binary system. Have you any guess about this enhancement? According to your results rather high mass loss rates
seem to be necessary for stars in double systems, at least
a factor of 5-10 larger than the estimate inferred from the
comparison of the occupation area of losing mass stars (near
the main sequence) with the theoretical models in core H-
burning.

Garmany: I may have missed this in your talk, but can
you say something about the time scale for the mass loss and
then for the Roche lobe overflow relative to the time spent
on the main sequence and post main sequence?

Vanbeveren: Roughly one can say that the hydrogen shell
burning stage (the post hydrogen core burning stage) is a
factor 10 smaller in time than the core hydrogen burning lifetime while the Roche overflow lifetime is approximately a
factor 10 smaller than the lifetime of the hydrogen shell
burning phase.

Henrichs: What are the reasons for assuming a constant
value of N during the whole evolution? Is it not preferable
to infer N as a function of $(L, \log T_{eff})$ from the observed
mass loss rates?

Vanbeveren: From the observations it is not clear whether
or not N is constant. Following the Barlow and Cohen values,
N should be constant. In any case, if we take a constant
value of N, this must be considered as an average value over
the whole main sequence. On the other hand, using different
functions for N will not change the general conclusions of
our calculations. Our purpose was to compare our results
with the results without a stellar wind and this comparison
is independent of the choice of the mass loss rate function.

COMMON ENVELOPE BINARIES AND MASS LOSS

A. Delgado
MPI für Physik und Astrophysik,
Föhringer Ring 6, 8 München 40, FRG

ABSTRACT

In this work we calculate the evolution of a binary system with a
common envelope, which consists of a blue supergiant and a neutron
star. We consider as a free parameter the effectivity with which the
energy liberated at the orbit produces mass loss from the system.
The evolutionary calculations were made, using various values of
this parameter, for a system with mass ratio 25:1. As initial state we
choose a model in the phase of Hydrogen-shell burning, before and
after the begin of Helium-burning in the core.
We found that, under certain conditions, it is possible for the
radius of the orbit and the period of the system to increase; the time
scale for the "spiral-in" would be of the order of 10^4-10^5 years. Mass
loss rates are between 10^{-3} $M_\odot$/y and 10^{-4} $M_\odot$/y.

INTRODUCTION

The aim of this work is to investigate the evolution of a massive close
binary system, consisting of a blue supergiant and a neutron star, during the phase of a common envelope. Following the normally accepted
scenario in the conservative case B (Kippenhahn and Weigert, 1967,
van den Heuvel and Heise, 1972), the system has undergone a supernova
explosion, before the first X-ray phase, leaving a main sequence O-
star and a neutron star, in a very excentrical orbit around it. The
orbit is circularised by tidal forces before the O-star envolves away
from the main sequence. For mass ratios small enough it becomes unstable and decays to the photosphere of the companion. The occurence
of a common envelope stage is made unavoidable (de Gréve et al. 1975).
Various authors have considered this problem more or less extensively
(Sparks and Stetcher, 1974, Chu et al., 1974, Paczynski, 1976,
van den Heuvel, 1976, Thomas, 1977, and, more recently, Taam et al.,
1978, Tutukov, 1978) but a good theoretical understanding is still
lacking. We present here a simple model which tries to make plausible
the possibility of getting a detached system as the end product of the

P. S. Conti and C. W. H. de Loore (eds.), Mass Loss and Evolution of O-Type Stars, 415–420.

common envelope stage. A more detailed treatment is left for future
work. In the next section our approach to the problem is described and
in section 3 the results are presented and some conclusions are outlined.

FORMULATION OF THE PROBLEM

We have calculated the evolution of a 25 $M_{\odot}$ star from the main sequence
up to a well advanced stage during Helium-burning in the core, using
the code described by Kippenhahn et al. (1976). Fig. 1 shows the variation of radius with time in the latter part of the evolution. Our calculations were made with two initial models, taken from this evolutionary sequence; one is in the phase of rapid expansion, corresponding to
the shell Hydrogen-burning phase with a neutral core. The other is in
the stage just after the beginning of Helium-burning in the core. It
should be pointed out that, from this diagramm, and also from observations of binary X-ray sources, the former case is the more probable
one. The initial period in this case is of the same order as that
given by the observations.

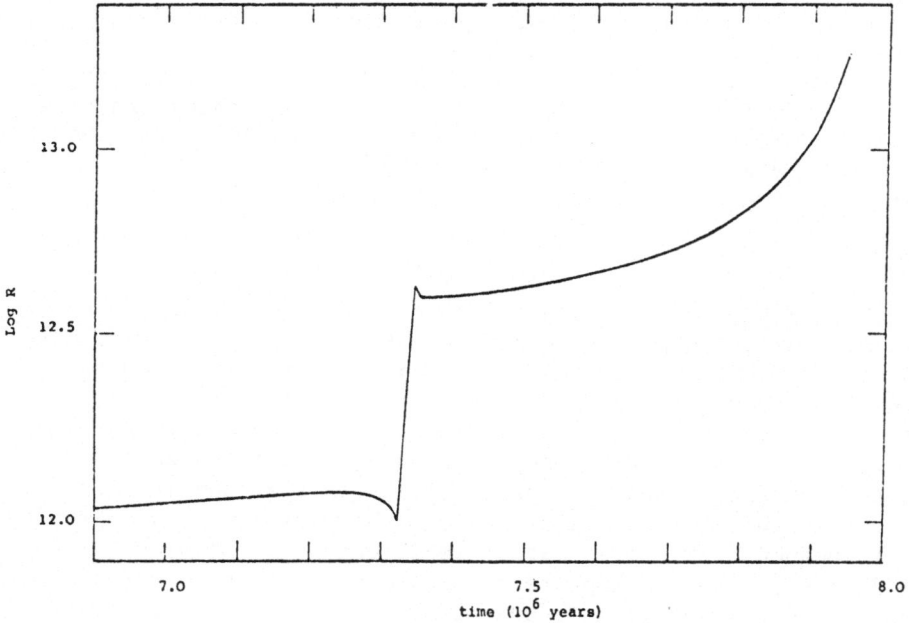

Fig.1: Evolution of the supergiant's radius with time.

We take as the initial state in our calculations the moment when
the neutron star penetrates the outer layers of its companion. It was
assumed that this star does not rotate, so that velocity of the neutron

star, relative to the medium, is equal to the orbital velocity. Balance of angular momentum determines the future evolution of the system.

From angular momentum conservation the equation of motion is

$$\frac{dr_B}{r_B} = - \frac{dM_B}{M_B} - \frac{2\dot{M}}{M_{ns}} dt \tag{1}$$

where r_B is the radius of the orbit, M_B the mass interior to it, M_{ns} the mass of the neutron star and $\dot{M}$ the rate at which mass is set into move by the neutron star and which is given orbital velocity. Equation (1) was obtained using the assumption that the time-scale of variation of the orbit is larger than the period, so that each orbit can be considered as Keplerian. Further we have assumed that the presence of the neutron star produces only a local perturbation on the companion which, therefore, can assimilate this perturbation without departing sensibly from hydrostatic and thermal equilibrium. This equation was integrated together with the stellar structure equations, using the initial conditions $r_{B_O} = r_+$ and $M_{B_O} = M_+$ at time $t = O$, where $+$ denotes the values for the supergiant at this time.

To calculate $\dot{M}$ we used the well-known schema of Bondi (Bondi and Hoyle, 1944). It assumes that a star moves supersonically relative to a homogeneous medium. It is reasonable to approximate the density as homogeneous if its scale hight is greater than about twice the accretion radius, R_A. This condition is well satisfied throughout the whole evolution. R_A and $\dot{M}$ are given by the following expressions

$$R_A = \frac{2GM_{ns}}{v^2} \tag{2}$$

$$\dot{M} = \pi R_A^2 \rho_B v \tag{3}$$

where ρ_B is the density at the position of the orbit and v is the relative velocity.

We take into account the effect of the accretion luminosity from the neutron star's surface by considering it to produce a force acting against the gravitational force of the neutron star. R_A and $\dot{M}$ then become

$$R_A = \frac{2GM_{ns}}{v^2} \left(1 - g \frac{L_{AC}}{L_{ED}} \right)$$

$$L_{ED} = \frac{4\pi c GM_{ns}}{\chi_{th}} \tag{4}$$

$$\dot{M} = 4\pi \rho_B \frac{\left(GM_{ns} - g \frac{\chi_{th} L_{AC}}{4\pi c}\right)^2}{v^3} \tag{5}$$

where L_{AC} is the accretion luminosity, L_{ED} the Eddington luminosity and g a factor varing between 0 and 1 which accounts for the possibility that L_{AC} might exceed the Eddington luminosity due to the asymmetry of the accretion flow. χ_{th} denotes the Thompson opacity.

In writing R_A and $\dot{M}$ in this form, it is assumed that the neutron star does not move more material than that which can be accreted, i.e., $\dot{M}$ is equal to the mass accretion rate. This is certainly an extreme assumption, but we only try to get an upper limit for the effect of this correction factor. Writing the accretion luminosity as

$$L_{AC} = \frac{GM_{ns}}{R_{ns}} \dot{M} \tag{6}$$

and using the transformations

$$x = \frac{v^3 \dot{M}}{4\pi \rho_B G^2 M_{ns}^2} \tag{7}$$

$$C = \frac{\chi_{th} \rho_B G^2 M_{ns}^2}{v^2 c R_{ns}} \tag{8}$$

equations (5) and (7) becomes

$$x = (1 - g Cx)^2 \tag{9}$$

$$x = \frac{2g C+1 - (4g C+1)^{1/2}}{2g^2 c^2} \tag{10}$$

from equations (7) and (10) one obtains the final form for the equation of motion

$$\frac{dr_B}{r_B} = - \frac{dM_B}{M_B} - 8\pi \rho_B G^{1/2} M_{ns} \left[\frac{r_B}{M_B}\right]^{3/2} x \, dt \tag{11}$$

Since $x \sim C^{-1}$ for $C \gg 1$ and $C \sim \rho_B$, the density dependence of the second term on the right-hand side of equation (11) is much weaker than in the case of unperturbed accretion radius.

Mass loss from the system is also considered. The energy liberated at the orbit is supposed to be immediately distributed on a spherical shell and transported from there to the photosphere, where it acts as an aditional source of radiation pressure and contributes to the mass loss from the system. The effectiveness of this energy in producing mass loss is measured by a factor, f, defined by

$$\frac{GM}{R} \dot{M}_L = f \dot{E}_{orb} \tag{12}$$

where $\dot{E}_{orb}$ is the rate of energy-generation at the orbit and $\dot{M}_L$ the mass-loss rate. f was considered to be a free parameter in the calculations. We have calculated evolutionary sequences for two values of g (1 and 0.2) and f ranging between 10^{-3} and 1.

RESULTS AND DISCUSSION

In table 1 are some relevant data from the evolutionary calculations made with initial models on the phase of rapid expansion. Mass loss rate, energy liberated at the orbit, and mass, bolometric magnitude, and period at the time given in the column on the right.

g	f	$\dot{M}_L (M_\odot/y)$	$\dot{E}_{orb} (L_\odot)$	$M_T (M_\odot)$	M_{B_T}	$P_T (d)$	$\tau (y)$
1	10^{-3}	10^{-5}	1.8×10^5	23	-8.50	1.81	5.2×10^4
1	10^{-2}	10^{-4}	1.8×10^5	14.41	-8.15	3.91	4.4×10^4
1	10^{-1}	10^{-3}	1.8×10^5	13.76	-8.00	3.40	1.0×10^4
1	2.3×10^{-1}	2×10^{-3}	1.8×10^5	8.5	-7.90	8.00	9.5×10^3
1/5	10^{-2}	7×10^{-4}	8.0×10^5	15.34	-8.01	3.81	1.4×10^4
1/5	5×10^{-2}	2×10^{-3}	8.0×10^5	8.4	-7.86	7.01	8.5×10^3

Table 1. Parameters for the evolutionary calculations with initial model in the phase of Hydrogen-shell burning with neutral core. The symbols are explained in the text.

Excepting the case with $f = 10^{-3}$, the supergiant has lost at this time a large part of its envelope and consists of a Helium-burning core, which contains about half of the total mass, and a hydrogen-poor envelope. The chemical composition at the surface is given by X = 0.52, Y = 0.48. At this point we could not calculate any further because of difficulties with the present code; the system should eventually become detached, when the supergiant has lost its whole hydrogen envelope, leaving a Helium star plus neutron star system with a period of the order of several days. Recent observations of the WR star HD 50896 (Firmani, 1978) could be explainable on the basis of a mechanism like that proposed in this paper.

We would like to finish with a remark about this possibility of increasing the period. As one can see from equation (1) two factors influence the variation of the orbit. The drag force (angular momentum loss) tries to bring the neutron star into the interior of the supergiant, whilst the expansion of the latter tries to increase the radius of the orbit. With these two processes we can associate two time-scales, $\tau_{orbital}$ and $\tau_{expansion}$. In our case the first time scale is determined by $\dot{M}$. The second one should be the thermal time-scale of the supergiant, since this is its expansion time scale in the evolutionary stage considered. The values $\tau_{orbital} = 10^6$ y and $\tau_{expansion} = 10^4$ y are obtained in the calculations, but, more generally, any factor reducing $\tau_{expansion}$ or increasing $\tau_{orbital}$, for instance, a faster expansion because of the heating at the orbit and, consequently, a decrease of the density at this position, could produce the effect of an increasing period. Calculations including this additional energy source in the energy equation and a more realistic formulation of angular momentum exchange were in the process of calculation at the date of this Symposium and will be presented in a future paper.

REFERENCES

Bondi, H. and Hoyle F., 1944. Monthly Notices Roy. Astr. Soc. 104, 273.

Firmani, C., this volume.

Chau, W.Y., Henriksen, R.N. and Alexander, M.E., 1974. Bull. of the American Astron. Soc. 6, 488.

de Gréve, J.P., de Loore, C., Sutantyo, W., 1975. Astroph. and Space Sci. 38, 301.

Heuvel, E.P.J., van den and Heise, J., 1972. Nature Phys.-Sci. 239, 67.

Heuvel, E.P.J. van den, 1976. IAU Symposium No 73 (ed. P. Eggleton, S. Mitton and J. Whelan), D. Reidel, Dordrecht-Holland.

Kippenhahn, R., Weigert, A. and Hofmeister, E. , 1976. Methods in Computational Physics 77, 129.

Kippenhahn, R. and Weigert, A., 1967. Z. Astrophys. 65, 251.

Paczynski, B., 1976. IAU Symposium No. 73 (ed. P. Eggleton, S. Mitton and J. Whelan), D. Reidel, Dordrecht-Holland.

Sparks, W.M., Stetcher, P.T. , 1974. Astroph. J. 188, 149.

Taam, R.E., Bodenheimer, P., Ostriker, J.P., 1978. Astroph. J. 222, 269.

Thomas, H.-C., 1977. Ann. Rev. of Astron. and Astroph. 15, 127.

Tutukov, A., this volume.

HD 50896: AN OTHER WR BINARY STAR

C. Firmani, G. Koenigsberger, G.F. Bisiacchi,
E. Ruíz and A. Solar
Instituto de Astronomía, Universidad Nacional Autónoma de México.

The current ideas concerning the evolution of close binary systems (van den Heuvel, 1976), accepting the hypothesis that the system is not disrupted by the first supernova (SN) explosion, predict that the Wolf-Rayet phase can occur twice. The first time the companion of the WR star is a normal OB star and the second time it is a collapsed object. In this context, the importance of searching for binary systems with collapsed companions among the "single" WR stars is evident. Due to its large distance from the galactic plane, z = 280 pc (Smith, 1968a), when compared with the average height (z = 60 pc, Cruz-González et al., 1974) of extreme Population I stars, HD 50896 was considered to be a likely candidate to this type of system.

It is important to note that this star belongs to a complete sample of WN stars brighter than v = 10^m and north of δ ≃ -25°. The statistical significance of the sample was suggested by Kuhi (1973). It contains seven binaries with OB companions and three supposedly single objects, two of which, HD 50896 and HD 192163, are associated with ring nebulae S 308 and NGC 6888, respectively. It is commonly believed that all WN stars associated with ring nebulae are single (Smith, 1968b). However, it must be noted that a ring nebula can be produced in a binary system by a spiral-in mass-loss mechanism, which could take place prior to the second WR phase.

In this framework, it would be very significant if HD 50896 were a binary system.

We have taken more than 100 spectra of this star in the wavelength range λλ4470-4800 A, with a very high signal-to-noise ratio, using the 1 meter telescope of the Tonantzintla Observatory and a SIT spectrophotometer. Figure 1 shows the spectra obtained from February to April 1977, and ordered according to a phase calculated with a period of 3.76 days. On this figure are indicated the phase, Julian Day - 2443000 for each spectrum, and the wavelength and identification of the more relevant features. The continuity of the profile variations of the NV blend at 4604-20 A and He II 4686 illustrates that the period is

P. S. Conti and C. W. H. de Loore (eds.), Mass Loss and Evolution of O-Type Stars, 421-424.
Copyright © 1979 by the IAU.

adequate. Upon including observations of October 1977 and February 1978, the periodical behavior as well as the period are confirmed.

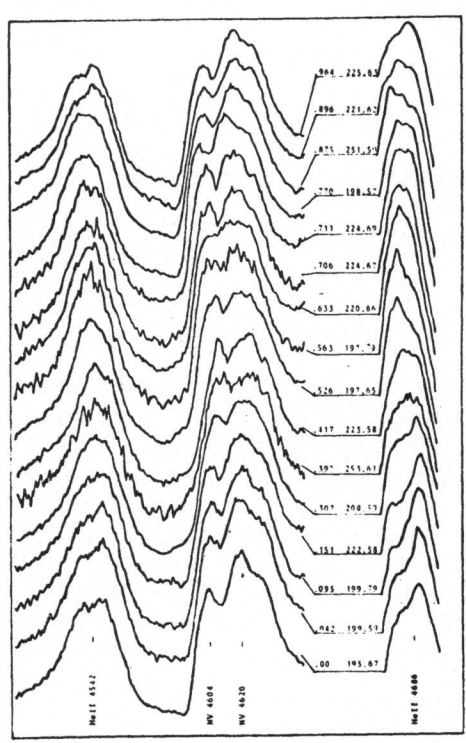

Differential photometry was made by E. de Lara and C.F. at the 1.5 meter telescope of the San Pedro Mártir Observatory from October 24 to November 7, 1977, with the 45 and 52 filters of the thirteen-color photometric system (Johnson and Mitchell 1975). The results ordered with the same phase used for the spectroscopic results are shown in Figure 2, and confirm the periodical nature of this star.

A 3.76-day period is too long to be compatible with radial pulsations (Stothers and Simon, 1970) and the periodical recurrence of similar spectral features over one year weakens a spot-rotation mechanism. Therefore, a binary nature for HD 50896 is strongly suggested.

Figure 1. Spectra of HD 50896, the phase obtained with a period P = 3.76 days. Only the top of HeII 4686 is shown.

The large line to continuum ratio, the absence of absorption lines and the small amplitude of the light curve favor a small-mass companion. Although there is a strong uncertainty due to the severe deformations in the varying profiles, the observations support radial velocity variations of about 1 A. This implies that the mass function is $\sim$ .1 M_O and for $M_{WR} \sim$ 10 M_O we obtain a mass for the companion between 2 and 4 M_O, depending on the inclination of the orbit (90° and 30°, respectively).

The large height above the galactic plane and the small mass of the companion support the idea that HD 50896 is a runaway and that the companion is probably a collapsar. This is, incidentally, also consistent with the energy involved in the variations of the He II line (10^{34} ergs s^{-1}). The absence of an X-ray source can be the result of the degradation of this radiation by the high density plasma of the WR wind, or the emission in the γ-ray region, which is plausible due to the high velocity of the material being accreted.

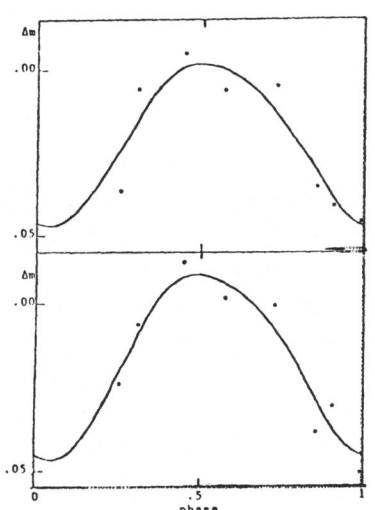

If we accept the hypothesis that HD 50896 is a WR + collapsar binary system, it remains to be explained why the number of these systems is so small compared to the WR + OB systems. The disruption of the binary system by the first SN explosion is an ad hoc explanation, and hence not attractive. We suggest that the explanation is related to the efficiency of the mass-loss mechanisms involved. That is, the first WR phase is a result of a Roche-lobe overflow mass-transfer mechanism, which is much more efficient than the spiral-in mechanism which produces the second WR stage. Therefore, because the He-core in both cases has presumably a similar life-time (Chiosi, private communication), the WR stage in its first occurrence lasts longer than in the second.

Figure 2. Differential photoelectric photometry of HD 50896 with the 45 and 52 filters. The phase is the same as in Fig. 1.

REFERENCES

Cruz-González,C., Recillas-Cruz,E., Costero, R., Peimbert, M. and
   Torres-Peimbert, S.: Rev. Mexicana Astron. Astrof., 1, 211.
Heuvel, E.P.J. van den: 1976, in Structure and Evolution of Close Binary
   Systems, IAU Symp. No. 73, P. Eggleton, S. Mitton and J. Whelan (eds.)
   D. Reidel, Dordrecht, p. 35.
Johnson, H.L. and Mitchell, R.I: 1975, Rev. Mexicana Astron. Astrof.,
   1, 299.
Kuhi, L.V.: 1973, in Wolf-Rayet and High-Temperature Stars, IAU Symp.
   No. 49, M.K.V. Bappu and J. Sahade (eds.) D. Reidel, Dordrecht, p. 205.
Smith, L.F.: 1968a, Monthly Notices Roy. Astron. Soc., 141, 317.
Smith, L.F.: 1968b, in Wolf-Rayet Stars, K.B. Gebbie and R.N. Thomas
   (eds.) U.S. Government Printing Office, Washington, D.C., p. 23.
Stothers, R., Simon, N.R.: 1970, Astrophys. J., 160, 1019.

DISCUSSION FOLLOWING FIRMANI

Niemela: Which mass did you use for the WN star?

Firmani: 8-10 solar masses.

Snow: How did you determine the energy required to per-
turb the He II λ emission line?

Firmani: We calibrated our instrument, which has a
linear response, by observing standard stars. We also adopted the absolute magnitude for HD 50896 determined by Morton.

Moffat: From my 1975 narrow band photoelectric observations (unpublished) of this star (HD 50896, WN5) on each of
37 continuous nights at La Silla, Chile, I found no clear
evidence for a real variation above that expected ($\lesssim 0^m.01$)
from the two constant comparison stars which were measured
simultaneously. The filters used had band passes of ~ 100 Å
centred on $\lambda\lambda \simeq$ 3635, 4680, 5640 the second of which is centred on the strong emission features of N III/V at λ ~ 4630
and He II at λ 4686.
(After returning home I plotted my photometric data in a 3.76
day phase diagram. The continuum shows a possible modulation
with peak-to-peak amplitude of 0.008 mag. while the quantity
m(4680)-m(5640) varies in antiphase with an amplitude of
0.012 mag. In addition, radial velocities os some lines on
12 Å/mm coudé plates taken in 1977 are compatible with a
period of 3.76 days).

Morton: Over howmany periods did you observe the light
variations?

Firmani: Three periods approximately.

Henrichs: Have you been able to determine the system
velocity?

Firmani: No, we believe that the variations of the line
profiles disguise the radial velocity effect.

GENERAL DISCUSSION

Underhill: The deduction of relative abundances of N, C, He and H from the strengths of the emission lines in Wolf-Rayet spectra is very tricky. We do not yet have a secure theory of how the spectral lines are formed and what controls the ionization balance under the different sets of state parameters (electron temperature, density, flow velocity, geometric size) that exist for Wolf-Rayet stars. Therefore, I urge caution in concluding that one or other element is over- or underabundant. In particular, some results which appeared recently in the literature (Willis and Wilson, 1978 M.N.) suggest that WN stars have a normal N/He ratio but that WN stars are deficient in C and H in contrast with Conti's statement that WN stars have an overabundance of N. The WC stars have an overabundance of C according to Willis and Wilson.

Vanbeveren: Did you use a spherical symmetric approximation for the radiation force? In that case I think that the shape of those critical surfaces (also computed by Kondo and McCluskey) is wrong as I said already earlier in this symposium. The picture changes totally if you include gravitation darkening.

Leung: Does your model take into account that the back side of the component cannot see the other star? A star is transparent to gravitation, but it is not transparent to radiation.

Van Blerkom: If a theoretician predicts all elephants have ten legs, but observers agree that they have four, can the theoretician insist that his model must be correct? This remark is a metaphor. For the nitrogen enhancements that the theoreticians insist must be present in the X-ray binaries, but the observers say is not at all apparent. Is there a problem here that should be discussed?
Conclusion at end of long discussion : All elephants have ten legs, but hide behind trees and only show four legs at any one time.

Dearborn: 1) The question that must be answered before it can be claimed that theory disagrees with observation, is how much ^{14}N enhancement is required to be observed. Walborn has indicated to me that the visible components of many X-ray binaries are not good candidates for observing ^{14}N enhancement (due to spectral type, rapid rotation, or poor data). The best candidate Cyg X-1 shows no ^{14}N enhancement, but the supposed mass loss is only marginally able to produce ^{14}N enhancement, a slightly lower initial mass, or higher observed mass is consistent with no enhancement. Composition

425

P. S. Conti and C. W. H. de Loore (eds.), Mass Loss and Evolution of O-Type Stars, 425–427.
Copyright © 1979 by the IAU.

can however give a significant clue to the amount of mass lost.

2) Dr. Chiosi is correct that the molecular weight gradient will drive thermohaline mixing and homogenize the envelope on a thermal time scale. The time required to convert ^{12}C to ^{14}N in the envelope of the now more massive mass accreting star is also short compared to the nuclear lifetime, so the point where $^{12}C = {}^{14}N$ moves outward in the mass accreting star. Therefore, when it begins to lose mass it does not have to lose all of the accreted mass plus its own original ^{12}C rich envelope. Mass transfer does not therefore allow a star to lose mass and not show ^{14}N.

van den Heuvel: For the X-ray binaries there is indeed the problem that people claim that they are very undermassive (have lost half of their mass by wind), which implies the prediction that they should show a nitrogen abundance anomaly which is not observed. I think that this may be a serious question.

Abbott: In regard to the relation of emission line strength of lines in the visible to mass loss rates, I would point out that the emission lines are measuring the gas density of the wind. The gas density depends not only on $\dot{M}$, but also on the velocity law, temperature, etc. Using the law $v(r) = v_\infty (1 - \frac{r_x}{r})^{1/2}$, the column density of material in the wind, $\int_0^\infty N_e dr$, scales as $\frac{\dot{M}}{v_\infty}$. For example even if ζ Pup and 9 Sgr had exactly the same $\dot{M}$, the column density of material in the envelope of ζ Pup would be ~ 30% larger. Since the emission line strength of a line like H_α scales

as $\int_0^\infty N_e^2 dr$, this dependence is exaggerated even more. Mass loss rates derived by Hutchings based on emission in the visible will be affected by this dependence on gravity. This will cause him to underestimate the rate of mass loss for the higher gravity main sequence stars.

Tutukov: We could explain overluminosity without too extensive mass loss at least of close binary components if we assume that duplicity promotes over mixing what increases the mass of the core enriched by helium. Observable mass loss rates give us no possibility to understand the absence of bright red supergiants.
That absence of very bright red supergiants ($L \geq 10^5 L_0$) is possible says us that for those supergiants the mass loss process is so extensive that stars are quite obscured by

circumstellar gas-dust envelopes. Probable candidates of
such stars could be OH/IR stars. When the hydrogen rich
envelope will be lost, the star quickly starts to move to
the WR region of the HR diagram. So, that scenario seems
now a probable one for the formation of single WR stars
and should be developed.

When the meeting was finally through
I asked "did you learn something new?"
After thought (at the bar)
He said that the star
Is γ Velorum not ².

SESSION 8

EVOLUTIONARY SCENARIO AND THE WR CONNECTION

Chairman: J.M. VREUX
Introductory Speaker: P.S. CONTI

1. A.F.J. MOFFAT and W. SEGGEWISS: The galactic WN7/WN8
   stars as massive O stars in advanced stages of
   evolution.

2. A.J. WILLIS and R. WILSON: The chemical nature and evolutionary status of the Wolf-Rayet stars.

3. E.M. LEEP: Of-WN evolution: spectral types and effective
   temperatures.

4. V.S. NIEMELA: Observations of velocity fields in WN and
   Of stars.

5. B. BOHANNAN: BE 381: WN9 or O8 Iafpe?

6. D. VANBEVEREN, J.P. DE GREVE, C. DE LOORE and E.L. VAN
   DESSEL: Conservative and non-conservative evolutionary computations in connection with Wolf-Rayet
   binaries.

Summary of Conference: E.P.J. VAN DEN HEUVEL

EVOLUTION OF O STARS AND THE WR CONNECTION

Peter S. Conti[*]
Joint Institute for Laboratory Astrophysics, University of
Colorado and National Bureau of Standards, and Department of
Astro-Geophysics, University of Colorado, Boulder, CO 80309

ABSTRACT

The stellar wind mass loss rates of at least some single Of type
stars appear to be sufficient to remove much if not all of the hydrogen-
rich envelope such that nuclear processed material is observed at the
surface. This highly evolved state can then be naturally associated
with classic Population I WR stars that have properties of high luminosity for their mass, helium enriched composition, and nitrogen or
carbon enhanced abundances. If stellar wind mass loss is the dominant
process involved in this evolutionary scenario, then stars with properties intermediate between Of and WR types should exist. The stellar
parameters of luminosity, temperature, mass and composition are briefly
reviewed for both types. All late WN stars so far observed are relatively luminous like Of stars, and also contain hydrogen. All early WN
stars, and WC stars, are relatively faint and contain little or no hydrogen. The late WN stars seem to have the intermediate properties required if a stellar wind is the dominant mass loss mechanism that trans-
forms an Of star to a WR type.

INTRODUCTION

It is with some irony that my talk today concerns mostly the status
of WR stars, whereas, as we are all aware, the Symposium that has brought
us all here is entitled "Mass Loss and Evolution of O-Type Stars." In
defense of this apparent disregard for the plans of the Scientific
Organizing Committee may I offer the following: At the IAU Symposium
#49 on "WR and High Temperature Stars" (Bappu and Sahade 1973) my paper
concerned Of stars; and at IAU Symposium #70 on "Be and Shell Stars"
(Slettebak 1976) my contribution with Stewart Frost discussed Oe stars.
Since we finally now have a Symposium devoted entirely to O-type stars
it then seems only fair to discuss WR stars.

It appears certain now that WR stars are the highly evolved remnants of the O stars that have evolved with mass loss of some kind.

431

P. S. Conti and C. W. H. de Loore (eds.), Mass Loss and Evolution of O-Type Stars, 431–445.

The populations and lifetimes (some 10^5 yrs) are <u>consistent</u> with the concept that <u>all</u> O stars more massive than 25 $M_\odot$ become WR stars (Smith 1973). I will first briefly review the status of O and Of stars. I will then consider what similar information we have about WR stars, and will emphasize the lack of quantitative data on these objects. Some preliminary highly suggestive results of newly acquired spectra of WN stars in the LMC will also be discussed.

Of TYPE STELLAR PARAMETERS

In an HR diagram the input parameters are luminosity and temperature, due essentially to various combinations of mass and composition. I will not discuss the composition in detail except to say it has been assumed to be "solar" for normal O-type stars. A few abundance analyses on <u>main-sequence</u> O types show this to be correct within a factor two. This is probably essentially correct for Of stars also. However, I have often worried that the H/He ratio could be different by factors of two or three from solar (e.g., enhanced helium) and we wouldn't know it from the spectra given the complications of line emission in these stars.

Masses can come directly only from those stars that are in binary systems; although it is possible to extract such information from independent measures of radius and surface gravity (spectroscopically) I do not think this method can ever be more promising than a factor two at best. Presently approximately 40 O-type binary systems are known, including the three involved with collapsed companions as X-ray sources (Conti 1978a). The O6.5III companion to the X-ray pulsar Cen X-3 has a very well determined mass of 18 $M_\odot$. In the other 37 systems masses can be determined exactly only for those few double-lined eclipsing systems. As Conti and Burnichon (1975) have pointed out, masses derived for those main sequence systems meeting these stringent observational criteria agree reasonably well with theoretical tracks.

So far, analyses have been completed for only five double-lined Of systems. The data are contained in Table 1. Only two of these systems eclipse so the values for the masses are lower limits in the other cases. The important fact about this table is the range in masses, from 9 $M_\odot$ to 58 $M_\odot$ for a similar spectral type: O7f. Clearly mass loss and evolutionary status are important parameters in these binary systems.

The luminosity-temperature relation for O and Of stars was extensively discussed by Conti and Burnichon (1975); a more recent diagram, which has been reproduced elsewhere in this Symposium, was constructed by Conti (1976). The O-type stars nicely fall into the expected region in the HR diagram compared to evolutionary tracks. The Of stars are invariably the more luminous stars and populate the upper portions of the diagram, being classified as late OB supergiants when their temperatures are sufficiently low that emission lines are not observed in the

Table 1.   Stellar masses -- Of types

Star	Types	$M\sin^3 i$	M_s/M_p	ΔM	Ref.
HD47129	O7If O7	58 64	1.1	$1^m.2$	1
29 CMa* UW CMa	O8.5If OB	20 24	1.2	$1^m.1$	2
HD149404	O8.5I O7III(f)	1.6 2.7	1.7	$\sim 0^m.5$	3
BD+40°4220* V729 Cyg	O7f O6f	31 9	0.29	$\sim 0^m$	4,5
HDE228766	O7.5 O5.5f	34 23	0.68	$\lesssim 0^m.5$	5

*Eclipsing system
References: 1. Hutchings and Cowley (1976); 2. Struve et al.
(1958); 3. Massey and Conti (1979 -- this Symposium);
4. Bohannan and Conti (1976); 5. Massey and Conti (1977)

blue region of the spectrum. The computational step in going from the
observed spectrum to temperature, and from the observed M_v to the luminosity are in reasonably good shape for O and Of stars.

Finally, what of the mass loss rates themselves? Extensive data
exist on only one Of star, ζ Pup. The four independent methods of UV
line analysis, Hα emission strength, IR free free measures and the radio
detection give a number of the order of $7 \times 10^{-6} M_\odot$ yr^{-1} (Conti 1978b).
I should remind all of you that it is believed that mass loss rate
scales with the Hα emission line strength: ζ Pup has one of the weaker
Hα profiles. As reported by Conti and Frost (1977) 6 out of 20 early
type O stars have Hα emission, and mass loss rates, of this order or
larger. Unfortunately, mass loss rates for O-type stars are not yet
available. It is expected, given the absence of emission lines in their
optical spectrum, that their rates will be below that of ζ Pup. However,
it is known (e.g., Snow 1979 -- this Symposium) that all of these stars
have stellar winds. The actual values for the rates of mass loss in
O-type stars is a very needed parameter for understanding their evolution, as the proceedings of this Symposium have stressed.

WR TYPE STELLAR PARAMETERS

As you all are aware, WR stars are believed to be helium burning
objects, in which the hydrogen rich envelope has been removed by an
extensive mass loss process, either due to mass exchange in a binary

(Paczynski 1973) or loss from a stellar wind (Conti 1976), or both.
The WN and WC subtypes are believed to be characterized by the presence
of enhanced nitrogen (WN) due to CNO cycle material being brought to
the surface, or enhanced carbon (WC) where the products of helium burning are exposed (Paczynski 1973). A recent abundance analysis of nine
WR stars by Willis and Wilson (1978) supports this general picture.

Stellar masses have been estimated from those eight double-lined
binary systems that are listed in Table 2. One problem is that invariably the radial velocity study concerns exclusively the emission
lines in the WR star and the absorption lines in the OB type. Although
this introduces an additional uncertainty in the values of K and the
masses, it cannot have a drastic effect. For the systems listed, the
WR star is probably always the fainter, based on eye estimates of line
intensities. The WR star would normally be called the secondary were
it not for its extraordinary spectrum. All WR stars in Table 2 are less
massive than their companions. Only three systems eclipse, so the

Table 2.   Stellar masses -- WR types

Star	Types	$M \sin^3 i$	M_{WR}/M_{OB}	Ref.
γ Vel	WC8 09I	17 32	0.53	1
HD90657	WN5 06	6.76 13.6	0.50	2
HD152270	WC7 08	1.85 6.85	0.27	3
HD168206* CV Ser	WC8 0	8.1 35.2	0.23	4
HD186943	WN4 B	3.36 7.94	0.42	5
HD190918	WN4 09I	0.21 0.78	0.27	5
HD193576* V444 Cyg	WN5 06	8.4 19.5	0.43	6
HD211853*	WN6 06I	11.5 33	0.35	7

*Eclipsing system

References: 1. Niemela and Sahade (1979 -- this Symposium);
2. Niemela (1976); 3. Seggewiss (1974); 4. Cowley et al.
(1971); 5. Bracher (1967); 6. Ganesh et al. (1967);
7. Stempien (1970).

masses are otherwise minimum values. Note that the WC8 companion to
γ Vel has a minimum mass of 17, nearly twice that of the Of secondary
in BD+40°4220 (Table 1). Considering the paucity of data in Tables 1
and 2 it would appear that an extensive observational program on more
systems would greatly improve our knowledge of their masses and
light ratios.

As for the luminosity and temperature of WR stars, these parameters are not as well known. What is reasonably well established is
the M_V for a number of stars, mostly those in the LMC (Smith 1968b)
where the distance modulus has been established from other objects.
Three cluster associations in our galaxy that contain WR stars --
Carina, Sco OB1, and TR 27 (Moffat, Fitzgerald and Jackson 1977) have
well-established distances. The single WR stars, classified by Smith
(1968a,b), which have distances from LMC or this cluster/association
membership, are plotted in Figure 1. This diagram represents what I
believe are the most reliable M_V data for <u>single</u> WR stars. I did
not use any stars with kinematic distances as discussed by Crampton
(1971), or any of the WR "pairs" of Smith (1968b), since I feel
these determinations are not as accurate as the galactic cluster
moduli. I did not plot any stars with companions (i.e., absorption
lines visible in their spectra) on the assumption that they are

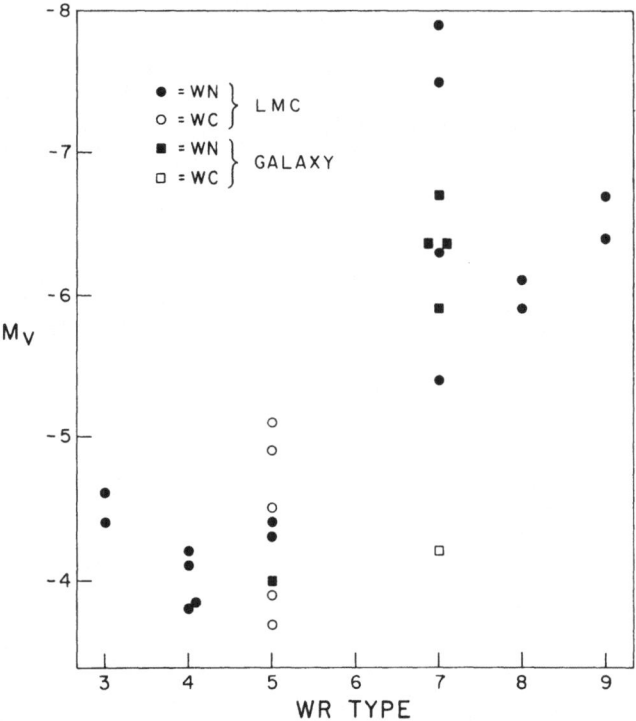

Figure 1. M_V for Wolf-Rayet stars. The values are for those
individual stars in the LMC, Carina, Sco OB1, or TR 27 which
have well-determined distances. See text.

binaries. I did plot those late WN types (Niemela 1979 -- this Symposium) in which absorption lines are identified in the WR star itself.
The two LMC WN stars plotted at type WN9 also have absorption lines.
Bruce Bohannan (1979) will discuss one of these two stars, BE 381,
later in this session.

The present M_V data on WR stars are admittedly sparse. There are
no WN6 stars identified in the LMC using Smith's classification,
although by Walborn's (1974) criteria three WN7 stars in Figure 1 would
be of this type. Three of the cluster WN7 stars are "transition" types
(Conti 1976). The WC star shown at type WC7 is a composite type WC7-N6
and needs to be investigated further. There are no late WC stars in the
LMC, for reasons that are not understood. Breysacher and Azzopardi
(1979 -- this Symposium) have also pointed out that all WR stars so far
identified in the SMC are early WN type. This is also not understood.

From Figure 1 the following statements can be made about the absolute visual magnitudes of WR stars (Underhill 1968; Smith 1973): The
WC stars all have an average M_V about -4.3, with a spread of about $1\overset{m}{.}2$.
The early WN stars also have M_V near this value, with a smaller range
among them. The late WN types are appreciably brighter with M_V averag-
ing about $-6\overset{m}{.}4$ and with a range in values of $2\overset{m}{.}5$. This large difference in M_V among the late WN stars requires comment. It would be nice
if there were some spectroscopic distinction for luminosity but a preliminary examination of image tube spectrograms of all of these stars
shows nothing clearcut. I frankly do not understand this large range
in M_V but it must be related to some combination of the mass and composition, i.e., how much of the star has been peeled away and from what
initial mass.

In Figure 1, the abscissa has been labeled with spectral type.
These have been adopted exclusively from Smith (1968a,b) who has done
the most extensive classification of WR stars in both hemispheres and
in the LMC. The WR star classification depends mostly, but not exclusively, on N III, N IV, and N V emission line ratios for WN stars and
C III, C IV emission line ratios for WC stars. This has been done only
qualitatively by eye estimates. Other classification schemes (e.g.,
Beals 1938; Hiltner and Schild 1966; Walborn 1974) have also generally
used line ionization criteria. A very disturbing feature of this entire classification problem is the lack of agreement among the different observers as to the types of identical stars.

What are the effective temperatures of WR stars? Clearly the
classification scheme itself describes an ionization and/or excitation
temperature (see, e.g., Bappu and Ganesh 1968). It is by no means certain that this envelope temperature is directly related to the effective
temperature as Anne Underhill has stressed many times (Underhill 1968,
1973). Without an envelope model, I believe that fitting continua
measurements is also fraught with difficulties. The reddening problem
at least appears to be in hand (Smith and Kuhi 1970) when relatively
narrow emission-line-free photometry is used.

Willis and Wilson (1978,1979 -- this Symposium), from a study of the UV lines and continua of nine WR stars, have evidence that they all have effective temperatures near 30,000°K. On the other hand, Ms. Leep (1979 -- this Symposium) has noted that those several late WN stars with absorption lines all could be consistently classified as early O type -- with effective temperatures near 45,000°K. If these authors are correct (no stars are in common) then it might be that early WN stars, and WC stars are cooler than later WN stars (i.e., there is no relation between the ionization/excitation value and the effective temperature -- Underhill 1973).

One other method has been used to determine effective temperatures of WN stars. This uses the Zanstra principle, extensively discussed by Morton (1969,1973) wherein the total flux in the H II region surrounding a single star can be estimated and the effective temperature inferred. This method alone avoids problems with the stellar envelope (to first order) but is, of course, not without its own difficulties. Following Smith (1973) and Morton (1973) I have adopted this scheme here but point out that only a few types have been measured: WN5, WN6, and WN8. No H II regions surrounding WC stars have yet been measured so no temperatures are available for them with this method.

I must now discuss the bolometric corrections (b.c.). As a zeroth order approximation it is probably reasonable to assume that the b.c. depends on the effective temperature alone. This seems to work reasonably well for O stars (Conti 1976) and probably Of stars also, but we have no independent confirmation of this for WR stars. There may well be drastic modifications to this simple assumption for WR stars with their extensive stellar winds. For mass loss rates as large as 10^{-4} $M_{\odot}$ yr^{-1}, the kinetic energy in the wind can be an appreciable fraction of the WR luminosity and adopting a b.c. from the effective temperature might well be an overestimate of the real value. In any case, for the purposes of discussion, I have assumed that the b.c. depends on the effective temperature only, adopting the values listed by Morton (1969) for OB stars.

Mass loss rates for WR stars seem to be typically a factor 10 larger than for Of stars, e.g., between 10^{-5} and 10^{-4} $M_{\odot}$ yr^{-1} (Conti 1978b; Willis and Wilson 1979), although only a few numbers are as yet available in the literature. Several have been detected as IR sources and as radio point sources due to emission from the surrounding envelope. If WR stars are burning helium in the core and have lifetimes on the order of a few 10^5 years (Smith 1973), then mass loss rates near the upper end of the above limits will be significant during their evolution.

EVOLUTIONARY STATUS

Figure 2 shows a theoretical HR diagram for Of and WN stars, with the location of the companions to the X-ray sources also indicated (from Conti 1978a). The Of stars are the same as discussed by Conti (1976).

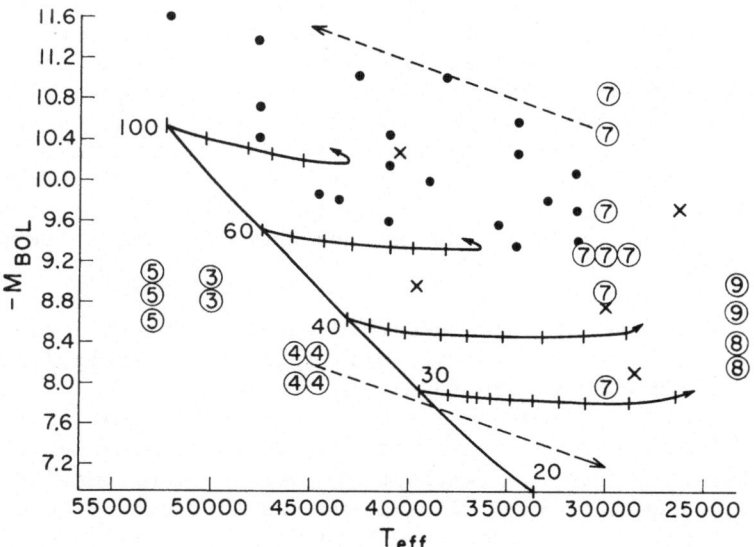

Figure 2. Theoretical HR diagram for evolution of massive
stars. The solid lines are for masses, as indicated, with
mass loss parameter N=300 (de Loore et al. 1978). The tic
marks on these tracks are every 5×10^5 yrs. The solid circles are the positions of Of stars (Conti 1976); the open
circles represent WN stars with an interior number showing
the type. The × represent the positions of binary companions
to X-ray sources (Conti 1978a). The dashed lines represent
different positions for the WN stars if alternative effective
temperatures are chosen. See text.

The M_{bol} for the WN stars come from those with M_v in Figure 1, and the
b.c. given from the T_{eff} values tabulated for the various types by
Smith (1973) and Morton (1973). The "WN9" stars are assumed to be
similar to WN8 stars for want of better information. The evolution
tracks in this figure are from de Loore et al. (1978) for various
stellar masses with a mass loss parameter N=300 to the end of core
hydrogen burning. With this parameter the endpoint masses are typically half the initial values. Such a mass loss rate is probably an overestimate of the real value for single stars with initial masses of
~40 $M_\odot$ or less but may not be too bad for binary systems or more massive objects. The companions to the X-ray sources are all undermassive
for their luminosities, consistent with this rate.

     The dashed lines in Figure 2 give the changes in the positions of
the WN7, 8, 9, stars if they have effective temperatures near 45,000°K,
following Leep (1979), or the positions of the WN3, 4, 5 stars if they
are at 30,000°K, following Willis and Wilson (1978). One may conclude
that the WN star effective temperatures are uncertain to 50%, a not
very satisfactory state of affairs.

A number of Of stars are appreciably brighter than theoretical
tracks for 100 $M_\odot$ with mass loss. They may well be evolved from higher
mass objects, or there may be some parameter amiss in the observations
or the theoretical tracks. In the former case the weakest link is the
bolometric correction: if this had been overestimated by up to a magnitude, then the brighter Of stars would fall lower in this diagram. A
number of simplifying assumptions have been made in the calculations,
but it is not obvious that any of these drastically affect the luminosity of the evolving stars (complete mixing might do it, but this is
highly speculative).

The early and late WN stars form distinct groupings: fainter
(and hotter?) and brighter (and cooler?), respectively. (The lack of
WN6 stars with known M_v is a real problem in their interpretation.)
Interestingly enough, there is one other <u>crucial</u> distinction between
the early and late WN stars. The former have little or no evidence of
hydrogen present in their spectra (Smith 1973), whereas the late types
all do. I have recently obtained image tube spectrograms of all the WR
stars in the LMC with M_v (the stars in Figure 1). <u>Without exception</u>,
all the WN7, 8, and 9 stars have evidence of hydrogen in their emission
line spectra. Figure 3 shows a preliminary reduction of a portion of
the blue region of the spectra of five late WN stars in the LMC. The
classification, it will be recalled, comes from the N III/N IV emission
line ratio (not shown). All stars have the Balmer lines appreciably
stronger than the alternating He II Pickering lines, hence some hydrogen

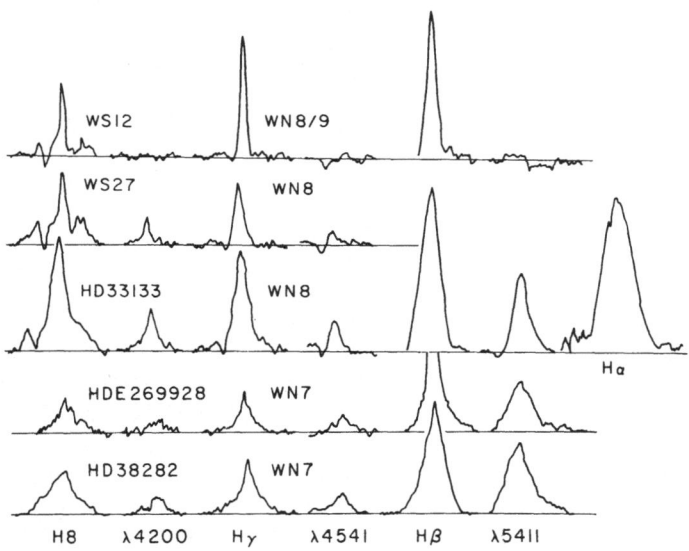

Figure 3. Selected emission line profiles of five stars in
the LMC. Without exception, all show evidence of hydrogen
as shown by the strength of the Balmer series with respect
to the Pickering series.

must be present, irrespective of the He I/He II ionization state. In
nearly all early-type WN stars, the Balmer and Pickering lines have
similar strength, indicating that He II dominates the blended lines
(Smith 1973).

It is probably no coincidence that it is these same late WN stars,
still containing hydrogen (on the surface at least), that appear to
have luminosities near those of Of objects, whereas the early WN stars,
without appreciable hydrogen and burning helium in the core, are considerably fainter. What is the relation between these late WN stars,
and Of stars (Underhill 1968)? As I have mentioned, a few of the former
have Balmer lines in absorption; but all have hydrogen and similar
luminosities to Of stars. The only spectroscopic distinction is
emission line width, which presumably means that the wind density is
higher and the mass loss rate is larger. Masses for late WN stars are
unknown at present (cf. Table 2) but at least one Of star has a mass
that is typical of WR stars (cf. Table 1). Could it be that these late
WN stars are also burning hydrogen in the core (or in a shell) and their
status is near the end of this phase of stellar evolution? I know of no
evidence specifically to the contrary for these WR stars. The actual
numbers of identified stars are few but Moffat and Seggewiss (1979 --
this Symposium) will indicate later that the statistics are consistent
with their being descendants of initially massive Of stars. It must
also be recalled that the effective temperatures are sufficiently uncertain that they cannot be used pro or con this argument.

There are no numbers for mass loss rates of late WN stars as yet
but values appreciably larger than 10^{-5} $M_\odot$ yr^{-1} would not be surprising
given the Hα and λ4686 emission line strengths. Late WN stars could
well evolve to more extreme WR types (e.g., those without hydrogen) in
their remaining lifetimes if they are near the end of core hydrogen
burning (or shell burning) or even if they are helium burning as other
WR stars are supposed to be.

DISCUSSION

Given that WR stars are the direct descendants of O and Of stars
that have suffered substantial mass loss, I now return to the question
of how they got there. In the (now) classical explanation this has
occurred by mass exchange in a close binary (Paczynski 1973). This
hypothesis requires all WR stars to be close binaries, a conclusion
which is becoming increasingly at odds with discovery of apparently
single stars (Castor and Van Blerkom 1970; Moffat and Seggewiss 1978;
Seggewiss and Moffat 1978; Conti, Niemela and Walborn 1979). Furthermore, if the mass loss process is an exchange, then not only must the
other star accept a great deal of material from the initial primary,
now the WR star, but the transfer process must be relatively rapid (on
the Kelvin-Helmholtz timescale, about ~3×10^4 yr for a 30 $M_\odot$ MS star --
Paczynski 1971). I know of no evidence that any companion of a WR
binary has accepted mass from its evolved companion; a most favorable

case for detection would be γ Vel, with an O9 supergiant star and a WC companion. If the former star has <u>accepted</u> appreciable material from the WC star, one would expect its composition to be anomalous (e.g., nitrogen and helium rich). There is no evidence for this in the spectrum of the supergiant. If mass exchange occurs, it must be on a thermal time scale, and one would not expect to see any WR stars with intermediate properties (e.g., hydrogen present) since evolution occurs relatively rapidly. As I have emphasized during this talk, all late WN stars have such properties. The late WN stars may therefore be a link in the evolution of O and Of stars to all WR types, or they may be endpoints in themselves. If many other WR type stars are found to be single, the former possibility seems more reasonable.

I should stress that discarding the concept of mass <u>exchange</u> as a process in the evolution of WR stars is not equivalent to saying the close binary nature of many WR stars has played no role. On the contrary, if one admits the possibility that such an interaction <u>enhances</u> the mass loss rate from the system via a stellar wind, and realizes that this process is a long-term one, on the time scale of the evolving star, stars with intermediate properties will be observed in binary systems also. BD+40°4220 (Bohannan and Conti 1976) would seem to be one such system but more work on known WN binaries (e.g., CQ Cep) would also prove useful. The present evidence concerning mass flow in early-type close binary systems (e.g., UW CMa, McClusky and Kondo 1976) suggests appreciable mass is leaving the system. Finally, in the case of those OB stars with X-ray companions, the present primary star is under-massive for its luminosity by about a factor two (Conti 1978a). For these four systems containing neutron stars, we are sure that the mass that has somehow been lost by the OB star had not been <u>exchanged</u> to the X-ray companion. These observations are suggestive that mass <u>exchange</u> in early-type systems is not important in their evolution, although mass loss certainly appears to be. It is suspected that mass loss phenomena are more important in stars that are members of close binary systems than in single stars of similar mass (Hutchings 1976). For the most luminous single stars, the stellar wind will control the evolution.

I am indebted to Dr. Nancy Morrison and Mr. Phil Massey for help with the line profile data. This research has been supported by the National Science Foundation under grant AST76-20842 through the University of Colorado.

REFERENCES

Bappu, M.K.V. and Ganesh, K.S.: 1968, Monthly Notices Roy. Astron. Soc. 140, p. 71.
Bappu, M.K.V. and Sahade, J., eds.: 1973, <u>Wolf-Rayet and High Temperature Stars</u> (Dordrecht: D. Reidel).
Beals, C.S.: 1938, Trans. IAU 6, p. 248.
Bohannan, B.: 1979, in <u>Mass Loss and Evolution of O-Type Stars</u>, eds. P. S. Conti and C. de Loore (Dordrecht: D. Reidel).
Bohannan, B. and Conti, P.S.: 1976, Astrophys. J. 204, p. 797.

Bracher, K.: 1967, Ph.D. Thesis, Indiana University.

Breysacher, J. and Azzopardi, M.: 1979, in Mass Loss and Evolution of O-Type Stars, eds. P.S. Conti and C. de Loore (Dordrecht: D. Reidel).

Castor, J.I. and Van Blerkom, D.: 1970, Astrophys. J. 161, p. 485.

Conti, P.S.: 1976, Mem. Soc. Roy. Sci. Liege 9, p. 193.

Conti, P.S.: 1978a, Astron. Astrophys. 63, p. 225.

Conti, P.S.: 1978b, Ann. Rev. Astron. Astrophys. 16 (in press).

Conti, P.S. and Burnichon, M.-L.: 1974, Astron. Astrophys. 38, p. 467.

Conti, P.S. and Frost, S.A.: 1977, Astrophys. J. 212, p. 728.

Conti, P.S., Niemela, V.S. and Walborn, N.R.: 1979, Astrophys. J. (in press).

Cowley, A.P., Hiltner, W.H. and Berry, C.: 1971, Astron. Astrophys. 11, p. 407.

Crampton, D.: 1971, Monthly Notices Roy. Astron. Soc. 153, p. 303.

Ganesh, K.S., Bappu, M.K.V. and Natarijan, V.: 1967, Kodaikanal Obs. Bull. #184.

Hiltner, W.A. and Schild, R.E.: 1966, Astrophys. J. 143, p. 770.

Hutchings, J.B.: 1976, Astrophys. J. 203, p. 438.

Hutchings, J.B. and Cowley, A.P.: 1976, Astrophys. J. 206, p. 490.

Leep, E.M.: 1979, in Mass Loss and Evolution of O-Type Stars, eds. P.S. Conti and C. de Loore (Dordrecht: D. Reidel).

de Loore, C., De Greve, J.P. and Van Beveren, D.: 1978, Astron. Astrophys. 67, p. 373.

Massey, P. and Conti, P.S.: 1977, Astrophys. J. 218, p. 431.

Massey, P. and Conti, P.S.: 1979, in Mass Loss and Evolution of O-Type Stars, eds. P.S. Conti and C. de Loore (Dordrecht: D. Reidel).

McClusky, G.E. and Kondo, Y.: 1976, Astrophys. J. 208, p. 760.

Moffat, A.F.J., Fitzgerald, M.P.M. and Jackson, P.D.: 1977, Astrophys. J. 215, p. 106.

Moffat, A.F.J. and Seggewiss, W.: 1978, Astron. Astrophys. (in press).

Moffat, A.F.J. and Seggewiss, W.: 1979, in Mass Loss and Evolution of O-Type Stars, eds. P.S. Conti and C. de Loore (Dordrecht: D. Reidel).

Morton, D.C.: 1969, Astrophys. J. 158, p. 629.

Morton, D.C.: 1973, in Wolf-Rayet and High Temperature Stars, eds. M.K.V. Bappu and J. Sahade (Dordrecht: D. Reidel), pp. 54-56.

Niemela, V.S.: 1976, Astrophys. Space Sci. 45, p. 191.

Niemela, V.S.: 1979, in Mass Loss and Evolution of O-Type Stars, eds. P.S. Conti and C. de Loore (Dordrecht: D. Reidel).

Niemela, V.S. and Sahade, J.: 1979, in Mass Loss and Evolution of O-Type Stars, eds. P.S. Conti and C. de Loore (Dordrecht: D. Reidel).

Paczynski, B.: 1971, Ann. Rev. Astron. Astrophys. 9, p. 183.

Paczynski, B.: 1973, in Wolf-Rayet and High Temperature Stars, eds. M.K.V. Bappu and J. Sahade (Dordrecht: D. Reidel), pp. 143-152.

Seggewiss, W.: 1974, Astron. Astrophys. 31, p. 211.

Seggewiss, W. and Moffat, A.F.J.: 1978, Astron. Astrophys. (in press).

Slettebak, A., ed.: 1976, Be and Shell Stars (Dordrecht: D. Reidel).

Smith, L.F.: 1968a, Monthly Notices Roy. Astron. Soc. 138, p. 109.

Smith, L.F.: 1968b, Monthly Notices Roy. Astron. Soc. 140, p. 409.

Smith, L.F.: 1973, in Wolf-Rayet and High Temperature Stars, eds.
    M.K.V. Bappu and J. Sahade (Dordrecht: D. Reidel), pp. 15-35.
Smith, L.F. and Kuhi, L.V.: 1970, Astrophys. J. 162, p. 535.
Snow, T.P.: 1979, in Mass Loss and Evolution of O-Type Stars, eds.
    P.S. Conti and C. de Loore (Dordrecht: D. Reidel).
Stempien, K.: 1970, Acta Astron. 20, p. 117.
Struve, O., Sahade, J., Huang, S.-S. and Zebergs, V.: 1958, Astrophys.
    J. 128, p. 328.
Underhill, A.B.: 1968, Ann. Rev. Astron. Astrophys. 6, p. 39.
Underhill, A.B.: 1973, in Wolf-Rayet and High Temperature Stars, eds.
    M.K.V. Bappu and J. Sahade (Dordrecht: D. Reidel), pp. 237-253.
Walborn, N.R.: 1974, Astrophys. J. 189, p. 269.
Willis, A.J. and Wilson, R.: 1978, Monthly Notices Roy. Astron. Soc.
    182, p. 559.
Willis, A.J. and Wilson, R.: 1979, in Mass Loss and Evolution of O-Type
    Stars, eds. P.S. Conti and C. de Loore (Dordrecht: D. Reidel).

*Visiting Astronomer, Cerro Tololo Inter-American Observatory.

DISCUSSION FOLLOWING CONTI

de Loore: Can you comment on the bolometric magnitudes?
How were they determined?

Conti: These were taken from Morton (Ap.J. 158, 629,
1969) and we make the assumption that the BC is dependent on
T_{eff} only, which is clearly a very serious limitation for WR
stars. The BC is probably allright for O and Of stars, but
may be highly suspect for WR stars.

Underhill: Placing the WR stars in the HR diagram is a
somewhat uncertain process and I am extremely sceptical that
the effective temperatures are as high as you have indicated.
They may all be of the order of 30 000 K or less, as an interpretation of spectrophotometric data on the visible continuum suggests. The uncertainty in M_{bol} is probably of the
order of $0^m.5$, for it is well known (Lindsey Smith, Crampton)
that the WN7 and WN8 stars lie in the range -6 to -7 for M_V,
while all the other WR subtypes lie in the range -4 to -5
for M_V. The bolometric correction is a function of T_{eff}.
Its value is not particularly uncertain if you know T_{eff},
for model atmospheres of all types, from black bodies to
complex NLTE atmospheres, produce about the same relation
between BC and T_{eff} once $T_{eff} > 25\ 000$ K. The real problem
is to relate the electron temperature in the emission-line
forming region to the formal effective temperature, T_{eff},
which describes the total flux of radiation flowing through
the atmosphere. From the pattern of emission-line intensities, one assigns a spectral type; one also may estimate an
electron temperature, T_e, by some type of spectroscopic diagnosis. From consideration of the physics of flowing ionized

gases one infers that T_e is a function not only of the radiative flux flowing through the atmosphere but also of the input of mechanical and/or magnetic energy which is injected.
R.N. Thomas has emphasized how necessary it is to use diagnostic techniques to infer the mechanical/magnetic heating
as well as radiative heating before relating spectral type
to T_{eff}. I suspect that the perturbation caused by mechani-
cal/magnetic heating is large in the case of WR stars and
that it far exceeds the perturbation which I told you of on
Monday for the B-type supergiants. In addition, it is possible that Zanstra temperatures found by comparing radio
emission with visible continuum emission are also in error,
for the electron temperature in the gas where the radio emission originates may be determined by the flux of mechanical/
magnetic energy while the visible continuum reflects the flux
of radiative energy only. Use of the Zanstra method implies
that the electron temperatures in both regions are determined
by the radiation field only. I doubt that this is true for
WR stars. Consequently I think your estimates of T_{eff} for
the Wolf-Rayet stars may be seriously overestimated. Lower
T_{eff} would resolve most of your problems.

Hearn: If you take a typical Wolf-Rayet star with mass
10 M_O and radius 10 R_O with a mass loss of a few $10^{-5} M_O yr^{-1}$,
then the optical depth of the matter streaming out of the
star is greater than 1. This means that we are not seeing
the surface of the star. The radius of the star could be
much less than the radius derived from optical depth 1 through
electron scattering. For a given radiative flux from the
star, the effective temperature depends on the radius of the
star ascribed. Thus the effective temperatures of Wolf-Rayet
stars could be much higher than the currently accepted values.

Morton: Electron scattering increases the radius of the
star but leaves the fluxes unchanged. Since the Zanstra temperature was determined by the ratio of visual flux to Lyman
continuum flux, which was estimated from the free-free emission of the surrounding nebula, the temperature should not
depend on whether electron scattering is present.

Carrasco: I would like to make a remark, the luminosities for WR stars would not be in such a large disagreement
if the adopted mass loss rate is reduced, and another theoretical track is plotted.
What is the base of your luminosity calibration for stars
not in the Magellanic Clouds?

Conti: Membership in Sco OB1, Carina or TR 27 (see
text).

Leung: I am in support of Ann Underhill's suggestion.
In the system of V729 Cyg (O7fEa + OfIa) the lines strength
are similar between the components if we interprete that the
two components may have similar luminosities (Bohannan and
Conti) from the lines strength. This lead to two difficulties: 1) the low mass component having radius much larger
than the Roche radius, and 2) the luminosity of the same component exceeds the Eddington limit. If we accept the alternative interpretation (Leung and Schneider), the low mass
does not encounter the above problem, but one new difficulty
arised. The luminosity of this component becomes too low to
be detached in the spectra. Thus, it leads to suggest that
the spectral lines detached are not formed in normal stellar
atmosphere but from a very rarely atmosphere. Under such
circumstance line intensity may not reflect the luminosity
of the star.

Chiosi: I have the impression that you cannot account
for the location of Of stars in your sample only because you
have compared it with theoretical models suffering a too
heavy mass loss during the core H burning phase, as it is in
the case of the N=300 set of de Loore et al. (1977). Conversely if we use slightly smaller mass loss rates as involved
in the α=0.90 or α=0.83 sets of Chiosi et al. (1978) the position of Of stars can be easily included in the core H-burning band.

THE GALACTIC WN7/WN8 STARS AS MASSIVE O STARS IN ADVANCED STAGES OF EVOLUTION

A.F.J. Moffat[1] and W. Seggewiss[2]
[1]Département de Physique, Université de Montréal
[2]Obs. Hoher List, Universitäts-Sternwarte Bonn

ABSTRACT

It is shown from various observational contraints that WN7/WN8 stars tend to occupy a class of their own set apart from other Wolf-Rayet subclasses. They probably evolved by stellar-wind mass loss from Of stars with original masses greater than ~35 $m_\odot$, whether single or component of a binary. Two WN7/WN8 stars may be runaways in the second WR phase. One, HD 197406, is situated over 1000 pc from the galactic plane and has a low-mass, overluminous companion which is eclipsed by the WN7 component.

1. COMPARISON OF WN7/WN8 WITH OTHER SUBCLASSES OF WR STARS

Observationally there are at least five factors which set WN7/WN8 stars apart from all other subclasses of Wolf-Rayet stars, whether WN4-6 or WC5-9:

(1) The main difference occurs in *intrinsic luminosity*: with $M_V = -6.8$ (-6.2) for WN7 (WN8) they are two full magnitudes brighter than other WR stars with mean $M_V = -4.4 \pm 0.4$ (max) according to the calibration of Smith (1973). The same trend persists for the bolometric absolute magnitudes while the effective temperatures of the WN7/WN8 stars lie at the hot end of the remaining WR stars (Conti, 1976).

(2) WN7/WN8 stars appear to be *younger* on the average than other WR stars. From the association of 19 galactic WR stars with 17 open clusters (Moffat and Seggewiss, 1978) six WN7/WN8 stars belong to clusters with mean earliest spectral type O4 ± 1 (s.d.) while 13 WR stars of other type belong to clusters of type O8 ± 2 (s.d.). This implies a total mean age of ~$3 \cdot 10^6$ a for WN7/WN8 stars and ~$5 \cdot 10^6$ a for other WR stars.

(3) Morphologically, the *spectra* of WN7/WN8 stars always show narrower lines than earlier WN stars. However, while

447

their lines are wider and more developed, the spectra of
WN7/WN8 stars resemble Of stars more than any other of the
WR subclasses. This is born out in Fig. 1 which shows photographic transmission tracings at coudé dispersion of the six
brightest WN7/WN8 stars in the sky along with a typical WN6
and an extreme Of star. Especially noticeable is the N IV
4058 emission line which is relatively narrow and unblended
in the WN7/WN8 spectra unlike the WN6 spectrum where it is
broad and asymmetric. It is absent in the Of spectrum.
Evidence is also accumulating for the existence of transition
types between Of and WN7/WN8 stars (e.g. the undermassive
Of star components in the spectroscopic binaries BD +40°4220
(Bohannan and Conti, 1976) and HD 228766 (Massey and Conti,
1977) and a sequence of five stars ranging from O through
Of to WN8 in a newly discovered southern cluster (Havlen
and Moffat, 1977)).

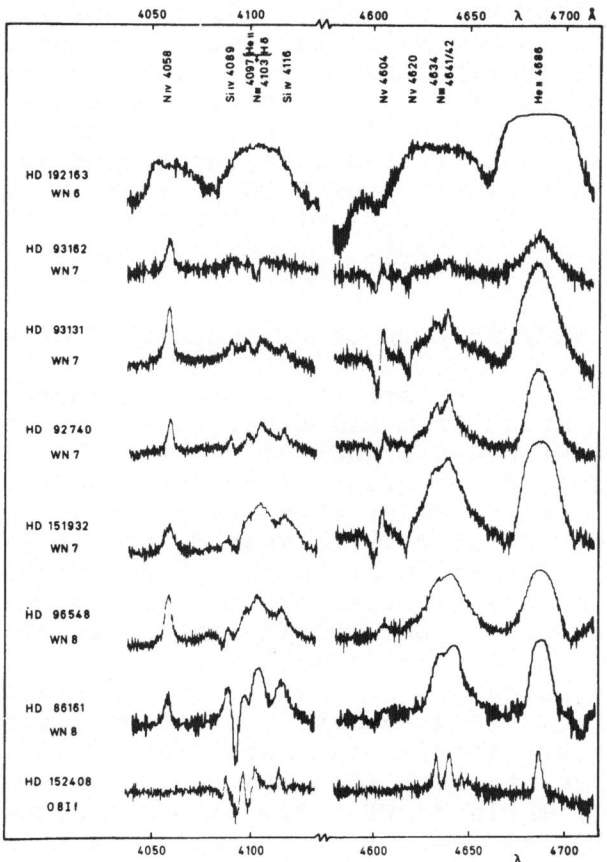

<u>Fig. 1.</u> Photographic transmission tracings (4040-4130 Å
and 4580-4710 Å) of a sample of stars ranging in spectral
type from WN6 through WN7, WN8 to an O8f supergiant.

(4) The *ionization/excitation structure of the envelopes* of
WN7/WN8 stars is dramatically set apart from the remaining
WN stars. This is demonstrated in Fig. 2 where we see a much
more rapid increase of expansion velocity with potential of
violet shifted absorption lines for the WN7/WN8 stars. Again
there appear to be no transition stars between WN6 and WN7.
(5) Another point of contrast concerns *abundance differences*.
The ratio by number of H/He, compared to the mean cosmic
value of 10, is ⪰1 for the WN7/WN8 stars and drops dramatically to 0.00-0.14 for the remaining WN subclasses (Smith,
1973; Rublev, 1975). WC spectra are too complex to allow
a useful estimate.

Taken all together, the above arguments strongly suggest
that the WN7/WN8 stars have evolved from luminous progenitors
of higher mass than all other WR subclasses. This agrees well
with the theoretical calculations and predictions of Chiosi

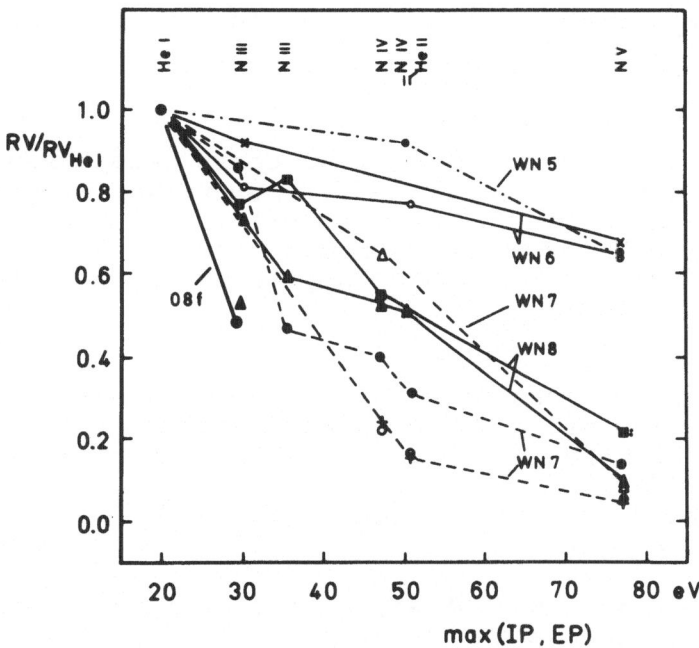

Fig. 2. Ionization/excitation structure of the envelopes of
WN5, WN6, WN7, WN8 stars and an O8If star. The radial velocities RV of the violet shifted absorption edges have been
reduced to unity for the He I lines which have the lowest
potential and are assumed to be formed in the outermost
region of the envelope where the maximum expansion velocity
is attained. The abscissa is the maximum of the ionization
or excitation potential for the line concerned (cf. Smith
and Aller, 1969).

et al. (1978) who claim parent masses for WN7/WN8 stars of
≥35-40 $m_\odot$ but only ~30 $m_\odot$ for the remaining WR classes.
In view of the distinct magnitude gap in the HR diagram and
the lack of transition WN6-WN7 types it appears doubtful
that WN7/WN8 stars evolve into earlier type WN or into WC
stars. If they in fact do, then the transition must occur
very rapidly.

2. STATISTICS OF WN7/WN8 STARS AND COMPARISON WITH O STARS

Using the catalogue of Smith (1968a) as a basis we present in Table 1 a complete list of known galactic WN7 and
WN8 stars (for notes to individual stars cf. Moffat and
Seggewiss, 1978). Some spectral types have been revised and
a few stars added or deleted. The distances are based on the
study of galactic WR stars by Smith (1968b) who assumed the
same absolute magnitudes for WN7 and WN8 stars as noted in
section 1.

Taking a distance of 3 kpc from the sun to represent
the radius of completeness for the discovery of galactic
WR stars (Stenholm, 1975) we find

n(WN7/WN8)/n(other WR) = 5/31 = 0.16

in which the statistics of the other WR stars were taken
again from Smith (1968b) with revised absolute magnitudes
where necessary (Smith, 1973). Theoretically, one can calculate the ratio to be expected from two assumptions:
(a) the number of progenitor stars obeys a universal initial
mass frequency function

$n(m) \sim m^{-2.35}$ (Salpeter, 1955) and

(b) the representative mass of a WN7/WN8 progenitor is
~50 $m_\odot$ and ~30 $m_\odot$ for other WR stars with mean duration of
the (He-burning) WR phase of ~3 and ~5·10^5 a, respectively
(Chiosi et al., 1978).

Thus one expects a ratio of $(3/5)\cdot(30/50)^{2.35} = 0.18$,
compatible with the observations.

Since Of stars and WN7/WN8 stars have nearly the same
absolute visual magnitudes in the mean (Conti, 1976) we can
compare their number density directly using apparent magnitudes. From the catalogue of 130 O stars with high quality
spectral types of Conti and Alschuler (1971) there are 46
galactic Of stars north of declination $\delta = -20^\circ$. For the
same limit of completeness (V ≤ 8.0) and declination range
there are four stars of type WN7/WN8; thus

n(WN7/WN8)/n(Of) = 0.09.

A ratio of this order is also expected if all Of stars
(i.e. O stars with m ≥ 35 $m_\odot$) become WN7/WN8 stars

Table 1. Catalogue of all known galactic WN7/WN8 stars

HD/name	MR	Sp	v	l	b	d	r	z
						kpc	kpc	pc
86161	19	WN8	$8^{m}_{.}43$	$281^{o}_{.}1$	$-2^{o}_{.}6$	4.0	10.0	- 181
92740	25	WN7	6.44	287.2	-0.8	2.9	9.6	- 40
93131	28	WN7	6.49	287.7	-1.1	3.5	9.5	- 67
93162	29	WN7+O7	8.17	287.5	-0.7	4.4	9.6	- 53
MS 3	-	WN7-8	12.67	288.6	-1.0	17.4	17.1	- 304
-	32	WN8+OB	10.88	289.8	-1.2	7.9	10.4	- 166
96548	34	WN8	7.85	292.3	-4.8	4.0	9.3	- 333
117688	49	WN7	10.87	307.8	+0.2	12.0	9.8	+ 42
134877	54	WN8	11.71	320.1	-1.8	7.6	6.4	- 238
151932	64	WN7	6.61	343.2	+1.4	2.3	7.8	+ 56
LS 11	-	WN7	12.42	341.9	-2.4	11.0	3.4	- 459
LSS 4064	-	WN8+OB:	(12.0)	348.7	-0.8	2.9	7.2	- 40
LSS 4065	-	WN8+OB	(11.0)	348.7	-0.8	2.9	7.2	- 40
-	89	WN7	(11.94)	27.8	+0.2			
177230	91	WN8	(11.1)	30.5	-4.8	9.4	5.1	- 787
M1-67	-	WN8	(11.1)	50.2	+3.3	4.3	8.0	+ 250
LS 16	-	WN8:	(13.7)	68.2	+1.0			
-	97	WN7	12.30	69.9	+1.7	1.0	9.7	+ 31
228766	105	WN7+O	9.33	75.2	+1.0	5.0	10.0	+ 87
-	111	WN7:	(10.5)	79.7	+0.7			
197406	113	WN7	10.50	90.1	+6.5	9.1	13.5	+1032
214419	118	WN7+O7	8.94	105.3	-1.3	5.8	12.8	- 130
-	119	WN8	11.18	109.8	+0.9	4.6	12.3	+ 72
-	122	WN7	11.49	115.0	+0.1	6.9	14.4	+ 12

Notes:

MR = no. in the catalogue of Roberts (1962)

d = distance from the sun

r = galactocentric distance

z = distance from the galactic plane

(assumed to be He stars of $m \gtrsim 10\ m_{\odot}$) and their respective mean life times are $\sim 3 \cdot 10^{6}$ a and $\sim 0^{.}3 \cdot 10^{6}$ a.

Another important observational quantity is the relative number of binary to single WN7/WN8 stars. From Smith's (1968a) catalogue of WR stars, the number ratio of single-line stars to spectroscopic binaries for all WR subclasses is nearly unity for $v \leq 10.5$ and increases dramatically (and therefore spuriously) for fainter stars whose spectroscopic data are progressively less complete (the ratio is

~10 for v ≥ 11). Therefore, we have made a list (Table 2) of
all galactic WN7/WN8 stars for v ≤ 10.5 separated according
to their spectroscopically proven single or binary nature
(for notes cf. Moffat and Seggewiss, 1978). One sees that
the binary frequency among galactic WN7/WN8 stars is close
to 50%. This compares well with the binary frequency among
O stars (including Of) which is 58% according to Conti et
al. (1977). This similarity is again strongly suggestive of
a common origin of all WN7/WN8 stars from massive Of stars;
whether single or binary appears to make little difference.

Table 2. List of single and binary stars among a complete sample
of WN7/WN8 stars with v ≤ $10^m\!.5$. The mass function f(m) refers to
the velocity of the WR component whose mass m_{WR} was assumed to be
10 $M_\odot$ except for HD 228766 where both masses were observed separately. (All masses are in solar units.)

Single stars (RV constant)

HD	Sp	Notes
86161	WN8	1
93131	WN7	2
93162	WN7	3
96548	WN8	1
151932	WN7	4

Proven binary stars (RV orbit available)

HD/name	Sp	Type	P	f(m)	$i = 90°$ m_{WR}	m_2	$i = 60°$ m_{WR}	m_2	Notes
92740	WN7	SB1	$80^d\!.35$	1.67	10	8	10	10	2
228766	WN7+O	SB2	10.7424	4.24	16	16	24	25	5
MR 111	WN7:	SB1	22:	7.7	10	18	10	24	6
197406	WN7	SB1	4.3207	0.251	10	3.6	10	4.3	7
214419	WN7+O7	SB1	1.64	4.38	10	14	10	17	8

3. KINEMATICS AND DISTRIBUTION OF GALACTIC WN7/WN8 STARS

It is fortunate that, unlike the other WR subclasses,
the WN7/WN8 stars have some narrow, symmetric emission
lines whose positions yield velocities close to the systemic
velocity. In particular, the emission line of N IV λ4058
has the additional advantage that it is unblended. This line
is generally not seen in Of stars whose emission spectrum
is not as strongly developed.

In Table 3 we present a compilation of 11 WN7/WN8 stars for which reliable radial velocities are available, mostly based on the emission line of N IV $\lambda4058$. Neglecting M1-67 and HD 197406 which may be high velocity stars, the mean difference between observed radial velocity RV(4058) and expected velocity RV(rot) due to differential, circular galactic rotation and peculiar solar motion for the nine remaining stars is

RV(O-C) = -15 ± 13 (s.d.) km s^{-1}.

This difference probably reflects the small systematic effect of line asymmetry in an expanding envelope. However, agreement is sufficiently good to indicate that the distances and absolute magnitudes on which the comparison is made are probably quite reliable. Applying this as a correction, we find that M1-67 has a residual, peculiar radial velocity of 184 km s^{-1} and HD 197406 has -35 km s^{-1}.

Table 3. WN7/WN8 stars with reliable velocities

HD/name	Sp	l	RV(4058)	RV(rot)	ΔRV
			km s^{-1}	km s^{-1}	km s^{-1}
86161	WN8	281°	- 23	+ 9	- 32
92740	WN7	287	- 23	- 6	- 17
93131	WN7	288	- 38	- 7	- 31
93162	WN7	288	- 26	- 4	- 22
96548	WN8	292	- 16	- 15	- 1
151932	WN7	343	- 48	- 28	- 20
M1-67	WN8	50	+207*	+ 38	+169
228766	WN7+O	75	- 10	- 12	+ 2
197406	WN7	90	-126	- 76	- 50
214419	WN7+O7	105	- 75	- 60	- 15
MR 119	WN8	110	- 50*	- 50	0

* N IV 4058 velocity not available; value based on other narrow lines.

The overall distribution of WR stars projected onto the galactic plane has recently been studied by Smith (1973) and Stenholm (1975). While there is some correlation with local spiral features, the overall correlation in the galaxy suffers from the lack of precision in single-star parallaxes.

In Fig. 3 we show a plot of perpendicular distance z
from the galactic plane versus galactocentric distance r for
WN7/WN8 stars of known distance. Restricting to r = 10 ±4 kpc
and omitting M1-67 and HD 197406 (MR 113) which are probably
high velocity stars and HD 96548 (MR 34) whose absolute z-
value is even larger than that of M1-67, we find

$$\overline{|z|} = 89 \text{ pc,}$$

much like the mean z found for the local, young population I
OB stars: 80 pc (Gunn and Ostriker, 1970).

With z = +1032 pc (i.e. >10 times larger than the aver-
age |z|) HD 197406 may be a runaway star with the main part
of its velocity vector directed perpendicular to the galactic plane; its systematic radial velocity component also
differs by ~35 km s^{-1} compared to that expected from normal
differential galactic rotation. On the other hand, M1-67
may be a runaway star with most of its velocity vector
directed in the plane. There may be one or more other run-
aways in the sample of WN7/WN8 stars but their velocities
and separation from the galactic plane do not appear to make
them stand out like HD 197406 and M1-67.

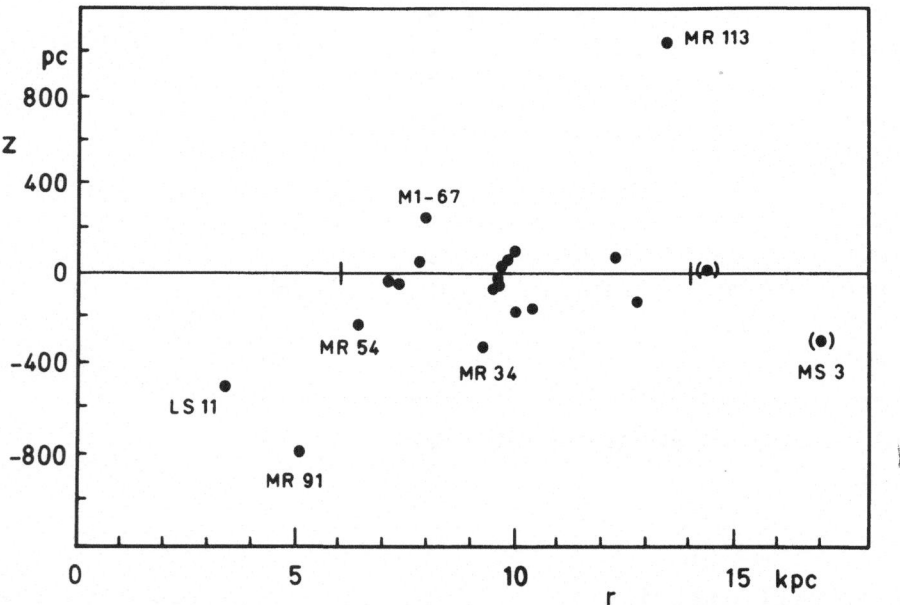

Fig. 3. Galactic distribution of WN7/WN8 stars: distance z
perpendicular to the plane versus galactocentric distance r.
The sun is assumed to lie 10 kpc from the galactic centre.
Brackets indicate less reliable data. The vertical bars at
r = 6 and 14 kpc indicate the limits posed for discussion.

In the context of possible runaway stars, it is interesting to compare with the O stars (Conti et al., 1977). Taking runaways to be those stars with RV(pec) > 40 km s^{-1} or |z| > 300 pc we find, from a sample of 87 O stars with reliable radial velocities 11% ± 4% runaway suspects (10 stars). Among our sample of 17 WN7/WN8 stars with r = 10 kpc ± 4 kpc we find the percentage to be 18% ± 10% (3 stars). These numbers overlap, again in support of the common origin of WN7/WN8 and (massive) O stars.

4. EVOLUTIONARY STATUS

From the preceding sections it appears very likely that most, if not all, WN7/WN8 stars evolve directly by stellar-wind mass loss from massive O-star progenitors. Perturbations due to binary mass transfer may speed up the process but are unlikely to be decisive. Support for this is found from direct observations of high mass loss rates among Of stars (Barlow and Cohen, 1977) and by the presence of expanding rings of thermally excited H II gas surrounding some single WN stars (cf. Smith, 1973).

Additional complications may occur in the evolution of massive stars to produce runaway WN7/WN8 stars, at least in the case of HD 197406 and M1-67. We postulate that these stars are in the second WR phase after having been accelerated to high velocities by a supernova explosion of the original primary in a massive binary system (cf. van den Heuvel, 1976). This is much like the explanation put forth to explain the runaway OB stars (cf. Bekenstein and Bowers, 1974) except that the massive runaway OB component has evolved by stellar-wind mass loss into a WN7/WN8 star.

Of particular interest is the WN7 star HD 197406 for which there is sufficient data available to make the above suggestion very plausible. Not only is it located further from the galactic plane than *any* other WR star (cf. Fig. 10 of Stenholm, 1975) and it has a moderately peculiar radial velocity, it is a single-line spectroscopic binary with a low mass function (cf. Table 2). Using the gravitational force law perpendicular to the galactic plane derived by Oort (1965) and a probable time ($\sim 5 \cdot 10^6$ a) elapsed since the supernova explosion occurred while the system was situated in the galactic plane, it is possible to calculate the original (Z_0) and present (Z) z-velocity components necessary to have reached the present distance from the galactic plane, z = 1032 pc. We obtain Z_0 = 207 km s^{-1} and Z = 198 km s^{-1} which are at the high end of the peculiar radial velocities observed for runaway OB stars (~ 200 km s^{-1}) by Bekenstein and Bowers (1974). For comparison, M1-67 has

an observed peculiar radial velocity of +184 km s^{-1} but
little is known about its possible binary nature.

A possible scenario for HD 197406 might be the following: Let as start with a binary WR system in the galactic
plane just before the supernova explosion. The system has
eccentricity e = O and period P = 2 d similar to the WN7+O7
binary HD 214419. Let the WR component have a mass of ~24 m$_\odot$
similar to the WN7 component of HD 228766 for a likely
orbital inclination i = 60° and the O star companion a mass
of 35 m$_\odot$, the minimum to produce the *present* WN7 star.
Then, following Sutantyo (1973) for an assumed instantaneous
symmetric supernova explosion we indicate schematically in
Fig. 4 a possible scenario. The final period and eccentricity are assumed to match the presently observed system, and
represent plausible values after tidal circulization of the
eccentric post-supernova orbit.

Stage	m/m$_\odot$	WR	O star	m/m$_\odot$	P(d)	e	z(pc)	Z(kms^{-1})
Start	24			35	2	O	O	O
Post SN	4	compact		35	7.2	0.51	O	200
After 5·10^{6}a (now)	4		WN7	10	4.3	O	1000	190

Fig. 4. A possible scenario for HD 197406

Further supporting evidence comes from the light
variations of HD 197406. Fig. 5 presents the light and
colour curves taken from the data of Bracher (1966), While
the colour b-v remains constant with phase, the magnitude v
varies systematically with phase (with some additional
intrinsic noise) with an amplitude ~0.045. Minimum light
occurs near phase 0.5 when the WR component is in front.

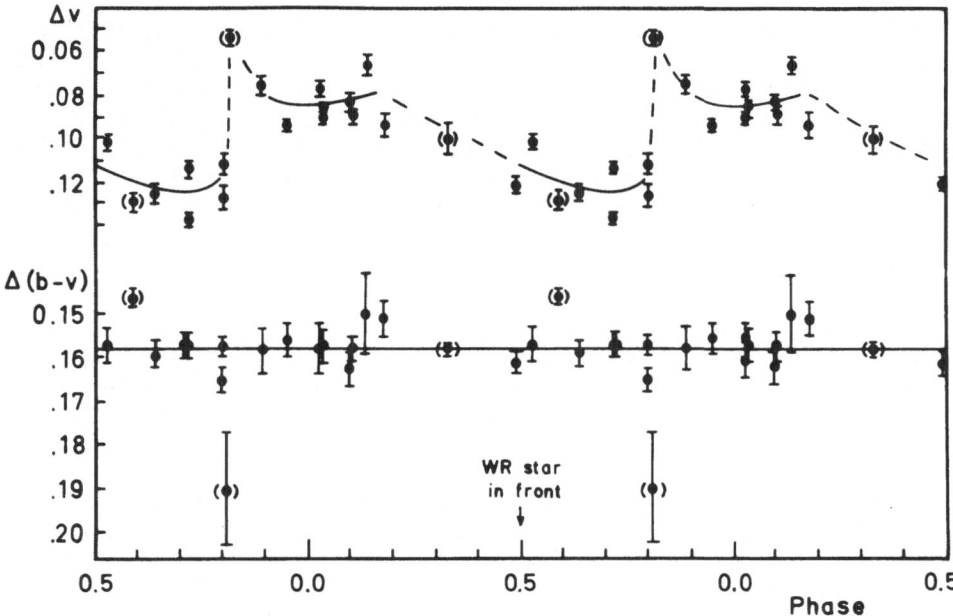

Fig. 5. Light curve of HD 197406 based on the original data of Bracher (1966). Visual magnitude and colour differences refer to HD 197406 minus BD +52° 2774 obtained during two runs in 1964 and 1965. Each point is the mean of from 9 to 20 individual measurements (3 or 4 measurements only for bracketed values) while the bars are the length of 2σ where σ is the internal standard deviation for each point. Fitted curves are best eye estimates.

The light variation is likely not due to ellipsoidal variation in view of the lack of two clear minima per cycle. Nor can it be due to oscillations of a dense He star (the WR component) which would be of the order of an hour (Stothers and Simon, 1970), not days as observed. This leaves as most plausible explanation the variation of light (directly or indirectly) mainly from a second star orbiting the WR component. In order to yield the observed visual amplitude, the unseen star must have an effective absolute magnitude $M_v \leq -3.4$ which is equivalent to a star at least as bright as an unreddened B1V star with luminosity $L_2 \sim 2 \cdot 10^{37}$ erg s^{-1}. Such a star would be much too massive to explain the observed low mass function. The most plausible explanation is that the secondary is a compact star which, by the process of mass accretion in the WR envelope, would appear overluminous for its mass. Such a star would be a result of the supernova explosion that accelerated the star to its present high z-value. From the orbit of HD 197406, this compact star

would be revolving well within the WR envelope that, even
when the star is on the near side, would almost entirely de-
grade the X-radiation into photons of longer wavelength (cf.
Moffat and Seggewiss, 1978). Although there is no information available on HD 197406 in the UV, it has no detectable
IR excess compared to other WN7 stars (Hackwell et al.,1974).
However, the optical light curve may be a result of the
modulation of the degraded source as it orbits in the dense
WR envelope. The constant colour would then be a result of
wavelength-independent electron scattering, the principle
source of opacity in the envelope.

The search for other candidate runaway WR stars is
continuing. In fact Firmani (1978) has recently found evidence that the WN5 star HD 50896 is probably such a star.

Acknowledgements: Both authors gratefully acknowledge
the allocation of observing time at the European Southern
Observatory. A.F.J.M. is indebted to Th. Schmidt-Kaler for
time on the Bochum 61 cm telescope in Chile which is finan-
ced by the Deutsche Forschungsgemeinschaft (DFG); to the
Ondřejov Observatory for observing time on the 2 m telescope;
to the Comité d'Attribution des Fonds Internes de Recherche
of the University of Montreal for a travel grant and to the
National Research Council of Canada for financial assistance.
W.S. expresses his thanks to the DFG for a travel grant.

REFERENCES

Barlow, M.J., Cohen, M.: 1977, *Astrophys. J.* 213, 737
Bekenstein, J.D., Bowers, R.L.: 1974, *Astrophys. J.* 190, 653
Bohannan, B., Conti, P.S.: 1976, *Astrophys. J.* 204, 797
Bracher, K.: 1966, *Thesis*, Indiana University
Chiosi, C., Nasi, E., Sreenivasan, S.R.: 1978, *Astron. Astrophys.* 63, 103
Conti, P.S.: 1976, *Mém. Soc. Roy. Sci. Liège, 6. Série*, 9, 193
Conti, P.S., Alschuler, W.R.: 1970, *Astrophys. J.* 171, 325
Conti, P.S., Leep, E.M., Lorre, J.J.: 1977, *Astrophys. J.* 214, 759
Firmani, C.: 1978, *IAU Symp. No. 83*, Reidel, Dordrecht (this volume)
Gunn, J.E., Ostriker, J.P.: 1970, *Astrophys. J.* 160, 979
Hackwell, J.A., Gehrz, R.D., Smith, J.R.: 1974, *Astrophys. J.* 192, 383
Havlen, R.J., Moffat, A.F.J.: 1977, *Astron. Astrophys.* 58, 351
Massey, P., Conti, P.S.: 1977, *Astrophys. J.* 218, 431
Moffat, A.F.J., Seggewiss, W.: 1978, *Astron. Astrophys.* (in press)
Oort, J.H.: 1965, *in: Stars and Stellar Systems*, Vol. 5, eds. A. Blaauw
    and M. Schmidt, University Press, Chicago, p. 455
Rublev, S.V.: 1975, *IAU Symp. No. 67*, Reidel, Dordrecht, p. 259
Salpeter, E.E.: 1955, *Astrophys. J.* 121, 161
Smith, L.F.: 1968a, *Monthly Notices Roy. Astron. Soc.* 138, 109
Smith, L.F.: 1968b, *Monthly Notices Roy. Astron. Soc.* 141, 317
Smith, L.F.: 1973, *IAU Symp. No. 49*, Reidel, Dordrecht, p. 15
Smith, L.F., Aller, L.H.: 1969, *Astrophys. J.* 157, 1245
Stenholm, B.: 1975, *Astron. Astrophys.* 39, 307
Stothers, S.R., Simon, N.R.: 1970, *Astrophys. J.* 160, 1019
Sutantyo, W.: 1973, *Astron. Astrophys.* 29, 104
van den Heuvel, E.P.J.: 1976, *IAU Symp. No. 73*, Reidel, Dordrecht, p. 35

DISCUSSING FOLLOWING MOFFAT and SEGGEWISS

<u>Bisiacchi</u>: If I remember well the calculation of Beken-
stein and Bowers, if the binary system is not disrupted by
the supernova explosion, you cannot obtain the velocities
necessary to put the star at that distance from the plane.

<u>Moffat</u>: Indeed Bekenstein and Bowers (1974) state that
most OB-runaway velocities will lie in the range 30-50 km s^{-1}
after receiving a kick from a SN explosion but exceptional
cases can lead to values up to ~ 200 km s^{-1} in very close,
very massive binaries, i.e. enough to get a star out to z ~
10^3 pc in ~ 5 10^6 years. Indeed, observed <u>radial</u> velocities
of OB runaways range up to ~ 200 km s^{-1}.

<u>Van Blerkom</u>: Tademaru has proposed a "tugboat" model in
which a newly formed, rapidly rotating neutron star is accelerated to high velocity by asymmetric emission of radiation
and pulls its companion along with it. This could account
for a runaway without disruption of the binary system.

<u>Moffat</u>: But admitting the presence of a neutron star
companion already implies a runaway system due to the violent
process of formation of the neutron star in the first place.
The "tugboat" will only help increase the acceleration if its
momentum vector is pointing in the right direction.

<u>Vanbeveren</u>: I just want to mention that starting with
a 35 M_O star including a stellar wind phase, will end up
with considerably higher masses ($\pm$ 18 M_O).

<u>Moffat</u>: Indeed, but this 18 M_O WN7 star will by virtue
of its very high mass loss rates, evolve to a smaller mass,
say ~ 10 M_O, the mean value observed for WR components in
known binaries.

<u>Massey</u>: Did you redo the orbit, or were your comments
on HD 197406 based on Katherine Bracher's orbit determination?
The reason for asking is that I think I recall that the mass
function was uncertain by a factor of ~ 4, depending on which
emission line was used.

<u>Moffat</u>: All data shown for HD 197406 are from Bracher's
thesis (1966). We took her mass function for the NIV 4058
line which is the most symmetrical, weak, unblended emission
line: f(m)=0,251 M_O. This is a conservative way not to push
the unseen star's mass down too far. Bracher had adopted
f(m)=0.07 M_O which would make our case of a low mass companion even stronger.

THE CHEMICAL NATURE AND EVOLUTIONARY STATUS OF THE WOLF-RAYET STARS

A.J.Willis and R.Wilson
Department of Physics and Astronomy
University College London, Gower Street, London WC1E 6BT

1   INTRODUCTION

     Much interest has been shown in recent years concerning the possible
link in stellar evolutionary terms between the O-type (particularly Of)
and the WR stars, arising from both observational studies of O stars
(Conti 1976, Bohannon and Conti 1977) and theoretical studies of the
evolution of massive hot stars with high mass loss (Chiosi and Nasi 1974,
de Loore et al. 1977, 1978). These investigations and ideas coincided
with a renewed attack on the abundance problem in the WR stars based on
the acquisition of new ultraviolet observations and the application of
recently developed techniques for treating line transfer in rapidly
expanding atmospheres (Willis and Wilson 1978a). The aim of this paper
is to summarize some of the salient aspects of this latter reference and
to examine the evolutionary status of the WR stars in the light of these
new results.

2   THE S2/68 OBSERVATIONS AND EFFECTIVE TEMPERATURES OF THE WR STARS

     The S2/68 experiment (described by Boksenberg et al. 1973) in the
ESRO satellite TD-1 has provided an extensive set of ultraviolet data for
nine WR stars - three WN, three WC and three WC+O binaries. These data
are in the form of low resolution ($\Delta\lambda \sim 35A$) spectrophotometric measurements over the wavelength range 1350-2550A, together with a broad band
($\sim 310A$) photometric measurement centred at 2740A. The data have been
calibrated to give absolute energy fluxes with an absolute photometric
accuracy believed to be better than twenty percent (Humphries et al. 1976).
The ultraviolet spectra of the six single stars are shown in Fig 1, where
we see that, as at visible wavelengths, the spectra are dominated by many
strong emission lines, identified in the figure.

     A combination of the S2/68 data with ground based measurements in
the visible (Smith 1968, Cohen et al. 1975) provides energy distributions
over the very extensive wavelength range of 1350A to 1 micron. Over this
large wavelength range the variation in magnitude of the interstellar

461

P. S. Conti and C. W. H. de Loore (eds.), Mass Loss and Evolution of O-Type Stars, 461–469.

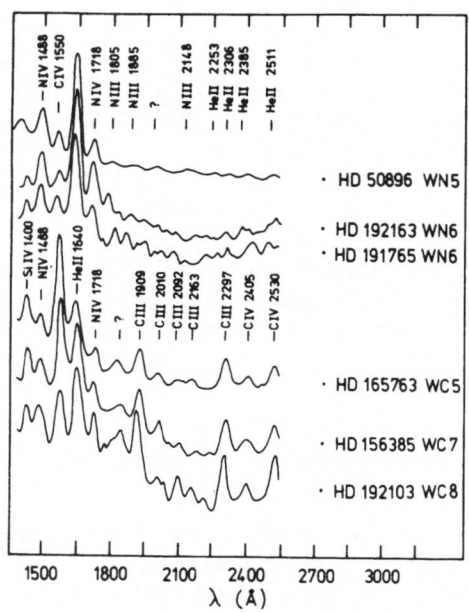

Fig 1   The S2/68 ultraviolet spectra of six single WR stars.

extinction is very large (Nandy et al. 1975) and thus accurate colour
excesses and subsequent intrinsic energy distributions for the WR stars
can be determined (Willis and Wilson 1978a). Colour temperatures based
on this long-wavelength baseline were determined using both black body
distributions and plane parallel model atmospheres (Kurucz et al. 1974)
and these are given in Table 1 for the nine WR stars observed. The black
body results are close to 33000 K whereas those derived from the model
atmosphere calculations lie near the somewhat lower temperature of 27000 K.
A detailed comparison of the observed and model distributions show severe
discrepancies, particularly in the infrared, where the observed fluxes
show an excess. These infrared excesses have been noted before (Kuhi
1966) and highlight the need for models which take into account the
extended nature of the WR atmospheres. Only a few such models, employing
spherical geometry, have been constructed to date, but the results of
Kunasz et al. 1975 indicate that in this temperature range, the true
effective temperature lies between the colour temperatures deduced by
comparison with black body and plane parallel model distributions. We
therefore take the average of the two colour temperatures listed in
Table 1 as indicative of the effective temperatures, which results in a
value close to 30000 K for the nine WR stars observed.

    Effective temperatures near this value are also indicated by a
Zanstra analysis of the HeII 1640A line observed strongly in emission in
the six single stars. These temperatures, denoted T_z, are also given in
Table 1. Additionally the effective temperature of the WC8 component in
the WR spectroscopic binary γ^2 Velorum, for which angular diameter

HD	Sp	T_k (K)	T_b (K)	T_z (K)
50896	WN5	26400	33000	31600
191765	WN6	28200	36000	29200
192163	WN6	28100	35000	30200
165763	WC5	27000	35000	27500
156385	WC7	25400	32000	27200
192103	WC8	26000	32000	26000
68273	WC8+09 I	27000	34000	
193793	WC6+0	25000	31000	
113904	WC6+0	27000	35000	

Table 1   WR colour temperatures T_k and T_b (see text) and Zanstra temperatures, T_z, determined from the HeII 1640A line.

information (Hanbury Brown et al. 1970) and a good distance estimate (Brandt et al. 1971) are available, is determined as 29000 K by Willis and Wilson (1978a). We can thus feel confident in ascribing effective temperatures of ~ 30000 K for the WR stars observed by S2/68. Morton (1973) has determined Zanstra temperatures for several WN stars exciting ring nebulae from observations of the radio emission. For the two WN stars in common with the S2/68 data he also determined temperatures close to 30000 K. However, for some of the other stars he estimates values in the range 40000-50000 K, which may indicate a real variation of effective temperature within each subclass. Further ultraviolet observations are probably needed to test this.

3   THE CHEMICAL NATURE OF THE WR STARS

Since the early separation of the WR stars into the WN and WC sequences (Beals 1934) the longstanding question has been whether the separation is the result of chemical or physical effects, i.e. is the apparent lack of nitrogen lines in WC spectra the result of an under-abundance of N and likewise the lack of carbon lines in WN spectra the result of an underabundance of C, or can these differences be explained in terms of different excitation effects. Simple analyses of the HeII Pickering decrement in WN and WC stars have shown that hydrogen is very deficient in both sequences (Smith 1973, Nugis 1975), but little information has been obtained on the C and N abundances because of the longstanding lack of suitable models of line formation and of observations in the ultraviolet where the appropriate low excitation lines occur. With the S2/68 observations and the developement of techniques for the treatment of line transfer in rapidly expanding atmospheres (Castor 1970) a renewed attack on the abundance problem in the WR stars became possible.

The model used is based on the Escape Probability Method (EPM) developed by Castor (1970) and simplified by Castor and van Blerkom (1970)

in their treatment of the HeII lines in the WN6 star HD 192163. The
model assumes that the emission lines are formed in a spherical,
homogenous region surrounding a continuum emitting core. The basic premise
of the EPM is that since the expansion velocity in the emission region
is very much larger than the thermal velocity, radiative interaction
with distant parts of the atmosphere is negligible, and therefore the
line transfer is locally constrained. The equations used in coupling the
statistical equilibrium equations for the ionic level populations with
the radiative transfer in the lines and continuum have been described by
Castor and van Blerkom (1970) and reviewed by Willis and Wilson (1978a).
It is possible through the EPM to set up these equations for the level
populations (and hence line source functions and strengths) in terms
of local values of electron temperature and density and ionic species
density, and thus to produce grids of computed line strengths for each
species in terms of these atmospheric parameters. The model grids are
then compared with the observed line strengths to determine the physics
and chemistry of the line emitting region. The EPM is best suited for
line transfer involving transitions arising amongst levels whose
populations are mainly determined by bound-bound processes; in general
these are low-lying and usually occur in the UV. It is in this context
that the S2/68 observations become important for abundance analysis of the
WR atmospheres. Equivalent widths for the transitions of HeII 1640,
CIII 2297, 1909, CIV 1550 and NIV 1718, 1488 observed in the six single
WR stars have been given by Willis and Wilson (1978a). It turns out that
the He modelling is best done using the visible Pickering lines, and our
analyses to date have concentrated on four stars which have both UV and
visible emission line measurements. These are HD 50896 (WN5), HD 192163
(WN6), HD 191765 (WN6) and HD 192103 (WC8). In order to determine the
NIII density in each star (and hence with NIV the total N) we have used
measurements for the excited lines of NIII 4640, 4100 in the three WN
stars (Smith and Kuhi 1976) and for the WC8 star the measured strength
of the NIII 991 resonance line arising in the WC8 component of the
binary star γ^2 Velorum as observed in a rocket spectrum of the system
(Burton et al. 1973, Willis 1976). The assumption of the similarity
of HD 192103 with the WC8 component in γ^2 Velorum is also used to fix
the radii of its continuum core and line emitting region, which are
inferred from the angular diameter measurements of the system given by
Hanbury Brown et al. (1970). The corresponding radii for the WN stars
have been derived using the observed strengths of three optically thick
HeII lines coupling levels n = 3, 4 and 5 using the method given by
Castor and van Blerkom (1970). Other parameters which enter each model
calculation are (i) the core continuum temperature, taken as 30000 K
and with a black body distribution, (ii) the expansion velocity inferred
from measurements of violet displaced absorption lines, (iii) the
electron temperature, T_e, (iv) the electron density, N_e and finally
(v) the ionic density of the species considered, $N(X^+)$. The last four
values are local values at the representative radius adopted for the
emission region. The model solutions, consisting of computed line strengths
are determined for grids of T_e, N_e and $N(X^+)$, and these atmospheric
parameters are determined for each star by fitting the observed line
strengths for each ion. For a detailed description of this model fitting

HD	N(CIII)	N(CIV)	N(NIII)	N(NIV)	N(He)	C/N	C/He	N/He
50896	5.0(6)	1.0(6)	1.0(9)	8.0(7)	2.0(11)	5.4(-3)	9.0(-5)	1.7(-2)
192163	5.0(6)	1.2(6)	1.4(9)	7.0(7)	2.0(11)	2.6(-3)	9.0(-5)	3.5(-2)
191765	5.0(6)	1.2(6)	1.1(9)	6.0(7)	2.0(11)	4.7(-3)	9.0(-5)	1.9(-2)
192103	3.0(8)	2.0(6)	7.0(7)	2.0(7)	1.0(11)	2.9(00)	9.0(-3)	3.2(-3)

Table 2   The abundances of He, C and N in four single WR stars
          observed by S2/68. The densities for each ionic species are
          in cm^{-3} with the values in () being the exponent.

see Willis and Wilson (1978a). It is noteworthy that the separate model
fitting for each ion results in similar values of T_e and N_e, providing
a good consistency check on the model employed.

The results are shown in Table 2 which lists the derived mass
abundance fractions of He, C and N in each star. The deduced electron
temperatures for the WN and WC stars are 50000 K and 30000 K respectively.
With He as the principle constituent and mainly doubly ionized, N_e is
simply twice the He density to good accuracy. Within the framework of the
EPM used it is concluded that the errors in these abundances resulting
from uncertainties in the radii and effective temperatures employed are
less than a factor of two, with the C/N ratios more accurate still. The
attainment of more accurate values will require the construction of more
sophisticated models of line transfer and also more observations in the
ultraviolet.

4   THE EVOLUTIONARY STATUS OF THE WR STARS

With the effective temperatures determined as outlined above and
the absolute magnitudes given by Smith (1968), the location of the WR
stars on the H-R diagram can now be reliably defined and this is shown
in Fig 2. The WR stars are seen to lie to the right of the MS and below
the Of-supergiant branch. A similar location was inferred by Conti (1976)
who also noted that the WN7,8 stars, which are more luminous lie close
to the realm of the Of stars to which they also bear some strong spectral
similarities. This prompted him to propose that the Of stars may be the
progenitors of the WR class, with the WN7,8 stars as an intermediate
stage. Bohannon and Conti (1977) have shown that in the binary BD +40 4220
both components are Of but the secondary has a mass of only 7 $M_\odot$ despite
its normal high Of luminosity which they suggest points to the secondary
as being " on its way to becoming a WR star ". The WR stars are known to
be overluminous for their mass and Smith (1973) has pointed out that their
masses and luminosities are consistent with those expected for helium
burning stars. The predominance of helium in their atmospheres would
support this assertion. Although very little information exists concerning
the abundances of H, He, C and N during the course of hydrogen and helium

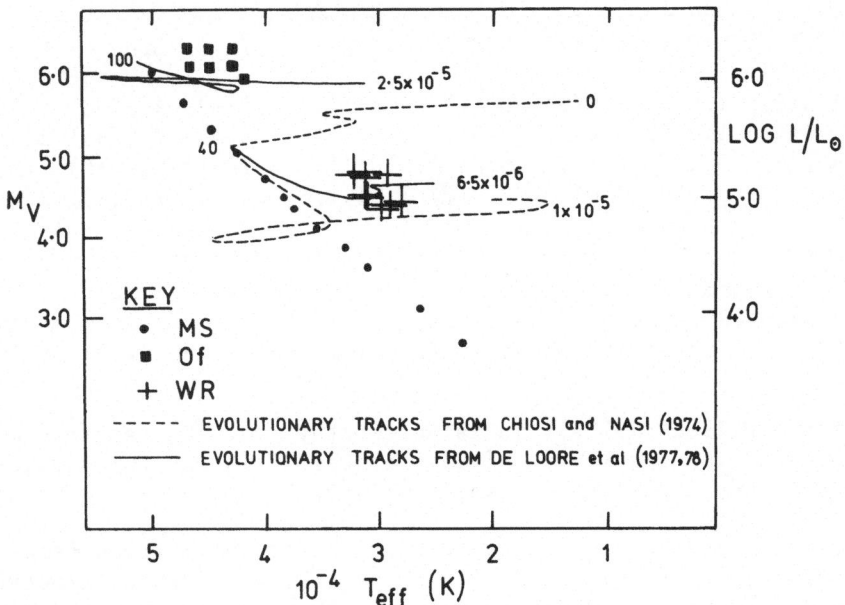

Fig 2   The upper region of the H-R diagram showing the location of the
          M-S, Of and WR stars and several evolutionary tracks for
          massive stars with high mass loss rates.

burning, we do find that the values deduced in the present study are
consistent with those expected to occur during these stages in massive
hot stars. Paczynski (1973) has tabulated in a semi-qualitative way the
abundance profiles to be expected during the course of hydrogen and
helium burning in massive stars where the former process operates through
the CNO-cycle and CNO-bicycle. These profiles are shown schematically in
Fig 3, starting with normal cosmic abundances. In the CNO-cycle and bi-
cycle, nearly all the hydrogen and much of the carbon is burned and in
the subsequent helium burning the carbon is quickly replenished and some
nitrogen is burned. The He, C and N abundances determined for the WR
stars are shown on the right of Fig 3, and it is clear that they match
the nuclear burning products at two specific stages which are marked at
the foot of the diagram. The WN point occurs at a very early stage of
helium burning, while the WC point occurs well into that process where
both C and N are quite abundant. Thus it appears that the WN stars are
at an earlier stage of evolution than the WC stars.

     This comparison between the observed abundances and the nuclear
burning products is only valid if the products of the nuclear burning
can be exposed to observation in the WR atmospheres. This can only occur
if there is extensive mixing and/or extensive mass loss, the latter
removing the outer unprocessed material. The WR stars are clearly losing
mass at a high rate and it is reasonable to suppose that the mass loss
process is the operative one. The atmospheric helium densities together

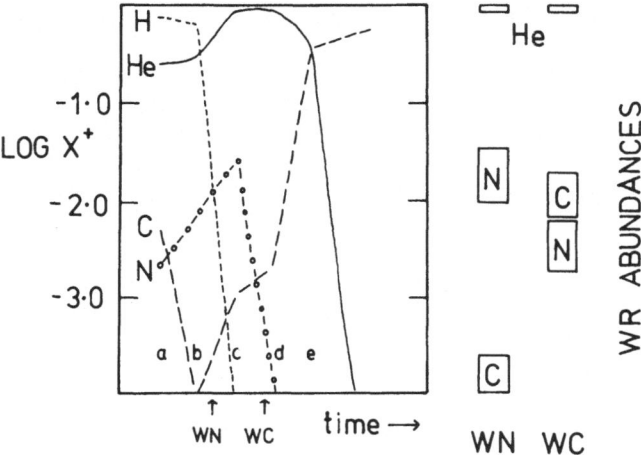

Fig 3   The abundance profiles of He, H, C and N during the course of
        hydrogen burning, points a to c, and helium burning, c to e, in
        massive hot stars, compared to deduced abundances for WR stars.

with the observed expansion velocities and line emitting radii deduced
in the present study imply mass loss rates of $\sim 10^{-4}$ $M_\odot$ y^{-1}. Willis and
Wilson (1978b) deduce a mass loss rate of 9×10^{-5} $M_\odot$ y^{-1} for the WC8 star
in γ^2 Velorum from observations with Copernicus of violet displaced
absorption in the intercombination CIII 1909 line. Mass loss rates of
this scale sustained over an extended period imply that the progenitors
of the WR class would have to be very massive stars, and Conti (1976)
has suggested that the Of stars with such large masses may have mass
loss rates sufficiently high enough to remove a large fraction of their
outer material during hydrogen burning and leave a star with WR characteristics. In the past few years several papers have dealt with the
evolution of massive stars with high mass loss rates. Chiosi and Nasi
(1974) studied the evolution of 20 and 40 $M_\odot$ models with a variety of
mass loss rates, and their track for a 40 $M_\odot$ star with an initial mass
loss rate of 1×10^{-5} $M_\odot$ y^{-1} is shown in Fig 2, together with the track
for the conserved case. When the mass loss rate is sufficiently high
they find that the star sheds most of its outer material and moves down
the M-S during the hydrogen burning phase and that during the subsequent
helium burning the star moves into the WR region. Although the masses
studied by Chiosi and Nasi (1974) are too small to give a starting point
among the Of stars, their results in many ways support the idea of a
Of-WR link through heavy mass loss in the hydrogen burning phase.
Similar results are reported by de Loore et al. (1977) who suggest that
the high mass loss rates needed to give the required evolution of a single
Of star to a WR star may not in fact occur. Subsequently de Loore et al.
(1978) extended their results to a star of 100 $M_\odot$ and an initial mass
loss rate of 2.5×10^{-5} $M_\odot$ y^{-1} and this track is also shown in Fig 2 in
which about 60 percent of the mass has been expelled by the end of the
hydrogen burning but the track has not proceeded down to the WR stars.

Additionally they find that the atmosphere is still comparatively rich
in hydrogen, in contrast to the observed situation in most classes of WR
star. Thus it would appear that the mass loss rates considered are still
not large enough to produce the required drop in luminosity from Of to
WR: perhaps rates of 3 to 4 $\times 10^{-5}$ $M_\odot$ y^{-1} are needed and one has to ask
whether such large mass loss rates can occur. This seems to be the case
for the WR stars themselves, and Hutchings (1976) has noted a few O-type
cases, and he has given evidence that mass loss is enhanced in binary
systems. De Loore et al. (1978) have argued that in massive binaries it
is very easy to remove 70 percent of the mass and leave a pure helium
star. It may therefore be that membership of a binary system is a
necessary condition for a WR star to develop, and that all WR stars are
thus in binary systems, a suggestion invoked by Kuhi (1973).

The evolution is clearly critically dependent on the mass loss rate
and it is not clear that the required very high mass loss rates for
single stars can be ruled out. It is possible that the mass loss rates
determined for O-type supergiants of 10^{-6} to 10^{-5} $M_\odot$ y^{-1} (Hutchings
1976) may refer to stars which are well into their hydrogen burning,
much higher rates having occured earlier in the evolution. In that case
it would not be inappropriate to consider higher mass loss rates than
those studied by de Loore et al. (1978) for the Of stars. Clearly there
is a need for more accurate and more extensive measurements of mass loss
rates for massive hot stars, together with more exhaustive models of
stellar evolution incorporating mass loss, in order to clarify the links
between the O-type and the WR stars.

REFERENCES

Beals,C.S., 1934, Pub.Dom.astrophys.Obs., 6, p95
Bohannon,B.,Conti,P.S., 1977, Astrophys.J., 204, p797
Boksenberg,A.,Evans,R.G.,Fowler,R.G.,Gardener,I.S.,Houziaux,L.,
    Humphries,C.M.,Jamar,C.,Macau,D.,Macau,J.P.,Malaise,D.,Nandy,K.,
    Thompson,G.I.,Wilson,R.,Wroe,H., 1973, Mon.Not.R.ast.Soc., 163, p297
Brandt,J.C.,Stecher,T.P.,Crawford,D.L.,Maron,S.P., 1971, Astrophys.J.,
    163, p199
Burton,W.M.,Evans,R.G.,Griffin,W.C.,Lewis,C.,Paxton,H.J.B.,Shenton,D.B.,
    Macchetto,F.,Boksenberg,A.,Wilson,R., 1973, Nature, 246, p37
Castor,J.I., 1970, Mon.Not.R.ast.Soc., 149, p111
Castor,J.I.,van Blerkom,D., 1970, Astrophys.J., 161, p485
Chiosi,K.,Nasi,E., 1974, Astron. Astrophys., 34, p355
Cohen,M.,Barlow,M.J.,Kuhi,L.V., 1975, Astron. Astrophys., 40, p291
Conti,P.S., 1976, Mem.Soc.R.Sci.Liege, Ser 6, 9, p193
De Loore,C.,de Greve,J.P.,Lamers,H.J., 1977, Astron. Astrophys., 40, p251
De Loore,C., de Greve,J.P.,Vanbeveren,D., 1978, Astron. Astrophys.,
    in press
Hanbury Brown,R.,Davis,J.,Herbison-Evans,D.,Allen,L.R., 1970,
    Mon.Not.R.ast.Soc., 148, p103
Humphries,C.M.,Jamar,C.,Malaise,D.,Wroe,H., 1976, Astron. Astrophys., 49

Hutchings,J.B., 1976, Astrophys.J., 203, p438
Kuhi.L.V., 1966, Astrophys.J., 143, p753
Kuhi,L.V., 1973, IAU Symposium 49, p205, D.Reidel Pub.Co., Holland
Kunasz,P.B.,Hummer,D.G.,Mihalas,D., 1975, Astrophys.J., 202, p92
Kurucz,R.L.,Peytremann,E.,Avrett,E.A., 1974, Blanketed Model Atmospheres
   for Early Type Stars, Washington KPA
Morton,D.C., 1973, IAU Symposium 49, p54, D.Reidel Pub.Co., Holland
Nandy,K.,Thompson,G.I.,Jamar,C.,Monfils,A.,Wilson,R., 1975,
   Astron. Astrophys., 44, p195
Nugis,T., 1975, IAU Symposium 67, p291, D.Reidel Pub.Co., Holland
Paczynski,B., 1973, IAU Symposium 49, p143, D.Reidel Pub.Co., Holland
Smith,L.F., 1968, Mon.Not.R.ast.Soc., 138, p109
Smith,L.F., 1973, IAU Symposium 49, p15, D.Reidel Pub.Co., Holland
Smith,L.F.,Kuhi,L.V., 1976, unpublished Atlas of WR Spectra
Willis,A.J., 1976, PhD Thesis, University of London
Willis,A.J.,Wilson,R., 1978a, Mon.Not.R.ast.Soc., 182, p359
Willis,A.J.,Wilson,R., 1978b, in preparation

DISCUSSION FOLLOWING WILLIS and WILSON

Hutchings: The very close grouping of WN7 stars in the
HR diagram and the similar close grouping of the 5 Of extreme
stars in the same place reinforces the connection between
them. It also suggests that something special happens to enhance mass loss, at this combination of temperature, luminosity and gravity.

Willis: The grouping of the WN7 stars in the extreme Of
region is the result of absolute magnitude determinations of
WN7 stars from LMC measurements and temperature determinations
from Zanstra analyses. We have not observed WN7 stars with
S2/68. However their apparent close grouping does suggest
an evolutionary link, as does the comparative similarity of
their spectra in the visible.

Castor: I am a little confused about the composition
determination for the WC stars. If the He/C ratio is normal,
thenthe mass fraction of carbon is enhanced by a factor 4 or
so, since the hydrogen is missing.

Willis: The carbon abundance determination is consistent
with the C being replenished in the helium burning phase and
so the He/C ratio should not really be compared with abundances relative to hydrogen in the hydrogen burning phase.

Of-WN EVOLUTION: SPECTRAL TYPES AND EFFECTIVE TEMPERATURES

E. Myckky Leep
Dept. of Astronomy, University of Washington,
Seattle, W.A. 98195

Because of the growing evidence (Conti 1976, Snow and Morton 1976, Chiosi et al. 1978, de Loore et al. 1977) that some Of stars are evolving into WN stars, we reexamine the classification of WN stars. There are four schemes for classifying WN stars, those of the IAU (1938), Hiltner and Schild (1966), Smith (1968), and Walborn (1974). All of these schemes involve estimating by eye the ratio of the nitrogen line strengths. Walborn's scheme is the most useful because he avoids problems that may arise from the use of helium lines. In other respects Walborn's scheme gives WN classifications similar to those found by Smith.

We have measured equivalent widths of lines of N III λλ4640–4642, N IV λ4058, N V λ4604 for 22 Of stars. These Of stars were then given WN subclasses according to the criteria of Walborn (1974). The equivalent widths of the absorption lines of He I λ4471 and He II λ4541 were measured in three WN transition stars (Conti 1976) HD 93162, HD 93131, HD 92740. (We know that only one of these stars, HD 92740, is a binary and here the absorption lines were determined to come from the WN component of the system (Conti, Niemela, Walborn 1978).) These three WN stars were given Of subclasses using the criteria of Conti and Frost (1977). Figure 1 shows the results. Note the inclusion of the subclass WN9 which Walborn (1977) proposes in connection with WN stars he observed in the Magellanic Clouds. Figure 1 shows that Of stars earlier than O6f all could be given a classical WN spectral type. Yet these stars have effective temperatures between 40000–55000 K (Conti and Frost 1977), much more than the usual 20000 K (Cassinelli and Hartman 1975) associated with late-type WN stars. The three WN stars are in turn classified with the hottest Of stars. From Figure 1 we can conclude that Of stars cannot be clearly separated from WN stars by the available classification criteria. Also, we see that at least some WN stars may have effective temperatures higher than the usual 20000 K. If the early Of stars do give us a true indication of the effective temperature of the WN6–WN8 stars and if WN3–WN5 stars have even higher temperatures (because more highly ionized lines are observed in these stars) the bolometric corrections for WN stars must increase, giving a higher absolute magnitude which would be closer to that observed for Of stars.

P. S. Conti and C. W. H. de Loore (eds.), Mass Loss and Evolution of O-Type Stars, 471–474.

Figure 1.

	03	04	05	05.5	06	06.5	07	07.5	08	08.5	09
WN3											
WN4											
WN5											
WN6	93129A	*93131 *93162									
WN7		190429A	*92740	228766							
WN8		15570 16691 ζPup	14947	152386							
WN9					14442 153919 λCep	148937 150958 602522	29 CMa 108 163758	166734 167971	151804 152408 9 Sge		

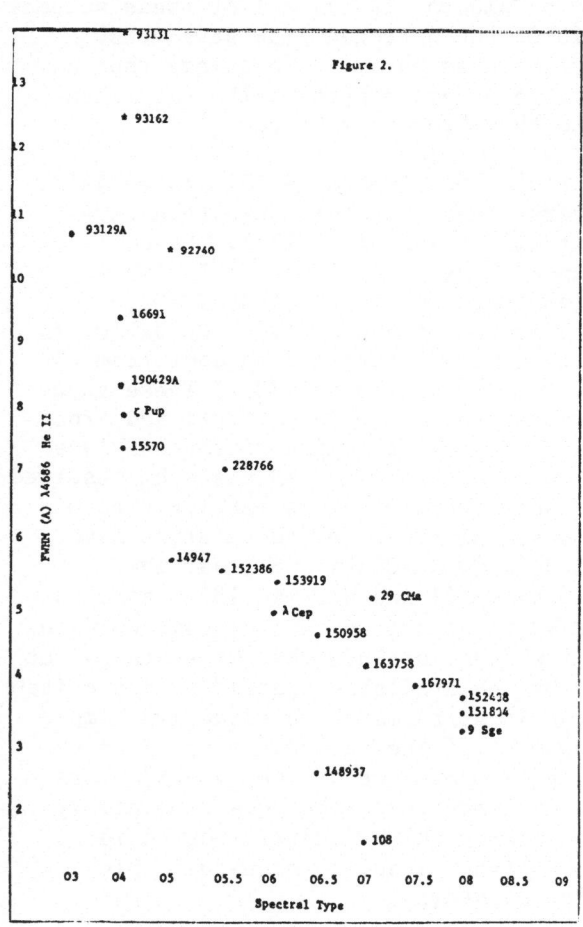

Figure 2.

Figure 2. Some stars do not fit the linear relation seen in Figgure 2. HDE 228766 and 29CMa are double lined spectroscopic binaries, where the secondary may influence the width of λ4686. For HDE 228766 the spectral type maybe as early as 05f. HD 148937 and HD 108 may also have incorrect spectral types because of P Cygni features in He I λ4471.

   In some stars it is observed that line widths correlate with
spectral type. Similar correlations were sought in Of stars, by
measuring the full width at half maximum (FWHM) of He II $\lambda4686$, which
is in emission in all Of stars. In Figure 2 we see the results, which
include the three WN stars discussed previously. One can see at once
the almost linear relation between spectral type (or effective temperature) and the FWHM of $\lambda4686$. Remember that $\lambda4686$ can be formed in emission
only in stars having envelopes (Mihalas, and Lockwood 1972, Mihalas 1974).
Then if the emission of $\lambda4686$ is due to the same mechanism in all Of stars,
one might expect that the width of $\lambda4686$ is an indication of the extent
of the envelope. If so, then the hotter Of stars appear to have the more
extended envelopes. Further, if the mechanism for emission of $\lambda4686$ is
the same for Of and WN stars, perhaps some WN stars differ from Of stars
primarily in the extent of their envelopes.

References

Cassinelli,J.P., Hartman,L: 1975, Ap.J. 202,718.
Chiosi,C.,Nasi,E.,Sreenivasan,S.R.: 1978,Astron.Ap.63,103.
Conti,P.S.: 1976,Mem.Soc.Roy.des Sci.de Liege, 6eSerie,Tome IX,193.
Conti,P.S.,Frost,S.A.: 1977, Ap.J.212,728.
Conti,P.S.,Niemela,V.S.,Walborn,N.R.: 1978,Preprint.
de Loore,C.,DeGreve,J.P.,Lamers, H.J.G.L.M.: 1977,Astron.Ap.61,251.
Hiltner, W.A.,Schild,R.E.: 1966,Ap.J.143,770.
I.A.U., 1938,Trans.Int.Astr.Union,6,248.
Mihalas,D.: 1974,A.J.79,1111.
Mihalas,D.,Lockwood,G.W.: 1972,Ap.J.175,757.
Smith,L.F.: 1968,M.N.R.A.S. 138,109.
Snow,T.P.,Morton,D.C.: 1976,Ap.J.Suppl.32,429.
Walborn,N.R.: 1974,Ap.J. 189,269.
Walborn,N.R.: 1977,Ap.J.215,53.

DISCUSSION FOLLOWING LEEP

Underhill: Your observations detail very nicely the
spectroscopic likenesses between Of and WN7/8 stars. The
next step will be to determine reliable flux effective temperatures for these stars and to see if they are as high as
we infer from the electron temperatures which we require to
account for the observed visible spectra. To obtain reliable flux effective temperatures for stars of spectral type
O8 and earlier we must have absolute spectrophotometry in
the range 1800 to 1000 Å. The range from 1800 Å to 4000 Å
is also needed, but these fluxes are available for a few of
the stars from S2/68 and OAO-2. The necessary observations
might be made successfully from the space shuttle. My determination from S2/68 spectra of T_{eff} for ζ Puppis is un-
reliable in the sense that my answer is constrained by the
model atmosphere effective temperature which I adopt, for
the estimated part of the integrated flux shortward of 1380
Å drives the solution. However, I am sure from the shape
of the spectrum in the range 6000 to 11000 Å that T_{eff} is
close to 47000 K. If it is as low as 32000 K as suggested
by Code et al. (1976, Ap.J.), the fit is not so good.

Conti: Ms. Leep's temperature scale puts some WN7 stars
at values about 45000 K, whereas Willis suggests that some
earlier types have effective temperatures near 30000 K. It
may be then that the effective temperatures are inverted
from the ionization values, in which the earlier types are
clearly hotter. The envelope temperature may not be exactly controlled by the effective temperature but by the heat
input.

OBSERVATIONS OF VELOCITY FIELDS IN WN AND Of STARS

Virpi S.Niemelä [*]

51 y 11,Villa Elisa,Buenos Aires,Argentina

Systematic wavelength shifts of series of spectral line centers observed in many early type stars,generally interpreted as due to large scale motions,can give us information about the velocity gradients in stellar atmospheres.However,it should be borne in mind that the velocity gradients inferred from the observed displacements of spectral lines may not correspond to a unique alternative (e.g.see Karp 1978).Also,and especially when we are dealing with stars which have emission lines in their spectra,the structure of the velocity field depends on the assumed temperature structure of the atmosphere,i.e.in which atmospheric region do the lines originate.

In this paper observations of mean wavelength displacements of absorption and emission line centers are presented for six stars,namely, HD 86161 (WN8),HD 92740 (WN7), HD 93129A (03If),HD 93131 (WN6),HD 93162 (WN6) and HD 163758 (06.5If).The spectral classifications quoted in parenthesis are from Walborn (1973,1974),and the spectra of the first five stars are illustrated in Walborn (1974).The observational data for HD 86161 and HD 163758 are based on 38 Å/mm Cassegrain spectra obtained and measured with a Grant engine at the Cerro Tololo Inter-American Observatory,Chile.The observational data for the other four stars are given in Conti,Niemelä and Walborn (1978),where the structure of their envelopes related to the radial velocities is discussed.

Earlier results of the velocity gradients of the hydrogen Balmer absorption lines in some of the stars studied here suggested that strong velocity fields exist at the atmospheric level where the absorptions originate (Niemelä 1975;Conti 1977).In Figures 1 to 4 the mean wavelength shifts of absorption and emission line centers are represented as a function of the excitation potential of the upper energy level of the corresponding line. These figures show that in all stars studied

---

[*] Visiting Astronomer,Cerro Tololo Inter-American Observatory,supported by the National Science Foundation under contract No. NSF-C866.

P. S. Conti and C. W. H. de Loore (eds.), Mass Loss and Evolution of O-Type Stars, 475–478.

here "velocity progressions" exist throughout the atmospheric regions
where both absorption and emission lines arise.

The kind of "velocity progressions",i.e.a correlation with increasingly negative radial velocity with decreasing excitation potential
of the upper energy level,shown in Figures 1 to 4 are generally interpreted as an outwards accelerated motion in a stellar atmosphere with
outwards decreasing temperature.Although this same type of correlation
seems to apply for the emission lines (Figure 4),however,the redshifts
may not be due to a Doppler effect (e.g.Auer and Van Blerkom 1972).We
note also,that the presence of a violet-shifted absorption edge in an
emission line does not always produce a redshift.For example in Fig.4
the lines of SiIV and NIII show P Cygni type profiles,but the emissions
are not redshifted.

The Figures 1 to 4 also suggest,and especially see Figure 1,a
smooth transition in the velocity gradients with the spectral type,in
the sense that for increasing ionization and decreasing strength of the
emission lines,the wavelength shifts seem smaller.Thus the relationship
between the Of and WN type spectra of these stars may be correlated to
the atmospheric velocity fields.An extrapolation to all other stars
showing Of and WN type spectra,however,is not obvious.

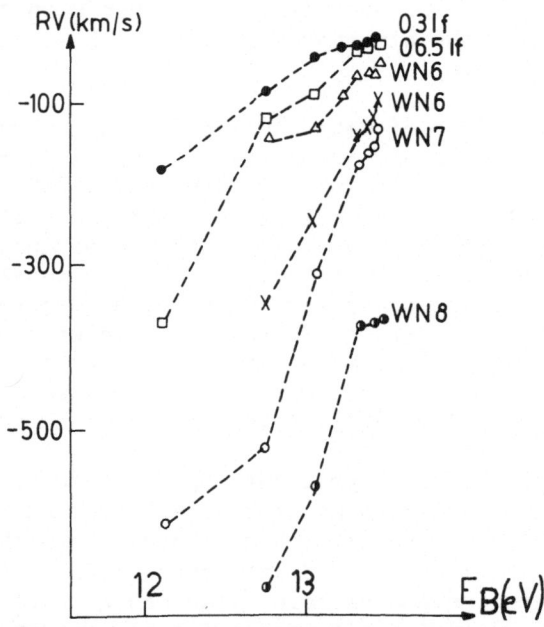

Figure 1. The mean displacements of the absorption line centers (in km/s) of the hydrogen Balmer series as a function
of the excitation potential of the upper energy level of the
corresponding line:HD 93129 (filled circles);HD 163758 (rec-
tangles);HD 93131 (crosses);HD 93162 (triangles);HD 92740
(open circles) and HD 86161 (half-filled circles).

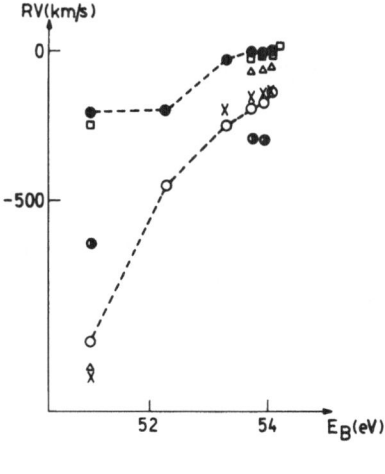

Figure 2. As in Fig.1 for the
HeII absorption lines.

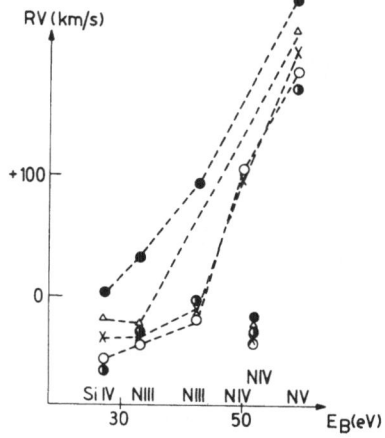

Figure 4. As in Fig.1 for the
narrow emission lines of SiIV,
NIII,NIV and NV.Note the distinct
value of the NIV 4057 singlet on
the lower right of the Figure.

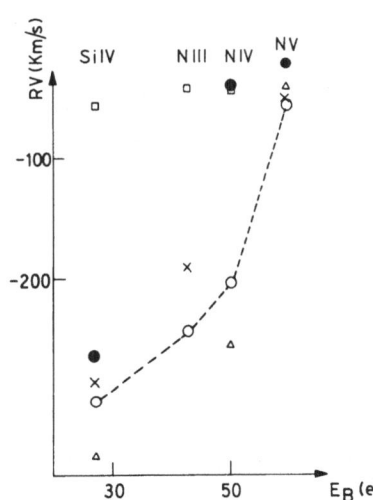

Figure 3. As in Fig.1 for the P Cygni absorptions of SiIV,
NIII,NIV and NV.

References:

Auer,L.H. and Van Blerkom,D.1972,Ap.J.,178,175.

Conti,P.S. 1977,in IAU Symp.No.80, in press.

Conti,P.S.,Niemelä,V.S. and Walborn,N.R.1978,Ap.J.,in press.

Karp,A.H. 1978,Ap.J.,222,578.

Niemelä,V.S. 1975,in Physics of Movements in Stellar Atmospheres,Coll.
        No.250 of C.N.R.S. of France,eds.R.Cayrel and M.Steinberg,p.467

Walborn,N.R. 1973,A.J.,78,1067.
        .1974 ,Ap.J.,189,269.

DISCUSSION FOLLOWING NIEMELA

Massey: I couldn't quite see from the graphs how did
the velocities of the Si IV, N III emission lines compare
with the uppermost Balmer absorption lines H 10, say. Were
they very different?

Niemela: The upper Balmer lines have more negative
velocity than the Si IV, N III emissions, which are not red
shifted. The difference between the velocities of the upper
Balmer absorptions and the narrow emissions depends on the
spectral type.

BE 381: WN9 or O8 Iafpe?

Bruce Bohannan
Department of Astro-Geophysics, University of Colorado
Boulder, Colorado 80309 USA

As John Hutchings summarized at the opening of this symposium the most easily detected symptom of mass loss in hot stars is emission at Hα. Some years ago I made a survey of the Large Magellanic Cloud for stars with Hα in emission (Bohannan and Epps 1974) and since then have obtained low resolution spectra of some of these stars to establish their spectral identity. I would like to talk today of one of these objects, BE 381, which displays spectroscopic features of both the extreme Of and low excitation WN Wolf-Rayet classifications.

The spectrum of BE 381 (Figure 1) exhibits well developed P Cygni profiles on the hydrogen transitions to Hε and on the neutral helium triplet lines (λ3889, λ4026, λ4471). N IIIλ4634-40 and He II λ4686 are in emission as in an Of - type star. The Si IV doublet around Hδ is strongly in emission. Ionized helium λ4542 and λ4200 are in absorption. In the table equivalent widths of the spectral lines in BE 381 are compared with those in HD 152408 (O8 Iafpe).

Identification	log Wλ (mÅ)	
wavelength	BE 381	HD 152408*
HeI λ3889 abs	2.67	
em	3.70E	
H ε λ3970 abs	1.90	
em	2.92E	
HeI λ4026 abs	2.98	
em	2.64E	
SiIV λ4116	3.18E	2.70E
H γ λ4340	3.70E	
N III λ4378	2.98E	
HeI λ4471 abs	2.85	2.98
em	3.28E	2.86E
HeII λ4542	2.83	2.81
NIII λ4634.40	4.25	
HeII λ4686	3.21E	3.52E

P. S. Conti and C. W. H. de Loore (eds.), Mass Loss and Evolution of O-Type Stars, 479-482.

The strength of He IIλ4542 in BE 381 is roughly equal to that in HD 152408, an extreme Of star (Hutchings 1968), which suggests comparable atmospheric conditions near the photosphere. The much stronger hydrogen and neutral helium P Cygni features in BE 381 indicate that the extended atmosphere is much more massive or the velocity law is slower. BE 381 is observed at an absolute visual magnitude fainter than that estimated for the extreme Of stars, a value of -6 is found if the reddening correction is only a few tenths in B-V, a typical value in the Large Magellanic Cloud.

The overall appearance of the spectrum of BE 381 is remarkably similar to HDE 269227, a Wolf-Rayet star in the LMC designated as WS 12 and classified as WN 8 by Westerlund and Smith (1964). The hydrogen, neutral helium and N III spectra are essentially the same, but Si IV emission is much stronger in BE 381. He II λ4542 absorption is slightly stronger in BE 381. N II λ3995 emission, present in HDE 269277 (Walborn 1977), is very weak or absent in BE 381.

It appears then that BE 381 should probably be considered a Wolf-Rayet type star rather than an extreme Of-type. If so classified, two arguments suggest that the type is later than WN 8. The typical WN 8-A star HD 86161 has N IV 4058 present (Walborn 1974), a line which is missing in both HD 269227 (WS 12) and BE 381. The excitation class vs. line width relation for WR stars suggests that BE 381 is of later type than HD 269227 because of the narrower lines in BE 381. The appropriate classification for BE 381 appears to be WN 9 or WN 10, (types previously undefined), similar to HDE 269227 (Walborn 1977). Further work is needed to define the sequence of low excitation Wolf-Rayet stars.

The relationship between WR and Of stars has been reviewed by Conti (1976, 1978). The WN 7-8 sub-types are much less pathological in their spectra than the earlier types. Peter Conti and Tony Moffat have independently described this afternoon how the low excitation WN stars may have evolved by mass loss from O-type stars with masses larger than 35 $M_\odot$. The location of the low excitation Wolf-Rayet stars near the low temperature limit of main sequence stellar evolution suggests that they represent massive stars which have completed core hydrogen burning and have just undergone gravitational contraction of the core. The subsequent envelope expansion causes a much more extended atmosphere with only a slight increase in mass-loss rate. The enhanced N III and Si IV are a consequence of the extended atmosphere rather than mixing of interior material. Helium, nitrogen or carbon may come to the surface later in the star's evolution.

I am especially grateful to P. S. Conti for obtaining a higher resolution spectrum of BE 381 and for his valuable conversations. This research was supported in part by National Science Foundation grant AST 76-20842.

* Conti and Alschuler 1971

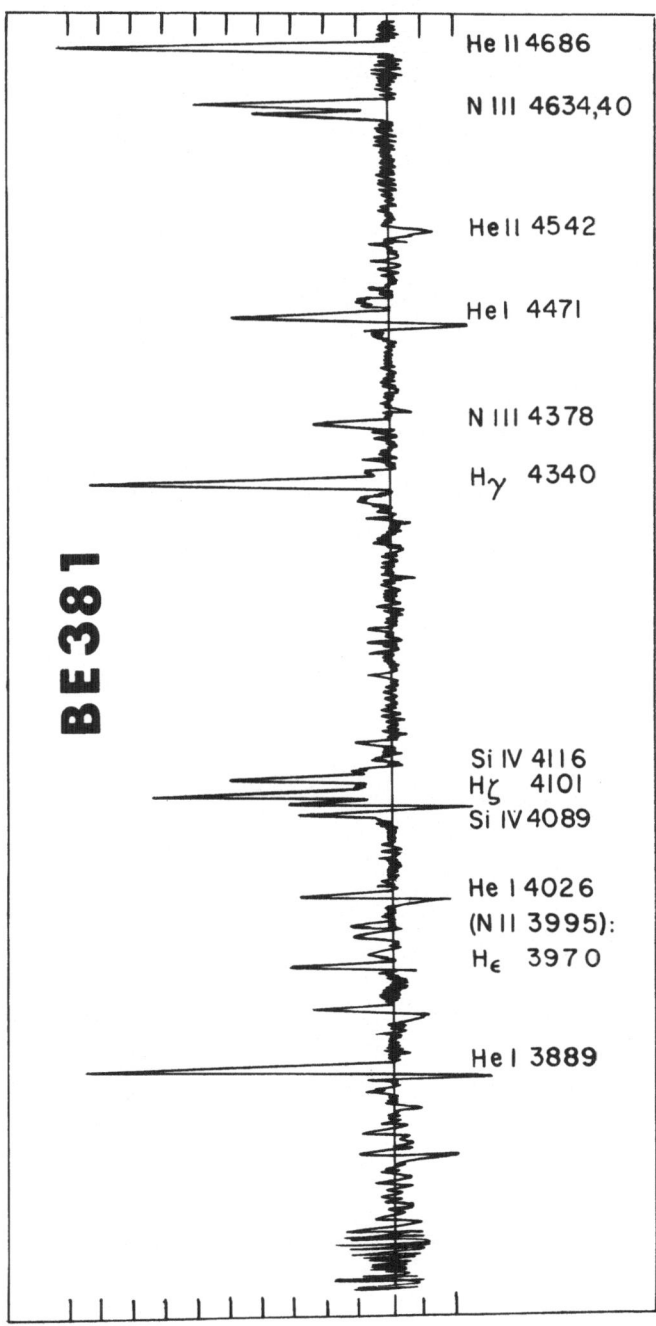

Figure 1: Spectrum of BE 381. This is an intensity tracing of a spectrogram taken with the image tube camera on the Four meter Cassegrain spectrograph at Cerro Tololo Inter-American Observatory. Original dispersion is 27 A/mm.

REFERENCES

Bohannan, B., and Epps, H. W.:1974, Astron. Astrophys. Suppl., 18, pp. 47-79.
Conti, P. S.:1976, Proc. 20th Liege International Astrophysical Collo-quium, Mem. Soc. Roy. Sci. Liege 9, pp. 193-212.
Conti, P. S.:1978, I.A.U. Symposium #80 (in press).
Conti, P. S., and Alschuler, W. R.: 1971, Ap. J. 170, pp. 325-344.
Hutchings, J. B.: 1968, M.N.R.A.S. 141, pp. 219-249.
Walborn, N. R.: 1974, Ap.J. 189, pp. 269-271.
Walborn, N. R.: 1977, Ap. J. 215, pp. 53-61.
Westerlund, B.E., and Smith, L. F.: 1964, M.N.R.A.S. 128, pp. 311-325.

DISCUSSION FOLLOWING BOHANNAN

Chiosi: Did you test whether your suggestion about the evolutionary status of the Of stars is consistent with the number of Of relative to main sequence stars, and the ratio of lifetimes? This ratio is in fact quite different in the two alternatives.

Sreenivasan: Would you care to indicate empirically theoretical mass loss rates for these WN stars? How does this compare with the rates for Of stars?

Bohannan: Only two to three times that of an Of star, say roughly 3×10^{-6} M_o/yr.

Breysacher: Does your star have a number in the LMC WR stars catalogue published by Fehrenbach et al.? I have ob-served their star No. 56 which is very similar to the one you described here.

Bohannan: I have a copy of my survey with me and could give you the identification. It is not one of those identi-fied by Fehrenbach.

CONSERVATIVE AND NON-CONSERVATIVE EVOLUTIONARY COMPUTATIONS IN
CONNECTION WITH WOLF-RAYET BINARIES.

D. Vanbeveren, J.P. De Grève, C. de Loore
Astrophysical Institute, Vrije Universiteit Brussel and
E.L. Van Dessel
Royal Belgian Observatory

1. INTRODUCTION

We consider the evolution of massive close binaries. The
reader is referred to the first contribution by the same authors during
this conference for the general ideas. The present contribution deals
with the implications for the Wolf-Rayet phase, i.e. that part of the
evolution where the primary is liable to show W-R characteristics (as
usual the primary is defined as the originally more massive component).
We refer to the evolution after the exchange of mass between the components, or even during the exchange phase (see section 3).

In Section 2 we recall some of the results from conservative
exchange calculations. These computations were extended in two ways.
Firstly, we include mass loss through stellar wind from ZAMS on for
both components as long as they evolve separately (first phase).
Secondly, we allowed for mass and momentum loss from the system during
the ensuing critical lobe overflow and mass exchange phase. For conve-
nience, we shall further specify here a second phase of violent mass
transfer and a third phase of slow exchange, during which the primary
slowly regains its previous luminosity.

2. COMPUTATIONS UNDER THE ASSUMPTION OF CONSERVATIVE MASS EXCHANGE

Our conservative computations have been described extensively by De Grève et al. (1978). Such computations can produce models
of W-R like stars, with $X_{atm} \approx 0.20$ (content of H by mass), suitable
luminosity and effective temperature and with masses within the range
of the observed values. It should be remarked, though, that in order
to obtain a remnant primary mass of e.g. 15 $M_\odot$, one has to start from
a star of $\approx$ 40 $M_\odot$; a large fraction of the mass is then exchanged in
about 1000 years. It is difficult to imagine an exchange rate of
0.02 $M_\odot$/year on the average as a conservative process. Anyway, one
cannot obtain systems with periods in the order of a few days (and some
of such are observed) from conservative computations; that is, if one

483

P. S. Conti and C. W. H. de Loore (eds.), Mass Loss and Evolution of O-Type Stars, 483–488.
Copyright © 1979 by the IAU.

wants to avoid contact systems. It has also been shown by De Grève et al. (1978) that systems with initial mass ratio q_i > 0.5 cannot produce the observed (final) periods and mass ratios. All in all, the conservative computations can only give a rough correspondence between observed and theoretical characteristics.

One of the important things we learned (or saw confirmed) from these computations, is that the remnant of the primary after the process of mass exchange is determined by the initial primary mass and composition. The choice of system parameters such as mass ratio and separation has no influence upon the final structure of the primary.

An example of evolutionary tracks in the H-R diagram is shown in Figure 1.

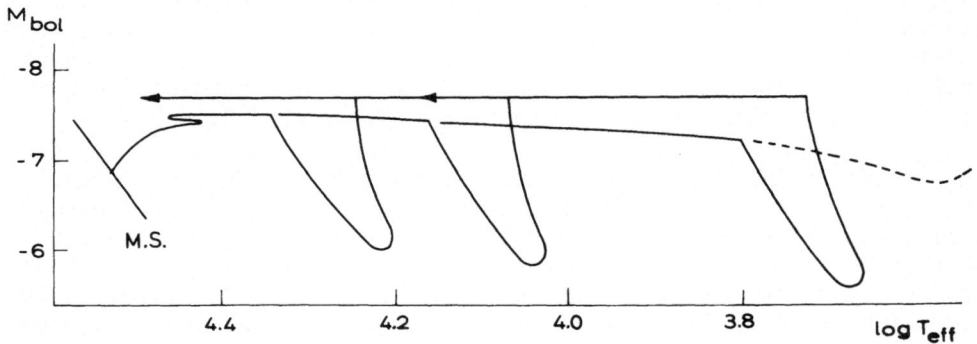

Figure 1. Evolutionary tracks for a 20 $M_{\odot}$ star with a 14 $M_{\odot}$ companion, conservative mass exchange. Initial separations : 40 $R_{\odot}$, 100 $R_{\odot}$, 490 $R_{\odot}$.

3. NON-CONSERVATIVE EVOLUTION AND THE WOLF-RAYET PHASE

The problems mentioned in section 2 disappear if the refinements described in the introduction are included in the computations.

For the stellar wind mass loss (first phase) the computations were performed as described by de Loore et al. (1977). The choice of the proportionality parameter N ($\dot{M}$ = N L c^{-2}) affects the value of the remnant mass (cf. Figure 1 in our first contribution). The size and structure of the primary after the first phase determine the final characteristics of the W-R star. Like in the conservative case, these are independent of separation and choice of secondary, but also of the assumptions about the amount of matter and momentum that is lost to the system or exchanged.

The computations for phase 2 and 3, including mass and angular momentum losses, were performed by means of two parameters β and α, where $1-\beta$ is the fraction of mass shed by the primary that is going out of the system and α describes the angular momentum losses through the relation

$$\Delta H/H = 1 - (1 - \frac{\Delta M}{M_{1i} + M_{2i}})^{\alpha} \quad , \quad \alpha \geq 0$$

The detailed derivation of this equation is given in Vanbeveren et al. (1978). In that paper calculations are given for systems with primary masses ranging from 30 $M_{\odot}$ to 60 $M_{\odot}$. Examples of these computations are shown in Figure 2. Typical values for mass exchange rate and time scale are :

phase 2 : 2 - 3 $\times$ 10^{-3} $M_{\odot}$/yr , 2000 - 4000 year

phase 3 : 1 - 2.5 $\times$ 10^{-4} $M_{\odot}$/yr, 7000 - 18000 year.

The systems we thus obtain at the end of phase 3 represent very well the characteristics of the observed W-R binaries. In other words, one can find a suitable initial system + evolutionary history for each of the observed systems. In particular, the observed combination of period and mass ratio can be obtained by an appropriate choice of mass and angular momentum loss (apart from the initial values for P and q one has to choose). For all these parameters a wide range of values is a priori possible; this leads to a variety of final configurations and gives the impression that the set of parameters is indetermined. We can, however, put certain restrictions on them.
(i) From the observed overluminosity of X-ray binaries N-values of $\approx$ 300 to 500 can be derived (Vanbeveren et al., 1978).
(ii) The observed combination of period and mass ratio for W-R binaries leads to restricted values for β and α. For the shortest periods (e.g. HD 214419 : P = 1.64 days; Khaliullin, 1972) one needs $\alpha \geq 3$. The fact that in general P < 100 days seem to exclude values of $\beta \approx 1$. From the analysis of some 10 W-R systems (De Grève et al., 1978, table 5) typical values of $\alpha \approx 3$ and $\beta \approx 0.5$ are found. Physically this means that about 50 % of the transferred mass leaves the system, carrying away about 50 % of the total orbital angular momentum. These rather high losses can be explained from the radial outflow of matter (because of the radiation pressure), rather than a directed flow in the vicinity of L_1, and from the expected occurence of a contact phase shortly after the beginning of phase 2 (cf. Kippenhahn and Meyer-Hofmeister, 1977). It was already found earlier (De Grève et al., 1978) that the later mass exchange sets in, the more violent the process is; in other words, the systems with large initial period will be subject to heavier mass loss during phase 2 and thus the period will be drastically reduced. This agrees with the observed scarcity of final periods larger than a few tens of days.

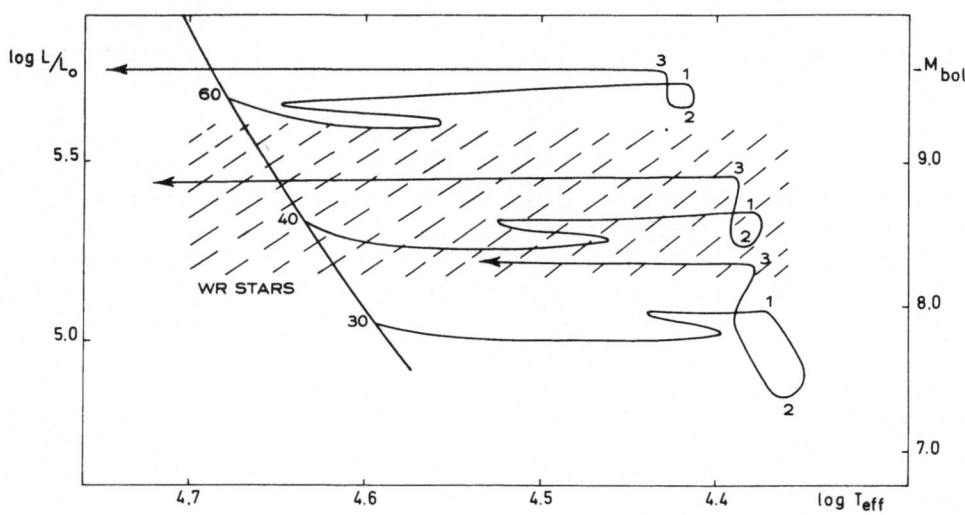

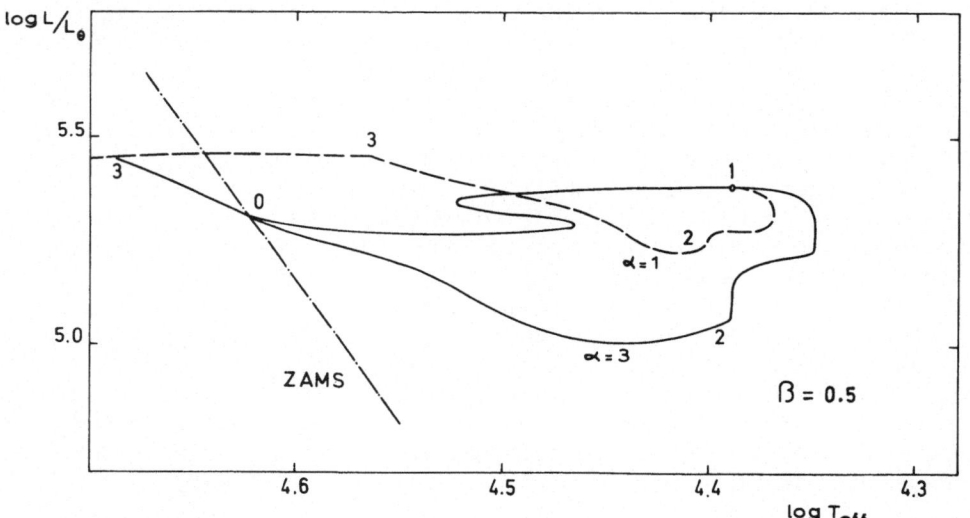

Figures 2a and 2b. Evolutionary tracks for the primary of a massive close binary system (companion star of half the primary mass). Stellar wind phase (N=300) : between points O (ZAMS) and 1. Mass transfer phase (2nd and 3rd phase) : between points 1 and 3.

Fig. 2a : Conservative mass exchange. The hatched area indicates the zone where W-R stars are found (Conti, 1976).

Fig. 2b : Non-conservative mass exchange with β = 0.5, α = 1 and
          α = 3.

(iii) Concerning the mass ratio q : consider the systems of Table 5 in
De Grève et al. (1978) and the mass estimates therein. If half of the
mass is lost from the system, the masses of the secondaries can only be
obtained if the initial mass ratio is close to unity. This means either
that q ≃ 1 is favoured during the formation of massive binary stars, or
that the companion star goes unnoticed if it is of smaller mass. The
latter explanation is sustained by the fact that the star which evolves
into a W-R becomes overluminous for its mass. Also, in some of the
systems there is a trace of the companion, but the determination of
orbital elements is not possible. Thus we expect that systems with small
initial q $^{(*)}$ will in general produce a W-R primary that apparently is
single. There may also be W-R stars with a compact companion (see our
first paper this volume), which would also appear to be single.

Finally, we want to examine if we can infer something from
the observed effective temperatures. The range of values one finds in
the litterature is very wide, but there seems to be an upper limit to
the estimations, situated around $\log T_{eff}$ = 4.75. Some of the estimates
used to be too high because of the use of plane parallel atmospheres
(cf. Cassinelli and Hartman, 1975). On the other hand, the effective
temperatures that come out of our evolutionary computations are also
systematically too high, because we had no models for extended atmospheres. Even then, it seems to us we can rule out the idea that the
observed W-R stars are stars on the He-M.S., because we are dealing
there with $\log T_{eff}$ ≥ 5.0. In fact, the box in the H-R diagram in which
the W-R stars are observed (see e.g. Conti, 1976) agrees very well with
(i) the primary stars in the stage of slow mass transfer (phase 3);
see e.g. in Figure 2b the track for α = 3; (ii) the ensuing evolutionary
phase, after termination of mass exchange, of rather fast contraction
towards the He-M.S.; a tentative explanation for an upper limit to
T_{eff} for the W-R phenomenon might be that the contraction of the star
puts an end to the existence of a large extended envelope. The star
reaches the $\log T_{eff}$ = 4.75 zone about 30 000 years after the end of
phase 3, i.e. 2 to 4 times the duration of slow mass exchange. It takes
another 30 000 years to attain $\log T_{eff}$ = 4.90. The evolutionary timescales may be considered as lower limits, because the effect of the
extended atmosphere will be to keep the star at smaller T_{eff}.

(*)
    q < 0.25 is probably excluded, because the massive star will then
    destroy its low mass companion during the formation (Hutchings
    1976).

REFERENCES

Cassinelli, J.P., and Hartman, L. : 1975, Astrophys. J. <u>207</u>, 718.

Conti, P.S. : 1976, Mem. Soc. Roy. Sci. Liège (6ème série) <u>9</u>, 193.

De Grève, J.P., de Loore, C., and van Dessel, E.L. : 1978, Astrophys.
        Space Sci., <u>53</u>, 105.

Hutchings, J.B. : 1976, I.A.U. Symp. N° 73 (ed. P. Eggleton, S. Mitton
        and J. Whelan), p. 9.

Khaliullin, Kh. F. : 1972, Sovj. Astron. <u>16</u>, 636.

Kippenhahn, R. and Meyer-Hofmeister, E. : 1977, astron. Astrophys.
        <u>54</u>, 539.

de Loore, C., De Grève, J.P., and Lamers, H.J.G.L.M. : 1977, Astron.
        Astrophys. <u>61</u>, 251.

Vanbeveren, D., De Grève, J.P., van Dessel, E.L., and de Loore, C. :
        1978, Astron. Astrophys., in press.

GENERAL DISCUSSION

<u>Plavec</u>: Concerning the shape of the critical Roche lobes when radiation pressure is included: for a purely gravitational case, you may indeed assume the attractive force emanating from a mass point. However, the radiation pressure felt by the other star emanates from the surface of its companion. In order to evaluate it, you must carefully integrate all incoming flux at a given point. You are dealing here with continuous radiation pressure. Naturally, part of the surface will be in the shadow cone and there will be no external radiation pressure there. If you want to study the effect of radiation pressure on gas motions in the circumstellar space, I think you must consider selective radiation pressure in individual lines which may be very different for different species and stages of ionization. Then it will be also necessary to include the interactions between individual particles.

<u>Niemela</u>: I would like to remark that the Roche-model is useful only when the force-field is mainly gravitational. But when e.g. radiation pressure has to be taken into account, the equipotential surfaces will change.

<u>Bolton</u> (to Willis): How well is the absolute calibration of your UV photometry determined? What effect does this uncertainty have on your effective temperature determination?

<u>Willis</u>: The absolute calibration of S2/68 was believed to be better than 20 percent in the absolute photometric sense. A comparison with S2/68 and other calibrated UV data (OAO-2, ANS, etc) shows that the calibration may be better than this, say 5-10 percent. I would not think this would introduce uncertainties of more than 5000 K.

<u>Bohannan</u>: The velocity gradients presented by Moffat and Niemela were steeper in WN7-8 than in WN5-6. What does this observation indicate about the extended atmosphere?

<u>Castor</u>: If the atmosphere is very dense, all lines may have the same or similar widths, while in less dense atmospheres the range in width is more. This effect goes in the right direction, if I understand your (Bohannan) comment correctly.

<u>Moffat</u>: Answer to Bohannan and Underhill concerning ionization/excitation structure of WN, Of envelopes : We assume that the violet shifted P-Cygni absorption edge of the optically observed He I lines (especially at λ 3889) represents the terminal wind velocity for each star. But since the ter-

489

P. S. Conti and C. W. H. de Loore (eds.), Mass Loss and Evolution of O-Type Stars, 489-490.
Copyright © 1979 by the IAU.

minal velocity of WR winds varies from one star to another by
up to a factor of ~ 2, we normalized our plot of velocity of
absorption edges versus ionization/excitation parameter to
RV/RV_{HeI} in order to facilitate a comparison for different
stars. Perhaps this is an oversimplification. The change of
velocity with excitation/ionization potential is less steep
for WN5/WN6 stars than WN7/WN8 stars possibly because most of
the observed absorption edges in the WN5/WN6 stars are formed
in the outer envelope close to the terminal velocity.

Moffat: Question to A. Willis : You mentioned that it
is plausible that the very luminous WN7/WN8 stars can and
should evolve downwards in the HRD to the less luminous WR
stars (WN4,5,6 and all WC). But if WR-mass loss is responsible for this, it cannot be of the normally observed wind
type - we require a more violent, rapid process to account
for the clear gap of stars in the HRD between the WN7/WN8's
and the rest.

Willis: An evolution from WN7-WN would indeed suggest
that we should see intermediate objects. The fact that we
don't may suggest a more violent, short time effect in the
mass loss which could cause the transition from the high to
low luminosity WR stars. The evolution models do seem to
show that if the mass loss rate is high enough, during the
hydrogen burning phase a single star can drop in luminosity
significantly. I don't think we can rule out mass loss rates
high enough to ignore the Of-WN5,WN6 possibility. What are
the mass loss rates for WN7,8 stars?

Sreenivasan: In answer to Willis' comment about the
evolutionary picture of these WR stars, I would like to remind him of the HR diagram that Conti showed wherein there
was argument regarding the location of the observed points.
The mass loss rate can of course be altered to produce an
evolutionary track with different time scales etc., but the
observers have control over ensuring that the points are in
the "right" places. If the observers could ensure that the
effective temperatures and magnitudes are consistent with
all the criteria used to check them, the model-makers could
do their share of producing reasonable evolutionary scenarios.

Chiosi: I wish to comment briefly on the problem of the
N-enhancement and its abundance in binary systems where it
seems normal. If during the first mass exchange CNO processed material (N-rich) falls to the surface of the companion, an inverse μ-gradient is built up. However, owing to
the Rayleigh-Taylor instability such a gradient is destroyed
on a very short time scale, diluting the N-rich material to
quite a normal abundance.

SUMMARY OF THE CONFERENCE - PROJECTS FOR FUTURE RESEARCH

E.P.J. van den Heuvel
Astronomical Institute, University of Amsterdam, Netherlands;
Astrophysical Institute, Vrije Universiteit Brussels, Belgium.

I have attempted to summarize what we have learnt during this most
fruitful week and to destill from this a number of possible target
points for future research.

1. MASS LOSS

One of the most important sessions of this symposium was the panel
discussion on tuesday afternoon, where the various stellar wind theories were confronted with one another and with reality. Apart from the
excellent chairman of that session - a Dutchman of course -, one of the
most brilliant stars of 'that day was Alpha Geminorum (also known by the
name of Castor), which produced a stable steady-state self-consistent
outflow of arguments precisely fulfilling the boundary conditions set
by the chairman. Although this outflow passed through a critical point
- when Dick Thomas jumped into the discussion - it never became turbulent or overheated, and it gave us a quite convincing picture of how
winds from early-type stars can be produced. Some of the other stars
in that discussion exhibited recurrent strong outbursts of counter
arguments, and several of them showed rapid profile changes, on timescales ranging from a few seconds to several minutes.

The discussion produced a considerable amount of acoustical heating,
the excess heat being radiated out into the audience, causing certain
individuals to reach very high states of excitation. Some opponents
could even be heard threatening one anothers lives. For example, Anne
Underhill to Joe Cassinelli in the discussion on line formation in hot
corona's: "You go in to at least optical depth two....", and when Joe
did not like to do so: "You will see, if you go in you will never meet
two points with the same temperature". Indeed, already at the first
point poor Joe would have been completely evaporated. Some other
highlights of that discussion:
- Cassinelli pointing out 580 problems with Lamers' warm coronal model;
- Thomas receiving zero out of ten for the confrontation of the predictions of his theory with the observations;

P. S. Conti and C. W. H. de Loore (eds.), Mass Loss and Evolution of O-Type Stars, 491–498.

and, very appropriately:
- the chairman of the local organising committee announcing a (Lucy and)
  Sol(o)mon barbecue for wednesday night.

Altogether, it appears from this discussion, that a reasonble
concensus is arising about the mechanisms driving the mass outflow. All
speakers seem to agree that once the flow is going, radiation pressure
will do the job of pushing it further to the high observed terminal
velocities. Also, all appear to agree that the highly ionized species
present in the wind, such as OVI, NVI, etc., indicate the presence of a
thin warm, possibly even hot, corona. Its temperature may be as low as
3.10^5 K, with a positive temperature gradient, as in Lamers's model; or
as high as 5.10^6 K, as Cassinelli has proposed. In the latter case there
are definite predictions about soft X-ray emission which will be observable within the near future with HEAO-II. The unknown source of the
heating may be Aflvén waves, meridional currents, turbulence in the
stellar envelope, etc. A number of us are convinced that this hot corona
provides the basic starting mechanism for the outflow. Others - although
agreeing about the presence of such a corona - are not sure whether it
is the cause, or just the result of the wind, or whether or not it has
anything to do with the driving mechanism. It seems that the time is not
yet ripe to resolve this problem, although as Olson has shown us, line-
strength ratios for various ions provide an important diagnostic means
for testing the coronal temperature and density structure and, hence,
of its importance in driving the wind.
It is encouraging to see that wind models including rotation are being
developed by Castor and his associates; this may be of great importance
for understanding the Oe and Be stars. And to see the application of
the radiation-pressure driven wind model to stars in other parts of
the Hertzsprung-Russell diagram, by Abbott; an especially interesting
suggestion by him was that possibly the wind problem is related to the
problem of the peculiar A and B stars, by causing selective depletion
of certain elements from their atmospheres.

The reviews by Snow and Barlow showed us that impressive progress
is being made in the gathering of empirical data on mass loss rates.
Especially the IR and radio parts of the spectrum promise to have
great potentials as was also illustrated by the contributions of Hyland,
Schwartz and Morton. Quite surprisingly, the mass loss rates from
Wolf-Rayet stars obtained in this way - and confirmed by the UV observations presented by Willis - are about 10^{-4} $M_\odot$/yr, which is some one
or two orders of magnitude larger than previously thought. Several
years ago Peter Noerdlinger already derived a mass loss rate of this
order for γ Velorum, by using the CIIIλ1909 Å line, as presented at
this meeting. This result was at that time found hard to believe, but
seems now fully confirmed.

Important areas for future investigation in the field of winds
and mass loss seem to me:
Theory:
1. The stability of radiation pressure driven winds (Castor is working

on this).
2. The origin of the time variations in spectrum and luminosity observed in almost all bright supergiants - of which Vreux and Andrillat as well as Wolf and Sterken gave us nice examples.
3. Models of winds in other parts of the HRD.
4. The origing of the mass outflow from Wolf-Rayet stars and especially: the duration of the phase of strong mass loss in these stars.
<u>Observations</u>:
5. Search for the possible presence of soft X-ray emission from coronas; measurements of temperature gradients.
6. Study of ratios of linestrengths of various ions in order to derive the temperature and density structure in the corona/wind (cf. Olson).

It seems to me that within one or two years we will have sufficient observational data available on stellar winds to enable one to make a critical review of observed stellar wind parameters as a function of the position of a star in the Hertzsprung-Russell diagram. Such a review would be very useful for our theorists who study the influence of mass loss on the evolution of early-type stars.

2. EVOLUTION

It seems quite clear from the work of Chiosi, De Loore et al., Sreenivasan and Wilson, Dearborn and Falk that the empirical properties of O and Of stars as well as of supergiants can - at least qualitatively - be understood in terms of mass loss rates of several times 10^{-6} $M_\odot$/yr right from the beginning of the evolution. For the hot stars such mass loss rates are indeed observed and we have in the CAK theory a plausible basis for understanding their origin. However, one point which still worries me is the rather ad hoc nature of the mass loss laws employed in the cooler parts of the HRD, where we know in fact very little about the physical mechanisms driving the mass loss - nor about observed mass loss rates. It is just the mass loss in this part of the diagram which determines whether or not a star will lose sufficient mass to return to the blue part of the HRD, where, as was stressed again by the beautiful observational work of Conti and of Leep presented today, the most luminous stars appear to turn into WN7 or WN8 stars. As many of these stars appear to be single, it appears that single stars can blow away enough matter to reach the WN7,8 region. Now, the observed mass loss rates in the blue part of the HRD are - as shown by De Loore's computations - not large enough to make the star directly go towards the WN7,8 box without first moving towards the red part of the HRD. Whether or not it will become a WN7,8 therefore depends completely on the mass loss rate which is - ad hoc - assumed in the red part of the HRD. One rather would like to have a more fundamental understanding of why and how the star becomes a WN7,8 object. Why does the star just shed the right amount of mass to become a WN7,8 star? May be an important clue to the answer is given by the beautiful observations presented by Mrs. Pismis, which show that some of the WR and Of stars are surrounded by massive expanding shells which recently must have been ejected by them. Such WR stars appear to be always of the WN type and appear - as far as one can tell - to be always single. The massive shells indicate that there must have been a short-lasting phase of very heavy mass loss, perhaps as much as 10^{-1} or 10^{-3} $M_\odot$/yr. Why? The answer to this question has perhaps nothing at all to do with stellar winds and atmospheric phenomena, but may perhaps be situated in the more fundamental stability properties of the star as a whole.

In Eddington's (1926) book on stellar structure, there is a most interesting passage which perhaps is relevant to this problem.   [*]) Eddington shows that the structure of a gaseous sphere with an internal radiation field is that of a polytrope of index 3; in such a star the ratio of gas pressure to radiation pressure is completely determined by two parameters: the mass M and the mean molecular weight μ.

"We can imagine a physicist on a cloud-bound planet who has never heard tell of the stars, calculating the ratio of gas pressure to radiation pressure for a series of globes of gas of various sizes, starting, say, with a globe of mass 10 gm., then 100 gm., 1000 gm., and so on, so that the nth globe contains 10^n gm. The table shows the more interesting part of his results.

No. of Globe	Radiation Pressure	Gas Pressure
30	·00000016	·99999984
31	·000016	·999984
32	·0016	·9984
33	·106	·894
34	·570	·430
35	·850	·150
36	·951	·049
37	·984	·016
38	·9951	·0049
39	·9984	·0016
40	·99951	·00049

Table 1. Fractional gas pressure and radiation pressure in globes of gas of increasing mass (Eddington 1926).

The rest of the table would consist mainly of long strings of 9's and 0's. Just for the particular range of mass about the 33rd to 35th globes the table becomes interesting, and then lapses back into 9's and 0's again. Regarded as a tussle between gas and radiation the contest is overwhelmingly one-sided except between numbers 33-35, where we may expect something interesting to happen. What 'happens' is the stars.
We draw aside the veil of cloud beneath which our physicist has been working and let him look up at the sky. There he will find a thousand million globes of gas all of mass between his 33rd and 35th globes - that is, between ½ and 50 times the sun's mass."

The reason why we don't see hydrogen-rich stars more massive than roughly 100 to 150 solar masses must be, indeed, that stars which are completely dominated by radiation pressure are dynamically unstable, since according to the virial theorem a star with Γ = 4/3 (radiation)

[*]) My attention to this passage was drawn by Chandrasekhar (1975), International School of Physics "Enrico Fermi" nr.65, Varenna).

cannot be in stable hydrostatic equilibrium. The upper limit M_L at which
the ratio of radiation pressure to total pressure becomes virtually
unity depends only on the mean molecular weight μ (and on some basic
constants of physics), and is given by

$$M_L = M_0/\mu^2 \tag{1}$$

where M_0 is constant. Because of the rather schematic theoretical model
underlying this equation (a polytrope with n = 3), the best procedure
seems to determine M_0 from the observations, which for hydrogen-rich
stars ($\mu = \frac{1}{2}$) suggest that M_L is about 150 $M_\odot$. This would imply that
for helium-stars ($\mu = 4/3$), M_L must be around 22 $M_\odot$.

Equation (1) together with the fact that in hydrogen-burning stars
with $M \gtrsim 40$ $M_\odot$ the burning core contains more than half of the stellar
mass may be the key to the problem of the large rates of mass ejection
from evolved massive stars. Because, consider for example a hydrogen-
rich star of 100 $M_\odot$, which has a burning core of about 60 $M_\odot$. At its
birth the star is stable against radiation pressure. However, neglecting
for the moment mass loss by stellar wind, it would at the end of
hydrogen burning have a helium core of 60 $M_\odot$, which is far above the
upper limit for stable helium stars. So, the star must become vibrationally unstable and shed its excess mass to regain stability. Even if
during hydrogen burning this star were to lose half of its mass by
stellar wind (implying $\dot{M} = 10^{-5}$ $M_\odot$/yr) it would still be left with an
unstable helium core of more than 25 $M_\odot$, and therefore, presumably,
would still eject its outer layers by vibrational instability.

In such a case one would always be left with a (single) Wolf-Rayet
star with a mass slightly below 22 $M_\odot$, and surrounded by a massive
expanding shell. This would, at the same time, explain why the WN7,8
stars have such high luminosities, as one would always expect them to
be located at the upper end of the stable helium main sequence.

Of course, this picture is a rather rough one, and instability may
already set in before the end of hydrogen burning, as during hydrogen
burning μ increases and a previously stable star may gradually move
into the unstable region, thereby continuously ejecting its excess mass
in order to regain stability. This may, in fact, be the reason why we
see all kinds of transition types between Of stars and WN7,8 stars,
always located in the top of the HR diagram, as was shown today by
the beautiful work of Conti and of Leep. At the same time it may be the
reason for the irregular light and spectrum variations exhibited by
almost all of the most luminous stars, as was shown by Wolf and Sterken,
Vreux and Andrillat, and for the large mass loss rate from P Cygni, as
discussed by Luud.

If the above speculations have any value, the path of employing
atmospheric stellar winds to "explain" the existence of WN7,8 stars
and the transition Of stars may well lead to a dead end.

After this detour let me resume the summary and turn to the work
presented by Lamb on pre-supernova mass loss from the progenitor of

Cas A, which showed us how continuous mass loss may leave its traces even after a supernova event. A further highlight was the statistical work on O-type binaries by Garmany, which presents us a number of interesting evolutionary problems, notably the absence of low-mass secondaries. This may also have considerable consequences for our understanding of the characteristics of Wolf-Rayet and X-ray binaries.

Detailed studies of individual binaries are of great value as they are the basic source of fundamental stellar parameters. It is therefore stimulating to see the work going on at several places, notably in Boulder under the inspiring guidance of Peter Conti (on what aspect of the O-stars is he *not* working?), at Goddard (Kondo and Rahe), at Buenos Aires (Niemela and Sahade), and at the University of Nebraska (Leung). Nancy Morrison's striking close triple system presents us with a mystery – probably also regarding stability – but makes triple-star models for Cygnus X-1 seem less unlikely now.

The first results of the systematic survey for Wolf-Rayet stars in the Magellanic Clouds presented by Breysacher showed us a doubling of the known number in the Small Cloud, and an additional dozen new ones in the Large Cloud. We are most eager to see his further results. Concerning the evolution of binaries the work of van Beveren showed us that continuous mass loss during the entire lifetime of massive binaries has to be taken into account if we wish to understand the Wolf-Rayet binaries. The works of Tutukov and of Delgado, on the other hand, showed us the possibilities for the final evolution of massive binaries in which one of the components is already a compact star. The problem of whether or not the compact star will spiral down completely into the center of its companion appears to depend on the initial conditions, but – as convincingly argued by Tutukov – the absence of large numbers of red supergiants shows us that the resulting objects cannot be very long lived.

The work of Moffat and Seggewiss showed us that there are several runaway Wolf-Rayet stars, one of which is a binary with an unseen low-mass component. Some four to five years ago it was conjectured by Tutukov and Yungelson (1974) and by ourselves (1973) that massive X-ray binaries might evolve into exactly such systems. The fact that such objects have now been found is of great importance for our understanding of the origin of the binary pulsar as well as for testing the spiral-in scenarios.

Some important projects or problems in the field of evolution that deserve further investigation are (numbers continue from those of the problem-list in section 1):
7. Application of the fine-grid (Iben-Lamb) evolutionary calculations to mass-losing stars.
8. Reasons for the high incidence of double-lined systems among the O-type spectroscopic binaries.
9. Further study of the evolution of common-envelope binaries; observational search for such systems.

10. Study of duplicity among LMC and SMC O-binaries and Wolf-Rayet binaries.
11. Study of mass outflow properties from runaway and halo O-type stars, in order to test the hypothesis of Carrasco et al., that part of these stars may belong to population II. Sarah Heap's work suggests that such stars may be distinguished from massive O-stars through UV spectroscopy.
12. Study of possible unresolved multiplicity of lumious OB stars and WR stars in the LMC and SMC, by using Space Telescope. Bolton mentioned that 30 Doradus may be an unresolved cluster of five stars, and – as pointed out by Underhill – there may be many more such systems.
13. Further work on determining the position of WR stars in the HR diagram.

Before terminating, may I ask you to memorate with me one of the pioneers in the field of O-stars, Su-Shu Huang, who died last September in his homeland, China. Much of the work presented here this week is based on foundations laid by him and Struve several decades ago (one minute silence).

Additional Topics for future investigations

In the subsequent discussion the following list of additional topics for future investigations was suggested by the audience.

14. Accurate determination of energy distributions of OB stars through absolute photometry from space.
15. Determination of mass loss rates also for stars less massive than 15 $M_\odot$.
16. Further concentration on measurements of radio emission from stellar winds; this is probably the most powerful technique for determining mass loss rates; VLA will be very suitable.
17. Study of the evolution of rotating massive protostars.
18. Incorporation of hydrodynamical atmospheric boundary conditions in evolutionary programs.
19. Study of the correct outer boundary conditions for massive red stars, in order to understand the discrepancies between the evolutionary tracks obtained by different investigators.
20. Study of changes in orbital periods of WR binaries in order to determine the mass-loss rates from their components.
21. Search for non X-ray emitting O stars with low-mass (presumably compact) companions.
22. Study of abundance anomalies in mass-losing stars, to determine how much mass has been lost.
23. Study in the UV of the mass-loss properties of luminous stars in the Magellanic Clouds.

REFERENCES

Eddington, A.S.: 1926, *The Internal Constitution of the Stars*, Cambridge
    University Press, p. 16.
Tutukov, A. and Yungelson, L.: 1973, *Nautsn. Informatsii of the Astron.
    Council of the USSR Acad. of Sci.* 27, 70.
van den Heuvel, E.P.J.: 1976, *Structure and Evolution of Close Binary
    Systems*, Proc. of IAU Symp. No. 73 (P. Eggleton et al., ed.).